Monographs in Mathematical Economics

Volume 7

This series, consisting of research monographs and advanced textbooks for graduate students, is designed to bring together mathematicians interested in obtaining challenging new stimuli from economic theories and economists seeking effective mathematical tools for their research.

The scope of the series includes but is not limited to:

- Economic theories in various fields based on rigorous mathematical reasoning
- Mathematical methods (e.g., analysis, algebra, geometry, probability) motivated by economic theories
- Mathematical results of potential relevance to economic theory
- Historical study of mathematical economics

Comparable, existing monographs series are mainly organized from the viewpoint of "users" of mathematics, and are thus of limited interest to mathematicians. This series in contrast aims at genuine interaction between economists and mathematicians.

Most economic phenomena are described by (1) optimizing behaviors of agents (consumers, firms, governments, etc.) and (2) equilibria generated through the interactions of these agents. Consequently, the most basic mathematical subjects for economics fall into two categories:

- Optimization theory, static and dynamic. Basic linear / nonlinear programming, calculus of variations, optimal control, dynamic programming etc. provide key mathematical tools. Modern developments in convex analysis, nonlinear functional analysis, set-valued analysis, and non-smooth analysis form indispensable foundations.
- Equilibrium theory. Existence in finite- / infinite-dimensional settings, characterization, computational algorithms, dynamic adjustment processes, correspondence with economic structures.

Some fixed-point theorems and variational inequalities were created to solve the existence problem of equilibria. The stability theory and the viability theory of differential equations play important roles in the dynamic aspects of the equilibrium theory. Further, differential topology is a basic tool for the geometric analysis of equilibrium manifolds.

Economic phenomena significant in the real world often have a dynamic character. For instance, business cycles and movements of asset prices are governed by dynamic laws expressed in terms of differential or difference equations. Nonlinear analysis, harmonic analysis, and stochastic calculus play significant roles in solving dynamic economic problems. Econometric methods, including time series analysis and forecasting, also require sophisticated mathematical tools. Finally, significant developments in game theory and interaction with mathematical logic should be mentioned.

All of these topics and phenomena come within the scope of this series, aimed at cooperation between economists and mathematicians and the development of their respective disciplines.

M. Ali Khan · Nobusumi Sagara ·
Alexander J. Zaslavski
Editors

Matching, Dynamics and Games for the Allocation of Resources

Essays in Celebration of David Gale's 100th Birthday

Springer

Editors
M. Ali Khan
Department of Economics
Johns Hopkins University
Baltimore, MD, USA

Nobusumi Sagara
Faculty of Economics
Hosei University
Machida, Tokyo, Japan

Alexander J. Zaslavski
Department of Mathematics
Technion—Israel Institute of Technology
Haifa, Haifa, Israel

ISSN 2364-8279 ISSN 2364-8287 (electronic)
Monographs in Mathematical Economics
ISBN 978-981-96-6660-7 ISBN 978-981-96-6661-4 (eBook)
https://doi.org/10.1007/978-981-96-6661-4

This Springer imprint is published by the registered company Springer Nature Singapore Pte Ltd.
The registered company address is: 152 Beach Road, #21-01/04 Gateway East, Singapore 189721, Singapore

Contents

Economic Dynamics and Growth

Matching, Contracts and Markets

Mathematics, Game Theory and Economics: Provisional Observations on David Gale's 75-Year Career (1949–2024)[1]

Preface to a 100th Birthday Anniversary Volume[2]

M. Ali Khan and Alexander J. Zaslavski

Abstract This preface introduces Gale's oeuvre and furnishes a current perspective on it, fifteen years after his passing. It also attempts a well-rounded picture of David Gale as one pioneer of post WWII economic theory and operations research. Finally, it introduces contributions by acknowledged leaders in mathematical economics, game theory, matching theory and dynamics and computation.

Keywords Gale-Nikaido-Kuhn-Debreu lemma · Gale-Shapley algorithm · Optimal growth

---

[1]David Gale was born on December 13, 1921 and died on March 7, 2008. His 75-year career began with the completion of his Princeton Ph. D. in 1949 and for the authors, lives on in increasing citations and influence. We ask forbearance from our more arithmetically inclined readers who think that a scholar's career ends with a scholar's passing.

[2]Nobusumi Sagara, one of the co-editors of this volume, passed away when the volume was ready: he was fully involved in all aspects of the project other than this preface: a scholar of the highest values and of unimpeachable integrity, we miss him deeply. Khan acknowledges with deep gratitude conversations and correspondence over the years with Mukul Majumdar, Marilda Sotomayor, Joel Sobel, and the late Monique Florenzano and Tapan Mitra. For their comments on this preface specifically, we thank Bob Anderson, Roger Backhouse, Gabrielle Demange, Dave Kellen, Roger Koenker, Andrew Mackenzie, Richard McLean, Arthur Paul Pedersen, Adriana Piazza, Liuchun Deng, Ani Ghosh, Simeon Reich, Al Roth, Olivia Schartman, David Schrittesser, Metin Uyanik, and Xinyang Wang. Khan's work on this preface was supported by the 2022 JHU Discovery Award "Deception and Bad-Faith Communication." Khan's work towards this revision by the 2022 JHU Discovery Award "Deception and Bad-Faith Communication," and in contract from "Blue Green Future & Rebalance Earth."

M. A. Khan (✉)
Department of Economics, Johns Hopkins University, Baltimore, MD, USA
e-mail: akhan@jhu.edu

A. J. Zaslavski
Department of Mathematics, Technion—Israel Inst of Technology, Haifa, Israel
e-mail: ajzasl@technion.ac.il

M. A. Khan et al. (eds.), *Matching, Dynamics and Games for the Allocation of Resources*, Monographs in Mathematical Economics 7,
https://doi.org/10.1007/978-981-96-6661-4_1

> ... he had the ability to look at each new problem as a *babe*—and he posed himself a large variety of questions in a large variety of domains - shedding all of his impressive knowledge and technical abilities in favor of the problem itself. Balinski-Laraki-Sorin (2009)[3]

> He was able to see simplicity and clarity in the middle of an awesome muddle. He found problems whose solutions are utterly non-trivial but which seem obvious after they are set forth. I've never quite seen anything like this.[4] Scarf (2009)

In this volume celebrating the hundredth birthday anniversary of David Gale, the editors are pleased to present 20 exemplary testimonials, categorized under three headings and in alphabetical order under each, to honor and celebrate Gale's memory and Gale's achievement. A measure of this achievement is not easily had, even after more than 15 years of his passing. It is still far too early. Nevertheless, in an attempt to get some little overview of the *oeuvre,* nothing if not richly polyvalent, this introduction rather poses the question "Who was David Gale?" This too is not an easy question to answer, and our attempt, surely subjective and surely rooted in the limited vantage point of our time and our location, not to mention our own limited abilities, leads us to begin with previous attempts as a point of departure: how future scholarship will answer, it is for future scholarship to determine and discover.[5]

There are at least four notable testimonials in the past: (i) a 1992 anthology edited by Majumdar, prefaced by him and Brock, and titled *Equilibrium and Dynamics*; (ii) a 2008 tribute in the form of a special issue of the journal *International Journal of Game Theory (IGJT)* edited by Sotomayor; (iii) a special 2009 memorial issue of the journal *Games and Economic Behavior (GEB)* edited by Sobel and von Stengel; and finally, (iv) Sobel's own 2017 *Palgrave* entry.[6] On a first cursory look, it is a little surprising how the tone, and the substance, of the four texts differ. The Majumdar anthology includes contributions from a veritable *Who's Who* of mainstream economics focused on their specialties, a tribute offered to Gale while he was still in his research prime;

---

[3] Balinski-Laraki-Sorin (2009) identify this ability as Gale's principal characteristic: "For the three of us—of different mathematical generations and backgrounds—David Gale evokes different memories and different impacts, though joint admiration and strong agreement on his principal characteristic. For me, Michel Balinski, David Gale was one of life's constants—like π or e—or so it seems." This is done in a memorium titled "David Gale in Paris" in a special issue of the journal *Games and Economic Behavior*. Throughout this preface, the date 2009 associated with an unreferenced author is uniformly associated with that particular author's contribution to Volume 66, Issue 2, pp. 581–1004 (July 2009) of this said journal. This issue also contains a bibliography of Gale's publications, and the reader is referred to it for all of Gale's articles not referenced in this preface.

[4] Sobel (2009) also has recourse to the word "awe" when he prefaces his article "ReGale: Some memorable results" by the single sentence abstract "This article reviews, with admiration and awe, the scientific contributions of David Gale."

[5] This is hardly empty rhetoric: as we shall have occasion to emphasize in the sequel, both "seminal" and "definitive apply to the *oeuvre*; see, for example, Babecki et al. [4] or Edelman et al. [24]. It would not be an exaggeration to say that the work is still opening lines and fields not yet fully conceived.

[6] For the first, see Majumdar [75]; for the second, see Sotomayor [101]; for the third, especially Sobel (2009) in the GEB special issue referred to in Footnote 2 in Chap. 4; and for the fourth, see Sobel [100].

while the Sotomayor collection essentially revolves around the 1962 Gale–Shapley paper in the *American Mathematical Monthly (AMM).*[7] Sobel's *Palgrave* entry is an abbreviated version of his contribution to the GEB special issue, a unique testimony that only a student can give of his or her dissertation advisor.[8] We begin with this 2009 special issue.

A focus on the latter reveals a certain playful gregariousness and a lack of earnestness, leave alone pomposity, in the way the contributors characterize his approach to his subject; and if we may dare to extrapolate, to the way he approached life. Gale comes out as a *bon vivant* gamester.[9] Thus, in his contribution, Osserman (2009) refers to Gale's "MathSite: An Interactive Source for Seeing, Hearing, Doing Mathematics," and the entry into his life of the critic, writer, and poet, Sandra Gilbert, an ideal personification of a bridge between C.P. Snow's "Two Cultures," a fitting closure to David's brilliant and kaleidoscopic career."[10] Osserman elaborates further with reference to himself:

> Our principal research interests were far apart and different, with mine more verbal and David's visual. I was especially impressed by his ability to produce actual hands-on models that beautifully encapsulated the ideas he was trying to convey. Among those, the ones that I found particularly striking were the physical models of *pseudospheres* of different sizes—the most beautiful and ingenious devices I had ever seen to illustrate some of the key properties of a central notion in differential geometry—that of the Gauss curvature of a surface.[11]

The computer scientist Karp, referring to his own collaboration, writes of how "David would weigh in on topics of interest to the public" in their wide-ranging conversations.

> David seldom mentioned his famous achievements in mathematical programming, polyhedral theory, game theory, convex analysis and economics. Instead the conversation would usually revolve around a provocative question that he would raise about a finite or infinite combinatorial game of perfect information. I loved hearing those ingenious problems but don't recall ever solving any of them.[12] Just a few months before his death, he received some attention from the media for his irrefutable mathematical argument debunking the widely held belief that, on the average, men have more partners in heterosexual liaisons than women do.[13]

---

[7] Of the 17 papers in this collection, 11 revolve around the AMM paper, 5 also cite his further 1985 investigation of the paper with Sotomayor in the direction of contributions of Dubins–Freedman and Knuth. Also see Footnote 16.

[8] We also make reference in Footnote 19 to another student's testimony.

[9] This is even after discounting the fact that this is a memorial issue. Even a cursory perusal of his *Mathscinet* pages brings this out—see, for example, [37, 41] and his various pieces on the game of Hex and other mathematical entertainments.

[10] In this and the subsequent quote, we have slightly re-arranged Osserman's prose. For Sandra Gilbert's poems, see [51].

[11] Osserman explains these pseudospheres: "items of "clothing" that were flexible but not stretchable, so that they would conform perfectly to the right size pseudosphere, but crinkle up and fit badly on any other surface."

[12] This connects to our first epigraph: David Gale's propensity to begin each problem anew and approach it on its own technical and substantive terms, in keeping with its own integrity and autonomy.

[13] See Karp (2009): a tribute titled "My memories of David Gale."

This is of course with reference to the result (and its proof) that Davis (2009) furnishes in his own two-paragraphed memoriam titled "Sex and the mathematician: The high school prom theorem." For his own part, Karp discusses a problem in linear algebra that he and Gale worked on.

> We came up with a number of entertaining variations on this theme, and David thought that the results would appeal to a broad mathematical audience, so we submitted a playfully written version of the paper to the *American Mathematical Monthly* under the title "The Non-Messing-Up Theorem." Perhaps the paper was too playful for the editors, as they rejected it. A more sober version was later published under the title "A Phenomenon in the Theory of Sorting."[14]

It is not so much a matter of Gale himself cultivating this image, but rather that his humility and innate curiosity of all things mathematical was evident to all those who were privileged to interact with him. In his 2002 review of the film *A Beautiful Mind,* he observes with a marked Wittgensteinian sensibility:

> I became increasingly aware of how each profession has its own rules and objectives and outlook on the world, its own idea of what the game is all about. Perhaps I should extend my earlier list[15] to include
>
> - Games mathematicians play: Proving theorems, e.g., the *Equilibrium-Point Theorem*
> - Games film makers play. Grabbing audiences, e.g., *A Beautiful Mind*
>
> For both these endeavors, the payoff turns out to be a prize.

This playful self-effacement, only half-serious and half-flippant, did not deter his shying away from "high-brow" philosophical issues, but always without any trace of pomposity.[16] In his 1989 article on "The truth and nothing but the truth," an article that perhaps deserves to be better-known, he quotes Saunders Mac Lane on the belief of logicians in Zermelo–Fraenkel set theory being based on a sort of Platonic imagination that there is out there in the real world some universe of sets to which their axioms will apply. Holding up Mac Lane's statement in which he submits "that the Platonic notion of a real universe of sets is absolute balderdash," Gale focuses on the adjective to say

---

[14] It is of interest that David Gale published the greatest number of his papers in the *Monthly*: his page in *Mathscinet* gives the number 11 out of 86. The second place is shared by *Mathematical Intelligencer* and the *Journal of Mathematical Economics* (tied at 7); and the third (fourth?) by *Econometrica* and the *Journal of Economic Theory* (tied at 5). We shall have occasion to refer to the *Mathscinet* archive in the sequel: it suffices it to say here that even this possibly incidental fact of publication is not irrelevant to answering the question that we continually pose in this preface.

[15] See p. 78 in Gale [43], italics the author's own. This is a continuation of a list that we reproduce below; see Footnote 33. On laying this aside, let us nevertheless note that in our judgment, this neglected piece is revelatory, and his passing comments on Nash reveal Gale's strength as a reliable portraitist and historian; in this connection, see, for example, Footnote 22 and the text it footnotes.

[16] One cannot help seeing these two sides of David Gale in the mirroring of the two issues in his honor: the tone of the *IJGT* tribute is serious and earnest as opposed to that of the *GEB* issue.

> Absolute balderdash is presumably a strong form of balderdash by analogy with absolute and ordinary convergence.[17]

Not everyone is aware, and we refer here to the community of economists rather than that of mathematicians, that this article features a Socratic dialogue between Plato and MOS (Mathematician on the Street) on the continuum hypothesis.[18]

Needless to say, this ability to puncture pomposity was undergirded by utmost and obvious compassion for minds lesser than his own,[19] and with a total disregard of questions of "scientific" credit and priority.[20] In his memorial article, the computer scientist Berlekamp (2009) discusses the Galetron.

> Eventually Neal Sloane solved the Galetron's minimax problem, and published the result under the title "The Solution to Berlekamp's Switching Game." This caused me some embarrassment, since I felt it should have been called "The Switching Game of Berlekamp & Gale." But, as always, David Gale didn't mind at all. His mathematical interests were unusually pure. Issues of credit and authorship were always of much lesser concern to him.

Insofar as the question of authorship is concerned, the preeminent example was already furnished at the beginning of Gale's career. In what he refers to as his own "brush with history," Gale does not refer to what Nash specifies in a footnote in his 1950 paper, and repeats in his own 2009 memoriam.

> My first short note credits David Gale with making me aware of the Kakutani fixed-point theorem. [He] happened to be very influential in relation to my original publications on "equilibrium points" and on my thesis at Princeton; and [he] was also influential in relation to other game-theoretic connections, such as the game of "Hex."[21]

David Gale only alludes to his contribution in the justification of a tangential (and generous) claim about Nash: "tactless, blunt, very ambitious certainly, but I don't think he had a swelled head."[22]

---

[17] In private correspondence, Al Roth writes "David Gale would have been amused that he posthumously shared the 2013 Golden Goose award (for unlikely results with impact) with Lloyd Shapley and me." The interested reader can see https://www.goldengooseaward.org/01awardees/market-design and http://vimeo.com/74905870.

[18] Gale [40] introduces the dialogue with the following sentence, "I happened to be present on the street and overheard the following dialogue." The dialogue concludes with an investigation of π and thereby connects to Balinski's assertion in Footnote 2 in Chap. 4. The sectioning of this article is worth noting as to whether Gale was serious or it was all being written "tongue in cheek": barring the title itself, the section titles are (i) The meaning of mathematical results, (ii) True, false, or meaningless, (iii) Reality. However, see the discussion of Gale–Stewart games in the sequel: Footnote 80, and the text it footnotes. With regard to Gale's facility to address different communities, see Footnote 42. An exegetically inclined reader may wish to track the word "community" in this very preface.

[19] In this connection see Brock's (2009) piece titled "More Gale's tales," a piece that follows his earlier "Gale's tales" in [75]. The portrait that Brock paints brings out the inevitable skepticism that one felt as to how Gale looked at one.

[20] The economics profession is perhaps now overly concerned with priority issues and with the marketing of one-self; see the reception of a 2014 volume reviewed by one of the authors of this preface in [61].

[21] See Nash (2009) where he adds "David Gale was among the group of Gale, Kuhn, and Tucker at Princeton who had a seminar studying connections between game theory and linear programming." We shall have more to say on the latter in the sequel.

[22] See [43] where Gale takes exception to the claim of Nash's arrogance in both Nassar's book and the film based on it. Also see Footnote 15.

Continuing on in this vein, we note that Scarf (2009) titles his contribution to the 2009 special *GEB* issue, "My introduction to top-trading cycles," and discusses Gale's alternative proof of the main result of his paper with Shapley on the nonemptiness of the core of housing markets.

> I did think about his [David Gale's] argument on the plane back to New Haven. When I returned, I sent David a letter with the phrase "I'm terribly sorry to have to tell you that there is nothing wrong with your argument."[23]

Sotomayor (2009) provides yet another anecdote.

> As soon as the manuscript was ready we sent a copy of it to David Gale. He replied with a letter in which he simplified my proof of the lemma and reduced it to ten lines! We invited him to be a co-author of our paper, which was published in Econometrica in 1989, but he didn't accept.

Mukul Majumdar, in conversation with one of the authors, also reveals a similar anecdote concerning the paper that presented the Kuhn–Tucker theorem, now referred to as the Karush–Kuhn–Tucker theorem.[24]

Curtis Eaves did not contribute to the *GEB* collection, but he too is clear about the genesis of the algorithm that has had, and continues to have, such a wide interdisciplinary impact. Eaves [22] has only three references, two of which are to Gale.[25]

> The linear pure exchange model was evidently first studied as such and in detail in Gale [2]. The concept of *reducibility* was introduced there, and assuming *irreducibility,* it was shown using the Kakutani fixed-point theorem that an equilibrium exists. Stymied in an effort to compute an equilibrium of the linear pure exchange model using Lemke's algorithm, the author approached David Gale, ... with the following question. If [the matrices] W and U are rational, does there exist a rational equilibrium? The success of the present paper rests upon the argument given in Gale [3] which supplied an affirmative answer.

Eaves [23] continues to anchor his algorithm in Gale's memorandum, but this time includes his name with Afriat, Dorfman–Samuelson–Solow (DOSSO, an acronym of names for a volume referred to below), Scarf and Karlin, all in the context of Leontief models.

Turning from others to Gale himself, from questions of persona and personality to the style and signature of his own writing, one cannot but begin with his *magnum opus* on "The Theory of Linear Economic Models." Published in 1960, a period especially

---

[23] Dinner remarks at the Gale Memorial in Stony Brook, July 12, 2007; these connect to our second epigraph. The authors of [84] write, "And then, stunningly, and explicitly as an afterthought, [Scarf-Shapley] also provided a constructive proof of the nonemptiness of the core by introducing Top Trading Cycles (TTC)." They then proceed to cite from Scarf (2009). For an up-to-date survey, see [1]: the fourth paragraph of their introduction discusses Gale's contribution.

[24] We briefly go into this below: see Footnote 77 and the text it footnotes. For a history of the contribution of Kuhn–Tucker to the Karush–Kuhn–Tucker theorem, see Kjeldsen [68] and his references. Gale (1967) returns to this theorem.

[25] In the quotation below, Gale [2] is [32], a paper that is not easily accessible in the Rand archive, and Gale [3] is a "personal communication." For the influence of the linear exchange model in French circles, see, for example, Florenzano (2009) and Bonnisseau-Florig-Jofré [9]. We shall have more to say on this model in the sequel.

fruitful for texts in mathematical economics, it can now be grouped with four clearly identifiable benchmark contributions of the period.[26] A comparison between these four incomparables brings out the tone of Gale's voice, and the care and thought it lavishes on the choice of notation. Gale's text was a second attempt at a consolidation of his early-career successes: the first attempt was already enthusiastically received by Robert Solow who noted that "These notes are based on lectures given as part of a graduate course in the Mathematics Department of Brown University... Their purpose is to give a unified mathematical treatment of... linear economic models."[27]

> The result is an extraordinarily neat and compact survey of zero-sum two-person games, linear programming, pure exchange models (and, implicitly, the elementary properties of finite Markoff chains), and pure production models of the Leontief and von Neumann types. Economic interpretations are usually given, but on the whole the treatment is abstract. Although the book is in principle self-contained, the author wisely recommends that it ought not to be used as a text at any lower level than graduate mathematics. On the other hand, some mathematical maturity could be replaced by economic maturity, and these notes should be of real interest to economic theorists.[28]

One may note here that Gale's book, presumably a polished version of these notes, contained central results on the theory of linear inequalities, including the 1951 paper of Gale, Kuhn and Tucker, and Gale's own 1962 extension of von Neumann's model of an expanding economy (both reconsidered below). It discusses Dantzig's simplex algorithm and gives an economic interpretation to problems now seen as canonical. The book also contains a concise and elementary introduction to the theory of linear inequalities (including a proof of the separating hyperplane theorem for convex polytopes), a chapter containing essential results on non-negative matrices, and a clean treatment of dynamic linear models of growth.[29]

In his review of the book, Jorgenson [59] singles out "rigor, parsimony, elegance, scope of applications as the characteristic features of modern mathematics," and sights Gale's volume as a worthy exemplar of these features, comparing it with Feller's treatment of discrete probability theory and its applications. He notes the author's "rare expository skill." Hahn's review [53] is much less restrained, and especially as it pertains to the two chapters on the theory of games, regarding which he poses a question.

> Once again the style is limpid, the exposition splendid. Why should the student learn about game theory? Surely not because it has proved of direct relevance to problems of economic

---

[26] Gale's book was published in the same year as his co-author Nikaido's text [87]; 2 years after "Linear Programming and Economic Analysis" by Dorfman–Samuelson–Solow, known to the *cognoscenti* as DOSSO, and a year after Debreu's "Theory of Value" and Karlin's two-volumed work [60]. We return to Gale's text below.

[27] Solow does not give a date for these notes—his own review is available as MR-0089111.

[28] Not to carp, but a 1960 survey on linear theory by a leading economic theorist of the time has no reference to any writing of David Gale's; see [57].

[29] Largely in recognition of their joint work, Gale, Kuhn and Tucker won the 1980 von Neumann prize for work they began in the late 1940s. As of this writing, the book has 2196 citations in *Google Scholar*.

analysis, but because it has indeed proved a most important source of methods for solving such problems. It is also splendid exercise.[30]

This review, dated as it is, is noteworthy in how Gale was so much ahead of his time in effecting the transition from Walrasian general equilibrium theory and economic dynamics toward establishing the current mainstream relevance of game theory. However what is missed in these, and in other, reviews is the nuanced way that Gale thinks of rigor, and especially if one compares his style with that of his influential contemporary, Gerard Debreu.[31] In his own piece on "truth" already referred to earlier, Gale[40], Gale writes:

In this essay I have written about a particular aspect of mathematics. The exposition, however, is deliberately informal and non-rigorous. Readers who are so inÂclined will find plenty of nits to pick, but if I were to proceed throughout with complete rigor, it would imÂpede the flow of ideas, so I hope the reader will not begrudge me some limited use of poetic license.

The novelty of the notion of rigor impeding the flow and fluency of ideas is brought out most sharply when one compares Gale's style and signature with that of his long-time Berkeley colleagues: Debreu most certainly but also Lucien LeCam, Dan McFadden, Roy Radner, and Ronald Shephard.

Beyond the question of rigor, however, Gale was deeply conscious that the register of mathematics was not the same as the register of economics. In his masterful 2002 exposition of Nash's theorem, he was blunt and oblique at the same time.[32]

Because he is a mathematician rather than an economist, Nash did what mathematicians do. He proved a theorem. Here it is. Nash's Nobel-Prize-Winning Theorem: *Every finite n-player game has an equilibrium point in mixed strategies.* What does all that mean, and why does it matter? I will approach the question indirectly.

And so he did. And after he was finished with answering the question of the difference between an economy and a game, he could conclude:

With this degree of generality much of social behavior fits into the game framework. Thus we have (i) Games politicians play— Impeaching a president, (ii) Games businessmen play— The Enron Game, (iii) Games countries play—The War on Terrorism.[33]

---

[30] Hahn's somewhat circumspect defense of the relevance of game theory in the sixties can be usefully complemented by Hick's [57] skepticism in his survey of the linearity postulate in economics. It is of interest that the master of linear structures among postwar mathematical economists does not get a single mention.

[31] Compare, for example, the preface of Debreu's *Theory of Value.* At this point, we should note the idiosyncratic nature of the referencing. This little preface is hardly a comprehensive survey of the references relevant to Gale's *oeuvre,* leave alone those of the post-WWII period. The bibliography already contains 109 items spread over six pages, and as a consequence, we ignore giving details of classical texts, one that any half-qualified reader of the subject would know well. Debreu's book, and DOSS) referred to below are obvious two exemplars of this idiosyncrasy.

[32] In his address to a joint session of the *American Mathematical Society* and the *Econometric Society,* Gale posed the question, "What is mathematical economics?" and proceeded to "convey the general flavor of the subject—a sort of answer to the question for the general mathematician" motivated largely by curiosity." In addition to Gale [34], the reader is referred to the fifth paragraph of Gale [39]. We shall have occasion to return to both papers below.

[33] This is the initiation of the list referred to above; see Footnote 15. Incidentally, this is also an answer to Hahn's rhetorical question as to why economists should study game theory.

In this (our) continuing attempt to answer, however provisionally, the question "Who was David Gale?"—a working portrait to serve as a backdrop to the 20 contributions of this volume—we could be criticized for having so far slighted pure mathematics. It is worth noting that Gale's contribution [31] to the subject of "compact sets of functions and function rings" represented his very first publication, a year after the completion of his thesis. It is a paper prominently featured in Willard's [105] text on *General Topology* and discussed in the context of k-spaces. We remind the reader that a k-space is a Hausdorff topological space X in which $S \subset X$ is closed if and only if S intersects every compact subset of X in a compact subset; and the class of k-spaces includes all locally compact and all first countable spaces.[34] Willard ascribes compactly generated topologies to Whitehead (1939), and notes that the term k-space first appears in Gale's paper where, not uncharacteristically, he first attributes the concept to Hurewicz, though without any reference.[35]

To be sure, the line between pure and applied mathematics is even less sharp than that between mathematical economics and economic theory, but it is of interest in this connection to see how the editors of *Mathscinet,* the functioning archive of the *American Mathematical Society,* perceive David Gale and identify his work. They classify his *oeuvre* into 16 categories, the first three of them being (i) operations research and mathematical programming, (ii) convex and discrete geometry, (iii) game theory, economics, finance, and other social and behavioral sciences. Of the bulk of the output, which *Mathscinet* sees to be of 86 papers, 52 are put in the first category, 8 in the second, and only 6 classified as falling in the category under which David Gale is primarily known in the economics profession, that of game theory and mathematical economics. As far as the categories are concerned, the Editors list four papers in the next category, and two each in the next two, and one in each of the ten subsequent ones.[36] David Gale's *Mathscinet* page lists thirty co-authors, and notes his Erdös number to be two. It is the professional mathematical community's portrait of one they regard as their own: an emergent portrait that is different from the one that we have so far culled from the GEB special issue. The diversity and depth of the mathematical contributions comes through clearly, but after all is said, the classifications tell as much about the Editors of *Mathscinet* as they tell about Gale.[37]

---

[34] See Gale [31], and especially the review by Myers MR-0036503 to frame Gale's contribution.

[35] See Sectiom 43 of the historical notes on pp. 320–321 in Willard [105]. This section gives references to the literature antecedent to Gale's paper, including the papers of Fox, Arens and Whitehead. We shall have occasion to discuss Gale's contribution to mathematical logic and set theory below; see Footnote 80 and the text it footnotes. Also see [55] for what the authors term as the "Gale property; also see Footnote 56.

[36] The categories are, respectively, (iv) General and overarching topics and collections; (v) Computer science, (vi) Combinatorics, (vii) Biology and other natural sciences, (viii) Mechanics, (ix) Probability theory and stochastic processes, (x) Topology, (xi) Real functions, (xii) Topological groups, Lie groups, (xiii) Linear and multilinear algebra; matrix theory, (xiv) Number theory, (xv) Classical algebra, (xvi) History and biography.

[37] As we shall have occasion to point out below, some papers of consequence for economists are not listed at all, or if listed, are not seen fit to be sent to be reviewed. We are all surely prisoners of our time and location, and our judgments reflect the cages in which we find ourselves.

This is precisely the point that we are out to make: Gale's *oeuvre* has enough of a diversity and rich maturity that one cannot but see in it what one wants to see, and what one is capable of seeing. Even within the mathematical community, by way of contrast with *Mathscinet,* the 1989 Editors of *Mathematical Intelligencer* have their own view of David Gale. They introduce the author of the article "The truth and nothing but the truth" with the following description.

> David Gale's thesis under A. W. Tucker may have been the first doctoral dissertation in the subject of game theory. Most of his work has been in mathematical ecoÂnomics, with occasional side excursions into convex geÂometry and combinatorics.[38]

Even a cursory perusal of the "occasional side excursions into combinatorial geometry" reveals him to be a founding figure for novel lines of inquiry as a result of his 1956 paper on "neighboring vertices," and his 1959 paper on "transient flows in networks." It is clear that the impact of this work has grown with the interest and advances in computation, and is now in full flight.[39]

We have focused on the GEB special issue to bring a somewhat neglected side of David Gale, and it is only fitting that in addition to the contributors, we also give a little space to how its own editors conceive of him. Thus Sobel-von Stengel (2009) write:

> David Gale played a leading role in developing theory of fundamental importance to economic and operations research. He authored papers in game theory and mathematical programming that have had a lasting impact. In addition, he was renowned for his ability to present his results in short papers combining precision and rigor with elegant style.[40]

Turning to the IJGT, we can couple these remarks with those of the editor of the special issue of a collection of 17 items: Sotomayor [101] notes:

> It only reflects a small part of his work in the following important areas of Game theory: the stable matching problem, auction theory, housing markets, combinatorial games and economics of fair division. In addition to original work, it has been notorious the ability of

---

[38] See the author's profile in Gale [40]. This journal is important for Gale studies since, along with the *Journal of Mathematical Economics,* he published seven papers in it. The only journal in which he published more papers was the *American Mathematical Monthly* with eleven papers. Incidentally, Gale includes this piece in his 1998 anthology "Tracking the Automatic Ant"; see Gale [41]. In the sequel, we shall have occasion to refer to the 1953 paper of Gale and Stewart as having a major impact on mathematical logic and other foundational issues in mathematics; see Footnote 80 and the text it footnotes.

[39] See the reviews of Coxeter and Kuhn of these papers in MR-0085552 and MR-0114669, respectively. Also see, for example, Chap. 5 in [102], Chap. 6 in [109], the Edmonds and Orlin papers in [89]. Indeed, much as in the use of Kuhn-Tucker theory in economics, Gale duality, Gale diagrams, Gale transforms and Gale Oiks are relied upon without citation; in addition to the references in Footnote 5, see, for example, [3, 18, 19, 25, 27, 59, 79, 96, 106]. Edelson-Popescu provide an interesting history: after citing Pascal in the 17th-century, and Castelnuovo, Hesse and Sturm writing in the 19th, they cite independent discovery of Whitney and Gale for the affine case. The authors are grateful to Joel Sobel for conversations leading to this footnote; also see Footnote 83.

[40] For the interested reader, this is to be complemented by Sobel (2009) and [100]. As we have pointed out above, like that of Brock's, his is a judgment of a distinguished student.

> Gale (*sic*) to combine precision and rigor with an elegant style of exposition and to provide simpler alternatives to complicated proofs.[41]

In their preface to the earlier 1992 volume, Brock–Majumdar provide their own listing of the subjects to which they ascribe his principal contributions, a listing that gives prominence to the mainstream economic theory of its time. Their open-ended list includes "revealed preference theory; Walrasian equilibrium theory; global univalence of mapping; intertemporal equilibrium with overlapping generations; money and interest; optimal development in a multi-sector economy; technological change; transfer problems and trade theory; economics of fair division; marriage and matching problems; auction theory." If one steps back and aggregates these labels into those now somewhat more conventional in the economics profession, it is clear that Brock-Majumdar see Gale's work as transcending boundaries between partial and general equilibrium analysis, between microeconomic and macroeconomic theory, between static and dynamic economics, and finally, between auction and matching theory. Their silence on game theory and on numerical analysis and computation is loud: the *oeuvre* is not easy to categorize and commodify, package and sell.[42] It is also then no wonder that the contributors to their volume are from a small set of the more distinguished economic theorists of their time, surely all fluent in mathematics, but not a single professional mathematician or a computer scientist among them.[43] The point again is one that we have made before, and one that Brock-Majumdar are explicit in making: "The collection of studies in this volume reflects the small subset of David Gale's work that has had profound influence in areas of economics that overlap with our own interests and competence."

So far the angles from which we have viewed David Gale and his work concern perspectives not primarily our own—we have relied on the voices of others. We now turn to our own summation before turning to the contributions of this volume.

First, the basic and boilerplate facts: they are clear and available at several places, and most of this paragraph reproduces for the convenience of the reader what is already in the public domain, and some of which we have already mentioned. David Gale was born in New York in 1921, and received an undergraduate degree from Swarthmore and a Master's degree from the University of Michigan before earning

---

[41] We have already had occasion to refer to Gale's expository gifts: Sotomayor reads her 1985 paper with Gale as providing a "survey of the existent theory for the Marriage and College Admissions models, complementing it with new results, providing new and shorter proofs for results already known in the literature."

[42] The Brock-Majumdar reference to "transfer problems and trade theory" stands out as a possible hagiographic exaggeration, but it is assuredly not. The skeptical reader is referred to [6, 7, 86] and their references. Indeed, Bhagwati discusses Gale's important position as a bridge between the two different communities, those of international trade theory and of mathematical economics. Also see Footnote 53 in the context of comparative-static exercises. As to the lack of a mention to game theory, the theories of matching and of auctions, are now seen as preeminent successes of game theory, though not necessarily regarded as such then.

[43] To be sure, the authors have no intention of pushing the distinction between economics and mathematics to the point where it becomes fruitless and a fool's errand—empty name-calling between "ideas and techniques."

a Ph.D. in Mathematics at Princeton. He worked at Brown University from 1950 through 1965 and then joined the faculty at the University of California, Berkeley. His principal appointment was in the Mathematics Department, but he maintained affiliations with the departments of Economics and Industrial Engineering. He won wide recognition for his research, and his awards included a Fulbright research fellowship, two Guggenheim fellowships, elections to the American Academy of Arts and Sciences and the National Academy of Sciences, USA, the Lester Ford Prize (for outstanding mathematical exposition), the John von Neumann Theory Prize (for fundamental contributions to operations research), and the Pirelli International Award (for the Internet Mathematics Museum 'MathSite'). He is an author of about ninety research publications and an author of three books, and has supervised sixteen doctoral students.

Even though Gale was elected to the membership of the National Academy of Sciences in the Mathematics section, an argument can be mounted ex-post that he could equally well have been in the section of Economic Sciences. His 1962 paper with Lloyd Shapley on "College Admissions and the Stability of Marriage," now holds a defining position in current mainstream economic theory that the 1950–1951 papers of Nash and the 1954 papers of Arrow–Debreu–McKenzie held for the first three post-WWII decades. It created a field as their papers did. Thus, in his preface to the 1990 book by Roth–Sotomayor [92], a book that Aumann sees as reporting "one of the outstanding success stories of the theory of games: theÂory and practice of matching markets," and which gives the Gale–Shapley paper pride of place.[44] He writes:

> The theoretical part of the story begins in 1962, with the publication of the famous Gale–Shapley paper. Since then, a large theoretical literature has grown from this paper. [T]he most dramatic development came in 1984, when Roth published his discovery that the Gale–Shapley algorithm had in fact been in practical use already since 1951 for the assignment of interns to hospitals in the United States; it had evolved by a trial-and-error process that spanned more than half a century.[45] The book describes this story in detail, as well as many other fascinating developments, on both the theÂoretical and practical side, and on the interface between the theoretical and the practical.

This success, one through which Gale's name is primarily known to the younger members of the economics profession,[46] should not be allowed to obscure the fact that he was one of the founding figures of at least three branches of modern economic theory:

---

[44] As of this writing, this paper is Gale's most cited paper: 9294 in *Google Scholar*, and 828 in *Mathscinet.* The reader interested in subsequent developments, including Gale's own follow-up work, can see Gale–Sotomayor [92, 93], even though their bibliographies are already more than three decades old. For its position in pure and applied mathematics, see [28], and Gale's own perspective on his result in [42]. Also see Footnotes 7, 17 and 78.

[45] Fleiner [28] dates it even further back to a theorem of Knaster that was a precursor of the Knaster–Tarski fixed-point theorem.

[46] Footnotes 7, 44 and 78, and the texts they footnote, all concern the 1962 Gale–Shapley paper.

(i) in addition to Arrow, Debreu, McKenzie, Nikaido, Uzawa and Negishi, he pioneered Walrasian equilibrium theory both in its original incarnation[47] as well as in its modern manifestation;
(ii) in addition to Ramsey, von Neumann, Samuelson, Malinvaud, Solow, Koopmans, McKenzie and Radner, he is an indispensable reference for capital theory, both aggregated and disaggregated; and finally,
(iii) in addition to von Neumann, Dantzig, Kuhn and Tucker, he gave shape to the interface of Walrasian and game theory on the one hand, and the theory of linear and concave programming on the other.

To put the matter in a better way, little of economic theory as documented in the standard textbooks exists beyond these branches as they emerge in his *oeuvre*. What follows is a brief elaboration on these three themes: the authors' own answer to the question "Who was David Gale?"

The contributions of Gale to Walrasian general equilibrium theory, even in its static atemporal formulation, has possibly been obscured by the success of Debreu's 1959 book, and now, also by popular accounts of the professional reception of the Arrow–Debreu–McKenzie papers of 1954.[48] This reception has revolved around fixed-point theory, and it is not fully appreciated to the extent that it perhaps ought, that the proof of the existence of a competitive economy in the 1954 Arrow–Debreu paper did not use a fixed-point theorem. It was based on converting the economy to a game, and then on the application of Debreu's 1952 generalization of Nash's theorem to the resulting game. To be sure, Debreu's "extension" of Nash's theorem relied on a fixed-point theorem, but the genealogy of this reliance can be directly and surely traced to Gale's suggestion of an alternative (first) proof of the Nash theorem already discussed above. This background is a necessary prelude to a fuller appreciation of Gale's 1955 contribution, and of Debreu's 1982 distinction between the simultaneous optimization and excess demand approaches to the existence question.

A clear and clarion testimony to the importance of Gale's 1955 paper is available in Florenzano (2009) where, under the title "Two lemmas that changed general equilibrium theory," she writes:

> Gale (1955) put the development of general equilibrium existence proofs initially based on a simultaneous optimization approach due to Arrow–Debreu (1954) on a new trail. The link between equilibrium existence and the Principal Lemma (typically referred to as the Gale–Nikaido–Debreu lemma) is simple, but no more than was, in Arrow–Debreu's paper, the link between equilibrium of an economy and equilibrium of the associated abstract economy.

In other words, Gale's paper provided an alternative whereby no connection need be made to a game (an abstract economy) and to the existence of an equilibrium of

[47] For an earlier defense of the claim that these magnificent seven names constitute an equivalence class as far as first wave of received Walrasian theory is concerned, see one of the authors [61].

[48] This is equally true of the contributions of the other names in Footnote 47; see, for example, an attempt at a corrective in [61]. How Debreu's book captured the imagination of the profession, and altered the very way that economics was done and written for at least three subsequent decades, is one of the foremost open questions of post-WWII economics. The style of the four rival texts mentioned above is of consequence for this question.

such a game, but rather only to the identification of a correspondence from prices to commodities generated by the economy itself and satisfying the properties of continuity, convex-valuedness and an inward-pointing property referred to as Walras's law. Gale's proof relied on a combinatorial lemma that goes under the names of Knaster, Kuratowski and Mazurkiewicz.[49]

To reiterate, the point is that the Gale–Nikaido–Kuhn–Debreu (GNKD) lemma is not a result on an economy or a game but a result on a correspondence that can be seen as encapsulating the underlying essence of the existence problem of the so-called private ownership economy.[50] So much for the first lemma, but what about the second? Florenzano writes:

> Almost 20 years after the publication of "The law of supply and demand," the second lemma, jointly authored with Mas-Colell,[51] reverses in some sense the direction stimulated by the first lemma to equilibrium existence proofs.

We might well ask what is the thrust of the phrase "reverses in some direction"? This is not so much for considerations relating to Gale's priority, but rather for a clearer understanding of the maturity and depth of Walrasian general equilibrium theory as is currently available. To see the matter from Gale's side, if his results are to be understood and to receive the reception that they merit, this reverse thrust needs to be clarified. Gale's suggested proof of Nash's theorem, referred to above, involved a collective fixed-point theorem for a correspondence taking the Cartesian product of a finite number of sets to itself, and the GNKD short-circuited this procedure by a correspondence obtained through aggregation. The Gale–Mas-Colell (GM) lemma

---

[49] This little lemma has since been extended by Fan and Shapley and given rise to acronyms such as the KKMS or FKKM. As we argue below, a case can also be made for the acronym KKMG; see [5, 17, 56, 73]. The last refers to the "Rainbow-KKM problem," a total search problem based on a generalization of KKMG lemma. Gale's own 1979 paper titled "The game of Hex and the Brouwer fixed-point theorem," reviewed by Fraenkel as MR-0551501, and Scarf [94, 95] are relevant in this connection; also see [88, 104], and Footnotes 60 and 84. As was pointed out to the authors by Simeon Reich, the issue goes back to what is now termed the Poincaré-Miranda theorem; see the Dinca–Mahwin [17] text and also [50]; to be sure there is no presumption of a unique fixed point, as brought out in [90, 91].

[50] It is interesting to ask how the names of Nikaido, Kuhn and Debreu get coupled with that of Gale in regard to this lemma. The *Mathscinet* archive is clear, and is worth investigating for the eye-witness accounts already there. Arrow's 1956–1957 testimony is crucial. It is he who reviewed Gale (1955), Nikaido (1956) and Debreu (1956); see MR-0077070, MR-0080571, MR-0081811. As to Kuhn (1956), Gale himself reviewed his contribution as a "new and shorter proof of the following lemma due to the reviewer"; see MR-0082411. The idea for the lemma was clearly in the air, and Karlin's book [60] already has the proof that is used in Debreu's text, though he does not reference, or discuss Debreu's 1956 PNAS piece; see Florenzano (2009) and Arrow's review for Debreu's contribution to the lemma, and also her own extension of it in [49]. In summary, Florenzano (2009) has yet to be supplemented by a fuller story of this reception and priority of Gale's paper: suffice it to say that a more or less complete narrative of the reception it would be based on the texts referenced in Footnote 26. The insertion of the letter K for Kuhn is due to [64] who can be consulted for a novel reading and formulation of the $GNKD$ lemma in game-theoretic terms. Also see Bryant [13] for an accessible exposition of the existence problem.

[51] Florenzano references the work as follows, "Formally, the fixed-point theorem (later referred to as Gale and Mas-Colell's lemma), stated in Sect. 2 of Gale and Mas-Colell (1975) and corrected in Gale and Mas-Colell (1979), gives a simpler and beautiful alternative to the complicated arguments given in Mas-Colell (1974)."

reverses this direction by going back to the Cartesian product of the individual action sets even for an economy without externalities as a consequence of non-ordered preferences. It is in this sense of the collective domain that the GM lemma reverses the direction of the GKND lemma.[52]

The existence theorem reported by Debreu in 1959, and then reworked in his 1982 Handbook chapter, is the *piéce de resistance* of neoclassical Walrasian general equilibrium theory, but two observations regarding Gale as one of the founding fathers of this theory deserve underscoring. The first is Gale's acute awareness that any existence theorem ought to be complemented by results on the cardinality, computablity, comparative-static properties and stability of an equilibrium whose existence is at issue so as to be useful.[53] The second is Gale's lack of defensiveness in going against the grain of static Walrasian theory and gleaning all that can be gleaned out by investigating simpler special cases: typically, linear functional forms or settings with indivisible commodities and objects of choice.[54]

Thus, in regard to the first observation, Arrow's review of Gale (1955) already makes clear his having gone beyond the question of existence of equilibrium.

> An elementary result on uniqueness is also obtained. Some other special results on Leontief models, the relation between supply functions and preference orderings, and the relation between competitive equilibrium and optimal distribution are derived.

Arrow felt it necessary to add that "all the results and the proofs are essentially contained in previously published work," a claim that we shall not question here, but simply point out that 15 years later, Arrow–Hahn [2] in their turn, also felt it necessary to single out Gale's contributions to both cardinality and stability issues. In their celebrated Chap. 9 on the uniqueness of competitive equilibria, They write:

> The study of uniqueness by examining the Jacobian of an appropriate mapping was initiated by Samuelson (1953–54). [His] proposition paid insufficient attention to the domain of a mapping and Gale and Nikaido (1965) showed it to be false by means of a counterexample.[55] They then considered a sufficient condition, here called "Gale property," which is the foundation of much of this chapter.[56]

---

[52] See Hildegard Dierker's review of Gale–Mas-Colell (1975) in MR-0381651 for this lemma. Also see [64] for ongoing work.

[53] To be sure, these *desiderata* go beyond Walrasian general equilibrium theory, and for one thing, investigation of comparative-static and stability properties of models is one of the hallmarks of the trade-theory community, and of MIT-Rochester economics as encapsulated in Samuelson's "correspondence principle." For this, see Footnote 42 and the text it footnotes.

[54] It is only fair to say that this strength and emphasis on discrete objects of choice led to the emergence of Gale as one of the founding figures of combinatorial optimization and geometry. On indivisibilities in the context of Walrasian theory, see [95]. For more recent work, see [15, 16] and their references. Also see Footnotes 5 and 39.

[55] For reference to Gale–Nikaido (1965), see the bibliography referred to in Footnote 4 and also Iyer's review MR-0204592. As of this writing, this paper has 931 citations in *Google Scholar*, and 96 in *Mathscinet.*

[56] See Definition 1 on p. 208, and its discussion on pp. 242–243 in [2]. The reader should beware that the Gale property of Arrow–Hahn is different from the same term used to characterize the compact-open topology of functional analysis: see Footnote 35.

Chapters 11–13 of Arrow–Hahn move beyond cardinality concerns to those of stability of processes studied by Walras and Edgeworth. They write "The first example to show the possibility of global instability was that of Scarf [1960]. The simpler case presented in Sect. 12.6 is due to Gale [1963]."[57] As to the comparison of competitive equilibria, in their Chap. 10, Arrow–Hahn identify the *Gale property* as crucial for the results that had been obtained uptil 1971.[58]

With regard to the second observation pertaining to static Walrasian theory with explicit functional forms, pride of place must surely be given to Gale's (1957) Rand memorandum: it is undoubtedly in the same equivalence class as Paul Samuelson's 1949 Rand memoranda on linear programming. It remains unpublished, and Gale begins his 1976 JME extension of it by viewing Eaves' algorithm as a constructive proof of his 1957 result. This is the elegant result with the simple statement that "any *irreducible* model has an equilibrium." We have had occasion to refer to this work above, and now offer the considered claim that Gale's influence in computer science can be directly traced to this memorandum. We can also couple it with Gale's IJGT 1984 paper "Equilibrium in a discrete exchange economy with money."[59] As it happened, Gale himself returns to the linear exchange model[60] in 1976.

It took only a year after his signal contribution to static Walrasian theory in 1955 for Gale to graduate from atemporal to intertemporal issues, and to what we have earlier identified as the second branch of his contributions to mainstream economic theory. In this, he very much kept company with his Illustrious contemporaries Malinvaud, McKenzie and Radner.[61] There are at least four different registers of postwar mainstream economic theory in which this transition from statics to dynamics manifests itself:

(a) the theory of adjustment processes to Walrasian equilibrium for which the 1963 Gale elaboration and simplification of Scarf's counterexample played such a significant role,

---

[57] See p. 322 in their Chap. 12. It is a little ironic that Gale [33], a simplification of an example of Scarf that was neither reviewed by *Mathscinet*, nor referenced by [2], put the naive attempts of Walrasian stability theory to rest.

[58] See the second paragraph on page 262 in [2] where the "Gale property" (GP) plays a salient role.

[59] These papers are not mentioned in the *Mathscinet* archive (also see Footnote 37). The point is that this work has given rise to a burst of recent interest in the computer science literature as well as that in computational economics and auction theory, especially in its combinatorial form. In addition to Moulin's [85] survey, the work an interested reader needs to track down is that of Chaudhuri, Duan, Garg, Shmyrev, Ye, among others; see [14–16, 20, 47, 98, 99, 103, 107]. Gale's 1979 paper with Eisenberg already mentioned plays a crucial role for this literature; see Footnote 76 and the text it footnotes.

[60] These two papers are, respectively, reviewed by Brock as MR-0441259 and by Mukherji as MR-0465100. Inspite of both being published in a mainstream mathematical economics journal, the second is largely ignored with only four *Google Scholar* citations. The first has the respectable count of 126. For the impact of these papers in the computer science community, see, for example, Shmyrev [98, 99]. As to Gale 1984, we return to it below; see Footnote 84. Also see Footnote 25 and the text it footnotes.

[61] To be sure, the trajectories of their entry into intertemporal Walrasian economics were different: both McKenzie and Radner entered the subject through the turnpike conjecture. For the analogy of atemporal and intertemporal allocation of resources, see [77].

(b) the theory of maximal balanced multi-sectoral growth associated with von Neumann and dating to 1936,
(c) the theory initiated by Ramsey in 1928 for the optimal intertemporal allocation of aggregate resources, and finally,
(d) the theory of overlapping generations due to Allais in 1947 and Samuelson in 1958.[62]

Gale's counterexample has been already referred to above, and we turn to the other three registers pertaining to allocation over time.

It is in his 1956 paper that Gale generalized and simplified the von Neumann model of an expanding economy by two essential additions to the original: he substituted von Neumann's requirement that each production process involve each good in the economy (as either input or output), with a weaker, more plausible condition; and he generalized it to a nonlinear setting.[63] A decade later, he continued with optimal multi-sectoral capital theory, but shifted from von Neumann's criterion of maximal balanced growth to allow for consumption that could yield strictly concave Rameysian felicities. Gale's 1967 paper "On optimal development in a multi-sector economy" was a seminal advance, and quickly recognized as such.[64] This definitive treatment of the multi-good Ramsey problem can also be seen as a generalization of the linear von Neumann–Gale model on the one hand, and of the continuous time extension of the papers of Samuelson and Solow on the other. This paper attained further refinement and maturity at the hands of Brock. In the 1968 *festschrift* for J. R. Hicks, Lionel McKenzie had already moved in this direction.[65] In the authors' judgment, this Gale–Malinvaud–McKenzie model now represents the modern theory of intertemporal resource allocation, a culmination of work that was initiated in the early fifties by the turn to infinite-dimensional spaces by Malinvaud, Debreu and Hurwicz, and continued by Radner in the very same 1967 issue.[66]

Following on the heels of the rediscovery and the reformulation of Ramsey aggregate capital theory in 1965 by Cass, Koopmans, Samuelson and Malinvaud—what

[62] For Allais' contribution to the so-called OLG model, see [78]. Also see [35].

[63] For von Neumann's model, see Champernowne's note in [54], Koompmans' exposition [70] and his references. In this connection to the von Neumann–Gale model, also see Gale's 1971 note [37]. This note, along with [44] and many other examples in the archive, bring out Gale's generous reception to criticisms and correction of his work.

[64] As of this writing, this paper has 827 citations in *Google Scholar*. The fact that it was not noted, leave alone reviewed, by the Editors of *Mathscinet* is a continuing warning to all investigators not to get fixated on one particular archive. Hahn's [54] anthology is blemished not only by the fact that it is limited to the articles in the journal *Review of Economic Studies* but also by the fact that he misses the famous 1967 issue of the journal, and thereby misses Gale's contributions, but also a paper culled from Brock's dissertation completed under Gale, and published in 1970. This criticism is not irrelevant to the current investigations of Ramsey theory by Duarte [21].

[65] His 1968 address already referred to above (see Footnote 32), serves as a useful complement to McKenzie's paper. It is a *tour de force*.

[66] The model has been applied to the economics of forestry and to the choice of technique in development planning, to take only two isolated examples. The associated literature, even if one confines one-self to economic theory, is immense: we refer the reader to [62, 63, 66, 108] and their references.

is currently referred to in the mainstream macroeconomic literature as the CKR model—the professions' attention shifted from the multi-sectoral to the Harrod–Domar–Solow–Swan aggregate setting of a single commodity that can be consumed or translated into capital. This shift was not lost on Gale: already his 1968 address [34] contained an analysis of the one-good case, and the next 5 years were to prove especially productive in this regard. The work is scattered in the applied mathematics literature.[67] In his investigation of the one-good model of economic development, Gale was naturally led to a conception whereby the consumption good was distinguished from the investment good, a special case of the two-sector Uzawa–Srinivasan model, one due to Joan Robinson and now current as the Robinson–Solow–Srinivsan (RSS) model studied by Mitra and his co-authors; see Gale [39].[68] But this was only one direction: by 1972, Gale [38] had already moved to the Allais–Samuelson overlapping generations (OLG) model. We turn to this.

Gale's 1972 consideration of the OLG model was a contribution aimed at applied mathematicians,[69] but in his two 1973 and 1977 JET papers "Pure exchange equilibrium of dynamic economic models" and "Dynamic behavior of a simple neoclassical model with wealth and capital unequal," he addressed the economics community. Already his turn to the simple aggregate model of growth had brought Gale under the sway of Samuelson's influence, and these two JET papers can be seen as Gale writing very much in the Samuelsonian mode. In terms of the search engines used here, these two papers were received in dissonant ways.[70] Burmeister's review is worth quoting at some length to see how deeply embedded Gale had become as a valued member of the neoclassical capital-theory community.

> The author constructs an unusual version of the standard one-commodity economic growth model in which consumer-workers are paid in IOU's issued by firms. However, firms own the entire capital stock, while the wealth of consumers consists entirely of accumulated IOU's. Except in cases where there exists an inefficient quantity of saving, the model converges to the so-called Golden Rule point where steady-state per capita consumption is maximal. This dynamic behavior is in marked contrast to the standard results obtained from one-commodity models both with and without money. The economic reason that the author obtains stable results, whereas most monetary growth models are featured by saddle-point equilibria, is that in his model economic agents are precluded from holding wealth in more than one form so that there do not exist any "unstable portfolio equilibrium equations" equating rates of return on alternative assets.

---

[67] In addition to [34], see the neglected exposition in [36], a paper with only 17 citations in *Google Scholar* and zero citations in *Mathscinet*—it was not sent for review by the editors. Majumdar [76] is an exception: this paper is one of the two references to Gale's work in his book. In this connection, also see Gale–Sutherland [45]. The Brock–Gale paper [10] is an exception in that it has had a good reception in the economics community with 96 citations in *Google Scholar* as of this writing.

[68] The reader is again referred to the Khan-Piazza references in Footnote 66.

[69] The contributions were presented at a "Symposium on Mathematical Economics" at the 1971 Fall meeting of the *Society for Industrial and Applied Mathematics (SIAM)*. In addition to Gale, the contributors included Chipman, Hansen, Koopmans, Radner and Samuelson.

[70] The first paper has been very well-received in economics with 729 *Google Scholar* citations, but not reviewed in *Mathscinet*. MR-0449529 limits itself to quoting the author's summary. On the other hand, though the Burmeister review of the 1977 paper, MR-0452525, is still worth reading, *Google Scholar* has only two citations to this paper.

We leave the subject with the judgment that a fuller integration of Gale's work on the OLG model still remains to be accomplished.[71] Indeed, one can go even further, and say that Gale's contribution to capital theory, and to the economics of time, has yet to be fully absorbed and received by the economics profession.

But let us backtrack to static Walrasian general equilibrium theory, and to the first branch of mainstream theory considered above. In addition to a theorem on the existence of competitive equilibrium in a private ownership economy, Debreu also presented in his 1959 text what later came to be called the two fundamental theorems of welfare economics.[72] A question that we have not asked, much less addressed, is whether Gale had anything to say about these theorems of normative analysis that were the very engine for the initiation of the Walrasian theory. As elsewhere in the *oeuvre*, the significance of his contributions is subtle and understated, and the answer to this question builds on the excursion into capital theory undertaken above. The genesis of modern welfare economics is the so-called socialist calculation debate, which in its most primitive form can be expressed by the aphorism that anything a competitive market can do, a planned economy can do equally well.[73] Less cryptically, and with a focus on the second rather than the first theorem, any "good" allocation of resources can be sustained under the convexity postulate as a suitable price-equilibrium. It is against this backdrop that Gale's work in capital theory can be most productively read: it was his contribution to the Koopmans–Kantorovich emphasis on prices, a simultaneous integration of the work of Ramsey, von Neumann and Malinvaud into Walrasian theory. We are well aware that these sentences require considerable unpacking for which it would not be appropriate to devote space in this preface. This must remain work for the future. We simply note that the word "good" acquires a technical meaning in his 1967 contribution, and the emphasis on decentralizing prices associated with efficient programs (and allocations) runs through the entirety of his work in intertemporal economic theory.[74]

We have been seeing Gale as one of the founding fathers of three branches of economic theory: we have considered him as a pioneer of both Walrasian competitive analysis and of Walrasian (and non-Walrasian) capital theory: we now turn to what we identify as the third branch of his work, that on the interface of Walrasian and game theory and the theory of linear and concave programming. Games against nature are two-person games of special interest to statistical theory, and within 2 years of completing his dissertation in 1949, Gale had five publications that were

---

[71] The dissonance referred to in Footnote 70 is not absent in Geanakoplos' definitive *Palgrave* entry on the OLG model; see [48] where Gale's 1972 and 1973 papers are not cited.

[72] See Koopmans' 1961 note [69] on these three theorems of Walrasian general competitive analysis. Also Bright [12] for an accessible exposition.

[73] See Hodgson [58], Levy-Peart [72] and their references. The authors thank Roger Backhouse for correspondence on this.

[74] This is the theme on which a further narrative could be constructed, and a connection to Lagrange multipliers forged. In this connection also see Gale's (1975) paper with Beveniste on the Cass criterion of intertemporal efficiency, and the salience given to it in [48]. This is also the point where Gale's indebtedness to Malinvaud's fundamental 1953 contribution is clear. As of this writing, Gale (1975) has 729 *Google Scholar* citations.

especially noted in statistical theory: the first three were concerned with two-person zero-sum games, and the other two with the mutual imbrication of programming and game theory based on Gale's investigation of convex polyhedral cones.[75] Eight years later, he published his only paper in the *Annals of Mathematical Statistics* that has proved very influential in computer science.[76] However, Gale's 1951 chapter jointly authored with Kuhn and Tucker stands out: it contained the first complete proof of the duality theorem of linear programming, and the theorem was used to prove the minimax theorem of zero-sum, two-person game theory. This paper uses convex analysis rather than fixed-point arguments to prove the minimax theorem, and implicitly provided a computational foundation for equilibrium points in zero-sum games.[77]

Whereas Gale's name is well known among professional and practicing economists as a result of the burgeoning Gale–Shapley matching literature,[78] mathematical logicians know him as a result of the literature on Borel determinacy associated with Gale–Stewart games.[79] Gale's 1949 Ph.D. dissertation was titled "Solutions of finite two-person games," and within 4 years he had generalized his result to infinite two-person games. In particular, this work led to the identification of a general class of games that do have *values,* but perhaps even more importantly, they have stimulated developments in set theory. Gale–Stewart games have had broad implications for mathematics, and what is remarkable in this line of work is that a theorem of "ordinary" mathematics makes indispensable use of the power set and replacement axioms, and thereby contributes to foundational set-theoretic issues.[80]

---

[75] The papers on two-person games constituted three chapters in the 1950 *Annals of Mathematics Studies* No. 24, and were reviewed by Wolfowitz: see MR-0039217, MR-0039219 and MR-0039221. The other two were chapters in the 1951 *Cowles Commission Monograph* No. 13, and were reviewed by Keifer: see MR-0042144 and MR-0046018.

[76] This is his 1959 paper "Consensus of subjective probabilities: the pari-mutuel method," reviewed in *Mathscinet* by Howard Raifa as MR-0102119, and as of this writing with 553 *Google Scholar* citations. For the influence of this paper, see, for example, [14, 85] and their references. Also see the Blackwell reference in Footnote 80.

[77] As of this writing, this paper has 521 *Google Scholar* citations; see Keifer's review in the footnote above for a reading of this paper that is as much a contribution to the theory of linear programming as it is to two-person games. Also see Footnote 24 and the text it footnotes.

[78] See Roth-Sotomayor [92] for its one-to-one and one-to many variants, as well as its one-sided and two-sided categorizations. We also refer the reader to Gale's involvement, and his papers with Alkan, Demange, Sotomayor and others referenced in the 2009 GEB bibliography, and also in a 2005 Ph.D. dissertation [97].

[79] As of this writing, this 1953 paper has 690 citations in *Google Scholar*, and 99 in *Mathscinet*; also see Sherman's review MR-0054922.

[80] Transfinite induction and set-theoretic issues arise naturally in other areas of game theory: the dialectical treatment of the *KRS* example relating to Radner–Rosenthal finite-player games with private information, as in [67] and their references, is an example. Game theorists and economists have so far ignored this direction of Gale's work. The interested reader can begin with the work of Friedman [30], Martin [80–83] and their references. The third Martin reference connects to Blackwell games as in [8]: this paper includes a reference to the von Neumann minimax theorem in its six references. Bryant's Master's dissertation [11] has a useful chapter on Gale–Stewart games. The Gale–Stewart paper is surely as much a contribution to pure mathematics as it is to game theory,

Indeed, 1953 was an important year in terms of Gale establishing himself as a game-theorist of the first rank. In addition to the Gale–Stewart paper, he also proposed a new solution for non-cooperative n-person games with perfect information. His co-author Kuhn gives a lucid account of the result.

> If Γ is the normal form of a game with perfect information, the successive application of the operations $\mathfrak{D}$ (simultaneously deleting all dominated strategies) and $\mathfrak{A}$ (simultaneously averaging equivalent strategies with equal weights) will reduce Γ to a game Γ^k with exactly one pure strategy for each player after a finite number of stages. These unique pure strategies for Γ^k are mixed strategies for Γ and are defined to be its "solution." They form an equilibrium point.[81]

Within 4 years, Gale's result had already made it into the standard textbook of the time. In their Chap. 7 on n-person games in normal form, Luce-Raiffa [74] emphasize the importance of *interchangeability* and the *equivalence* of equilibrium points and present Gale's result as a "plausible condition which may be added to isolate a single equilibrium point as more acceptable than the others."[82]

By now, we are in better position to answer the question posed in this preface, namely "Who was David Gale?" For the authors of these lines, bypassing his mastery over polyhedral cones and linear algebra and of issues of computation more generally, he is a mathematical scientist whose name figures prominently in several registers of the current mainstream of economics and game theory: in two fundamental lemmata of Walrasian theory, in two fundamental models of growth theory, one descriptive and the other normative, in two fundamental theorems of game theory, and in at least two algorithms.[83] Under the first, we refer to Gale–Nikaido–Kuhn–Debreu and to Gale–Mas-Colell; under the second, with the models associated with von Neumann–Gale and with Gale–Malinvaud–McKenzie; under the third, with the Gale–Stewart and Gale–Shapley theorems of game theory, both of which initiated two seminal subfields of inquiry, and finally, the Eisenberg–Gale and Gale–Shapley algorithms.[84]

---

and is an exemplar of the futility of drawing boundaries between pure and applied mathematics; also see Footnote 38 in this connection. The authors are especially grateful to David Schrittesser for introducing them to this literature.

[81] See Kuhn's review MR-0054921. As of this writing, this 1953 paper has 87 citations in *Google Scholar*, and 11 in *Mathscinet*.

[82] See p. 172 in [74]. The authors note that "failure of the general equilibrium notion to have these two properties raises much more serious questions as to its merits than could be raised against the minimax concept," and whereas they see it as a valuable start, they are not completely sold on what they refer to as the Gale procedure. Also see [65] in this connection.

[83] We ignore the use of Gale's name in combinatorial geometry and discrete optimization; see Footnote 39 for a sample of such naming. This whole issue of proper names denoting universal concepts and defining communities takes us into deep anthropological and philosophical waters that are beyond the scope of this little preview: the interested reader can begin with [71].

[84] We leave it to the reader to compare the Knaster–Kuratowski–Mazurkiewicz–Gale lemma with that of the Knaster–Kuratowski–Mazurkiewicz–Shapley lemma, and fit it in some slot that they think appropriate. The inclusion of Gale must rest on Gale 1984, a paper already referred to above in the context of computation; see Footnote 60. We now single it out in the context of the KKM lemma. Referring to the substantive theorem in this paper, Gale writes, "Our proof of the equilibrium theorem is based on a generalization of the well-known lemma of Knaster, Kuratowski and Mazurkewicz (KKM) in combinatorial topology." As of this writing, this paper has 208 *Google Scholar* citations. For the Eisenberg–Gale algorithm, see Footnotes 76 and 59.

We now turn to the 20 contributions of the volume to which this is a preface. We divide the volume into three parts: "On Economies and Games," "On Economic Dynamics and Growth," and "On Matching, Contracts, and Finance." All of these parts are devoted to those areas of mathematical economics to which David Gale gave his name.

The first part "On economies and games" contains nine chapters. The first chapter by Yuhki Hosoya is on Gale's contribution in revealed preference. More precisely, the author examines Gale's seminal paper published in 1960 which contains an example of a demand function that satisfies the weak axiom of revealed preference but cannot be shown to have been generated by some weak order. The author identifies a gap in Gale's proof and presents three complete proofs of Gale's claim.

In the second chapter, Roman A. Polyak studies nonlinear production–consumption equilibrium (NPCE) which is a combination and generalization of both classical linear programming (LP) and classical input–output (IO) models. In contrast to LP and IO the NPCE has both production and consumption components. Moreover, the production cost, the consumption and the factors (resources) availability are not fixed. Instead they are correspondent functions of the production output, prices of goods and prices of factors. Finding NPCE is equivalent to solving a variational inequality with a particular nonlinear operator and a simple feasible set.

In the third chapter of this part, John Stachurski improves the result of Eisenberg and Noe [26] who analyze systemic risk for financial institutions linked by a network of liabilities and showed that the solution to their model is unique when the financial system satisfies a regularity condition involving risk orbits. The author shows that this condition is not needed: under a mild convention used to handle edge cases, a unique solution always exists.

In the fourth chapter, Nengfa Wang and Zhe Yang discuss some coalitional transferable utility (TU) extensions of Gale–Mas-Colell economy. By the allocation rule of wealth, they introduce the notion of coalitional TU equilibria and prove the existence theorem under the regular conditions. Furthermore, the authors obtain a fuzzy extension of coalition TU equilibria.

In the fifth chapter, Gabrielle Demange studies visibility design in network games. In her chapter, individuals interact through social media networks by posting contributions, comments and the like. These actions entail complementarities and generate spillovers. A social media platform is carefully designed often with the objective of increasing users' activity, and the author studies visibility strategies which promote some individuals by making their contributions more visible than others. It is shown that under some specifications, optimal strategies are simple, making visible the individuals whose total impact, measured in terms of the total number of followers, is maximal.

In the sixth chapter, Janos Flesch and Eilon Solan study repeated games with tail-measurable payoffs. More precisely, they consider multi-player Blackwell games, extensions of the two-player Gale–Stewart perfect-information games that are repeated games where the payoff of each player is a bounded and Borel-measurable function of the infinite stream of actions played by the players during the game. The authors give an overview of recent work on Blackwell games by proving, in four different ways, that such games with a finite number of players, finite action sets, and tail-measurable payoffs admit an ϵ-equilibrium, for all positive ϵ.

In the seventh chapter, Sergiu Hart presents a formal write-up of the simple proof (1995) of the existence of calibrated forecasts by the minimax theorem, which moreover shows that N^3 periods suffice to guarantee a calibration error of at most $1/N$. This is work that builds on the author's work with Dean Foster and relies on Kuhn's (1953) result on extensive form games, and Blackwell's (1956) *approachability* result; see, for example, [29] and their references.

In the eighth chapter, Peicong Hu and Joel Sobel analyze the problem of getting permission when options are partially ordered. In their model a manager has access to expert advisers. The manager selects at most one project and can implement it only if one expert provides support. The game in which the manager consults experts simultaneously typically has multiple equilibria including one in which at least one expert supports the manager's favorite project. The authors describe the set of outcomes that survive iterative deletion of weakly dominated strategies. These outcomes typically exclude the manager's most preferred equilibrium outcome. They introduce sequential procedures and compare their performance to the simultaneous game.

Kakutani's 1941 extension of the Brouwer fixed-point theorem occupies a canonical position in both mathematical economics and in game theory. In the ninth chapter, Toru Maruyama follows Rudin's use of a fixed-point theorem of Kakutani's to extend the existence of a Haar measure on a compact topological group to a locally compact one.

The second part "On economic dynamics and growth," contains seven chapters. In the first chapter, Joel Blot and Hasan Yilmaz build a generalization of the Pontryagin maximum principle in infinite horizon in the framework of the continuous time. The authors improve a classical result of Halkin by lightening the assumptions and provide conditions to ensure that the multiplier associated to the criterion is equal to 1. They extend these results to the multiobjective problems. In this framework, a new sufficient condition of Pareto optimality for multiobjective optimal control problems in infinite horizon is obtained.

In the second chapter of this part, Liuchun Deng, Minako Fujio, and M. Ali Khan synthesize recent findings of the Robinson–Shinkai–Leontief (RSL) model of optimal growth, a miniature two-sector model that has been used as a workhorse in the literature of economic models of economic growth. This is a generalization of the two-sector (RSS) model due the Robinson, Solow and Srinivasan; a background of this literature is available in [62, 63, 66].

In the third chapter, Natali Hritonenko and Yuri Yatsenko analyze dynamic CES production functions with continuum of capital vintage inputs in a wide range of elasticity of substitution among vintages. The authors identify additional constraints for the original CES function and prove that the constructed modified CES function converges to a Cobb–Douglas production function with continuum of vintage inputs when the elasticity of substitution approaches unity.

In the fourth chapter, Kazumichi Iwasa and Kazuo Nishimura investigate a two-country dynamic model that allows international borrowing and lending. It is known that international trade patterns become undecidable when international borrowing and lending is allowed. In order to avoid this, the authors assume a consumable capital good to be nontradable. A key feature of their model is the existence of a continuum of steady-state levels of capital stocks, which enables them to examine how the initial amount of physical capital and assets in each country affects the amount of capital and assets in the steady state.

In the fifth chapter Mukul Majumdar, reminds the reader of the expanse of Gale's research and discusses some of his influential papers on static and dynamic equilibrium in the Walrasian tradition. Then the author considers the duality theorems of linear programming, and characterization of 'dual' variables as prices, saddle points and two-person zero-sum matrix games. The duality theorem of linear programming is used to prove that every two-person zero-sum matrix game has a solution in mixed strategies. The existence of a saddle point is proved by invoking the minimax theorem of Kakutani. The use of dual prices and the value loss technique to prove turnpike theorems are elaborated in this chapter.

In the sixth chapter of this part, Huiqi Wang, Lin Meng and George Xianzhi Yuan establish an interpretable prototype system by describing the development and growth mechanisms of small and micro enterprises under a tech-economic paradigm by applying the stochastic dynamical system approach with the associated stochastic resonance theory. They also develop an associated quantitative method of risk management and growth optimization under the measurement of systematic response risk, bilateral cooperation risk, and value at response risks performance. Based on the real-world supporting profiles from manufacturing sector in China, the authors investigate the validity by developing different strategies from micro profiles adjustment, mesoscopic structure control and macro climate regulation. It is found that any arbitrary strategy is not always effective, and some non-monotonic phenomena, counter to intuition, are the result of stochastic resonance features caused by the stochastic dynamical system.

In the final chapter of this part, Alexander J. Zaslavski studies the turnpike phenomenon for a generalized von Neumann–Gale model and obtains two turnpike results which describe the structure of approximate optimal trajectories. This paper builds on the author's work consolidated in [108], and earlier initiated in literature surveyed in [66].

The third part "On matching, contracts, and finance," contains four chapters. In the first chapter, Keisuke Bando and Toshiyuki Hirai study a relationship between stable outcomes and weakly setwise stable outcomes in two-sided many-to-many matching with contracts. They show that the set of stable outcomes coincides with the set of weakly setwise stable outcomes if the choice function of every agent on both sides is unitary, and the choice function of every agent on one side is substitutable, though the latter is generally larger than the former. The authors also show that both conditions are essential by demonstrating counterexamples each of which violates one of the conditions.

In the second chapter of this part, Guilherme Carmona and Krittanai Laohakunakorn discuss a version of Gale and Shapley's (1962) marriage market featuring a continuum of women and men based on Greinecker and Kah [52]. The authors define a deferred acceptance algorithm for this setting and show that it terminates after a finite number of iterations to yield a side-optimal stable matching.

In the third chapter of this part, Vladimir Danilov and Gleb Koshevoy study the broader problem of the existence of stable systems of contracts between agents of two complementary groups, say workers and firms. The authors introduce the concept of semi-stable pairs, define a dynamic process (a lá Gale–Shapley algorithm) on a set of semi-stable pairs and show that for agents' preferences defined by path-independent choice functions, the steady states yield stable sets of contracts. In addition to existence, this process makes it possible to show that the set of stable sets forms a complete lattice.

In the fourth chapter of this part, Shan Li, Xinpeng Li, and George Xianzhi Yuan study the general notions of upper and lower variances, which were initially introduced by Peter Walley for the bounded random variables. The authors obtain properties of upper and lower variances and new formulas for calculation of them. They take US markets for case study, which shows that the upper variance would be a useful tool to provide the risk management in the state or national level for the predication of (financial) systemic risk management by providing worst indicators to prepare necessary implementation for the worst markets to coming in the practice.

In conclusion, the 20 chapters that are presented in this volume are authored by a select group of experts in the area of Mathematical Economics, all of whom have been influenced by scientific work of David Gale. However, we should like to propose an exercise for the reader attempting to get a measure of David Gale's impact as a scholar: it is to slot each of the 20 contributions in the implicit categories that subtext the first part of the preface. It would be each reader's own answer to the question "Who was David Gale?" We have every hope and expectation that the difficulty of this exercise will play some important part not only in furthering Gale's legacy to mathematical economics but also gauging the polyvalence of his achievement.

References

1. Afacan, M.O., Hu, G., Li, J.: Housing markets since Shapley and Scarf, J. Math. Econ. **111**, online (2024)
2. Arrow, K.J., Hahn, F.H.: General Competitive Analysis. Holden Day Inc., San Francisco (1971)
3. Aurenhammer, F.: A relationship between Gale transforms and Voronoi diagrams. Discrete Appl. Math. **28**(1), 83–91 (1990)
4. Babecki, C., Thomas, R.R.: Graphical designs and Gale duality. Math. Program. **200**(2), 703–737 (2023)
5. Bapat, R.B.: A constructive proof of a permutation-based generalization of Sperner's lemma. Math. Program. **44**(1), 113–120 (1989)
6. Bhagwati, J.N., Brecher, R., Hatta, T.: The generalized theory of transfers and welfare: bilateral transfers in a multilateral world. Am. Econ. Rev. **73**(4), 606–618 (1983)

7. Bhagwati, J.N., Brecher, R., Hatta, T.: The paradoxes of immiserizing growth and donor-enriching "recipient-immiserizing" transfers: a tale of two literatures. Weltwirtschaftliches Archiv **120**(2), 228–243 (1984)
8. Blackwell, D.: Infinite G_δ-games of infinite length. Zastoswania (Matematuki) Applicationes Mathemticae Hugo Steinhaus Jubilee **10**, 99–101 (1969)
9. Bonnisseau, J.M., Florig, M., Jofré, A.: Continuity and uniqueness of equilibria for linear exchange economies. J. Optim. Appl. **109**(2), 237–263 (2001)
10. Brock, W.A., Gale, D.: Optimal growth under factor augmenting progress. J. Econ. Theory **1**(2), 229–243 (1969)
11. Bryant, R.: Borel determinacy and metamathematics. Unpublished Master's thesis, University of Toronto (2001)
12. Bryant, W.: Misinterpretations of the second fundamental theorem of welfare economics: barriers to better economic education. J. Econ. Educ. **25**(1), 75–80 (1994)
13. Bryant, W.: Conditions for the existence of market equilibrium. J. Econ. Educ. **28**(3), 230–254 (1997)
14. Chaudhuri, B.R., Kroer, C., Mehta, R., Nan, T.: Competitive equilibrium for chores: from dual Eisenberg-Gale to a fast, greedy, LP-based algorithm (2024). arXiv: arXiv:2402.10439v1
15. Danilov, V.I., Koshevoy, G.A., Murota, K.: Gross substitution, discrete convexity, and submodularity. Math. Soc. Sci. **41**(3), 251–273 (2001)
16. Danilov, V.I., Koshevoy, G.A., Lang, C.: Discrete convexity and equilibria in economies with indivisible goods and money. Discrete Appl. Math. **131**(3), 283–298 (2003)
17. Dinca, G., Mahwin, J.: Brouwer Degree: The Core of Nonlinear Analysis. Birkhäuser, Berlin (2012)
18. Doignon, J.-P.: The best-worst-choice polytope on four alternatives. J. Math. Psychol. **114**, 102769 (2023)
19. Donoho, D.L., Tanner, J.: Counting faces of randomlly projected polytopes when the projection radically lowers dimension. J. Am. Math. Soc. **22**(1), 1–53 (2009)
20. Duan, R., Mehlhorn, K.: A combinatorial polynomial algorithm for the linear Arrow-Debreu market. Inform. Comput. **243**(1), 112–132 (2015)
21. Duarte, P.G.: The growing of Ramsey's growth model. Hist. Polit. Econ. **41**(Winter Supplement), 161–181 (2009)
22. Eaves, B.C.: A Finite algorithm for the linear exchange model. Cowles Foundation Discussion Paper No. 622. Published in 1976 in the J. Math. Econ. **3**(2), 197–203 (1975)
23. Eaves, B.C.: Finite solution of pure trade markets with Cobb-Douglas utilities. Math. Program. Study **23**, 226–239 (1985)
24. Edelman, P.H., Peterson, S.W., Reiners, V., Stout, J.H.: Geochemical phase diagrams and Gale diagrams. SIAM J. Appl. Math. **64**(1), 231–251 (2003)
25. Edmonds, J.R.: Euler complexes (Oiks). Electron. Notes Discrete Math. **36**, 1289–1293 (2010)
26. Eisenberg, L., Noe, T.H.: Systemic risk in financial systems. Manag. Sci. **47**, 1289–1293 (2001)
27. Eisenbud, D., Popescu, S.: The projective geometry of the Gale transform. Manag. Sci. **230**(2), 127–133 (2000)
28. Fleiner, I.: A fixed-point approach to stable matchings and some applications. Math. Oper. Res. **28**(1), 103–126 (1988)
29. Foster, D., Hart, S.: 'Calibeating': beating forecasters at their own game. Longer Version of the Publication in Theor. Econ. **18**(2023), 1441–1474 (1922)
30. Friedman, H.: Higher set theory and mathematical practice. Ann. Pure Appl. Logic **2**(3), 326–357 (1971)
31. Gale, D.: Compact sets of functions and function rings. Proc. Am. Math. Soc. **1**(2), 303–308 (1950). Reviewed in Mathematical Reviews MR0036503 by S. B. Myers
32. Gale, D.: Price equilibrium for linear models of exchange. Rand Corporation Technical Report No. MP-M56 (1957)
33. Gale, D.: A note on global stability of competitive equilibrium. Naval Res. Logist. Quart. **10**(1), 81–87 (1963)

34. Gale, D.: A mathematical theory of optimal economic development. Bull. Am. Math. Soc. **74**(2), 207–223 (1968). Reviewed in Mathematical Reviews MR-0221835 by A. C. Williams
35. Gale, D.: A Note on the nonexistence of optimal price vectors in the general balanced-growth model of Gale: comment. Econometrica **40**(2), 391–392 (1970). Reviewed in Mathematical Reviews MR-0378727 by M. Intrilligator
36. Gale, D.: Nonlinear duality and qualitative properties of optimal growth. In: Abadie, J. (ed.) Integer and Nonlinear Programming, pp. 309–319. North Holland, Amsterdam (1970)
37. Gale, D.: Optimal strategy for serving in tennis. Math. Mag. **44**(4), 197–199 (1971)
38. Gale, D.: On equilibrium growth of dynamic economic models. In: Day, R.H., Robinson, S.M. (eds.) Mathematical Topics in Economic Theory and Computation, Philadelphia: Society for Industrial and Applied Mathematics (1972). Reviewed in Mathematical Reviews MR0406406 by G. Tintner
39. Gale, D.: On the theory of interest. Am. Math. Month. **80**(8), 853–868 (1973). Reviewed in *Mathematical Reviews* MR0378725 by V. Strnad
40. Gale, D.: The truth and nothing but the truth. Math. Intell. **11**(3), 62–67 (1989)
41. Gale, D.: Tracking the Automatic Ant and Other Mathematical Explorations. Springer, New York (1998)
42. Gale, D.: The two-sided matching problem. Origin, development and current issues. Int. Game Theory Rev. **3**(2–3), 237–252 (2001)
43. Gale, D.: Review of "A Beautiful Mind: A Biography". Math. Intell. **24**(3), 75–78 (2002)
44. Gale, D., Rockwell, R.: The Malinvaud eigenvalue lemma: correction and amplification of "Note and comment on the interest rate theorems of Malinvaud and Starrett". Econometrica **44**(6), 1323–1324 (1976)
45. Gale, D., Sutherland, W.R.: Analysis of a one good model of economic development. In: Dantzig, G.B., Veinott, A.F., Jr. (eds.) Mathematics of the Decision Sciences. American Mathematical Society, Providence (1968)
46. Garg, J., Végh, L.A.: A strongly polynomial algorithm for linear exchange markets. In: Proceedings of the 51st Annual ACM SIGACT Symposium on the Theory of Computing (STOC '19), June 23–26 (2019)
47. Garg, J., Husić, E., Végh, L.A.: An auction algorithm for market equilibrium with weak gross substitute demands. ACM Trans. Econ. Comput. **11**(3–4), 1–24 (2023)
48. Geanakoplos, J.: Overlapping generations of general equilibrium model. In: Jones, R. (ed.) The New Palgrave Dictionary of Economics, 3rd edn. Palgrave MacMillan, London (2017)
49. Geistdoerfer-Florenzano, M.: The Gale-Nikaido-Debreu lemma and the existence of transitive equilibrium with or without the free-disposal assumption. J. Math. Econ. **9**(2), 113–134 (1982)
50. Ghosh, A., Ali Khan, M., Uyanik, M.: The intermediate value theorem and decision-making in psychology and economics: an expositional consolidation. Games **11**(3–4), 51–75 (2022)
51. Gilbert, S.: Four poems from when she was kissed by the mathematician. Math. Intell. **31**(1), 18–20 (2009)
52. Greinecker, M., Kah, C.: Pairwise stable matching in large economies. Econometrica **89**(6), 2929–2974 (2021)
53. Hahn, F.H.: Review of "The Theory of Linear Economic Models". Econ. J. **72**(285), 175–176 (1962)
54. Hahn, F.H.: Readings in the Theory of Growth. MacMillan, London (1971)
55. Henry, M., Reynolds, D., Trapp, G.: A note on Gale's property. Topol. Proc. **7**(Research Announcement), 193–196 (1982)
56. Herings, P.J.:, An extremely simple proof of the K-K-M-S Theorem. Econ. Theory **10**(2), 361–367 (1997)
57. Hicks, J.R.: Linear theory. Econ. J. **70**(280), 671–709 (1960)
58. Hodgson, G.M.: Some limitations of the socialist calculation debate. Schmollers Jahrbuch **136**(1), 33–57 (2017)
59. Jorgenson, D.: Review of "The Theory of Linear Economic Models". J. Am. Stat. Assoc. **56**(294), 426–427 (1961)

60. Karlin, S.: Mathematical Methods and Theory in Games. Addison-Wesley, Programming and Economics, Reading (1959)
61. Khan, M.A.: On the finding of an equilibrium: Düppe-Weintraub and the problem of scientific credit. J. Econ. Literat. **59**(2), 590–633 (2021)
62. Khan, M.A., Piazza, A.: An overview of turnpike theory: towards the discounted deterministic case. Adv. Math. Econ. **14**, 39–68 (2011)
63. Khan, M.A., Piazza, A.: Turnpike theory: a current perspective. In: Jones, R.J.G. (ed.) The New Palgrave Dictionary of Economics, 3rd ed. Palgrave MacMillan, London (2017)
64. Khan, M.A., McLean, R.P., Uyanik, M.: On game-theoretic underpinnings of the Gale-Nikaido-Kuhn-Debreu lemma: generalized formulations (2024). Versions of this forthcoming in Economic Theory and in Economic Theory Bulletin
65. Khan, M.A., Pedersen, A.P., Schrittesser, D.: Two-person adversarial games are zero-sum: an elaboration of a folk theorem (2024). A version forthcoming in Economics Letters
66. Khan, M.A., Zaslavski, A.J.: On two classical turnpike results for the Robinson-Solow-Srinivasan model. Adv. Math. Econ. **13**(2010), 47–97 (2010)
67. Khan, M.A., Zhang, Y.: On sufficiently-diffused information in Bayesian games: a dialectical formalization. Adv. Math. Econ. **21**(2017), 47–73 (2017)
68. Kjeldsen, T.H.: A contextualized historical analysis of the Kuhn-Tucker theorem in nonlinear programming: the impact of World War II. Historia Mathematica **27**(4), 331–361 (2000)
69. Koopmans, T.C.: Convexity assumptions, allocative efficiency, competitive equilibrium. J. Polit. Econ. **69**(5), 478–479 (1961)
70. Koopmans, T.C.: Economic growth at a maximal rate. Quart. J. Econ. **78**(3), 143–160 (1964)
71. Kripke, S.: Naming and Necessity. Harvard University Press, Cambridge (1980)
72. Levy, D., Peart, S.J.: The socialist calculation debate. In: Jones, R.J.G. (ed.) The New Palgrave Dictionary of Economics, 3rd ed. Palgrave MacMillan, London (2017)
73. Lock, E., Zephyr, Q, Teytelboym, A.: The computational complexity of the housing market (2024). arXiv: arXiv:2402.08484v2
74. Luce, D., Raiffa, H.: Games and Decisions: Introduction and Critical Survey. Wiley, New York (1957)
75. Majumdar, M. (ed.): Equilibrium and Dynamics: Essays in Honor of David Gale. Palgrave MacMillan, New York (1992)
76. Majumdar, M.: Sustainability and Resources: Theoretical Issues in Dynamic Economics. World Scientific, Singapore (2020)
77. Malinvaud, E.: The analogy between atemporal and intertemporal theories of resource allocation. Rev. Econ. Stud. **28**(3), 143–160 (1961)
78. Malinvaud, E.: The overlapping generations model in 1947. J. Econ. Literat. **25**(1), 103–105 (1987)
79. Marcus, D.A.: Gale diagrams of convex polytopes and positive spanning sets of vectors. Discrete Appl. Math. **9**(1), 47–67 (1984)
80. Martin, D.A.: Borel determinacy. Ann. Math. **102**(2), 363–371 (1975)
81. Martin, D.A.: A purely inductive proof of Borel determinacy. In: Nerode, A., Shore, R.A. (eds.) Recursion Theory: Proceedings of the Symposium in Pure Mathematics, vol. 42, pp. 303–308. American Mathematical Society, Providence (1982)
82. Martin, D.A.: An extension of Borel determinacy. Ann. Pure Appl. Logic **49**(2), 279–293 (1990)
83. Martin, D.A.: A simple proof that determinacy implies Lebesgue measurability, Rendiconti del Seminario Matematico - Politecnico di Torino**61**(4), 393–397 (2003). Reviewed in Mathematical Reviews MR2040199 by J. Dupare
84. Morrill, T., Roth, A.E.: Top trading cycles. J. Math. Econ. **112**, online (2024)
85. Moulin, H.: Fair division in the internet age. Ann. Rev. Econ. **11**, 407–41 (2019)
86. Moulin, J.H., Thompson, W.: Can everyone benefit from growth? Two difficulties. J. Math. Econ. **17**(4), 339–345 (1988)
87. Nikaido, H.: Gendai Keitzaigaku No Sugakuteki Hoho (in Japanese) Iwanami Shoten, Tokyo (1960). English translation 1970, Introduction to Sets and Mappings in Modern Economics. Amsterdam, North Holland

88. Petri, H., Voorneveld, M.: No bullying! A playful proof of Brouwer's fixed-point theorem. J. Math. Econ. **78**(1), 1–5 (2018)
89. Pulleyblank, W.R. (ed.): Progress in Combinatorial Optimization. Academic Press, New York (1984)
90. Reich, S., Zaslavski, A.J.: Most continuous and increasing functions have two different fixed points. Carpathian J. Math. **39**(1), 231–236 (2023)
91. Reich, S., Zaslavski, A.J.: Most continuous and increasing functions on a compact real interval have infinitely many different fixed points. J. Fixed Point Theory Appl. online (2023)
92. Roth, A.E., Sotomayor, M.: Two-Sided Matching: A Study in Game-Theoretic Modelling and Analysis. Cambridge University Press, New York (1990)
93. Roth, A.E., Sotomayor, M.A.O.: Two-sided matching, Chapter 16 In: Aumann, R.J., Hart, S. (eds.) Handbook of Game Theory, vol 1. Elsevier Science Publications, New York (1992)
94. Scarf, H.E.: The origins of fixed point methods. In: Lenstra, J.K., Rinnooy Kan, A.H.G., Schrijver, A. (eds.) History of Mathematical Programming, A Collection of Personal Reminiscences, pp. 126–134, Amsterdam, North Holland (1991)
95. Scarf, H.E.: The allocation of resources in the presence of indivisibilities. J. Econ. Perspect. **8**(4), 111–128 (1994)
96. Schneider, R.: Random Gale diagrams and neighborly polytopes in high dimensions. Beiträge zur Algebra und Geometrie **62**(3), 641–650 (2021)
97. Scott, S.: A study of stable marriage problems with ties. University of Glasgow, Ph.D. dissertation (2005)
98. Shmyrev, V.I.: An algorithm for finding equilibrium in the linear exchange model with fixed budgets. J. Appl. Indus. Math. **3**(4), 505–518 (2009). Original Russian text published in 2008, Sibirskii Zhurnal Industrial'noi Matematiki **XI**(2), 139–154
99. Shmyrev, V.I.: An iterative approach for searching an equilibrium in piecewise Linear exchange model. In: Kochetov, Y., Khachay, M., Beresnev, V., Nurminski, E., Pardalos, P. (eds.) Proceedings Discrete Optimization and Operations Research, pp. 61–74. Springer, Berlin (2016)
100. Sobel, J., Gale, D. (1921–2008), Jones, R.J.G. (eds.): The New Palgrave Dictionary of Economics, 3rd ed. Palgrave MacMillan, London (2017)
101. de Oliviera Sotomayor, M.A.: A collection of papers dedicated to David Gale on the occasion of his 85th birthday. Int. J. Game Theory **36**(3/4), 317–319 (2008)
102. Thomas, R.R.: Lectures in Geometric Combinatorics. American Mathematical Society, Providence (2006)
103. van der Laan, G., Talman, D., Yang, Z.: Existence of an equilibrium in a competitive economy with indivisibilities and money. J. Math. Econ. **10**(2), 361–367
104. Voorneveld, M.: An elementary direct proof that the Knaster-Kuratowski-Mazurkiewicz lemma implies Sperner's lemma. Econ. Lett. **158**(1), 1–2 (2012)
105. Willard, D.: General Topology. Addison-Wesley, New York (1970)
106. Wotzlaw, R.F., Ziegler, G.M.: A lost counterexample and a problem on illuminated polytopes. Am. Math. Month. **118**(6), 534–543 (2011)
107. Ye, Y.: A path to the Arrow-Debreu competitive market equilibrium. Math. Program. **111**, 315–358 (2008)
108. Zaslavski, A.J.: Optimal Control Problems arising in Mathematical Economics. Monograph in Mathematical Economics, vol. 5. Springer, Singapore (2022)
109. Ziegler, G.: Lectures on Polytopes, 7th edn. Springer, Berlin (2007)

Revealed Preference, Economies and Games

On Gale's Contribution in Revealed Preference Theory

Yuhki Hosoya

Abstract We investigate Gale's important paper published in 1960. This paper contains an example of a candidate of the demand function that satisfies the weak axiom of revealed preference and that is doubtful that it is a demand function of some weak order. We examine this paper and first scrutinize what Gale proved. Then we identify a gap in Gale's proof and show that he failed to show that this candidate of the demand function is not a demand function. Next, we present three complete proofs of Gale's claim. First, we construct a proof that was constructible in 1960 by a fact that Gale himself demonstrated. Second, we construct a modern and simple proof using Shephard's lemma. Third, we construct a proof that follows the direction that Gale originally conceived. Our conclusion is as follows: although, in 1960, Gale was not able to prove that the candidate of the demand function that he constructed is not a demand function, he substantially proved it, and therefore it is fair to say that the credit for finding a candidate of the demand function that satisfies the weak axiom but is not a demand function is attributed to Gale.

Keywords Consumer theory · Revealed preference · Weak axiom · Strong axiom · Shephard's lemma · Integrability theory

1 Introduction

Since the weak axiom of revealed preference was created by Samuelson [20], there had been an active debate in consumer theory regarding what type of choice behavior can be expressed by utility maximizing behavior. When a candidate of the demand function is given, what are the conditions for it to behave as if it were a consequence of some weak order maximization problem? In the 1950s, several researchers argued that the weak axiom of revealed preference is a necessary and sufficient condition for this problem. Although Houthakker [12] presented the strong axiom of revealed preference and argued that it was the answer to the above question, Houthakker

Y. Hosoya (✉)
Faculty of Economics, Chuo University, Tokyo, Japan
e-mail: ukki@gs.econ.keio.ac.jp

M. A. Khan et al. (eds.), *Matching, Dynamics and Games for the Allocation of Resources*, Monographs in Mathematical Economics 7,
https://doi.org/10.1007/978-981-96-6661-4_2

himself does not present any example of a function that obeys the weak axiom but violates the strong axiom, and some economists seemed to believe that the weak axiom of revealed preference is sufficient.

Gale [5] is a landmark paper that offered an answer to this controversy. He presented a candidate of the demand function and showed that, although it satisfies the weak axiom of revealed preference, it is not a consequence of any weak order maximizing behavior. This made it clear that the weak axiom of revealed preference is not sufficient to answer the question in the previous paragraph. Eventually, Richter [18] rigorously showed that Houthakker's claim is correct, and that settled this issue.

However, some economists argue that this thesis presented by Gale is insufficient. According to them, Gale only showed that the candidate of the demand function he constructed satisfies the weak axiom of revealed preference, but he did not rigorously show that it does not correspond to any weak order. The primary objective of the present paper is to investigate this claim; that is, to investigate what Gale's contribution is and whether the example Gale constructed is really not a demand function.

We first scrutinize Gale's constructed candidate of the demand function and his discussion on this function. Gale showed that this candidate satisfies the weak axiom of revealed preference and the corresponding inverse demand function does not satisfy Jacobi's integrability condition. However, in Gale's paper, there is no sufficient proof for this function to be not able to be represented by the weak order maximizing behavior. Thus, the assertion in the previous paragraph is correct.

Next, we present three complete proofs of Gale's claim. The first is a proof using another fact that Gale himself presented in 1960. In addition to the argument above, Gale showed that the constructed candidate of the demand function does not satisfy the strong axiom of revealed preference. As mentioned above, the strong axiom of revealed preference is a necessary and sufficient condition for a function to be a demand function that corresponds to a weak order. This was proved by Richter in [18]; thus, this was not yet proved in 1960. So can Gale's claim be proved without using Richter's result but only from what was known in 1960? We answer this question affirmatively. From this point of view, we think that it is fair to say that while the assertions in the two previous paragraphs are indeed correct, the proof of Gale's claim can be easily reconstructed from what Gale himself showed, and thus it is Gale who showed that the weak axiom of revealed preference is insufficient for rationalizability.

Second, we provide a modern and sophisticated proof of Gale's claim. The above proof is correct, but Gale's proof that the function does not satisfy the strong axiom of revealed preference is too genius to be easily conceived. In this section, we provide a more mechanical, easy, and simple proof using Shephard's lemma. Perhaps, this proof is the most standard one at the present time.

Third, we prove the above result using Jacobi's integrability condition as addressed by Gale. Although both of the above proofs are simple and rigorous, they are fundamentally different from the proof scheme that Gale himself would have had in mind. Therefore, we start from the inverse demand function, construct a method for calculating a binary relation, and prove that Jacobi's integrability condition is equivalent to the transitivity of this binary relation. Because Gale's function can be seen as a consequence of the maximizing behavior of this binary relation, the fact that this binary relation is not weakly ordered means that this function cannot be rationalized. Although this proof is heavily complicated, this is probably the closest to the proof that Gale had in mind.

We prove our main claims in the main text, but we have chosen to place some of the proof in the appendix. In particular, the result of Richter [18] (Theorem 1), Shephard's lemma (Theorem 3), and the method for constructing a binary relation from an inverse demand function (Theorem 4) are all results that are too heavy to read the proofs in situ, and thus we place the proofs of these results in the appendix.

This paper is organized as follows. In Sect. 2, we introduce many symbols and terms that are necessary for this paper. In Sect. 3, we scrutinize Gale's paper and the result that Gale showed. Section 4 contains three proofs of Gale's claim listed above. Section 5 is the conclusion.

2 Preliminaries

In this section, we provide basic knowledge of consumer theory. First, we define $\mathbb{R}^M_+ = \{x \in \mathbb{R}^M | x_i \geq 0 \text{ for all } i\}$ and $\mathbb{R}^M_{++} = \{x \in \mathbb{R}^M | x_i > 0 \text{ for all } i\}$. The former is called the **nonnegative orthant** of $\mathbb{R}^M$, and the latter is called the **positive orthant** of $\mathbb{R}^M$, respectively. If $M = 1$, then this symbol is abbreviated, and these sets are simply written as $\mathbb{R}_+$ and $\mathbb{R}_{++}$. For $x, y \in \mathbb{R}^M$, we write $x \geq y$ if and only if $x - y \in \mathbb{R}^M_+$, and $x \gg y$ if and only if $x - y \in \mathbb{R}^M_{++}$.

In this paper, Ω denotes the consumption space, and we assume that Ω is a subset of $\mathbb{R}^n$. An element $x \in \Omega$ is called a **consumption plan**, and x_i denotes the amount of consumption for the ith commodity.

Choose any binary relation R on Ω.[1] We write xRy instead of $(x, y) \in R$. We say that R is[2]

- **reflexive** if xRx for all $x \in \Omega$,
- **complete** if for all $x, y \in \Omega$, either xRy or yRx,
- **transitive** if xRy and yRz imply xRz,
- **p-transitive** if $\dim(\text{span}\{x, y, z\}) \leq 2$, xRy and yRz imply xRz,
- **continuous** if R is closed in Ω^2,
- **symmetric** if xRy implies yRx,
- **asymmetric** if xRy implies $\neg(yRx)$,
- **antisymmetric** if xRy and yRx imply $x = y$,
- **monotone** if $x \gg y$ implies xRy and $\neg(yRx)$,
- **convex** if xRy, yRx and $t \in [0, 1]$ imply $[(1 - t)x + ty]Rx$, and

[1] That is, $R \subset \Omega^2$.

[2] Note that, reflexivity, completeness, transitivity, symmetry, asymmetry, and antisymmetry can be defined even when R is a binary relation for some abstract set X.

- **strictly convex** if xRy, yRx, $x \neq y$, and $t \in]0, 1[$ imply $[(1-t)x + ty]Rx$ and $\neg(xR[(1-t)x + ty])$.

We call a complete and transitive binary relation a **weak order**, a transitive and asymmetric binary relation a **strong order**, and a reflexive, transitive, and symmetric binary relation an **equivalence relation**. Suppose that $\succsim$ is a binary relation, and define

$$x \succ y \Leftrightarrow x \succsim y \text{ and } y \not\succsim x,$$

$$x \sim y \Leftrightarrow x \succsim y \text{ and } y \succsim x.$$

It is known that if $\succsim$ is a weak order, then $\succ$ is a strong order and $\sim$ is an equivalence relation.[3]

Choose any binary relation $\succsim$ on Ω. A function $u : \Omega \to \mathbb{R}$ is said to **represent** $\succsim$, or to be a **utility function** of $\succsim$, if

$$x \succsim y \Leftrightarrow u(x) \geq u(y).$$

It is easy to show that if $\succsim$ is represented by a function, then it is a weak order. Debreu [3] showed that if $\succsim$ is a continuous weak order and Ω is connected, then there exists a continuous function that represents $\succsim$.

Choose $p \in \mathbb{R}^n_{++}$ and $m > 0$. Define

$$\Delta(p, m) = \{x \in \Omega | p \cdot x \leq m\},$$

and for a binary relation $\succsim$ defined on Ω, define

$$f^{\succsim}(p, m) = \{x \in \Delta(p, m) | y \not\succ x \text{ for all } y \in \Delta(p, m)\}.$$

We call this multi-valued function $f^{\succsim}$ a **semi-demand function** corresponds to $\succsim$, and if $\succsim$ is a weak order, then we call $f^{\succsim}$ a **demand function** corresponds to $\succsim$. If $\succsim$ is represented by u, then $f^{\succsim}$ is also written as f^u.

Note that, for given $(p, m) \in \mathbb{R}^n_{++} \times \mathbb{R}_{++}$, $f^u(p, m)$ is the set of solutions to the following maximization problem:

$$\begin{aligned} \max \quad & u(x) \\ \text{subject to.} \quad & x \in \Omega, \\ & p \cdot x \leq m. \end{aligned} \tag{1}$$

[3] Only transitivity of $\succ$ is not trivial, and thus we show this. Suppose not. Then, there exist $x, y, z \in \Omega$ such that $x \succ y$ and $y \succ z$ but $x \not\succ z$. Because $x \succ y$ and $y \succ z$, $x \succsim y$ and $y \succsim z$, and thus $x \succsim z$. Hence, we must have $z \succsim x$. By the transitivity of $\succsim$, we have that $z \succsim y$, which contradicts $y \succ z$.

We call this problem (1) the **utility maximization problem** for the utility function u. Therefore, the demand function f^u is explained as the (possibly multi-valued) solution function to the utility maximization problem (1).

Suppose that $f : P \to \Omega$, where P is a nonempty cone that is included in the following set:

$$\{(p, m) \in \mathbb{R}^n_{++} \times \mathbb{R}_{++} | \Delta(p, m) \neq \emptyset\}.$$

We call this function f a **candidate of demand** (CoD) if $f(p, m) \in \Delta(p, m)$ for all $(p, m) \in P$. If $f = f^{\succsim}$, then we say that f corresponds to $\succsim$ or $\succsim$ corresponds to f. Moreover, if $\succsim$ is represented by u, then we say that f corresponds to u or u corresponds to f. We call a CoD f a **semi-demand function** if it is a semi-demand function corresponds to some binary relation $\succsim$, and a **demand function** if it is a demand function corresponds to some weak order $\succsim$.

For a CoD f, let $R(f)$ be the range of f: that is,

$$R(f) = \{x \in \Omega | x = f(p, m) \text{ for some } (p, m) \in P\}.$$

We say that a CoD f is **homogeneous of degree zero** if

$$f(ap, am) = f(p, m)$$

for any $a > 0$. Note that, because P is a cone, the above definition has no problem. Next, we say that a CoD f satisfies **Walras' law** if

$$p \cdot f(p, m) = m$$

for any $(p, m) \in P$. Note that, if $f = f^{\succsim}$ for some weak order $\succsim$, then f is automatically homogeneous of degree zero. Moreover, if $\succsim$ is monotone, then f automatically satisfies Walras' law.

For a CoD f, define a binary relation $\succ_r$ on Ω such that

$$x \succ_r y \Leftrightarrow \exists (p, m) \in P \text{ s.t. } f(p, m) = x, \; y \in \Delta(p, m) \setminus \{x\}.$$

We call this relation $\succ_r$ the **direct revealed preference relation**, and we say that f satisfies the **weak axiom of revealed preference** (or simply, the **weak axiom**) if $\succ_r$ is asymmetric.

Choose any binary relation R on Ω, and define

$$R^* = \cap \{\bar{R} \subset \Omega^2 | R \subset \bar{R} \text{ and } \bar{R} \text{ is transitive}\}.$$

This binary relation R^* is called the **transitive closure** of R. Because the intersection of any family of transitive binary relations is also transitive, we have that R^* is transitive. In the appendix, we show that xR^*y if and only if there exists a finite sequence $x_1, \ldots, x_k$ such that $x_1 = x$, $x_k = y$ and $x_i R x_{i+1}$ for all $i \in \{1, \ldots, k-1\}$.

Specifically, let $\succ_{ir}$ be the transitive closure of $\succ_r$. We call this relation $\succ_{ir}$ the **indirect revealed preference relation**, and we say that f satisfies the **strong axiom of revealed preference** (or simply, the **strong axiom**) if $\succ_{ir}$ is asymmetric. Clearly, the strong axiom implies the weak axiom.

Finally, suppose that $f : P \to \Omega$ is a CoD that is differentiable at (p, m). Define

$$s_{ij}(p, m) = \frac{\partial f_i}{\partial p_j}(p, m) + \frac{\partial f_i}{\partial m}(p, m) f_j(p, m).$$

The $n \times n$ matrix $S_f(p, m)$ whose (i, j)th component is $s_{ij}(p, m)$ is called the **Slutsky matrix**. We sometimes call the matrix-valued function S_f itself the Slutsky matrix. This matrix has the following alternative notation:

$$S_f(p, m) = D_p f(p, m) + D_m f(p, m) f^T(p, m),$$

where D_p and D_m denote the partial differential operators, and f^T denotes the transpose of f.[4]

3 Gale's Example

Gale [5] presented a CoD that satisfies the weak axiom of revealed preference but is doubtful to be a demand function. In this section, we introduce Gale's example. Throughout this section, we consider $\Omega = \mathbb{R}^n_+$ and $P = \mathbb{R}^n_{++} \times \mathbb{R}_{++}$, and usually assume that $n = 3$.

Let

$$A = \begin{pmatrix} -3 & 4 & 0 \\ 0 & -3 & 4 \\ 4 & 0 & -3 \end{pmatrix},$$

and define

$$h_A(p) = \frac{1}{p^T A p} A p$$

on the set

$$C = \{p \in \mathbb{R}^3_{++} | Ap \geq 0\}.$$

Note that, if $p \in C$, then $Ap \neq 0$ and $p^T Ap > 0$, and thus $h_A(p)$ is well-defined. Indeed, if $p \in \mathbb{R}^3_{++}$ and $Ap \leq 0$, then

$$4p_2 \leq 3p_1,\ 4p_3 \leq 3p_2,\ 4p_1 \leq 3p_3,$$

[4] Throughout this paper, A^T denotes the transpose of the matrix A.

and thus, $64p_1 \le 27p_1$, which is a contradiction. Hence, if $p \in C$, then $Ap \ge 0$ and $Ap \ne 0$, which implies that $p^T Ap > 0$.

Now, choose any $p \in \mathbb{R}^3_{++}$, and define $\bar{p}$ as follows.

(I) If $p \in C$, then we define $\bar{p} = p$.

(II) Suppose that for some $(i, j, k) \in \{(1, 2, 3), (2, 3, 1), (3, 1, 2)\}$, $-3p_i + 4p_j \le 0$, and $-3p_j + 4p_k \le 0$. By our previous argument, we have that $-3p_k + 4p_i > 0$. In this case, define $\bar{p}_i = \frac{16}{9}p_k$, $\bar{p}_j = \frac{4}{3}p_k$, and $\bar{p}_k = p_k$.

(III) Suppose that for some $(i, j, k) \in \{(1, 2, 3), (2, 3, 1), (3, 1, 2)\}$, $-3p_i + 4p_j \le 0$, $-3p_j + 4p_k \ge 0$, and $-3p_k + 4p_i \ge 0$. We separate this case into two subcases.

   (i) If $16p_j - 9p_k \ge 0$, then define $\bar{p}_i = \frac{4}{3}p_j$, $\bar{p}_j = p_j$, and $\bar{p}_k = p_k$.
   (ii) If $16p_j - 9p_k \le 0$, then define $\bar{p}_i = \frac{4}{3}p_j$, $\bar{p}_j = p_j$, and $\bar{p}_k = \frac{16}{9}p_j$.

In the appendix, we check that $\bar{p}$ is well-defined and the mapping $p \mapsto \bar{p}$ is continuous. Note that, in any case, $\bar{p} \le p$ and $\bar{p} \in C$. Moreover, $\bar{p}$ is homogeneous of degree one with respect to p: that is, $\overline{ap} = a\bar{p}$ for all $a > 0$. Define

$$f^G(p, m) = h_A(\bar{p})m. \tag{2}$$

We call this CoD f^G **Gale's example**.

It is easy to show that Gale's example f^G is a continuous CoD that is homogeneous of degree zero and satisfies Walras' law. We show that f^G satisfies the weak axiom. Suppose not. Then, there exist $x, y \in \Omega$ such that $x \succ_r y$ and $y \succ_r x$. Therefore, $x \ne y$, and there exist $(p, m), (q, w) \in P$ such that $x = f^G(p, m),\ p \cdot y \le m$ and $y = f^G(q, w),\ q \cdot x \le w$. Because Gale's example is homogeneous of degree zero, replacing $(p, m), (q, w)$ with $\frac{1}{m}(p, m), \frac{1}{w}(q, w)$, we can assume without loss of generality that $m = w = 1$. Moreover, because $\bar{p} \le p$ and $\bar{q} \le q$, we can assume without loss of generality that $p = \bar{p},\ q = \bar{q}$. Then,

$$q^T Ap \le p^T Ap,\ p^T Aq \le q^T Aq.$$

Define

$$\lambda = \frac{p^T Ap}{q^T Ap}.$$

By definition, $\lambda \ge 1$. Define $r = \lambda q$. Then, $(r - p)^T Ap = 0$, and

$$p^T Ar = \lambda p^T Aq \le \lambda^2 p^T Aq \le \lambda^2 q^T Aq = r^T Ar,$$

which implies that

$$0 \le (r - p)^T Ar = (r - p)^T A(r - p).$$

Let $z = r - p$. By the above arguments,

$$z^T Az \geq 0,\ z^T Ap = 0.$$

If $z \neq 0$, then for some $(i, j, k) \in \{(1, 2, 3), (2, 3, 1), (3, 1, 2)\}$, $z_i \geq 0$, $z_k \leq 0$ and $z_i - z_k > 0$. If $z_j \geq 0$, then

$$\begin{aligned} 0 \leq z^T Az &= -3z_i^2 - 3z_j^2 - 3z_k^2 + 4z_iz_j + 4z_jz_k + 4z_kz_i \\ &= -(3z_i^2 - 4z_iz_j + 3z_j^2) - 3z_k^2 + 4z_k(z_i + z_j) < 0, \end{aligned}$$

which is a contradiction. If $z_j \leq 0$, then

$$0 \leq z^T Az = -(3z_j^2 - 4z_jz_k + 3z_k^2) - 3z_i^2 + 4z_i(z_j + z_k) < 0,$$

which is a contradiction. Therefore, we have that $z = 0$. This implies that q is proportional to p, and thus $x = y$, which is a contradiction. Hence, f^G satisfies the weak axiom.

Gale claimed that f^G is not a demand function. The reason he explained is as follows. First, suppose that $f : P \to \Omega$ is a CoD, and $g : \mathbb{R}^n_{++} \to \mathbb{R}^n_{++}$ satisfies

$$x = f(g(x), g(x) \cdot x)$$

for every $x \in \mathbb{R}^n_{++}$. We call this function g an **inverse demand function** of f. Gale claimed that if f is a demand function, then $g(x)$ must satisfy the following **Jacobi's integrability condition:**[5]

$$g_i\left(\frac{\partial g_j}{\partial x_k} - \frac{\partial g_k}{\partial x_j}\right) + g_j\left(\frac{\partial g_k}{\partial x_i} - \frac{\partial g_i}{\partial x_k}\right) + g_k\left(\frac{\partial g_i}{\partial x_j} - \frac{\partial g_j}{\partial x_i}\right) = 0. \tag{3}$$

We define

$$B \equiv A^{-1} = \frac{1}{37}\begin{pmatrix} 9 & 12 & 16 \\ 16 & 9 & 12 \\ 12 & 16 & 9 \end{pmatrix}.$$

Then, $Bx \geq 0$ for all $x \geq 0$, and $Bx \gg 0$, $x^T Bx > 0$ if $x \neq 0$. Define

$$g(x) = \frac{1}{x^T Bx} Bx. \tag{4}$$

Then, for all $x \in \mathbb{R}^3_{++}$,

---

[5] Here, we abbreviate variables to avoid the expression becoming too long. Note that, we say that $g(x)$ satisfies (3) if and only if g satisfies (3) for all x and $i, j, k \in \{1, ..., n\}$, and $g(x)$ violates (3) if there exist x and $i, j, k \in \{1, ..., n\}$ such that (3) is not satisfied.

$$h_A(g(x)) = \frac{(x^T Bx)^2}{x^T B^T ABx} \frac{1}{x^T Bx} ABx = x,$$

which implies that $g(x) \in C$, and thus

$$f^G(g(x), g(x) \cdot x) = x.$$

However, this $g(x)$ violates (3), and thus, if Gale's claim is correct, then f^G is not a demand function.[6]

We point out the problem included in Gale's explanation. First, Gale's explanation is deeply related to the following theorem.

Frobenius' Theorem. *Suppose that $U \subset \mathbb{R}^n$ and $g : U \to \mathbb{R}^n \setminus \{0\}$ is C^1. Then, the following two statements are equivalent.*

1. *$g(x)$ satisfies* (3).
2. *For every $x \in U$, there exist a neighborhood $V \subset U$ of x and a pair of a C^1 function $u : V \to \mathbb{R}$ and a positive continuous function $\lambda : V \to \mathbb{R}$ such that*

$$\nabla u(y) = \lambda(y)g(y) \tag{5}$$

for all $y \in V$.

For a proof of this theorem, see Theorem 2 of Hosoya [10] and Theorem 10.9.4 of Dieudonne [4].

If $f^G = f^u$ for some C^1 nondegenerate function u,[7] then by Lagrange's multiplier rule, Eq. (5) must hold, which implies that $g(x)$ satisfies (3). As we have checked, $g(x)$ actually violates (3). By the contrapositive of the above argument, we have that there is no C^1 nondegenerate function u such that $f^G = f^u$. From this fact, Gale stated that f^G is not a demand function.[8]

However, this does not mean that there is no weak order $\succsim$ such that $f^G = f^{\succsim}$, because there exists a weak order that cannot be represented by a utility function.[9] Therefore, we cannot determine whether f^G is a demand function by Gale's explanation alone. With this consideration, we believe that Gale only showed that f^G is not a demand function corresponding to some C^1 nondegenerate utility function u.

In the next section, we consider this problem, and treat three approaches to this problem. First, we use Gale's another contribution to show that f^G is not a demand

[6] We verify the fact that $g(x)$ violates (3) in the appendix. See subsection "Property of the Inverse Demand Function".

[7] A C^1 function u is said to be **nondegenerate** if $\nabla u(x) \neq 0$ for all x.

[8] Note that, Gale's construction to this example is not a guesswork. Actually, Gale defined $h_A(p) = \frac{1}{p^T Ap} Ap$ for an arbitrary 3×3 matrix A, and proved that $g(x)$ defined by (4) satisfies (3) if and only if A is symmetric. See his Lemma. And then, he tried to construct a matrix A that is asymmetric but $f(p, m) = h_A(p)m$ satisfies the weak axiom, and obtained the above matrix A.

[9] In fact, Gale stated that "f^G is not a demand function corresponds to some utility function u." However, even this claim is not obvious if u is not differentiable.

function. Second, we construct a proof that f^G is not a demand function using Shephard's lemma. Third, we use Hosoya's [7] integrability result to show that a restriction of f^G into $(f^G)^{-1}(\mathbb{R}^3_{++})$ is actually not a demand function but a semi-demand function corresponds to a complete, p-transitive, and continuous preference relation, and construct an alternative proof that f^G is not a demand function.

4 Three Proofs of Gale's Claim

4.1 First Proof: Direct Method

In this subsection, we prove that f^G is not a demand function using a fact that Gale found. Gale considered the following four price vectors and consumption vectors.

$$p^1 = (9, 16, 12),\ x^1 = f^G(p^1, 9) = (1, 0, 0),$$

$$p^2 = (340, 440, 330),\ x^2 = f^G(p^2, 303) = (0.6, 0, 0.3),$$

$$p^3 = (410, 400, 300),\ x^3 = f^G(p^3, 303) = (0.3, 0, 0.6),$$

$$p^4 = (16, 12, 9),\ x^4 = f^G(p^4, 9) = (0, 0, 1).$$

We can easily check that

$$p^1 \cdot x^2 = 9,\ p^2 \cdot x^3 = 300,\ p^3 \cdot x^4 = 300,$$

and thus

$$x^1 \succ_r x^2,\ x^2 \succ_r x^3,\ x^3 \succ_r x^4.$$

Because $\succ_r \subset \succ_{ir}$,

$$x^1 \succ_{ir} x^2,\ x^2 \succ_{ir} x^3,\ x^3 \succ_{ir} x^4,$$

and because $\succ_{ir}$ is transitive, we conclude that

$$(1, 0, 0) \succ_{ir} (0, 0, 1).$$

Gale stated that, by almost the same arguments, one can show that $(0, 0, 1) \succ_{ir} (0, 1, 0)$ and $(0, 1, 0) \succ_{ir} (1, 0, 0)$, which means that $\succ_{ir}$ is not asymmetric and f^G violates the strong axiom. Actually, his claim is correct. Define

$$p^5 = (440, 330, 340),\ x^5 = f^G(p^5, 303) = (0, 0.3, 0.6),$$

$$p^6 = (400, 300, 410),\ x^6 = f^G(p^6, 303) = (0, 0.6, 0.3),$$

$$p^7 = (12, 9, 16),\ x^7 = f^G(p^7, 9) = (0, 1, 0),$$

$$p^8 = (330, 340, 440),\ x^8 = f^G(p^8, 303) = (0.3, 0.6, 0),$$

$$p^9 = (300, 410, 400),\ x^9 = f^G(p^9, 303) = (0.6, 0.3, 0),$$

$$x^{10} = x^1 = (1, 0, 0).$$

Then, we can easily check that $x^i \succ_r x^{i+1}$ for $i \in \{1, ..., 9\}$, and thus

$$(0, 0, 1) \succ_{ir} (1, 0, 0),$$

which implies that the strong axiom does not hold.

Gale claimed that because f^G is not a demand function, it must violate the strong axiom by Houthakker's [12] theorem, and Gale verified this result by constructing the above sequence. However, we claim that the converse relationship is more important. Suppose that $f^G = f^{\succsim}$ for a weak order $\succsim$, and $x \succ_r y$. By the definition of $f^{\succsim}$, we have that $x \succ y$. Because $\succ$ is transitive,

$$x^1 \succ x^4,\ x^4 \succ x^{10}.$$

However, because $x^{10} = x^1$, this contradicts the asymmetry of $\succ$. This implies that such a weak order $\succsim$ is absent, and thus f^G is not a demand function. This is one approach for proving that Gale's example does not become a demand function.

In this context, Richter [18] presented the following elegant theorem.

Theorem 1 *Let $P = \{(p, m) \in \mathbb{R}^n_{++} \times \mathbb{R}_{++} | \Delta(p, m) \neq \emptyset\}$, and suppose that $f : P \to \Omega$ is a CoD. Then, f is a demand function if and only if f satisfies the strong axiom of revealed preference.*

We present the proof of this theorem in the appendix.

We make several remarks on the above arguments. First, some readers might think that using Houthakker's result, Gale actually proved that f^G is not a demand function. However, we believe this consideration is wrong. Houthakker's paper itself contains ambiguities characteristic of older economics papers, and it is debatable as to what his theorem is. As far as we can see, Houthakker's paper seems to calculate a so-called indifference hypersurface, and in this argument, he treated some kind of differential equation. Therefore, we believe that Houthakker's paper, like Gale's, implicitly assumed the differentiability of u. If this consideration is correct, then Houthakker's theorem cannot be used to show f^G is not a demand function. However, even if our consideration is wrong, it is reasonable to assume that Gale could not show his result using Houthakker's theorem. Gale stated that "Since the demand function in our example does not come from a preference relation it follows from Houthakker's theorem that it must fail to satisfy the Strong Axiom." A straightforward understanding of this statement suggests that Gale considered Houthakker's theorem to be the assertion that "if a CoD is not a demand function, then it does not satisfy the strong axiom." And, as we saw above, the claim needed to complete the proof of Gale's result is the opposite of this, namely, that "if the CoD does not satisfy the strong axiom, then it is not a demand function."

Theorem 1 resolves this problem perfectly, because there is no vague claim. However, Richter's paper was published in 1966. Therefore, in 1960, this result was not known, and thus Gale could not apply it to f^G.

By the way, Gale mentioned Uzawa's [24] result. His result is as follows.

Theorem 2 *Suppose that* $\Omega = \mathbb{R}^n_+$, $P = \mathbb{R}^n_{++} \times \mathbb{R}_{++}$, *and a CoD* f *satisfies the following assumptions.*

(I) $R(f) = \Omega \setminus \{0\}$.
(II) f *satisfies Walras' law.*
(III) For all $p \in \mathbb{R}^n_{++}$, *there exist* $\varepsilon > 0$ *and* $L > 0$ *such that if* $\|q - p\| < \varepsilon$ *and* $m, w > 0$, *then* $\|f(q, m) - f(q, w)\| \le L|m - w|$.[10]
(IV) f *satisfies the strong axiom of revealed preference.*

Define

$$\succsim = \{(x, y) \in \Omega^2 | y \not\succ_{ir} x\}.$$

Then, $\succsim$ *is a continuous weak order and* $f = f^{\succsim}$.

We omit the proof of this theorem. Note that Gale's example f^G satisfies (I)–(III) of this theorem. Therefore, in 1960, Gale knew that if f^G satisfies the strong axiom, then f^G is a demand function. However, as we noted above, the converse relationship is important for proving the absence of a weak order $\succsim$ such that $f^G = f^{\succsim}$. Therefore, this result is also not able to be used to prove that f^G is not a demand function.

In the twenty-first century, we can easily show that Gale's example is not a demand function using Theorem 1. In 1960, proving this result was not easy. However, at least this result can be shown using Gale's sequence $x^1, ..., x^{10}$, and the fact that $\succ_r \subset \succ$. We think that Gale could have constructed this logic even in 1960. In conclusion, we can say the following. First, there is a gap in Gale's proof that Gale's example is not a demand function. Second, Gale would have been able to prove this in a different manner, even in 1960. Therefore, it is fair to say that the credit for finding a CoD that satisfies the weak axiom but is not a demand function is attributed to Gale.

4.2 Second Proof: Using Shephard's Lemma

In the previous subsection, the sequence $x^1, ..., x^{10}$ was given. The construction of such a sequence is not easy in general, and probably this could be found because

[10] Uzawa originally stated this condition as follows: "for all $(p, m) \in P$, there exist $\varepsilon > 0$ and $L > 0$ such that if $\|q - p\| < \varepsilon$ and $|w_i - m| < \varepsilon$ for $i \in \{1, 2\}$, then $\|f(q, w_1) - f(q, w_2)\| \le L|w_1 - w_2|$." However, to the best of our understanding, Uzawa's condition is too weak to prove this theorem and there is a gap, unless our strengthened condition is used.

Gale is a genius.[11] However, we and almost all readers are not geniuses, and thus finding such a sequence is an incredibly hard task.

Hence, we want to find another proof that can be constructed by usual economists. One way to find such a proof is to use Shephard's lemma and the Slutsky matrix.

To introduce the Shephard's lemma, we first consider the following minimization problem.

$$\begin{aligned} \min \quad & p \cdot y, \\ \text{subject to.} \quad & y \in \Omega, \\ & u(y) \geq u(x). \end{aligned} \tag{6}$$

This problem (6) is the dual problem for the utility maximization problem (1). The problem (6) is called the **expenditure minimization problem**. We define the value of this problem as $E^x(p)$, and call the function E^x the **expenditure function**.

It is useful to consider another definition of the expenditure function. Suppose that a weak order $\succsim$ is given. Then, the expenditure function E^x is defined as follows.

$$E^x(p) = \inf\{p \cdot y | y \succsim x\}. \tag{7}$$

Note that, if $\succsim$ is represented by u, then $E^x(p)$ becomes the value of the problem (6), and thus these two definitions of the function E^x coincide.

The following theorem was shown in Lemma 1 of Hosoya [9].

Theorem 3 *Suppose that either $\Omega = \mathbb{R}^n_+$ or $\Omega = \mathbb{R}^n_{++}$, $P = \mathbb{R}^n_{++} \times \mathbb{R}_{++}$, and $f : P \to \Omega$ is a continuous demand function. Suppose also that $\succsim$ is a weak order such that $f = f^{\succsim}$, and define E^x by (7). Then, $E^x : \mathbb{R}^n_{++} \to \mathbb{R}_+$ is concave and continuous. Moreover, if f satisfies Walras' law and $x \in R(f)$, then the following facts hold.*

(1) $E^x(p) > 0$ for all $p \in \mathbb{R}^n_{++}$.
(2) If $x = f(p^*, m^*)$, then $E^x(p^*) = m^*$.
(3) For every $p \in \mathbb{R}^n_{++}$, the following equality holds.

$$\nabla E^x(p) = f(p, E^x(p)). \tag{8}$$

We provide the proof of this result in the appendix. The Eq. (8) is famous and called **Shephard's lemma**. Note that, because f is continuous, we have that E^x is C^1. If f is continuously differentiable at (p^*, m^*), then by (2) and (3) in Theorem 3, we can easily verify that E^x is twice continuously differentiable at p^*, and

$$D^2E^x(p^*) = S_f(p^*, m^*),$$

where $D^2E^x(p^*)$ denotes the Hessian matrix of E^x at p^*. By Young's theorem, $D^2E^x(p^*)$ is symmetric, and thus we conclude that $S_f(p^*, m^*)$ is also symmetric.

[11] Gale himself stated that the finding of this sequence is a "considerable labour."

Consider applying this result to Gale's example f^G. It is easy to show that $R(f^G)$ includes $\mathbb{R}^3_{++}$, and if $(p, m) \in f^{-1}(\mathbb{R}^3_{++})$, then f^G is continuously differentiable at this point. Hence, choose $p^* = (1, 1, 1)$, $m^* = 3$, and $x = (1, 1, 1)$. Then, $x = f^G(p^*, m^*)$. If f^G is a demand function, then by the above consideration,

$$S_f(p^*, m^*) = D^2E^x(p^*)$$

is symmetric. However,

$$s_{12}(p^*, m^*) = \frac{11}{3} \neq -\frac{1}{3} = s_{21}(p^*, m^*)$$

and thus, f^G is not a demand function.

This method of proof does not require any inspiration of a genius and shows that Gale's example is not a demand function in the shortest way. To construct this proof, we need only Theorem 3 and a simple calculation for s_{ij}.

4.3 Third Proof: On the "Open Cycle" Theoretic Approach

In this subsection, we investigate another aspect of Gale's example. In previous subsections, we proved that Gale's example is not a demand function. Consider the restriction $\tilde{f}$ of Gale's example to the inverse image of the positive orthant. Then, we can construct a complete, p-transitive, and continuous binary relation $\succsim$ such that $\tilde{f} = f^{\succsim}$. Thus, the restriction of Gale's example is in fact a semi-demand function. Moreover, our construction method for $\succsim$ has two features. First, for our construction method, $\tilde{f}$ is a demand function if and only if our $\succsim$ is transitive. Second, (3) for an inverse demand function is equivalent to the transitivity of $\succsim$. Hence, we can conclude that Gale's example is not a demand function because $g(x)$ defined in (4) does not satisfy (3). This explanation seems to be very close to Gale's original explanation, and thus we think that this is the form of proof that Gale originally intended to present. Furthermore, this proof is deeply related to the mysterious "open cycle theory" due to Pareto in 1906.

We start from recalling the notion of the **inverse demand function**. Choose any function $g : \mathbb{R}^n_{++} \to \mathbb{R}^n_{++}$. This function g is called an inverse demand function of a CoD $f : P \to \Omega$ if and only if $x = f(g(x), g(x) \cdot x)$ for all $x \in \mathbb{R}^n_{++}$. If $f = f^u$ for some continuous utility function $u : \Omega \to \mathbb{R}$ that is increasing, C^1, and nondegenerate on $\mathbb{R}^n_{++}$,[12] then by Lagrange's multiplier rule, g is an inverse demand function of f if and only if

$$\nabla u(x) = \lambda(x)g(x)$$

[12] A function u is said to be **increasing** if $x \gg y$ implies $u(x) > u(y)$. Note that, if $f = f^u$ for some increasing function u, then f must satisfy Walras' law.

for some $\lambda(x) > 0$.

Define $\tilde{\Omega} = \mathbb{R}^n_{++}$. Let $g : \tilde{\Omega} \to \mathbb{R}^n_{++}$ be a given C^1 function. Choose any $(x, v) \in \tilde{\Omega}^2$, and consider the following ordinary differential equation (ODE):

$$\dot{y}(t) = (g(y(t)) \cdot x)v - (g(y(t)) \cdot v)x, \; y(0) = x. \tag{9}$$

Let $y(t; x, v)$ denote the nonextendable solution for the above parametrized ODE. Define

$$w^* = (v \cdot x)v - (v \cdot v)x,$$

$$t(x, v) = \inf\{t \geq 0 | y(t) \cdot w^* \geq 0\}.$$

In the appendix, we show that if x is not proportional to v, then $t(x, v)$ is well-defined and positive, and $y(t; x, v)$ is proportional to v if and only if $t = t(x, v)$. Define

$$u^g(x, v) = \frac{\|y(t(x, v); x, v)\|}{\|v\|},$$

$$\succsim^g = (u^g)^{-1}([1, +\infty[).$$

The following result is a slight modification of Theorem 1 of Hosoya [7].

Theorem 4 *Suppose that $k \geq 1$ and $g : \tilde{\Omega} \to \mathbb{R}^n_{++}$ is a C^k function. Then, the following results hold.*[13]

(I) $u^g(x, v)$ is a well-defined continuous function on $\tilde{\Omega}^2$ and $\succsim^g$ is a complete, p-transitive, continuous, and monotone binary relation on $\tilde{\Omega}$.
(II) $x \in f^{\succsim^g}(g(x), g(x) \cdot x)$ for all $x \in \tilde{\Omega}$ if and only if $w^T Dg(x) w \leq 0$ for all $x \in \tilde{\Omega}$ and $w \in \mathbb{R}^n$ such that $w \cdot g(x) = 0$. Moreover, if $w^T Dg(x) w < 0$ for all $x \in \tilde{\Omega}$ and $w \in \mathbb{R}^n$ such that $w \neq 0$ and $w \cdot g(x) = 0$, then $f^{\succsim^g}(p, m)$ contains at most one element, and if $f^{\succsim^g}(p, m) \neq \emptyset$, then p is proportional to $g(x)$ for $x = f^{\succsim^g}(p, m)$.
(III) Consider a CoD $f : P \to \tilde{\Omega}$ such that g is an inverse demand function of f, and suppose that $f = f^{\succsim^g}$. Then, the following statements are equivalent.
   (i) f is a demand function.
   (ii) $\succsim^g$ is transitive.
   (iii) For every $v \in \tilde{\Omega}$, $u^g_v : x \mapsto u^g(x, v)$ is C^k and there exists $\lambda : \tilde{\Omega} \to \mathbb{R}$ such that $\nabla u^g_v(x) = \lambda(x) g(x)$ for all $x \in \tilde{\Omega}$.
   (iv) For every $v \in \tilde{\Omega}$, $u^g_v : x \mapsto u^g(x, v)$ is C^k and represents $\succsim^g$.
   (v) $g(x)$ satisfies (3).

We provide the proof of this theorem in the appendix. Applying this theorem to Gale's example, we can present the following arguments. First, recall that in Gale's

[13] Throughout this paper, $Dg(x)$ denotes the Jacobian matrix of g at x.

example, the consumption set Ω is $\mathbb{R}^3_+$ and the domain $P = \mathbb{R}^3_{++} \times \mathbb{R}_{++}$. Let $\tilde{\Omega} = \mathbb{R}^3_{++}$ and $\tilde{P} = (f^G)^{-1}(\tilde{\Omega})$. In the appendix, we check that $\tilde{P} = \{(mg(x), m)|x \in \tilde{\Omega}, m > 0\}$, where $g(x)$ is given by (4).[14] Define $\tilde{f} : \tilde{P} \to \tilde{\Omega}$ as the restriction of f^G to $\tilde{P}$. Then, $\tilde{f}$ is also a CoD, where the consumption set is $\tilde{\Omega}$ and the domain is $\tilde{P}$.

Second, suppose that f^G is a demand function. Then, $f^G = f^{\succsim}$ for some weak order $\succsim$ on Ω. Let $\succsim' = \succsim \cap (\tilde{\Omega})^2$: that is, $\succsim'$ is the restriction of $\succsim$ to $\tilde{\Omega}$. Then, $\succsim'$ is a weak order on $\tilde{\Omega}$ and $\tilde{f} = f^{\succsim'}$, which implies that $\tilde{f}$ is a demand function.

Third, recall the function g defined in (4). We found that

$$x = f^G(g(x), g(x) \cdot x) = \tilde{f}(g(x), g(x) \cdot x)$$

for all $x \in \tilde{\Omega}$, which implies that $g(x)$ is an inverse demand function of $\tilde{f}$. In the appendix, we show that $w^T Dg(x)w < 0$ for all $x \in \mathbb{R}^3_{++}$ and $w \in \mathbb{R}^3$ such that $w \neq 0$ and $w \cdot g(x) = 0$.[15] By Theorem 4, $\tilde{f} = f^{\succsim^g}$ and $g(x)$ satisfies (3) if and only if $\tilde{f}$ is a demand function.[16] Because $g(x)$ defined in (4) violates (3), we have that $\tilde{f}$ is not a demand function.

In conclusion, we obtain two results. (1) if f^G is a demand function, then $\tilde{f}$ is also a demand function. (2) $\tilde{f}$ is not a demand function. Therefore, by the contrapositive of (1), f^G is also not a demand function. This is the third proof that Gale's example is not a demand function.

Notes on Theorem 4. This result is deeply related to the "open cycle theory" in Pareto [16]. In 1906, Pareto published his famous monograph called "Manuale" [15]. It is said that Volterra wrote a review of this "Manuale" [25], and pointed out a mathematical error in Pareto's discussion of the problem of consumer choice. In fact, there are various arguments about this story, and some say that Volterra criticized Pareto for not writing (3), whereas others say that Volterra was not actually criticizing Pareto, but rather referring to another work of Pareto and recommending that it be incorporated into the book.[17] In any case, Pareto's response to Volterra's review was to try to construct the consumer theory for the case in which (3) does not hold [16], and eventually, this paper was incorporated into the mathematical appendix of the French edition of his book [17], known as "Manuel".

We explain why this theory is called "open cycle theory" using "three-sided tower" arguments due to Samuelson [21].[18] Samuelson explained what occurs when the integrability condition (3) is violated using three linearly independent vectors x, y, z.

---

[14] See subsection "Proof of a Fact on the Domain of Gale's Example".

[15] See subsection "Property of the Inverse Demand Function."

[16] Note that, the claim $\tilde{f} = f^{\succsim^g}$ means that $\tilde{f}(p, m) = f^{\succsim^g}(p, m)$ for all $(p, m) \in \tilde{P}$ and $f^{\succsim^g}(p, m) = \emptyset$ if $(p, m) \notin \tilde{P}$. Therefore, we needed to verify that $\tilde{P} = \{(mg(x), m)|x \in \tilde{\Omega}, m > 0\}$ in the above argument.

[17] If the reader can read Japanese, we recommend Suda [22], which includes a very detailed survey on this problem.

[18] We note that Samuelson consistently argued that Pareto's argument was wrong. Although Pareto connects this problem with the order of consumption, Samuelson did not think that this explanation is correct.

First, consider the plane spanned by x, y, and draw the indifference curve passing through x. This curve intersects the straight line passing through 0 and y only once, and this point can be written as ay for some $a > 0$. Similarly, consider the plane spanned by y, z, and draw the indifference curve passing through ay. Then, this curve intersects the straight line passing through 0 and z only once, and this point can be written as bz for some $b > 0$. Finally, consider the plane spanned by z, x, and draw the indifference curve passing through bz. Then, this curve intersects the straight line passing through 0 and x only once, and this point can be written as cx for some $c > 0$. Samuelson said that $c = 1$ must hold if (3) holds, but if (3) does not hold, then $c \neq 1$ in some cases. In other words, if (3) is violated, then the above cycle constructed from three indifference curves is not necessarily a closed curve. Pareto's theory tried to treat the case in which the cycle constructed using indifference curves does not become a closed curve, and thus this theory is called "open cycle theory."

By the way, Samuelson called the consumer who violates (3) "a man that can be easily cheated." To illustrate why, consider the above example with $c < 1$. In this case, when the consumer is offered to exchange x for ay, then he/she accepts this exchange because these are indifferent. Next, if he/she is offered to exchange ay for bz, then he/she also accepts this because these are indifferent. Finally, when someone offers him/her to exchange bz for cx, he/she accepts this because these are indifferent. As a result of these exchanges, his/her initial consumption vector x is reduced to cx. Samuelson described such an individual as "easily cheated." Of course, a similar exchange can be done in the case $c > 1$.

This discussion can be further understood using Theorem 4. First, the ODE (9) is intentionally designed so that the right-hand side is orthogonal to $g(y(t))$. In microeconomics, the budget hyperplane is orthogonal to the price vector, and at the optimal consumption plan, the indifference hypersurface is tangent to the budget hyperplane. Therefore, the supporting hyperplane of the indifference hypersurface is orthogonal to the price vector at this point. Because $g(y(t))$ denotes the price vector under which $y(t)$ is optimal, the trajectory of the curve $y(t; x, v)$ can be seen as the indifference curve passing through x in the plane spanned by x, v. Hence, translating Samuelson's arguments into our symbols,

$$ay = y(t(x, y); x, y),\ bz = y(t(ay, z); y, z),\ cx = y(t(bz, x); z, x).$$

By an easy calculation, we obtain that

$$a = u^g(x, y),\ b = u^g(ay, z),\ c = u^g(bz, x).$$

It is easy to verify that

$$x \sim^g ay,\ ay \sim^g bz,\ bz \sim^g cx.$$

Therefore, if $\succsim^g$ is transitive, then $c = 1$. However, if $\succsim^g$ is not transitive, then there exist x, y, z such that $c \neq 1$. Hence, Theorem 4 can be viewed as the result of a mathematical expression of Samuelson's explanation.

Probably, Gale had this Samuelson's argument in mind and thus focused on the fact that $g(x)$ in (4) does not satisfy (3). In this sense, the proof in this subsection can be considered to be the closest to the argument that Gale originally had in mind.[19]

Samuelson discussed using three linearly independent vectors, and he did not have a problem with the fact that an indifference curve can be drawn. In other words, he thought that such a problem would not occur for the case in which $n = 2$. In this connection, we mention two facts. First, (3) automatically holds if any two of i, j, k coincide, and therefore it holds unconditionally when $n = 2$. Second, Rose [19] showed that the weak axiom implies the strong axiom when $n = 2$. Therefore, in light of Theorem 1, there is no CoD that satisfies the weak axiom but is not a demand function. Indeed, Gale's example treated the case in which $n = 3$, which is an essential assumption.

We must mention Hurwicz and Richter [13, 14]. They considered an axiom called **Ville's axiom of revealed preference** (or simply, **Ville's axiom**) for a function $g : \mathbb{R}^n_{++} \to \mathbb{R}^n_{++}$. A function $g : \mathbb{R}^n_{++} \to \mathbb{R}^n_{++}$ is said to satisfy Ville's axiom if there is no piecewise C^1 closed curve $x : [0, T] \to \mathbb{R}^n_{++}$ such that

$$g(x(t)) \cdot \dot{x}(t) > 0 \tag{10}$$

for almost all $t \in [0, T]$. They showed that when g is C^1, g satisfies this axiom if and only if g satisfies (3). For a modern proof of this result, see Hosoya [8]. Because $g(x)$ defined by (4) violates (3), this $g(x)$ also violates Ville's axiom. Therefore, there exists a closed curve $x(t)$ that satisfies (10) for almost all t.

If $f^G = f^u$ for some C^1 utility function u, then by Lagrange's multiplier rule, $\nabla u(x) = \lambda(x)g(x)$ for all x. Therefore,

$$\frac{d}{dt}u(x(t)) > 0$$

for almost all $t \in [0, T]$, which contradicts $x(0) = x(T)$. Ville's axiom is built on this idea, and it is worth noting that this idea and the "open cycle" argument are very similar. In fact, we can easily show the following result.

Theorem 5 *Suppose that $g : \tilde{\Omega} \to \mathbb{R}^n_{++}$ is a C^1 function. Then, g satisfies Ville's axiom if and only if $\succsim^g$ is transitive.*

We present a proof in the appendix.

Finally, we mention a fact. We have already shown that the restriction $\tilde{f}$ of Gale's example is a semi-demand function. Actually, Hosoya [11] showed that any CoD that satisfies Walras' law and the weak axiom is a semi-demand function corresponds to some complete binary relation. Hence, Gale's example itself is also a semi-demand

[19] However, it cannot be assumed that Gale himself knew or substantially proved Theorem 4. This can be seen in the proof of Theorem 4. As we see in the appendix, the proof of Theorem 4 is extremely long and difficult. If Gale had proved such a result, we cannot believe that he would not have left a record of it anywhere. In this regard, although it is possible to prove that f^G is not a demand function using Gale's own idea, it is inconceivable that Gale would have done so.

function. We think that Gale's example is actually a semi-demand function corresponds to a complete, p-transitive, and continuous binary relation, and such a binary relation can be constructed in the same manner as Theorem 4. However, there are several technical difficulties, and thus it is an open problem.

5 Conclusion

We have scrutinized Gale's paper written in 1960 to see how far Gale had shown. As a result, we found that Gale showed that his constructed CoD satisfies the weak axiom of revealed preference, and that the corresponding inverse demand function does not satisfy Jacobi's integrability condition, but he did not show that this CoD is not a demand function. Next, we proved that this CoD is not a demand function in three ways. The first is the proof that Gale was able to construct in 1960, the second is the simplest modern proof, and the third is the proof constructed in the direction Gale originally intended.

We paid particular attention to the first proof, and found that, although what Gale showed in his paper was insufficient, we could easily confirm Gale's claim from what he showed. Thus, it is fair to say that the credit for discovering a CoD that is not a demand function while satisfying the weak axiom of revealed preference belongs to Gale.

Appendix: Proof of Results

Proof of the Well-Definedness of $\bar{p}$ in Gale's Example

In this subsection, we prove the well-definedness of $\bar{p}$.

Recall the definition of $\bar{p}$. For $p \in \mathbb{R}^3_{++}$, the definition of $\bar{p}$ is as follows.

(I) If $p \in C$, define $\bar{p} = p$.

(II) Suppose that for some $(i, j, k) \in \{(1, 2, 3), (2, 3, 1), (3, 1, 2)\}$, $-3p_i + 4p_j \le 0$, and $-3p_j + 4p_k \le 0$. By our previous argument, we have that $-3p_k + 4p_i > 0$. In this case, define $\bar{p}_i = \frac{16}{9}p_k$, $\bar{p}_j = \frac{4}{3}p_k$, and $\bar{p}_k = p_k$.

(III) Suppose that for some $(i, j, k) \in \{(1, 2, 3), (2, 3, 1), (3, 1, 2)\}$, $-3p_i + 4p_j \le 0$, $-3p_j + 4p_k \ge 0$, and $-3p_k + 4p_i \ge 0$. We separate this case into two subcases.

(i) If $16p_j - 9p_k \ge 0$, then define $\bar{p}_i = \frac{4}{3}p_j$, $\bar{p}_j = p_j$, and $\bar{p}_k = p_k$.

(ii) If $16p_j - 9p_k \le 0$, then define $\bar{p}_i = \frac{4}{3}p_j$, $\bar{p}_j = p_j$, and $\bar{p}_k = \frac{16}{9}p_j$.

Our problem is the existence of p for which multiple definitions are applicable. For example, if $p = (4, 3, 4)$, then $-3p_1 + 4p_2 = 0$, $-3p_2 + 4p_3 > 0$ and $-3p_3 + 4p_1 > 0$, which implies that (I) and (III)-(i) are applicable for this p. We need to show that, in such a case, both definition in (I) and (III)-(i) lead to the same $\bar{p}$.

We separate the proof into seven cases. Let $I = \{(1, 2, 3), (2, 3, 1), (3, 1, 2)\}$. Recall that if $-3p_i + 4p_j \le 0$ and $-3p_j + 4p_k \le 0$, then $-3p_k + 4p_i > 0$.

Define $C_1 = C$ and

$$\begin{aligned}
C_2 &= \{p \in \mathbb{R}^3_{++} \mid -3p_i + 4p_j \le 0,\ -3p_j + 4p_k \le 0 \text{ for some } (i, j, k) \in I\},\\
C_3 &= \{p \in \mathbb{R}^3_{++} \mid -3p_i + 4p_j \le 0,\ -3p_j + 4p_k \ge 0,\\
&\qquad -3p_k + 4p_i \ge 0,\ 16p_j - 9p_k \ge 0 \text{ for some } (i, j, k) \in I\},\\
C_4 &= \{p \in \mathbb{R}^3_{++} \mid -3p_i + 4p_j \le 0,\ -3p_j + 4p_k \ge 0,\\
&\qquad -3p_k + 4p_i \ge 0,\ 16p_j - 9p_k \le 0 \text{ for some } (i, j, k) \in I\}.
\end{aligned}$$

Then, $p \in C_1$ if and only if (I) is applicable, $p \in C_2$ if and only if (II) is applicable, $p \in C_3$ if and only if (III)-(i) is applicable, and $p \in C_4$ if and only if (III)-(ii) is applicable.

Case 1. $-3p_i + 4p_j > 0,\ -3p_j + 4p_k > 0,\ -3p_k + 4p_i > 0.$

In this case, $p \in C_1$ but $p \notin C_2 \cup C_3 \cup C_4$. Thus, only definition (I) is applicable, and thus $\bar{p} = p$.

Case 2. $-3p_i + 4p_j = 0,\ -3p_j + 4p_k > 0,\ -3p_k + 4p_i > 0.$

In this case, $-9p_k + 16p_j > 0$, and thus $p \in C_1 \cap C_3$ but $p \notin C_2 \cup C_4$. Hence, definitions (I) and (III)-(i) are applicable, and in both definitions, $\bar{p} = p$.

Case 3. $-3p_i + 4p_j < 0,\ -3p_j + 4p_k > 0,\ -3p_k + 4p_i > 0.$

In this case, if $-9p_k + 16p_j > 0$, then $p \in C_3$ but $p \notin C_1 \cup C_2 \cup C_4$. If $-9p_k + 16p_j < 0$, then $p \in C_4$ but $p \notin C_1 \cup C_2 \cup C_3$. In both cases, $\bar{p}$ is well-defined. If $-9p_k + 16p_j = 0$, then $p \in C_3 \cap C_4$ and thus definitions (III)-(i) and (III)-(ii) are applicable, and in both definitions, $\bar{p}$ are the same.

Case 4. $-3p_i + 4p_j = 0,\ -3p_j + 4p_k = 0,\ -3p_k + 4p_i > 0.$

Define $i^* = j$, $j^* = k$, $k^* = i$. In this case, $-9p_k + 16p_j > 0$ and $-9p_{k^*} + 16p_{j^*} = 0$. This implies that $p \in C_1 \cap C_2 \cap C_3 \cap C_4$, and thus definitions (I), (II), (III)-(i), and (III)-(ii) are applicable. And in all cases, $\bar{p} = p$.

Case 5. $-3p_i + 4p_j = 0,\ -3p_j + 4p_k < 0,\ -3p_k + 4p_i > 0.$

Define $i^* = j$, $j^* = k$, $k^* = i$. Then, $-9p_{k^*} + 16p_{j^*} < 0$, and thus $p \in C_2 \cap C_4$ but $p \notin C_1 \cup C_3$. Hence, definitions (II) and (III)-(ii) are applicable, and in both cases, $\bar{p}$ are the same.

Case 6. $-3p_i + 4p_j < 0,\ -3p_j + 4p_k = 0,\ -3p_k + 4p_i > 0.$

In this case, $-9p_k + 16p_j > 0$, and thus $p \in C_2 \cap C_3$ but $p \notin C_1 \cup C_4$. Hence, definitions (II) and (III)-(i) are applicable, and in both cases, $\bar{p}$ are the same.

Case 7. $-3p_i + 4p_j < 0,\ -3p_j + 4p_k < 0,\ -3p_k + 4p_i > 0.$

In this case, $p \in C_2$ but $p \notin C_1 \cup C_3 \cup C_4$, and thus only definition II) is applicable, and thus $\bar{p}$ is defined.

Hence, in all cases $\bar{p}$ is well-defined.

Using this argument, we can easily show that the mapping $p \mapsto \bar{p}$ is continuous. Actually, the restriction of this mapping to C_i is trivially continuous, and if $p \in C_i \cap C_j$, then the value of these functions coincide. The continuity of this mapping immediately follows from this fact.

Proof of the Basic Property on the Transitive Closure

In this subsection, we prove the following.

Lemma 1 *Let $X \neq \emptyset$ and suppose that R is a binary relation on X. Let R^* be the transitive closure of R. Then, xR^*y if and only if there exists a finite sequence $x_1, ..., x_k$ such that $x_1 = x$, $x_k = y$ and x_iRx_{i+1} for all $i \in \{1, ..., k-1\}$.*

Proof Define R^+ as the set of all $(x, y) \in X^2$ such that there exists a finite sequence $x_1, ..., x_k$ such that $x_1 = x$, $x_k = y$ and x_iRx_{i+1} for all $i \in \{1, ..., k-1\}$. It suffices to show that $R^+ = R^*$.

It is obvious that R^+ is transitive, and thus $R^* \subset R^+$.

Conversely, suppose that xR^+y. Then, there exists a finite sequence $x_1, ..., x_k$ such that $x_1 = x$, $x_k = y$ and x_iRx_{i+1} for all $i \in \{1, ..., k-1\}$. Because $R \subset R^*$, we have that $x_iR^*x_{i+1}$ for all $i \in \{1, ..., k-1\}$. Since R^* is transitive, we have that xR^*y, which implies that $R^+ \subset R^*$. This completes the proof. □

Proof of Theorem 1

Suppose that $f = f^{\succsim}$ for some weak order $\succsim$. We have already shown that $x \succ_r y$ implies that $x \succ y$. Hence, Lemma 1 and the transitivity of $\succ$ mean that $x \succ_{ir} y$ implies that $x \succ y$. Hence,

$$x \succ_{ir} y \Rightarrow x \succ y \Rightarrow y \not\succ x \Rightarrow y \not\succ_{ir} x,$$

which implies that $\succ_{ir}$ is asymmetric, and thus f satisfies the strong axiom.

To prove the converse relationship, we need the following lemma.

Lemma 2 (Szpilrajn's extension theorem) *Suppose that X is a nonempty set and $\succ$ is a strong order on X. Then, there exists an antisymmetric weak order $\succsim^*$ that includes $\succ$.*[20]

Proof Let L be the set of all strong orders on X that includes $\succ$. Define

$$\trianglerighteq = \{(\succ_1, \succ_2) \in L^2 | \succ_2 \subset \succ_1\}.$$

Then, $\trianglerighteq$ is a reflexive and transitive binary relation on L, and clearly every chain C of $\trianglerighteq$ has an upper bound. By Zorn's lemma, there exists a maximal element $\succ^* \subset L$ with respect to $\trianglerighteq$. Define

$$\succsim^* = \succ^* \cup \{(x, x) | x \in X\}.$$

Clearly $\succsim^*$ is antisymmetric, and it is easy to show that $\succsim^*$ is transitive. It suffices to show that $\succsim^*$ is complete.

Suppose not. Then, there exists $x, y \in X$ such that $x \not\succsim^* y$ and $y \not\succsim^* x$. By the definition of $\succsim^*$, we have that $x \neq y$. Define

$$\succ^+ = \succ^* \cup \{(z, w) \in X^2 | z \succsim^* x,\ y \succsim^* w\}.$$

We first show that $\succ^+$ is asymmetric. Suppose not. Then, there exist $z, w \in X$ such that $z \succ^+ w$ and $w \succ^+ z$. If $z \not\succ^* w$ and $w \not\succ^* z$, then $z \succsim^* x$ and $y \succsim^* z$. This implies that $y \succsim^* x$, which is a contradiction. Therefore, either $z \succ^* w$ or $w \succ^* z$. Without loss of generality, we assume that $w \succ^* z$. Then, $z \not\succ^* w$, and thus $z \succsim^* x$ and $y \succsim^* w$. Because $w \succ^* z$, we have that $w \succsim^* z$, and by the transitivity of $\succsim^*$, $y \succsim^* x$, which is a contradiction.

Second, we show that $\succ^+$ is transitive. Suppose that $z \succ^+ w$ and $w \succ^+ v$. If $z \succ^* w$ and $w \succ^* v$, then $z \succ^* v$, and thus $z \succ^+ v$. Suppose that $z \not\succ^* w$. Then, $z \succsim^* x$ and $y \succsim^* w$. If $w \not\succ^* v$, then $w \succsim^* x$. This implies that $y \succsim^* x$, which is a contradiction. Hence, we have that $w \succ^* v$. This implies that $y \succsim^* v$, and thus $z \succ^+ w$. Next, suppose that $z \succ^* w$ and $w \not\succ^* v$. Then, $w \succsim^* x$ and $y \succsim^* v$. Because $z \succ^* w$, we have that $z \succsim^* x$, which implies that $z \succ^+ v$. Thus, in any case, we have that $z \succ^+ v$, which implies that $\succ^+$ is transitive.

Therefore, $\succ^+$ is a strong order that includes $\succ^*$. Because $x \succ^+ y$ and $x \not\succ^* y$, we have that $\succ^* \subsetneq \succ^+$. This implies that $\succ^*$ is not maximal with respect to $\trianglerighteq$, which is a contradiction. This completes the proof. □

Now, suppose that f satisfies the strong axiom. Then, $\succ_{ir}$ is a strong order on Ω, and thus there exists an antisymmetric weak order $\succsim^*$ on Ω that includes $\succ_{ir}$. Suppose that $x = f(p, m)$ and $y \in \Delta(p, m)$. If $x = y$, then $x \succsim^* y$. If $x \neq y$, then $x \succ_{ir} y$, and thus $x \succsim^* y$. Moreover, because $\succsim^*$ is antisymmetric, $y \not\succsim^* x$. This implies that $x = f^{\succsim^*}(p, m)$, and thus $f = f^{\succsim^*}$. Hence, f is a demand function. This completes the proof of Theorem 1. □

[20] This theorem was shown by Szpilrajn [23]. For another modern proof, see Chambers and Echenique [1].

Proof of Theorem 3

Choose any p, q and $t \in [0, 1]$, and define $r = (1-t)p + tq$. Choose any $\varepsilon > 0$. Then, there exists $y \in \Omega$ such that $y \succsim x$ and $r \cdot y \le E^x(r) + \varepsilon$. Thus,

$$E^x(r) + \varepsilon \ge r \cdot y = (1-t)p \cdot y + tq \cdot y \ge (1-t)E^x(p) + tE^x(q),$$

which implies that E^x is concave. Because any concave function defined on an open set is continuous, E^x is continuous.

In the reminder of the proof, we assume that f satisfies Walras' law and $x \in R(f)$. Suppose that $x = f(p^*, m^*)$. Choose any $p \in \mathbb{R}^n_{++}$. Then, there exists $\varepsilon > 0$ such that $p^* \cdot y < m^*$ for all $y \in \Delta(p, \varepsilon)$. In this case, $x \succ y$ for all $y \in \Delta(p, \varepsilon)$, and thus $E^x(p) > \varepsilon$. Hence, we have that $E^x(p) > 0$, and thus (1) holds.

Next, suppose that $y \in \Delta(p^*, m^*)$ and $x \neq y$. Then, $x \succ y$. By the contrapositive of this, we have that if $y \succsim x$, then either $y = x$ or $p^* \cdot y > m^*$, which implies that $E^x(p^*) = m^*$. Hence, (2) holds.

Third, choose any $p \in \mathbb{R}^n_{++}$. Define

$$D_{i,+}E^x(p) = \lim_{h \downarrow 0} \frac{E^x(p + he_i) - E^x(p)}{h}, \quad D_{i,-}E^x(p) = \lim_{h \uparrow 0} \frac{E^x(p + he_i) - E^x(p)}{h},$$

where e_i denotes the ith unit vector. Because E^x is concave, both limits are defined, and

$$D_{i,+}E^x(p) \le D_{i,-}E^x(p).$$

For any $q \in \mathbb{R}^n_{++}$ and $\varepsilon > 0$, define $X(q) = f(q, E^x(q))$ and $X^\varepsilon(q) = f(q, E^x(q) + \varepsilon)$. By definition of $E^x(q)$, there exists $y \in \Omega$ such that $y \succsim x$ and $q \cdot y \le E^x(q) + \varepsilon$. Because $X^\varepsilon(q) \succsim y$, we have that $X^\varepsilon(q) \succsim x$, and thus $p \cdot X^\varepsilon(q) \ge E^x(p)$. Because f is continuous, letting $\varepsilon \to 0$, we obtain

$$p \cdot X(q) \ge E^x(p) = p \cdot X(p).$$

Therefore, if we set $q = p + he_i$, then

$$\begin{aligned} E^x(q) - E^x(p) &= q \cdot X(q) - p \cdot X(p) \\ &= p \cdot (X(q) - X(p)) + hX_i(q) \\ &\ge hX_i(q). \end{aligned}$$

By the above inequality,

$$D_{i,+}E^x(p) \ge f_i(p, E^x(p)) \ge D_{i,-}E^x(p),$$

and thus,

$$\frac{\partial E^x}{\partial p_i}(p) = f_i(p, E^x(p)),$$

which implies that (8) holds. This completes the proof. □

Property of the Inverse Demand Function

In the main text, we claimed several property of the inverse demand function $g(x)$ defined in (4). In this subsection, we present proofs of these facts.

First, we present a lemma.

Lemma 3 *Suppose that $U \subset \mathbb{R}^n$ is open, $g : U \to \mathbb{R}^n \setminus \{0\}$ is C^1, and $\lambda : U \to \mathbb{R}_{++}$ is also C^1. Define $h(x) = \lambda(x)g(x)$. Then, the following holds.*

(i) $g(x)$ satisfies (3) if and only if $h(x)$ satisfies (3).
(ii) $w^T Dg(x)w \le 0$ for all $x \in U$ and $w \in \mathbb{R}^n$ such that $w \cdot g(x) = 0$ if and only if $w^T Dh(x)w \le 0$ for all $x \in U$ and $w \in \mathbb{R}^n$ such that $w \cdot h(x) = 0$.
(iii) $w^T Dg(x)w < 0$ for all $x \in U$ and $w \in \mathbb{R}^n$ such that $w \neq 0$ and $w \cdot g(x) = 0$ if and only if $w^T Dh(x)w < 0$ for all $x \in U$ and $w \in \mathbb{R}^n$ such that $w \neq 0$ and $w \cdot h(x) = 0$.

Proof By Leibniz's rule,

$$\begin{aligned} & h_i\left(\frac{\partial h_j}{\partial x_k} - \frac{\partial h_k}{\partial x_j}\right) + h_j\left(\frac{\partial h_k}{\partial x_i} - \frac{\partial h_i}{\partial x_k}\right) + h_k\left(\frac{\partial h_i}{\partial x_j} - \frac{\partial h_j}{\partial x_i}\right) \\ &= \lambda^2\left[g_i\left(\frac{\partial g_j}{\partial x_k} - \frac{\partial g_k}{\partial x_j}\right) + g_j\left(\frac{\partial g_k}{\partial x_i} - \frac{\partial g_i}{\partial x_k}\right) + g_k\left(\frac{\partial g_i}{\partial x_j} - \frac{\partial g_j}{\partial x_i}\right)\right] \\ &\quad +\lambda\left[g_i\left(g_j\frac{\partial \lambda}{\partial x_k} - g_k\frac{\partial \lambda}{\partial x_j}\right) + g_j\left(g_k\frac{\partial \lambda}{\partial x_i} - g_i\frac{\partial \lambda}{\partial x_k}\right) + g_k\left(g_i\frac{\partial \lambda}{\partial x_j} - g_j\frac{\partial \lambda}{\partial x_i}\right)\right] \\ &= \lambda^2\left[g_i\left(\frac{\partial g_j}{\partial x_k} - \frac{\partial g_k}{\partial x_j}\right) + g_j\left(\frac{\partial g_k}{\partial x_i} - \frac{\partial g_i}{\partial x_k}\right) + g_k\left(\frac{\partial g_i}{\partial x_j} - \frac{\partial g_j}{\partial x_i}\right)\right], \end{aligned}$$

and thus, $h(x)$ satisfies (3) if and only if $g(x)$ satisfies (3). Hence, (i) holds.

Next, we note that $w \cdot g(x) = 0$ if and only if $w \cdot h(x) = 0$. If $w \cdot g(x) = 0$, then

$$w^T Dh(x)w = w^T g(x)D\lambda(x)w + \lambda(x)w^T Dg(x)w = \lambda(x)w^T Dg(x)w,$$

and thus, (ii) and (iii) hold. This completes the proof.

We show that $g(x)$ defined in (4) violates (3). Choosing $\lambda(x) = 37(x^T Bx)$ and applying Lemma 3, we have that $g(x)$ satisfies (3) if and only if

$$h(x) = \begin{pmatrix} 9 & 12 & 16 \\ 16 & 9 & 12 \\ 12 & 16 & 9 \end{pmatrix} \begin{pmatrix} x_1 \\ x_2 \\ x_3 \end{pmatrix} \tag{11}$$

satisfies (3). However, if $x_1 = x_2 = x_3 = 1$, then

$$\begin{aligned}&h_1\left(\frac{\partial h_2}{\partial x_3}-\frac{\partial h_3}{\partial x_2}\right)+h_2\left(\frac{\partial h_3}{\partial x_1}-\frac{\partial h_1}{\partial x_3}\right)+h_3\left(\frac{\partial h_1}{\partial x_2}-\frac{\partial h_2}{\partial x_1}\right)\\&=37(12-16)+37(12-16)+37(12-16)=-444\neq 0,\end{aligned}$$

and thus $h(x)$ violates (3).

Define

$$C=\begin{pmatrix}9 & 14 & 14\\ 14 & 9 & 14\\ 14 & 14 & 9\end{pmatrix}.$$

Then, $w^T Dh(x)w = w^T Cw$, where $h(x)$ is defined in (11). Let $t = \frac{h_1(x)}{h_2(x)}$. Then, $\frac{9}{16} \le t \le \frac{4}{3}$, and thus

$$\begin{vmatrix}9 & 14 & h_1(x)\\ 14 & 9 & h_2(x)\\ h_1(x) & h_2(x) & 0\end{vmatrix}=(h_2(x))^2(-9t^2+28t-9)>0.$$

Next, let $s_1 = \frac{h_1(x)}{h_3(x)}$ and $s_2 = \frac{h_2(x)}{h_3(x)}$. Then, $\frac{3}{4} \le s_1 \le \frac{16}{9}$, $\frac{9}{16} \le s_2 \le \frac{4}{3}$, and $\frac{9}{16} \le \frac{s_1}{s_2} \le \frac{4}{3}$, and thus

$$\begin{aligned}&\begin{vmatrix}9 & 14 & 14 & h_1(x)\\ 14 & 9 & 14 & h_2(x)\\ 14 & 14 & 9 & h_3(x)\\ h_1(x) & h_2(x) & h_3(x) & 0\end{vmatrix}\\&=(h_3(x))^2[115s_1^2-140s_1s_2+115s_2^2-140s_1-140s_2+115]<0.\end{aligned}$$

By Theorem 5 of Debreu [2], we have that $w^T Cw < 0$ for all $w \in \mathbb{R}$ such that $w \neq 0$ and $w \cdot h(x) = 0$. Therefore, by Lemma 3, $w^T Dg(x)w < 0$ for all $x \in \mathbb{R}^3_{++}$ and $w \in \mathbb{R}^3$ such that $w \neq 0$ and $w \cdot g(x) = 0$.

Proof of a Fact on the Domain of Gale's Example

In Subsect. 4.3, we first introduce Theorem 4 and then provide a proof that f^G is not a demand function. In this proof, we defined $\tilde{\Omega} = \mathbb{R}^3_{++}$ and $\tilde{P} = (f^G)^{-1}(\tilde{\Omega})$, and claimed that $\tilde{P} = \{(mg(x), m) | x \in \tilde{\Omega},\ m > 0\}$, where $g(x)$ is defined by (4). In this subsection, we prove this claim rigorously.

First, we present three lemmas.

Lemma 4 *Suppose that $U \subset \mathbb{R}^n$ is open, and $g : U \to \mathbb{R}^n \setminus \{0\}$ is C^1. Suppose also that $g_n(x) \neq 0$. Then,*

$$g_i\left(\frac{\partial g_j}{\partial x_n}-\frac{\partial g_n}{\partial x_j}\right)+g_j\left(\frac{\partial g_n}{\partial x_i}-\frac{\partial g_i}{\partial x_n}\right)+g_n\left(\frac{\partial g_i}{\partial x_j}-\frac{\partial g_j}{\partial x_i}\right)=0 \qquad (12)$$

for all $i, j \in \{1, ..., n-1\}$ *if and only if* g *satisfies (3) for all* $i, j, k \in \{1, ..., n\}$.

Proof Clearly, Eq. (3) implies (12). Conversely, suppose that (12) holds for all $i, j \in \{1, ..., n-1\}$. Choose any $i, j, k \in \{1, ..., n\}$. If two or three of them are n, then trivially (3) holds. If one of them is n, then by (12), (3) holds. Therefore, without loss of generality, we can assume that $i, j, k \in \{1, ..., n-1\}$. Then,

$$g_i\left(\frac{\partial g_j}{\partial x_n}-\frac{\partial g_n}{\partial x_j}\right)+g_j\left(\frac{\partial g_n}{\partial x_i}-\frac{\partial g_i}{\partial x_n}\right)+g_n\left(\frac{\partial g_i}{\partial x_j}-\frac{\partial g_j}{\partial x_i}\right)=0,$$

$$g_j\left(\frac{\partial g_k}{\partial x_n}-\frac{\partial g_n}{\partial x_k}\right)+g_k\left(\frac{\partial g_n}{\partial x_j}-\frac{\partial g_j}{\partial x_n}\right)+g_n\left(\frac{\partial g_j}{\partial x_k}-\frac{\partial g_k}{\partial x_j}\right)=0,$$

$$g_k\left(\frac{\partial g_i}{\partial x_n}-\frac{\partial g_n}{\partial x_i}\right)+g_i\left(\frac{\partial g_n}{\partial x_k}-\frac{\partial g_k}{\partial x_n}\right)+g_n\left(\frac{\partial g_k}{\partial x_i}-\frac{\partial g_i}{\partial x_k}\right)=0.$$

Multiplying g_k by the first line, g_i by the second line, and g_j by the third line and summing them, we obtain

$$g_n\left[g_i\left(\frac{\partial g_j}{\partial x_k}-\frac{\partial g_k}{\partial x_j}\right)+g_j\left(\frac{\partial g_k}{\partial x_i}-\frac{\partial g_i}{\partial x_k}\right)+g_k\left(\frac{\partial g_i}{\partial x_j}-\frac{\partial g_j}{\partial x_i}\right)\right]=0,$$

and because $g_n(x) \neq 0$, (3) holds. This completes the proof. □

Lemma 5 *Suppose that* $U \subset \mathbb{R}^n$ *is open, and* $g : U \to \mathbb{R}^n \setminus \{0\}$ *is* C^1. *Moreover, suppose that* $g_n \equiv 1$. *Define*

$$a_{ij}(x)=\frac{\partial g_i}{\partial x_j}(x)-\frac{\partial g_i}{\partial x_n}(x)g_j(x),$$

and let $A_g(x)$ *be the* $(n-1)\times(n-1)$ *matrix whose* (i, j)*th component is* $a_{ij}(x)$. *Then, the following holds.*[21]

(i) $g(x)$ *satisfies (3) if and only if* $A_g(x)$ *is symmetric for all* $x \in U$.
(ii) $w^T Dg(x)w \le 0$ *for all* $x \in U$ *and* $w \in \mathbb{R}^n$ *such that* $w \cdot g(x) = 0$ *if and only if* $A_g(x)$ *is negative semi-definite for all* $x \in U$.
(iii) $w^T Dg(x)w < 0$ *for all* $x \in U$ *and* $w \in \mathbb{R}^n$ *such that* $w \neq 0$ *and* $w \cdot g(x) = 0$ *if and only if* $A_g(x)$ *is negative definite for all* $x \in U$.

Proof By Lemma 4, $g(x)$ satisfies (3) if and only if $g(x)$ satisfies (12). Therefore, $g(x)$ satisfies (3) if and only if for every $x \in U$ and $i, j \in \{1, ..., n-1\}$,

[21] The matrix-valued function A_g is sometimes called the **Antonelli matrix** of g.

$$g_i\left(\frac{\partial g_j}{\partial x_n}-\frac{\partial g_n}{\partial x_j}\right)+g_j\left(\frac{\partial g_n}{\partial x_i}-\frac{\partial g_i}{\partial x_n}\right)+g_n\left(\frac{\partial g_i}{\partial x_j}-\frac{\partial g_j}{\partial x_i}\right)=0.$$

Because $g_n \equiv 1$, we have that $\frac{\partial g_n}{\partial x_i}=\frac{\partial g_n}{\partial x_j}=0$, and thus the above formula can be transformed as follows:

$$g_i\frac{\partial g_j}{\partial x_n}-g_j\frac{\partial g_i}{\partial x_n}+\frac{\partial g_i}{\partial x_j}-\frac{\partial g_j}{\partial x_i}=0,$$

where the left-hand side is $a_{ij}-a_{ji}$. Therefore, (3) is equivalent to the symmetry of A_g, and (i) holds.

Next, define $\hat{g}=(g_1,...,g_{n-1})$. By an easy calculation,

$$Dg(x)=\left(\begin{array}{c|c} A_g(x) & \frac{\partial \hat{g}}{\partial x_n}(x) \\ \hline 0^T & 0 \end{array}\right)+\left(\begin{array}{c|c} \frac{\partial \hat{g}}{\partial x_n}(x)\hat{g}^T(x) & 0 \\ \hline 0^T & 0 \end{array}\right).$$

Choose any $w \in \mathbb{R}^n$ such that $w \cdot g(x)=0$, and define $\hat{w}=(w_1,...,w_{n-1})$. Then, $\hat{w}\cdot\hat{g}(x)=-w_n$, and thus

$$\begin{aligned} w^T Dg(x)w &= \hat{w}^T A_g(x)\hat{w}+\hat{w}^T\frac{\partial \hat{g}}{\partial x_n}(x)w_n+\hat{w}^T\frac{\partial \hat{g}}{\partial x_n}(x)\hat{g}^T(x)\hat{w} \\ &= \hat{w}^T A_g(x)\hat{w}, \end{aligned}$$

which implies that (ii) holds. Moreover, if $w \neq 0$ and $w \cdot g(x)=0$, then $\hat{w}\neq 0$, which implies that (iii) holds. This completes the proof. □

Lemma 6 *Suppose that $f: P \to \Omega$ is a CoD that satisfies Walras' law, and $g: \mathbb{R}^n_{++}\to\mathbb{R}^n_{++}$ is an inverse demand function of f such that $g_n \equiv 1$. Let $x=f(p,m)$, and suppose that f is differentiable at (p,m) and g is differentiable at x. Define $\tilde{S}_f(p,m)$ as the $(n-1)\times(n-1)$ matrix whose (i,j)th component is $s_{ij}(p,m)$. Then, the matrix $A_g(x)$ is regular and $(A_g(x))^{-1}=\tilde{S}_f(p,m)$.*

Proof Define[22]

$$\hat{S}=\begin{pmatrix} s_{1,1} & \cdots & s_{1,n-1} & \frac{\partial f_1}{\partial m} \\ \vdots & \ddots & \vdots & \vdots \\ s_{n,1} & \cdots & s_{n,n-1} & \frac{\partial f_n}{\partial m} \end{pmatrix},\quad F=\begin{pmatrix} \frac{\partial f_1}{\partial p_1} & \cdots & \frac{\partial f_1}{\partial p_{n-1}} & \frac{\partial f_1}{\partial m} \\ \vdots & \ddots & \vdots & \vdots \\ \frac{\partial f_n}{\partial p_1} & \cdots & \frac{\partial f_n}{\partial p_{n-1}} & \frac{\partial f_n}{\partial m} \end{pmatrix}.$$

In other words, $\hat{S}$ is the matrix whose nth column vector is $D_m f(p,m)$ and jth column vector is the same as that of $S_f(p,m)$ for $j \neq n$, and F is the matrix whose

[22] In this proof, we frequently abbreviate variables. Note that we use the symbol $s_{i,j}$ instead of s_{ij} because if $j=n-1$, then the latter expression is confusing.

nth column vector is $D_m f(p,m)$ and jth column vector is the same as that of $D_p f(p,m)$ for $j \neq n$. Define

$$H = \begin{pmatrix} \frac{\partial g_1}{\partial x_1} & \cdots & \frac{\partial g_1}{\partial x_{n-1}} \\ \vdots & \ddots & \vdots \\ \frac{\partial g_{n-1}}{\partial x_1} & \cdots & \frac{\partial g_{n-1}}{\partial x_{n-1}} \end{pmatrix},$$

$$b = \left(\frac{\partial g_1}{\partial x_n}, ..., \frac{\partial g_{n-1}}{\partial x_n}\right)^T,$$

$$c = \left(\frac{\partial}{\partial x_1}[g(x)\cdot x], ..., \frac{\partial}{\partial x_{n-1}}[g(x)\cdot x]\right),$$

$$\hat{f} = (f_1, ..., f_{n-1})^T,\ \hat{g} = (g_1, ..., g_{n-1})^T.$$

Because $g_n \equiv 1$, to differentiate $y = f(g(y), g(y)\cdot y)$ in y at $y = x$, we have that[23]

$$I_n = D_p f(p,m) Dg(x) + D_m f(p,m) D[g(x)\cdot x] = F \times \begin{pmatrix} H & b \\ c & \frac{\partial}{\partial x_n}[g(x)\cdot x]. \end{pmatrix}$$

Therefore, F is regular and

$$\begin{aligned} F^{-1} &= \begin{pmatrix} H & b \\ c & \frac{\partial}{\partial x_n}[g(x)\cdot x] \end{pmatrix} \\ &= \begin{pmatrix} I_{n-1} & 0 \\ \hat{f}^T & 1 \end{pmatrix} \times \begin{pmatrix} H & b \\ \hat{g}^T & 1 \end{pmatrix} \\ &= \begin{pmatrix} I_{n-1} & 0 \\ \hat{f}^T & 1 \end{pmatrix} \times \begin{pmatrix} H - b\hat{g}^T & b \\ 0 & 1 \end{pmatrix} \times \begin{pmatrix} I_{n-1} & 0 \\ \hat{g}^T & 1 \end{pmatrix}. \end{aligned}$$

Because $H - b\hat{g}^T = A_g(x)$ by definition, we have that $A_g(x)$ is regular, and

$$\begin{aligned} \hat{S} &= F \times \begin{pmatrix} I_{n-1} & 0 \\ \hat{f}^T & 1 \end{pmatrix} \\ &= \begin{pmatrix} I_{n-1} & 0 \\ \hat{g}^T & 1 \end{pmatrix}^{-1} \times \begin{pmatrix} A_g(x) & b \\ 0 & 1 \end{pmatrix}^{-1} \\ &= \begin{pmatrix} I_{n-1} & 0 \\ -\hat{g}^T & 1 \end{pmatrix} \times \begin{pmatrix} (A_g(x))^{-1} & -(A_g(x))^{-1}b \\ 0 & 1 \end{pmatrix} \end{aligned}$$

as desired. This completes the proof. □

[23] I_n denotes the identity matrix.

Now, let $g(x)$ be defined by (4) and choose any $x \in \tilde{\Omega}$. Let $k(x) = \frac{1}{g_3(x)} g(x)$, $p = k(x)$ and $m = k(x) \cdot x$. Then, $f^G(p, m) = x$. Clearly f^G is differentiable at (p, m), and k is differentiable at x.

Because f^G is homogeneous of degree zero, we have that

$$D_p f^G(p, m)p + D_m f^G(p, m)m = 0.$$

On the other hand, because f^G satisfies Walras' law,

$$m = (f^G)^T(p, m)p.$$

To combine these equations, we obtain

$$S_{f^G}(p, m)p = 0.$$

By Lemma 6, we have that $\tilde{S}_{f^G}(p, m)$ is regular, and thus the rank of $S_{f^G}(p, m)$ is 2.

Suppose that there exists $(q, w) \in \mathbb{R}^3_{++} \times \mathbb{R}_{++}$ such that $f^G(q, w) = x$ and q is not proportional to p. Choose any $t \in [0, 1]$, and define $(r, c) = (1 - t)(p, m) + t(q, w)$. Suppose that $f^G(r, c) = y \neq x$. By Walras' law, $r \cdot x = c$. Because $r \cdot y \leq c$, either $p \cdot y \leq m$ or $q \cdot y \leq w$, which contradicts the weak axiom. Therefore, we have that $f^G(r, c) = x$ for all $t \in [0, 1]$. Hence,

$$S_{f^G}(p, m)(q - p) = \lim_{t \downarrow 0} \frac{f^G((1 - t)p + tq, (1 - t)m + tw) - f^G(p, m)}{t} = 0.$$

This implies that the dimension of the kernel of $S_{f^G}(p, m)$ is greater than or equal to 2, which contradicts the rank-nullity theorem. Therefore, such a (q, w) is absent, and we conclude that $\tilde{P} = \{(mg(x), m) | x \in \tilde{\Omega},\ m > 0\}$.

By the way, suppose that f is a CoD that satisfies Walras' law, g is an inverse demand function of f, $x = f(p, m)$, f is differentiable at (p, m), and g is differentiable at x. Using Lemmas 3, 5, and 6, we can easily show that g violates (3) at x if and only if $S_f(p, m)$ is not symmetric. Let $x = (1, 1, 1)$ and $(p, m) = (1, 1, 1, 3)$. Because $g(x)$ defined by (4) violates (3) at x, $S_{f^G}(p, m)$ is not symmetric, and by Theorem 3, we can conclude that f^G is not a demand function. This is another proof that Gale's example is not a demand function.

However, this logic can be used because f is differentiable. To prove (III) of Theorem 4, we need a more complicated argument because $f^{\succsim^g}$ is not necessarily differentiable.

Proof of (I) of Theorem 4

First, we must define several symbols. In this definition, we abbreviate variables, but all symbols are actually functions on $\tilde{\Omega}^2$. For any $(x, v) \in \tilde{\Omega}^2$, we define[24]

$$a_1 = \frac{1}{\|x\|}x,$$

$$a_2 = \begin{cases} \frac{1}{\|v-(v\cdot a_1)a_1\|}(v-(v\cdot a_1)a_1) & \text{if } v \neq (v\cdot a_1)a_1, \\ 0 & \text{otherwise,} \end{cases}$$

$$Py = (y\cdot a_1)a_1 + (y\cdot a_2)a_2,$$

$$Rw = (w\cdot a_1)a_2 - (w\cdot a_2)a_1,$$

$$v_1 = \arg\min\{w\cdot a_1 | w \in P\mathbb{R}^n_+,\ \|w\| = 1,\ w\cdot a_2 \geq 0\},$$

$$v_2 = \arg\min\{w\cdot a_1 | w \in P\mathbb{R}^n_+,\ \|w\| = 1,\ w\cdot a_2 \leq 0\}.$$

$$\Delta = \{w \in \text{span}\{x, v\} | w\cdot Rv \leq 0,\ w\cdot v_1 \geq x\cdot v_1,\ w\cdot v_2 \leq x\cdot v_2\}.$$

$$C = \|x\|\|v-(v\cdot a_1)a_1\|.$$

Moreover, we define $y_1 = y_2 = x$ if x is proportional to v, and otherwise, y_i is the unique intersection of $\{s_1 v | s_1 \in \mathbb{R}\} \cap \{x + s_2 Rv_i | s_2 \in \mathbb{R}\}$.

Lemma 7 *All the above symbols are well-defined. Moreover, the following results hold.*

(i) If x is not proportional to v, then $\{a_1, a_2\}$ is the orthonormal basis of span$\{x, v\}$ derived from x, v by the Gram–Schmidt method, and P is the orthogonal projection from $\mathbb{R}^n$ onto span$\{x, v\}$.[25]

(ii) If x is not proportional to v, then R is the unique orthogonal transformation on span$\{x, v\}$ such that $Ra_1 = a_2$ and $Ra_2 = -a_1$. Moreover, if T is an orthogonal transformation on span$\{x, v\}$ such that $w\cdot Tw = 0$ for any $w \in$ span$\{x, v\}$, then we must have either $T = R$ or $T = -R = R^{-1} = R^3$.[26]

(iii) If x is not proportional to v, then both v_1 and v_2 are continuous and single-valued at (x, v). Moreover, $P\mathbb{R}^n_+ = \{c_1 v_1 + c_2 v_2 | c_1, c_2 \geq 0\}$.

(iv) Both y_1 and y_2 are continuous and single-valued. Moreover, for any $(x, v) \in \tilde{\Omega}^2$, $y_1(x, v),\ y_2(x, v) \in \tilde{\Omega}$.

---

[24] We restrict the operator R to span$\{x, v\}$.

[25] As a result, $y\cdot w = Py\cdot w$ for any $y \in \mathbb{R}^n$ and $w \in$ span$\{x, v\}$.

[26] In particular, if $z \in$ span$\{x, v\} \cap \tilde{\Omega}$ and $[x, z] \cap \{cv | c \in \mathbb{R}\} = \emptyset$, then $R(x, v) = R(z, v)$. Note that $[x, z]$ represents $\{(1-t)x + tz | t \in [0, 1]\}$.

(v) $\Delta = (x + RP\mathbb{R}^n_+) \cap \{w \in span\{x, v\} | w \cdot Rv \le 0\} = co\{x, y_1, y_2\}$.[27]
(vi) $(y \cdot x)v - (y \cdot v)x = C \cdot RPy$ *for any* $y \in \mathbb{R}^n$.

We omit the proof of Lemma 7. See the proof of Theorem 1 in Hosoya [7] for more detailed arguments.

We now prove (I) of Theorem 4. For any $(x, v) \in \tilde{\Omega}^2$, let $y(\cdot; x, v)$ be the nonextendable solution to the following initial value problem:

$$\dot{y}(t) = (g(y(t)) \cdot x)v - (g(y(t)) \cdot v)x, \ y(0) = x. \tag{13}$$

Note that, by Lemma 7, this equation can be rewritten as

$$\dot{y}(t) = CRPg(y(t)), \ y(0) = x. \tag{14}$$

Recall the following definition:

$$w^* = (v \cdot x)v - (v \cdot v)x = CRv,$$

$$t(x, v) = \inf\{t \ge 0 | y(t; x, v) \cdot w^* \ge 0\}.$$

We first show the well-definedness of $t(x, v)$. Actually, $t(x, v) = 0$ if x is proportional to v, and thus we can assume that x is not proportional to v. By the Cauchy–Schwarz inequality, we have that

$$x \cdot w^* < 0.$$

Moreover,

$$\dot{y}(t; x, v) \cdot w^* = (CRPg(y(t; x, v))) \cdot (CRv) = C^2 g(y(t; x, v)) \cdot v > 0.$$

Furthermore, because $\dot{y}(t; x, v) \in RP\mathbb{R}^n_+$, by (v) of Lemma 7, we have that $y(t; x, v) \in \Delta$ if and only if $t \ge 0$ and $y(t; x, v) \cdot w^* \le 0$. Because Δ is compact and $y(t; x, v)$ is nonextendable, we have that there exists $t^* > 0$ such that $y(t^*; x, v) \notin \Delta$, which implies that $y(t^*; x, v) \cdot w^* > 0$. This indicates that $t(x, v)$ is well-defined. Moreover, clearly $y(t; x, v) \cdot w^* = 0$ if and only if $t = t(x, v)$, which implies that $y(t; x, v)$ is proportional to v if and only if $t = t(x, v)$. Because $t(x, v)$ is the unique solution to the following equation:

$$y(t; x, v) \cdot w^* = 0,$$

by the implicit function theorem, the function $t : (y, z) \mapsto t(y, z)$ is C^k around (x, v), and thus u^g is also C^k around (x, v). By (iv) of Lemma 7, we can easily verify that u^g is continuous on $\tilde{\Omega}^2$,[28] and thus $\succsim^g$ is a continuous binary relation on $\tilde{\Omega}$.

[27] By this result, we have that Δ is a compact subset of $\tilde{\Omega}$.

[28] Recall that $u^g(x, v)v \in [y_1(x, v), y_2(x, v)]$ for any $(x, v) \in \tilde{\Omega}^2$.

It suffices to show that $\succsim^g$ is complete, p-transitive, and monotone. We introduce a lemma.

Lemma 8 *Suppose that $x, v, z, w \in \Omega$, x is not proportional to v, z is not proportional to w, and span$\{x, v\} =$ span$\{z, w\}$. If there exist $t_1, t_2 \in \mathbb{R}$ such that $y(t_1; x, v) = y(t_2; z, w)$, then the trajectory of $y(t; x, v)$ is the same as that of $y(t; z, w)$.*

Proof Since $P(x, v) = P(z, w)$, we abbreviate both $P(x, v)$ and $P(z, w)$ as P. We abbreviate $R(x, v)$ as R. By (ii), $R(z, w)$ is equal to either R or $-R$. We define $s = 1$ if $R(z, w) = R$ and $s = -1$ otherwise. Define

$$z_1(t) = y((C(x, v))^{-1}t; x, v), z_2(t) = y(s(C(z, w))^{-1}t; z, w).$$

For (vi),

$$\dot{z}_1(t) = RPg(z_1(t)), \ \dot{z}_2(t) = RPg(z_2(t)).$$

Hence, both z_1 and z_2 are solutions to the following autonomous differential equation:

$$\dot{z} = RPg(z).$$

Moreover, we can easily check that both z_1 and z_2 are nonextendable. By assumption, there exist t_3, t_4 such that $z_1(t_3) = z_2(t_4)$, and thus the trajectory of z_1 must be equal to that of z_2. This completes the proof. □

Now, we prove that

$$u^g(y, v) \ge u^g(z, v) \Leftrightarrow u^g(y, z) \ge 1 (\Leftrightarrow y \succsim^g z) \tag{15}$$

for all $x, v \in \tilde{\Omega}$ and $y, z \in \text{span}\{x, v\} \cap \tilde{\Omega}$.[29] If x is proportional to v, then (15) holds trivially. Hence, we assume that x is not proportional to v. If z is proportional to v, then $z = av$, and by Lemma 8, $y(t(y, z); y, z) = y(t(y, v); y, v)$. Therefore, $u^g(z, v) = a$ and $u^g(y, v) = au^g(y, z)$, which implies that

$$u^g(y, v) \ge u^g(z, v) \Leftrightarrow au^g(y, z) \ge a \Leftrightarrow u^g(y, z) \ge 1,$$

as desired.

Hence, we assume that z is not proportional to v. Suppose that y is proportional to z. If $u^g(y, z) = 1$, then $y = z$, and thus $u^g(y, v) = u^g(z, v)$. Suppose that $u^g(y, z) > 1$. Define

$$c(s) = \frac{\|y(t(y, (1 - s)y + sv); y, (1 - s)y + sv)\|}{\|y(t(z, (1 - s)z + sv); z, (1 - s)z + sv)\|}.$$

[29] That is, the function $y \mapsto u^g(y, v)$ represents $\succsim^g$ on $\tilde{\Omega} \cap \text{span}\{x, v\}$.

By assumption, $c(0) = u^g(y, z) > 1$. If $c(1) \leq 1$, then $c(s) = 1$ for some $s \in [0, 1]$ by the intermediate value theorem. Thus, there exist t', t'' such that $y(t'; y, v) = y(t''; z, v)$ and $c(s) \equiv 1$ by Lemma 8. This implies that $u^g(y, z) = c(0) = 1$, which is a contradiction. Hence, $c(1) > 1$, and thus, $u^g(y, v) > u^g(z, v)$. Finally, if $u^g(y, z) < 1$, then by the symmetrical argument as above, we can show that $u^g(y, v) < u^g(z, v)$, and thus (15) holds when y is proportional to z.

Next, suppose that y is not proportional to z. By Lemma 8, the trajectory of $y(t; y, v)$ is the same as that of $y(t; u^g(y, z)z, v)$, and thus

$$\begin{aligned} u^g(u^g(y, z)z, v) &= \frac{\|y(t(u^g(y, z)z, v); u^g(y, z)z, v)\|}{\|v\|} \\ &= \frac{\|y(t(y, v); y, v)\|}{\|v\|} \\ &= u^g(y, v). \end{aligned} \tag{16}$$

Because we have already shown that $u^g(u^g(y, z)z, v) \geq u^g(z, v)$ if and only if $u^g(y, z) \geq 1$, we have that (15) holds.[30]

Completeness and p-transitivity of $\succsim^g$ directly follow from (15).[31]

It suffices to show that $\succsim^g$ is monotone. Actually, we prove that

$$v \gneq x \Rightarrow v \succ^g x.$$

Let $(x, v) \in \tilde{\Omega}^2$ and $v \gneq x$. If x is proportional to v, then clearly $v \succ^g x$. Hence, we assume that x is not proportional to v. By the completeness of $\succsim^g$, it suffices to show that $u^g(x, v) < 1$. Define $z = v - x$. Then, $z \in \operatorname{span}\{x, v\}$ and $z \gneq 0$. By (i), (ii), and (vi) of Lemma 7,

$$\dot{y}(t; x, v) \cdot Rz > 0,$$

and thus,

$$\begin{aligned} u^g(x, v)v \cdot Rz &= y(t(x, v); x, v) \cdot Rz \\ &> y(0; x, v) \cdot Rz \\ &= x \cdot Rz \\ &= v \cdot Rz. \end{aligned}$$

Therefore, to prove that $u^g(x, v) < 1$, it suffices to show that $v \cdot Rz < 0$.

Because $x \cdot a_2 = 0$,

---

30 We mention that (16) holds unconditionally: that is, if $\dim(\operatorname{span}\{y, z, v\}) \leq 2$, then

$$u^g(u^g(y, z)z, v) = u^g(y, v)$$

even when y is proportional to z or z is proportional to v. The proof is easy and thus omitted.

31 For completeness, note that by (15), $u^g(x, v) \geq 1 \Leftrightarrow u^g(v, x) \leq 1$.

$$\begin{aligned}v \cdot Rz &= v \cdot [(z \cdot a_1)a_2 - (z \cdot a_2)a_1]\\&= (z \cdot a_1)(v \cdot a_2) - (v \cdot a_2)(v \cdot a_1)\\&= -(v \cdot a_2)(x \cdot a_1).\end{aligned}$$

Since $x \cdot a_1 > 0$ by definition and $v \cdot a_2 \geq 0$ by the Cauchy–Schwarz inequality, we have that $v \cdot Rz \leq 0$. If $v \cdot Rz = 0$, then v is proportional to z, and thus, x is proportional to v, which contradicts our initial assumption. Hence, we have that $v \cdot Rz < 0$, which completes the proof of (I). □

Proof of (II) of Theorem 4

Let $D_x u^g$ denote the partial derivative of u^g with respect to the first variable or its transpose.[32] We first present a lemma.

Lemma 9 *Suppose that $(x, v) \in \tilde{\Omega}^2$ and x is not proportional to v. Then, the following results hold.*

(i) $D_x u^g(x, v) \neq 0$ *and* $D_x u^g(x, v)x > 0$.
(ii) $PD_x u^g(x, v) = \lambda(x)Pg(x)$, *where* $\lambda(x) = \frac{D_x u^g(x,v)x}{g(x)\cdot x} > 0$.

Proof If $y, z, w \in \tilde{\Omega}$ are linearly dependent, then by (16), $u^g(y, w) = u^g(u^g(y, z)z, w)$. Therefore, for any $t > 0$,

$$t = u^g(tv, v) = u^g(u^g(tv, x)x, v),\ 1 = u^g(x, x) = u^g(u^g(x, v)v, x).$$

Taking derivatives with respect to t on both sides of the former equation at $t = u^g(x, v)$, we have that,

$$\begin{aligned}1 &= D_x u^g(u^g(u^g(x, v)v, x)x, v)x \times [D_x u^g(u^g(x, v)v, x)v]\\&= D_x u^g(x, v)x \times [D_x u^g(u^g(x, v)v, x)v],\end{aligned}$$

which implies that $D_x u^g(x, v) \neq 0$ and $D_x u^g(x, v)x \neq 0$. Since $u^g(tx, v) > u^g(x, v)$ for any $t > 1$, $D_x u^g(x, v)x > 0$ and (i) holds.

To prove (ii), define $z = \dot{y}(0; x, v)$. By (16), $u^g(y(t; x, v), v) = u^g(x, v)$. Taking derivatives with respect to t at $t = 0$, we have that $D_x u^g(x, v)z = 0$. Note that because $z = CRPg(x)$, $g(x) \cdot z = 0$, and thus both $Pg(x)$ and $PD_x u^g(x, v)$ belong to $\{w \in \text{span}\{x, v\} | w \cdot z = 0\}$, whose dimension is exactly 1. Hence, these are linearly dependent, and thus (ii) holds. This completes the proof. □

Suppose that $x \in f^{\succsim^g}(g(x), g(x) \cdot x)$ for all $x \in \tilde{\Omega}$. Choose any $x \in \tilde{\Omega}$ and $w \in \mathbb{R}^n$ such that $w \cdot g(x) = 0$. If $w = 0$, then clearly $w^T Dg(x)w = 0$. Otherwise, we

---

[32] We think that this abbreviation of the transpose symbol does not cause any confusion.

can assume without loss of generality that $x + w, x - w \in \tilde{\Omega}$. Let $v = x + w, x(t) = x + tw$ and $c(t) = u^g(x(t), v)$. Since $x \in f^{\succsim^g}(g(x), g(x) \cdot x)$, we have that $c(0) \geq c(t)$ for any $t \in [-1, 1]$. Note that $c'(t) = D_x u^g(x(t), v)w$ for all $t \in]-1, 1[$. By the mean value theorem, there exists a positive sequence (t_m) such that $t_m \downarrow 0$ as $m \to \infty$ and $c'(t_m) \leq 0$ for all m. Then,

$$\begin{aligned} 0 &\geq \limsup_{m\to\infty} \frac{D_x u^g(x(t_m), v)w}{t_m} \\ &= \limsup_{m\to\infty} \frac{\lambda(x(t_m))g(x(t_m)) \cdot w}{t_m} \\ &\geq M \limsup_{m\to\infty} \frac{g(x(t_m)) \cdot w}{t_m} \\ &= M \limsup_{m\to\infty} \frac{g(x(t_m)) - g(x)}{t_m} \cdot w \\ &= M[w^T Dg(x)w], \end{aligned}$$

where $M = \max_{t\in[0,\frac{1}{2}]} \lambda(x(t)) > 0$. Hence, $w^T Dg(x)w \leq 0$ for all $x \in \tilde{\Omega}$ and $w \in \mathbb{R}^n$ such that $w \cdot g(x) = 0$.

Second, suppose that there exists $x \in \tilde{\Omega}$ such that $x \notin f^{\succsim^g}(g(x), g(x) \cdot x)$. By p-transitivity, continuity, and monotonicity of $\succsim^g$, there exists $v \in \tilde{\Omega}$ such that $g(x) \cdot v < g(x) \cdot x$ and $v \sim^g x$. Clearly, v is not proportional to x. Define $x(t) = (1 - t)x + tv$. Then,

$$\begin{aligned} \frac{d}{dt}[u^g(x(t), v)]\bigg|_{t=0} &= D_x u^g(x, v)(v - x) \\ &= \lambda(x)g(x) \cdot (v - x) \\ &= \lambda(x)[g(x) \cdot v - g(x) \cdot x] < 0, \end{aligned}$$

and thus, $u^g(x(t), v) < u^g(x, v)$ for a sufficiently small $t > 0$, which implies that $x(t) \not\succsim^g x$, and thus $\succsim^g$ is not convex.

Hence, for proving the first claim of (II) of Theorem 4, it suffices to show that if $w^T Dg(x)w \leq 0$ for all $x \in \tilde{\Omega}$ and $w \in \mathbb{R}^n$ such that $w \cdot g(x) = 0$, then $\succsim^g$ is convex. Suppose not. Then, there exist $x, v \in \tilde{\Omega}$ and $t \in]0, 1[$ such that $v \sim^g x$ and $(1 - t)x + tv \not\succsim^g x$. Clearly, v is not proportional to x. Let $w = v - x, x(s) = x + sw$, and

$$s^* = \max[\arg\min\{u^g(x(s), v) | s \in [0, 1]\}].$$

Since $u^g(x(t), v) < 1$, we have $s^* \in]0, 1[$. Define $c(s) = u^g(x(s), v)$. Then, $c(s^*) \leq c(s)$ for any $s \in [0, 1]$, and thus, $c'(s^*) = 0$. Hence, $D_x u^g(x(s^*), v)w = 0$. Let $p = PD_x u^g(x(s^*), v)$. Then, $p \cdot w = 0$. By (1) of Lemma 9, $p \cdot x(s^*) > 0$, and thus $p \neq 0$.

Consider the following equation:

$$f(a,b)=u^g(bp+x(s^*+a),v)=u^g(x(s^*),v).$$

Note that $\frac{\partial f}{\partial b}(0,0)=\|p\|^2\neq 0$. Applying the implicit function theorem, we can prove the existence of the C^k function $b:]-\varepsilon,\varepsilon[\to\mathbb{R}$ such that $b(0)=0$ and $f(a,b(a))=u^g(x(s^*),v)$ for any $a\in]-\varepsilon,\varepsilon[$. Differentiating both sides with respect to a, we have

$$D_xu^g(b(a)p+x(s^*+a),v)[b'(a)p+w]=0.$$

Hence,

$$0=D_xu^g(x(s^*),v)[b'(0)p+w]=b'(0)\|p\|^2,$$

which implies that $b'(0)=0$. By Lemma 9, if $a>0$ is so small that $D_xu^g(b(a)p+x(s^*+a),v)p>0$, then

$$\begin{aligned}
0=&\limsup_{a'\downarrow a}\frac{1}{a'-a}[-D_xu^g(b(a)p+x(s^*+a),v)[b'(a')p+w]\\
&+D_xu^g(b(a)p+x(s^*+a),v)[b'(a')p+w]]\\
=&\limsup_{a'\downarrow a}\frac{1}{a'-a}[-\lambda(b(a)p+x(s^*+a))g(b(a)p+x(s^*+a))\cdot[b'(a')p+w]\\
&+D_xu^g(b(a)p+x(s^*+a),v)[b'(a')p+w]]\\
=&\limsup_{a'\downarrow a}\frac{1}{a'-a}[\lambda(b(a)p+x(s^*+a))\\
&\times[g(b(a')+x(s^*+a'))-g(b(a)p+x(s^*+a))]\cdot[b'(a')p+w]\\
&+D_xu^g(b(a)p+x(s^*+a),v)[(b'(a')p+w)-(b'(a)p+w)]]\\
=&\lambda(b(a)p+x(s^*+a))[b'(a)p+w]^TDg(b(a)p+x(s^*+a))[b'(a)p+w]\\
&+D_xu^g(b(a)p+x(s^*+a),v)p\times\limsup_{a'\downarrow a}\frac{b'(a')-b'(a)}{a'-a},
\end{aligned}$$

and thus, $\limsup_{a'\downarrow a}\frac{b'(a')-b'(a)}{a'-a}\geq 0$. By the same arguments, we can show that $\limsup_{a'\uparrow a}\frac{b'(a')-b'(a)}{a'-a}\geq 0$.

Now, choose any such $a>0$. Define $h(\theta)=b'(\theta)a-b'(a)\theta$. Then, $h(0)=h(a)=0$, and thus, there exists $\theta^*\in]0,a[$ such that either $h(\theta^*)\geq h(\theta)$ for any $\theta\in[0,a]$ or $h(\theta^*)\leq h(\theta)$ for any $\theta\in[0,a]$. If the former holds,

$$\begin{aligned}
0\geq&\limsup_{\theta\downarrow\theta^*}\frac{h(\theta)-h(\theta^*)}{\theta-\theta^*}\\
=&a\limsup_{\theta\downarrow\theta^*}\frac{b'(\theta)-b'(\theta^*)}{\theta-\theta^*}-b'(a),
\end{aligned}$$

and thus we have that $b'(a) \geq 0$. By the symmetrical arguments, we can show that $b'(a) \geq 0$ in the latter case. Hence, we have that $b'(a) \geq 0$ for a sufficiently small $a > 0$, and thus, $b(a) \geq 0$ for a sufficiently small $a > 0$.

On the other hand, $D_x u^g(x(s^*), v)p = \|p\|^2 > 0$, and thus,

$$D_x u^g(b'p + x(s^* + a), v)p > 0 \text{ for any } b' \in [0, b(a)],$$

for a sufficiently small $a > 0$. Therefore,

$$u^g(x(s^* + a), v) \leq u^g(b(a)p + x(s^* + a), v) = u^g(x(s^*), v),$$

which contradicts the definition of s^*. Hence, the first claim of (II) is correct.

Next, suppose that $w^T Dg(x)w < 0$ for all $x \in \mathbb{R}^n_{++}$ and $w \in \mathbb{R}^n$ such that $w \neq 0$ and $w \cdot g(x) = 0$. Choose any $(p, m) \in \mathbb{R}^n_{++} \times \mathbb{R}_{++}$. For any $x \in \tilde{\Omega}$ such that $p \cdot x = m$, if $g(x)$ is not proportional to p, then there exists $v \in \tilde{\Omega}$ such that $p \cdot v \leq p \cdot x$ and $v \succ^g x$, and thus, $x \notin f^{\succsim^g}(p, m)$. Therefore, if there is no x such that $p \cdot x = m$ and $g(x)$ is proportional to p, then $f^{\succsim^g}(p, m) = \emptyset$. Next, suppose that $p = ag(x)$ for some $a > 0$ and $m = p \cdot x$. We have already shown that $x \in f^{\succsim^g}(p, m)$. Suppose that $\succsim^g$ is strictly convex. Then, $x \succ^g v$ for any $v \in \tilde{\Omega}$ such that $p \cdot v \leq m$ and $v \neq x$, which implies that $f^{\succsim^g}(p, m) = \{x\}$. Hence, it suffices to show that $\succsim^g$ is strictly convex. Suppose not. Then, there exist $y, z \in \tilde{\Omega}$ and $t \in]0, 1[$ such that $y \sim^g z$, $y \neq z$ and $(1 - t)y + tz \not\succ^g y$. Define $y(s) = (1 - s)y + sz$. We have already shown that $\succsim^g$ is convex, and thus $y(s) \succsim^g y$ for any $s \in [0, 1]$. This implies that $u^g(y(t), y) \leq u^g(y(s), y)$ for any $s \in [0, 1]$. By the first-order condition and Lemma 9,

$$g(y(t)) \cdot (z - y) = 0,$$

and by the mean value theorem, there exists a sequence (s_m) such that $s_m \downarrow t$ as $m \to \infty$ and

$$g(y(s_m)) \cdot (z - y) \geq 0$$

for all m. Therefore,

$$(z - y)^T Dg(y(t))(z - y) = \lim_{m \to \infty} \frac{[g(y(s_m)) - g(y(t))] \cdot (z - y)}{s_m - t} \geq 0,$$

which contradicts our initial assumption. This completes the proof of (II). □

Proof of (III) of Theorem 4

We separate the proof into two steps.

Step 1. *Suppose that* $g : \tilde{\Omega} \to \mathbb{R}^n_{++}$ *is* C^k. *Then, (ii)-(v) of (III) of Theorem 4 are equivalent.*

Proof of Step 1. First, we show that (iii) is equivalent to (iv). Suppose that (iii) holds. Then, for any $x, z \in \tilde{\Omega}$,

$$\frac{d}{dt}[u^g_v(y(t; x, z))] = 0,$$

and thus $u^g_v(x) = u^g_v(u^g(x, z)z)$. Hence,

$$u^g_v(x) \ge u^g_v(z) \Leftrightarrow u^g_v(u^g(x, z)z) \ge u^g_v(z) \Leftrightarrow u^g(x, z) \ge 1 \Leftrightarrow x \succsim^g z,$$

which implies that (iv) holds.

Conversely, suppose that (iv) holds. Choose any $x \in \tilde{\Omega}$ and any linearly independent family $v_1, ..., v_{n-1} \in \mathbb{R}^n$ such that $v_i \cdot g(x) = 0$ and $x + v_i \in \tilde{\Omega}$ for all $i \in \{1, ..., n-1\}$. Since both v_i and $\dot{y}(0; x, x + v_i)$ are orthogonal to $P(x, x + v_i)g(x)$, $\dot{y}(0; x, x + v_i)$ is proportional to v_i. Since $y(t; x, x + v_i) \sim^g x$ for any t, we have that $\nabla u^g_v(x)$ is orthogonal to $\dot{y}(0; x, x + v_i)$, and thus $\nabla u^g_v(x) \cdot v_i = 0$ for all $i \in \{1, ..., n-1\}$. Therefore, $\nabla u^g_v(x) = \lambda(x)g(x)$ for some $\lambda(x) \in \mathbb{R}$, which implies that (iii) holds.

Second, we show that (iv) is equivalent to (ii). It is obvious that (iv) implies (ii). Conversely, suppose that (ii) holds. First, choose any $x, z, v \in \tilde{\Omega}$. Then,

$$u^g(x, v)v \sim^g x \sim^g u^g(x, z)z \sim^g u^g(u^g(x, z)z, v)v,$$

which implies that $u^g(x, v) = u^g(u^g(x, z)z, v)$.

Next, choose any $x, z, v \in \tilde{\Omega}$ and suppose that $u^g_v(x) \ge u^g_v(z)$. Then,

$$u^g(u^g(x, z)z, v) = u^g(x, v) \ge u^g(z, v),$$

which implies that $u^g(x, z) \ge 1$. Hence, $x \succsim^g z$. Conversely, suppose that $x \succsim^g z$. Then, $u^g(x, z) \ge 1$, and thus,

$$u^g_v(x) = u^g(u^g(x, z)z, v) \ge u^g_v(z).$$

Hence, u^g_v represents $\succsim^g$.

Fix any $v \in \tilde{\Omega}$ and choose any $z \in \tilde{\Omega}$ such that z is not proportional to v. Then, for any $x \in \tilde{\Omega}$, we have x is not proportional to either v or z. If x is not proportional to v, clearly u^g_v is C^k around x. Otherwise,

$$u^g_v(y) = u^g(u^g(y, z)z, v)$$

for any $y \in \tilde{\Omega}$ and the right-hand side is C^k in y around x. Therefore, u_v^g is C^k on $\tilde{\Omega}$, and thus (iv) holds.

It suffices to show that (iii) is equivalent to (v). By Frobenius' theorem, (iii) implies (v). Conversely, suppose that (v) holds. We introduce a lemma.

Lemma 10 *For any $v \in \tilde{\Omega}$, there exist open neighborhoods W^v, W_1^v of v, $\hat{u}^v : W_1^v \to \mathbb{R}$, and $\lambda^v : W_1^v \to \mathbb{R}_{++}$ such that*

(a) $u^g(x, z) = u^g(u^g(x, y)y, z)$ for any $x, y, z \in W^v$,
(b) $y([0, t(x, y)]; x, y) \subset W_1^v$ for any $x, y \in W^v$,
(c) $\hat{u}^v$ is C^k on W_1^v and $\nabla \hat{u}^v(x) = \lambda^v(x)g(x)$ for any $x \in W_1^v$.

Proof Fix any $v \in \tilde{\Omega}$. By Frobenius' theorem, there exists an open and convex neighborhood $W_1^v \subset \tilde{\Omega}$ of v, $\hat{u}^v : W_1^v \to \mathbb{R}$, and $\lambda^v : W_1^v \to \mathbb{R}_{++}$ such that $\hat{u}^v$ is C^k and $\nabla \hat{u}^v(x) = \lambda^v(x)g(x)$ for all $x \in W_1^v$. Since y_1, y_2 are continuous on $\tilde{\Omega}^2$ and $y_1(v, v) = v = y_2(v, v)$, $U_1^v = y_1^{-1}(W_1^v) \cap y_2^{-1}(W_1^v)$ is open and includes (v, v).[33] Let $W_2^v \subset W_1^v$ be an open, convex neighborhood of v such that $W_2^v \times W_2^v \subset U_1^v$, $U_2^v = y_1^{-1}(W_2^v) \cap y_2^{-1}(W_2^v)$, and $W^v \subset W_2^v$ be an open neighborhood of v such that $W^v \times W^v \subset U_2^v$.

Clearly, (c) holds. To show (b), choose any $x, y \in W^v$. If x is proportional to y, then $t(x, y) = 0$, and thus (b) holds. Otherwise, since $x, y \in W^v$, we have that $(x, y) \in U_2^v$, and thus, $y_1(x, y), y_2(x, y) \in W_2^v$. Since W_2^v is convex, we have that $\Delta(x, y) \subset W_2^v \subset W_1^v$, and thus, $y([0, t(x, y)]; x, y) \subset \Delta(x, y) \subset W_1^v$. Hence, (b) holds.

To show (a), suppose that $x, y, z \in W^v$. If these vectors are linearly dependent, then we have shown that $u^g(x, z) = u^g(u^g(x, y)y, z)$ holds in (16). Otherwise, we have already shown that $y([0, t(x, y)]; x, y) \subset W_2^v$, and thus, $u^g(x, y)y \in W_2^v$. Hence, $y_1(u^g(x, y)y, z), y_2(u^g(x, y)y, z) \in W_1^v$, and thus, $\Delta(u^g(x, y)y, z) \subset W_1^v$. Therefore, $y([0, t(u^g(x, y)y, z)]; u^g(x, y)y, z) \subset W_1^v$. Moreover, we have already proved that $y([0, t(x, z)]; x, z) \subset W_1^v$.

By easy computations,

$$\frac{d}{dt}\hat{u}^v(y(t; x, y)) = 0 \text{ for any } t \in [0, t(x, y)],$$

$$\frac{d}{dt}\hat{u}^v(y(t; u^g(x, y)y, z)) = 0 \text{ for any } t \in [0, t(u^g(x, y)y, z)],$$

and

$$\frac{d}{dt}\hat{u}^v(y(t; x, z)) = 0 \text{ for any } t \in [0, t(x, z)].$$

Hence, we obtain

$$\hat{u}^v(u^g(x, z)z) = \hat{u}^v(x) = \hat{u}^v(u^g(x, y)y) = \hat{u}^v(u^g(u^g(x, y)y, z)z).$$

[33] See Lemma 7.

Since $\nabla\hat{u}^v(w) \gg 0$ for all $w \in W_1^v$, we have that $u^g(x, z) = u^g(u^g(x, y)y, z)$. Hence, (a) holds. This completes the proof. □

Choose any $(x, v) \in \tilde{\Omega}^2$. If x is not proportional to v, then u^g is C^k around (x, v). Otherwise, there exists $s > 0$ such that $x = sv$. Let W^x be the set defined in Lemma 10, and choose any $z \in W^x$ that is not proportional to x. Let $W' = W^x \setminus \{tz|t \in \mathbb{R}\}$ and $W'' = s^{-1}W'$. Then, W' is an open neighborhood of x, W'' is an open neighborhood of v, and for any $(y, w) \in W' \times W''$,

$$u^g(y, w) = su^g(y, sw) = su^g(u^g(y, z)z, sw) = u^g(u^g(y, z)z, w),$$

where the right-hand side is C^k around (x, v). Hence, u^g is C^k on $\tilde{\Omega}^2$.

Define

$$X = \{x \in \tilde{\Omega}|\exists\mu(x) \in \mathbb{R} \text{ such that } D_x u^g(x, v) = \mu(x)g(x)\}.$$

To prove (iii), it suffices to show that $X = \tilde{\Omega}$.

First, we show that X includes some open neighborhood of $L = \{sv|s > 0\}$. Fix any $s > 0$. Let $x = sv$ and choose W^x, W_1^x, $\hat{u}^x$, λ^x as in Lemma 10. Fix any $y \in W^x$. Choose any $z \in W^x$ such that $\hat{u}^x(z) = \hat{u}^x(y)$. Since

$$\frac{d}{dt}\hat{u}^x(y(t; y, z)) \equiv 0$$

for all $t \in [0, t(y, z)]$, we have that $\hat{u}^x(y) = \hat{u}^x(u^g(y, z)z)$, and thus, $u^g(y, z) = 1$. Hence,

$$u^g(y, v) = su^g(y, x) = su^g(u^g(y, z)z, x) = su^g(z, x) = u^g(z, v).$$

Conversely, choose any $z \in W^x$ such that $u^g(y, v) = u^g(z, v)$. Then,

$$u^g(z, v) = u^g(y, v) = su^g(y, x) = su^g(u^g(y, z)z, x) = u^g(u^g(y, z)z, v),$$

which implies that $u^g(y, z) = 1$. Hence,

$$\hat{u}^x(y) = \hat{u}^x(u^g(y, z)z) = \hat{u}^x(z).$$

Therefore, $\hat{u}^x(z) = \hat{u}^x(y)$ if and only if $u^g(z, v) = u^g(y, v)$.

By the preimage theorem, the set $(\hat{u}^x)^{-1}(y)$ is an $n - 1$ dimensional C^k manifold and both $\nabla\hat{u}^x(y)$ and $D_x u^g(y, v)$ are in the orthogonal complement of the tangent space $T_y((\hat{u}^x)^{-1}(y))$.[34] Hence, there exists $c \in \mathbb{R}$ such that

$$D_x u^g(y, v) = c\nabla\hat{u}^x(y) = c\lambda^x(y)g(y).$$

[34] See Sect. 1.4 of Guillemin and Pollack [6].

Therefore, we have proved that $\cup_{s>0} W^{sv} \subset X$.

Second, fix any $x \in \tilde{\Omega}$. It suffices to show that $x \in X$. If x is proportional to v, then it is clear. Otherwise, define A_x as the trajectory of $y(\cdot; x, v)$ and $B_x = A_x \cap Y$, where Y is the interior of X. It suffices to show that $x \in B_x$. To show this, we prove that $B_x = A_x$. Since A_x is connected, it suffices to show that B_x is nonempty, open, and closed in A_x. Clearly, B_x is open in A_x. Since $u^g(x, v)v \in B_x$, B_x is nonempty. Hence, it suffices to show that B_x is closed in A_x.

Suppose that (z^m) is a sequence on B_x that converges to $z^* \in A_x$. It suffices to show that $z^* \in B_x$. Without loss of generality, we assume that $z^* \neq u^g(x, v)v$. Choose W^{z^*}, $W_1^{z^*}$, $\hat{u}^{z^*}$, λ^{z^*} as in Lemma 10.

Since W^{z^*} is open in $\tilde{\Omega}$, there exists m such that $z^m \in W^{z^*}$. By the definition of A_x and Lemma 8, there exists t^m such that $y(t^m; z^*, v) = z^m$. Since $z^m \in B_x$, there exists an open set $W_{z^m} \subset W^{z^*}$ and $\mu : W_{z^m} \to \mathbb{R}$ such that $D_x u^g(z, v) = \mu(z)g(z)$ for any $z \in W_{z^m}$. Let W_3 be an open neighborhood of z^* such that $W_3 \subset W^{z^*}$ and $y(t^m; z, v) \in W_{z^m}$ for any $z \in W_3$.

By the local submersion theorem,[35] there exist open W_4, $V \subset \mathbb{R}^n$, and $\phi : V \to W_4$ such that $z^* \in W_4 \subset W_3$, V is convex, ϕ is a bijection from V onto W_4, both ϕ and ϕ^{-1} are C^k, and $(\hat{u}^{z^*} \circ \phi)(w) = w^1$ for any $w \in V$.

Now, suppose that $y, z \in W_4$, and $\hat{u}^{z^*}(y) = \hat{u}^{z^*}(z)$. Define $z_1(s)$ and $z_2(s)$ such that

$$z_1(s) = \phi((1-s)\phi^{-1}(y) + s\phi^{-1}(z)),\ z_2(s) = y(t^m; z_1(s), v).$$

Then, $\hat{u}^{z^*}(z_2(s)) = \hat{u}^{z^*}(z_1(s)) = \hat{u}^{z^*}(y)$ for any $s \in [0, 1]$. On the other hand, $z_2(s) \in W_{z^m}$ for any $s \in [0, 1]$, and thus,

$$\frac{d}{ds}[u^g(z_2(s), v)] = 0,$$

which implies that

$$u^g(y, v) = u^g(z_1(0), v) = u^g(z_2(0), v) = u^g(z_2(1), v) = u^g(z_1(1), v) = u^g(z, v).$$

Symmetrically, we can show that if $u^g(y, v) = u^g(z, v)$, then $\hat{u}^{z^*}(y) = \hat{u}^{z^*}(z)$. Hence, $u^g(y, v) = u^g(z, v)$ if and only if $\hat{u}^{z^*}(y) = \hat{u}^{z^*}(z)$, and thus, we can prove $W_4 \subset X$ using the preimage theorem. Therefore, (v) implies (iii), as desired. This completes the proof of Step 1. □

Step 2. *Suppose that $f : P \to \tilde{\Omega}$ is a CoD such that g is an inverse demand function of f, and $f = f^{\succsim^g}$. Then, (i) and (ii) in (III) of Theorem 4 are equivalent.*

Proof of Step 2. Clearly (ii) implies (i). Conversely, suppose that (i) holds. Then, there exists a weak order $\succsim$ such that $f^{\succsim^g} = f^{\succsim}$. First, we note the following fact. Choose any $x, v \in \tilde{\Omega}$, and define

$$x_0^m = x,$$

---

[35] Again, see Sect. 1.4 of Guillemin and Pollack [6].

$$x_{i+1}^m = x_i^m + \frac{t(x,v)}{m} CRPg(x_i^m).$$

The finite sequence $x_0^m, ..., x_m^m$ is the explicit Euler approximation of (13), and thus $x_m^m \to y(t(x,v); x, v)$ as $m \to \infty$. Moreover, because $g(x_i^m) \cdot x_i^m = g(x_i^m) \cdot x_{i+1}^m$, we have that $x_i^m \succ x_{i+1}^m$, and thus $x \succ x_m^m$. This implies that if $0 < c < u^g(x, v)$, then $x \succ cv$.

Now, suppose that (ii) is violated. Then, there exist $x, y, z \in \tilde{\Omega}$ such that $x \succsim^g y$, $y \succsim^g z$, but $z \succ^g x$. This implies that $u^g(z, x) > 1$. Choose any $a \in]1, u^g(z, x)[$. Then, $z \succ ax$. Because $ax \gg x$, we have that $u^g(ax, y) > 1$, and thus there exists $b \in]1, u^g(ax, y)[$. Then, $ax \succ by$, and thus $z \succ by$. Because $by \gg y$, we have that $u^g(by, z) > 1$, and thus $by \succ z$, which is a contradiction. This completes the proof of Step 2. □

Clearly, Steps 1–2 imply (III) of Theorem 4. This completes the proof. □

Proof of Theorem 5

Suppose that $\succsim^g$ is transitive. By Step 1 in the proof of (III) of Theorem 4, it is represented by u_v^g, and u_v^g is C^1. By Lagrange's multiplier rule, we have that

$$\nabla u_v^g(x) = \lambda(x) g(x)$$

for some $\lambda(x) > 0$. Now, choose any piecewise C^1 closed curve $x : [0, T] \to \tilde{\Omega}$. Then,

$$u_v^g(x(0)) = u_v^g(x(T)),$$

and thus, there exists a non-null set $S \subset [0, 1]$ such that for all $t \in S$,

$$\frac{d}{dt} u_v^g(x(t)) \le 0.$$

This implies that for all $t \in S$,

$$g(x(t))\dot{x}(t) \le 0,$$

and thus g satisfies Ville's axiom.

Conversely, suppose that $\succsim^g$ is not transitive. Then, there exists $x, y, z \in \tilde{\Omega}$ such that $x \succsim^g y$, $y \succsim^g z$, and $z \succ^g x$. Because $\succsim^g$ is p-transitive, we have that x, y, z are linearly independent. Because $z \succ^g x$, we have that $u^g(x, z) < 1$. Consider the following differential equation

$$\dot{z}^1(t; a) = (g(z^1(t; a)) \cdot x)z - (g(z^1(t; a)) \cdot z)x + az, \ z^1(0; a) = x.$$

We have that $z^1(t;0) = y(t;x,z)$, and thus for sufficiently small $a > 0$, there exists $t_1 > 0$ such that $z^1(t_1;a) = \alpha z$ for some $\alpha \in]0,1[$. Note that,

$$g(z^1(t;a)) \cdot \dot{z}^1(t;a) = az \cdot g(z^1(t;a)) > 0$$

for all $t \in [0, t_1]$. Because $u^g(y,z) \geq 1$, by (15), we have that $u^g(z,y) \leq 1$, and thus $u^g(\alpha z, y) < 1$. Consider the following differential equation

$$\dot{z}^2(t;b) = (g(z^2(t;b)) \cdot \alpha z)y - (g(z^2(t;b)) \cdot y)\alpha z + by,\ z^2(0;b) = \alpha z.$$

We have that $z^2(t;0) = y(t;\alpha z, y)$, and thus for sufficiently small $b > 0$, there exists $t_2 > 0$ such that $z^2(t_2;b) = \beta y$ for some $\beta \in]0,1[$. Again note that,

$$g(z^2(t;b)) \cdot \dot{z}^2(t;b) = by \cdot g(z^2(t;b)) > 0$$

for all $t \in [0, t_2]$. Because $u^g(x,y) \geq 1$, we have that $u^g(y,x) \leq 1$, and thus $u^g(\beta y, x) < 1$. Consider the following differential equation

$$\dot{z}^3(t;c) = (g(z^3(t;c)) \cdot \beta y)x - (g(z^3(t;c)) \cdot x)\beta y + cx,\ z^3(0;c) = \beta y.$$

We have that $z^3(t;0) = y(t;\beta y, x)$, and thus for sufficiently small $c > 0$, there exists $t_3 > 0$ such that $z^3(t_3;c) = \gamma x$ for some $\gamma \in]0,1[$. Note that

$$g(z^3(t;c)) \cdot \dot{z}^3(t;c) = cx \cdot g(z^3(t;c)) > 0$$

for all $t \in [0, t_3]$. Finally, define $z^4(t) = [(1-t)\gamma + t]x$, and

$$x(t) = \begin{cases} z^1(t;a) & \text{if } 0 \leq t \leq t_1, \\ z^2(t-t_1;b) & \text{if } t_1 \leq t \leq t_1+t_2, \\ z^3(t-t_1-t_2;c) & \text{if } t_1+t_2 \leq t \leq t_1+t_2+t_3, \\ z^4(t-t_1-t_2-t_3) & \text{if } t_1+t_2+t_3 \leq t \leq t_1+t_2+t_3+1. \end{cases}$$

Then, for $T = t_1+t_2+t_3+1$, $x : [0,T] \to \tilde{\Omega}$ is a piecewise C^1 closed curve, and for almost all $t \in [0,T]$,

$$g(x(t)) \cdot \dot{x}(t) > 0,$$

which implies that g violates Ville's axiom. This completes the proof. □

References

1. Chambers, C.P., Echenique, F.: Revealed Preference Theory. Cambridge University Press, Cambridge (2016)
2. Debreu, G.: Definite and semi-definite quadratic forms. Econometrica **20**, 295–300 (1952)
3. Debreu, G.: Representation of a preference ordering by a numerical function. In: Thrall, R.M., Coombs, C.H., Davis, R.L. (eds.) Decision Processes, pp. 159–165. Wiley, New York (1954)
4. Dieudonne, J.: Foundations of Modern Analysis. Academic Press, London (1969)
5. Gale, D.: A note on revealed preference. Economica **27**, 348–354 (1960)
6. Guillemin, V., Pollack, A.: Differential Topology. Prentice Hall, New Jersey (1974)
7. Hosoya, Y.: Measuring utility from demand. J. Math. Econ. **49**, 82–96 (2013)
8. Hosoya, Y.: Revealed preference theory. Appl. Anal. Optim. **3**, 179–204 (2019)
9. Hosoya, Y.: Recoverability revisited. J. Math. Econ. **90**, 31–41 (2020)
10. Hosoya, Y.: Equivalence between Nikliborc's theorem and Frobenius' theorem. Pure Appl. Funct. Anal. **6**, 719–741 (2021a)
11. Hosoya, Y.: The weak axiom of revealed preference and inverse problems in consumer theory. Linear Nonlinear Anal. **7**, 9–31 (2021b)
12. Houthakker, H.S.: Revealed preference and the utility function. Economica **17**, 159–174 (1950)
13. Hurwicz, L., Richter, M.K.: Ville axioms and consumer theory. Econometrica **47**, 603–619 (1979a)
14. Hurwicz, L., Richter, M.K.: An integrability condition with applications to utility theory and thermodynamics. J. Math. Econ. **6**, 7–14 (1979b)
15. Pareto, V.: Manuale di Economia Politica con una Introduzione alla Scienza Sociale. Societa Editrice Libraria, Milano (1906a)
16. Pareto, V.: L'ofelimità Nei Cicli Non Chuisi. Giornale degli Economisti **33**, 15–30 (1960b). English translated by Chipman, J.S.: Ophelimity in nonclosed cycle. In: Chipman, J.S., Hurwicz, L., Richter, M.K., Sonnenschein, H.F. (eds.) Preferences, Utility, and Demand, pp. 370–385. Harcourt Brace Jovanovich, New York (1971)
17. Pareto, V.: Manuel d'Economie Politique. Giard et E, Briere, Paris (1909)
18. Richter, M.K.: Revealed preference theory. Econometrica **34**, 635–645 (1966)
19. Rose, H.: Consistency of preference: the two-commodity case. Rev. Econ. Stud. **25**, 124–125 (1958)
20. Samuelson, P.A.: A note on the pure theory of consumer's behaviour. Economica **5**, 61–71 (1938)
21. Samuelson, P.A.: The problem of integrability in utility theory. Economica **17**, 355–385 (1950)
22. Suda, S.: Vilfredo Pareto and the integrability problem of demand function. Keio J. Econ. **99**, 637–655 (2007)
23. Szpilrajn, E.: Sur l'Extension de l'Ordre Partiel. Fundamenta Mathematicae **16**, 386–389 (1930)
24. Uzawa, H.: Preferences and rational choice in the theory of consumption. In: Arrow, K.J., Karlin, S., Suppes, P. (eds.) Mathematical Models in the Social Science, 1959: Proceedings of the First Stanford Symposium, pp. 129–149. Stanford University Press, Stanford (1959). Reprinted. In: Chipman, J.S., Hurwicz, L., Richter, M.K., Sonnenschein, H.F. (eds.) Preferences, Utility, and Demand, pp. 7–28. Harcourt Brace Jovanovich, New York (1971)
25. Volterra, V.: L'economia Matematica ed il Nuovo Manuale del Prof. Pareto. Giornale degli Economisti **33**, 296–301 (1906). English translated by Kirman, A.P.: Mathematical economics and professor Pareto's new manual. In: Chipman, J.S., Hurwicz, L., Richter, M.K., Sonnenschein, H.F. (eds.) Preferences, Utility, and Demand, pp. 365–369. Harcourt Brace Jovanovich, New York (1971a)

Finding Nonlinear Production–Consumption Equilibrium

Roman A. Polyak

Abstract We introduce and study nonlinear production–consumption equilibrium (NPCE). The NPCE is a combination and generalization of both classical linear programming (LP) and classical input–output (IO) models. In contrast to LP and IO the NPCE has both production and consumption components. Moreover, the production cost, the consumption and the factors (resources) availability are not fixed. Instead they are correspondent functions of the production output, prices of goods and prices of factors. At the NPCE the total production cost reaches its minimum, while the total consumption, without factors expenses, reaches its maximum. At the same time, the production cost is consistent with the production output, the consumption is consistent with the prices for goods and the factors availability is consistent with prices for factors. Finding NPCE is equivalent to solving a variational inequality (VI) with a particular nonlinear operator and a simple feasible set. Under natural assumptions on the production, consumption and factor operators the NPCE exists and it is unique. Projection on the feasible set is a low-cost operation, therefore for solving the VI we use two projection methods. Each of them requires at each step few matrix by vector multiplications and allows, along with convergence and convergence rate, establishing complexity bound. The methods decompose the problem, so both the primal and the dual variables are computed simultaneously. On the other hand, both methods are pricing mechanisms for establishing NPCE, which is a generalization of Walras–Wald equilibrium (see [12]) in few directions.

Keywords Financial system

1 Introduction

For several decades, linear programming (LP), introduced by Leonid V. Kantorovich before WW2 was the main instrument for optimal resources allocation (see [5, 8, 10] and references therein). In 1975 L.V.Kantorovich and T.C.Koopmans shared the

R. A. Polyak (✉)
George Mason University, Fairfax, VA, USA
e-mail: rpolyak@gmu.edu

M. A. Khan et al. (eds.), *Matching, Dynamics and Games for the Allocation of Resources*, Monographs in Mathematical Economics 7,
https://doi.org/10.1007/978-981-96-6661-4_3

Nobel Prize in Economics "for their contribution to the theory of optimal allocation limited resources." Also before the WW2 Wassily W. Leontief introduced the input–output (IO) model (see [14]). The model has been widely used for planning production, analysis economic activities, international trade, just to mention a few. W. Leontief received the Nobel Prize in economics in 1973 "for development the input-output model and for its application to important economic problems." For quite some time, LP and IO models were used independently. Systematic effort to combine them has been undertaken by Michio Morishima more than 50 years ago. His Walras–Leontief's (WL) model consists of three systems of equations (see [16]). The first system requires balance between supply and demand for each product. The second system requires balance between availability and the required consumption of each factor. The third system requires balance between the price for each produced item and total expenses required for its production.

The solution of the systems defines the WL equilibrium. Unfortunately, the assumptions on the supply and demand functions, under which WL equilibrium exists, is hard to justify from the economic standpoint. According to M. Morishima "there is no basis for the supply and demand functions to satisfy the assumptions assumed." The NPCE is a generalization of Leontief–Walras model (see [19]). The assumptions, under which the NPCE exists, are justifiable from the economic stand point. The NPCE model has two parts: the production and the consumption. The production part transforms factors (resources) into products. The consumption part uses the production output to maximize the consumption.

We start by describing the building blocks of the NPCE.

1. Let $A : \mathbb{R}^n \to \mathbb{R}^n$ be the balance matrix with elements a_{ij}, which show how much of product $1 \leq i \leq n$ needed to produce one item of product $1 \leq j \leq n$. We assume that A has no zero rows and columns. For consumption vector y, we have

$$y = x - Ax = (I - A)x,$$

where $x = (x_1, \dots, x_n)^T$ is the production vector with components x_j—the number of items of product $1 \leq j \leq n$ produced. We assume that the economy is productive. It means that for any consumption vector $c \in \mathbb{R}^n_+$ the following system:

$$(x - Ax) = c, \qquad x \in \mathbb{R}^n_+ \tag{1}$$

has a solutions. In such case, matrix A is called productive matrix, (see [6, 13]). Let λ be the price vector, i.e., λ_j defines the price of one item of product $1 \leq j \leq n$, then $q = (I - A)^T \lambda$ is the profit vector, i.e., q_j, $1 \leq j \leq n$ is the profit out of one item of product $1 \leq j \leq n$, that is

$$q_j = \left((I - A)^T \lambda\right)_j, \ 1 \leq j \leq n, \ \lambda \in \mathbb{R}^n_+. \tag{2}$$

Solution of system (1) defines such production vector x_c, which guarantee consumption given by $c \in \mathbb{R}^n_+$. Solution of system (2) defines such price vector

$\lambda_q \in \mathbb{R}^n_+$, which guarantee profit given by $q \in \mathbb{R}^n_+$. For a productive matrix A, the inverse $(I - A)^{-1}$ exists and it is a non-negative matrix, so both systems (1) and (2) have a unique non-negative solutions x_c and λ_q (see, for example, [6, 13]). The productivity assumption we will keep throughout the paper.

2. Let $B : \mathbb{R}^n \to R^m$ be the technological matrix. The elements b_{ij} show how much of factor $1 \le i \le m$ needed to produce one item of product $1 \le j \le n$. We assume that matrix B has no zero row and zero column. The fundamental difference between LP, IO and NPCE is: instead of fixed production cost vector p, consumption vector c, and availability vector r, we used correspondent operators.
3. First, the fixed production cost vector $p \in \mathbb{R}^n_+$ is replaced by production operator $p : \mathbb{R}^n_+ \to \mathbb{R}^n_+$, which maps the output $x = (x_1, \ldots, x_n)^T$ into production per item cost vector $p(x) = (p_1(x), \ldots, p_n(x))^T$.
4. Second, the fixed consumption vector $c \in \mathbb{R}^n_+$ is replaced by consumption operator $c : \mathbb{R}^n_+ \to \mathbb{R}^n_+$, which maps the price for goods vector $\lambda = (\lambda_1, \ldots .\lambda_n)^T \in \mathbb{R}^n_+$ into consumption vector $c(\lambda) = (c_1(\lambda), \ldots, c_n(\lambda))^T$.
5. Third, the fixed resource vector $r \in \mathbb{R}^m_+$ is replaced by resource operator $r : \mathbb{R}^m_+ \to \mathbb{R}^m_+$ which maps the price per unit factor vector $v = (v_1, \ldots, v_m)^T$ into availability vector $r(v) = (r_1(v), \ldots, r_m(v))^T$.

2 Nonlinear Production–Consumption Equilibrium

The NPCE is defined as a triple $y^* = (x^*, \lambda^*, v^*) \in \Omega = \mathbb{R}^n_+ \times \mathbb{R}^n_+ \times \mathbb{R}^m_+$:

$$x^* \in \text{Argmin}\left\{\langle p(x^*), X\rangle \middle| (I - A)X \ge c(\lambda^*),\ BX \le r(v^*),\ X \in \mathbb{R}^n_+\right\}, \tag{3}$$

where X—production vector;

$$(\lambda^*, v^*) \in$$
$$\text{Argmax}\left\{\langle c(\lambda^*), \Lambda\rangle - \langle r(v^*), V\rangle \middle| (I - A)^T\Lambda - B^T V \le p(x^*),\ \Lambda \in \mathbb{R}^n_+,\ V \in \mathbb{R}^m_+\right\} \tag{4}$$

where Λ—consumption price vector and V—factors price vector. In other words the NPCE defines such production vector $x^* \in \mathbb{R}^n_+$, consumption price vector λ^*, and factor price vector $v^* \in \mathbb{R}^m_+$, that the total production cost $(p\,(x^*)\,, x^*)$ reaches its minimum, while the consumption without factors cost reaches its maximum. Moreover, the production cost per unit vector $p\,(x^*)$ is consistent with production output x^*. The consumption $c\,(\lambda^*)$ is consistent with the consumption prices λ^*. The factor availability $r\,(v^*)$ is consistent with factors price vector v^*. Keeping in mind that under fixed $y^* = (x^*, \lambda^*, v^*)$ problems (3) and (4) are just primal and dual LP, we obtain

$$\langle c(\lambda^*), \lambda^*\rangle = \langle p(x^*), x^*\rangle + \langle r(v^*), v^*\rangle. \tag{5}$$

Thus, the total consumption is equal to the sum of the production and the factors costs. The NPCE is a generalization of Leontief–Walras equilibrium (see [19]). It combines nonlinear equilibrium for resource allocation [17] with nonlinear input–output equilibrium [18]. From the complementarity conditions for all $1 \le i \le m$ and all $1 \le j \le n$ follows

$$\begin{aligned} x_j^* > 0 &\Rightarrow p_j(x^*) + \left(B^T v^*\right)_j - \left((I - A)^T \lambda^*\right)_j = 0 \\ x_j^* = 0 &\Leftarrow p_j(x^*) + \left(B^T v^*\right)_j - \left((I - A)^T \lambda^*\right)_j > 0, \end{aligned} \tag{6}$$

$$\begin{aligned} \lambda_j^* > 0 &\Rightarrow \left((I - A)x^*\right)_j - c_j(\lambda^*) = 0 \\ \lambda_j^* = 0 &\Leftarrow \left((I - A)x^*\right)_j - c_j(\lambda^*) > 0, \end{aligned} \tag{7}$$

$$\begin{aligned} v_i^* > 0 &\Rightarrow r_i(v^*) - \left(Bx^*\right)_i = 0 \\ v_i^* = 0 &\Leftarrow r_i(v^*) - \left(Bx^*\right)_i > 0. \end{aligned} \tag{8}$$

In other words, the NPCE leads to the market clearing. Thus, the main issue at this point is existence of NPCE and its uniqueness.

2.1 Existence of NPCE

We start with conditions on operators p, c, and r, which guarantee existence and uniqueness of NPCE. Let $p : \mathbb{R}^n_+ \to \mathbb{R}^n_+$ be strongly monotone increasing operator, that is there exist $\alpha > 0$, that for any x_1 and x_2 from $\mathbb{R}^n_+$ follows

$$\langle p(x_1) - p(x_2), x_1 - x_2 \rangle \ge \alpha \left\| x_1 - x_2 \right\|^2, \tag{9}$$

where $\|x\| = \langle x, x\rangle^{1/2}$ is the Euclidean norm. Assumption (9), in particular, implies: the increase of production of any good, when the production of the rest is fixed leads to the cost per item increases. Moreover, the lower bound of the margin is $\alpha > 0$. Let $c : \mathbb{R}^n_+ \to \mathbb{R}^n_+$ be strongly monotone decreasing operator that is there exists $\beta > 0$, that for any λ_1 and λ_2 from $\mathbb{R}^n_+$ follows

$$\langle c(\lambda_1) - c(\lambda_2), \lambda_1 - \lambda_2 \rangle \le -\beta \left\| \lambda_1 - \lambda_2 \right\|^2. \tag{10}$$

Assumption (10), in particular, implies: price increase for an item of any given product, when prices for the rest are fixed, leads to the consumption decrease of the product. Moreover, the upper bound of the margin is $-\beta < 0$. Finally, let $r : \mathbb{R}^m_+ \to \mathbb{R}^m_+$ be strongly monotone increasing operator, then there exists $\gamma > 0$ that for any v_1 and v_2 from $\mathbb{R}^m_+$ follows

$$\langle r(v_1) - r(v_2), v_1 - v_2 \rangle \geq \gamma \left\| v_1 - v_2 \right\|^2. \qquad (11)$$

Assumption (11), in particular, implies: the increase of the price for any given factor, when the prices for the rest are fixed, leads to availability increase of the factor. Moreover, the lower bound of the margin is $\gamma > 0$. We start by showing that finding y^* from (3) to (4) is equivalent to solving the following variation inequality (VI):

$$\left\langle g(y^*), y - y^* \right\rangle \leq 0, \ \forall y \in \Omega, \qquad (12)$$

where operator $g : \mathbb{R}^{2n+m} \to \mathbb{R}^{2n+m}$ is defined as follows:

$$g(y) = \left((I - A)^T \lambda - p(x) - B^T v; c(\lambda) - (I - A)x; Bx - r(v) \right). \qquad (13)$$

Theorem 1 *For $y^* = (x^*, \lambda^*, v^*)$ to be a solution of the dual LP pair (3) and (4), it is necessary and sufficient for y^* to be a solution of the VI (12) with operator g given by (13).*

Proof Let consider the Lagrangian for LP (3)

$$\mathcal{L}(y^*, X, \Lambda, V) = \langle p(x^*), X \rangle - \langle \Lambda, (I - A)X - c(\lambda^*) \rangle - \langle V, -BX + r(v^*) \rangle.$$

Then,

$$y^* \in \operatorname*{Argmin}_{X \in \mathbb{R}^n_+} \max_{\substack{\Lambda \in \mathbb{R}^n_+ \\ V \in \mathbb{R}^m_+}} \mathcal{L}(y^*; X, \Lambda, V). \qquad (14)$$

In other words,

$$x^* \in \text{Argmin} \left\{ \mathcal{L}\left(y^*; X, \lambda^*, v^*\right) \middle| X \in \mathbb{R}^n_+ \right\}$$

$$= \text{Argmin} \left\{ \left\langle p(x^*) - (I - A)^T \lambda^* + B^T \lambda^*, X \right\rangle \middle| X \in \mathbb{R}^n_+ \right\}$$

$$= \text{Argmax} \left\{ \left\langle (I - A)^T \lambda^* - p(x^*) - B^T \lambda^*, X \right\rangle \middle| X \in \mathbb{R}^n_+ \right\}.$$

Therefore,

$$(I - A)^T \lambda^* - p(x^*) - B^T \lambda^* \leq 0. \qquad (15)$$

On the other hand, from (4) follows

$$(\lambda^*, v^*) \in \text{Argmax} \left\{ \mathcal{L}\left(y^*; x^*, \Lambda, V\right) \middle| \Lambda \in \mathbb{R}^n_+, V \in \mathbb{R}^m_+ \right\}$$

$$= \text{Argmax} \left\{ \left\langle c(\lambda^*) - (I - A)x^*, \Lambda \right\rangle + \left\langle Bx^* - r(v^*), V \right\rangle \middle| \Lambda \in \mathbb{R}^n_+, V \in \mathbb{R}^m_+ \right\}.$$

Therefore,

$$c(\lambda^*) - (I - A)x^* \leq 0, \qquad Bx^* - r(v^*) \leq 0. \tag{16}$$

The complementarity condition for the dual LP pair (3) and (4) is given by

$$\langle y^*, g(y^*) \rangle = 0. \tag{17}$$

From (15) to (17) follows (12).

Conversely, let us assume existence $\bar{y} \in \Omega$, that

$$\langle g(\bar{y}), y - \bar{y} \rangle \leq 0, \quad \forall y \in \Omega,$$

then

$$\langle g(\bar{y}), y \rangle \leq \langle g(\bar{y}), \bar{y} \rangle, \quad \forall y \in \Omega. \tag{18}$$

From (18) follows $g(\bar{y}) \leq 0$, otherwise the left-hand side can be made as large as one wants by taking correspondent component of y large enough, while the right-hand side of (18) is fixed. Therefore,

$$(I - A)^T \bar{\lambda} - p(\bar{x}) - B^T \bar{\lambda} \leq 0, \quad \bar{x} \geq 0, \tag{19}$$

$$c(\bar{\lambda}) - (I - A)\bar{x} \leq 0, \quad \bar{\lambda} \geq 0, \tag{20}$$

$$B\bar{x} - r(\bar{v}) \leq 0, \quad \bar{v} \geq 0. \tag{21}$$

From (19) follows that $\bar{x}$ is a feasible solution of the primal LP (3) with y^* replaced by $\bar{y}$.

From (20) to (21) follows that $(\bar{\lambda}, \bar{v})$ is a feasible solution of the dual (4) with y^* replaced by $\bar{y}$.

Also for $y = 0$ from (18) follows

$$\langle g(\bar{y}), \bar{y} \rangle \geq 0,$$

which together with (19)–(21) leads to

$$\langle g(\bar{y}), \bar{y} \rangle = 0 \tag{22}$$

Hence, $\bar{x}$ is a feasible solution of the primal LP

$$\langle p(\bar{x}), \bar{x} \rangle = \min \left\{ \langle p(\bar{x}), X \rangle \middle| (I - A)X \geq c(\bar{\lambda}); -BX \geq -r(\bar{v}); X \in \mathbb{R}^n_+ \right\} \tag{23}$$

and $\langle \bar{\lambda}, \bar{v} \rangle$ is feasible solution of the dual LP

$$\langle c(\bar{\lambda}), \bar{\lambda}\rangle - \langle r(\bar{v}), \bar{v}\rangle$$
$$= \max\left\{\langle c(\bar{\lambda}), \Lambda\rangle - \langle r(\bar{v}), V\rangle \middle| (I-A)^T\Lambda \le B^T V + p(\bar{x}), \Lambda \in \mathbb{R}^n_+, V \in \mathbb{R}^m_+\right\}. \tag{24}$$

Thus, for the primal feasible vector $\bar{x}$ and the dual feasible vector $(\bar{\lambda}; \bar{v})$, the complementarity conditions satisfied, therefore $\bar{x}$ solves the primal LP (23), $(\bar{\lambda}; \bar{v})$ solves the dual LP (24) and $\bar{y} = y^*$ □

Finding a saddle point (14) of the Lagrangian $\mathcal{L}(y^*, X, \Lambda, V)$ is equivalent to finding an equilibrium of two person concave game with following payoff functions.

For the first player, the payoff function is

$$\varphi_1(y; X, \lambda, v) = -\mathcal{L}(y; X, \lambda, v) = \langle -p(x) + (I-A)\lambda - B^T v, X\rangle + \langle v, r(v)\rangle$$

with strategy $X \in \mathbb{R}^n_+$

For the second player, the payoff function is

$$\varphi_2(y; x, \Lambda, V) = \mathcal{L}(y; x, \Lambda, V) = \langle c(\lambda) - (I-A)x, \Lambda\rangle + \langle Bx - r(v), V\rangle + \langle p(x), x\rangle$$

with the strategy (Λ, V).

Now let us consider the normalized payoff function $\Phi : \Omega \to \mathbb{R}$, given by the following formula:

$$\Phi(y, Y) = \varphi_1(y; X, \lambda, v) + \varphi_2(y; x, \Lambda, V).$$

We also consider the following map

$$y \to \omega(y) = \operatorname{Argmax}\left\{\Phi(y, Y) \middle| Y \in \Omega\right\}. \tag{25}$$

Then, finding $y^* = (x^*, \lambda^*, v^*)$ is equivalent to finding a fixed point of the map $y \to \omega(y)$.

The normalized payoff function $\Phi(y, Y)$, however, is linear in Y and $\Omega = \mathbb{R}^n_+ \times \mathbb{R}^n_+ \times \mathbb{R}^m_+$ is an unbounded set, therefore, one cannot use the map (25) for proving existence and uniqueness of y^*.

We will do it later by using assumption (9)–(11). Meanwhile, we would like to point out that the nonlinear operator g defined by (13) is, in fact, the pseudo-gradient of $\Phi(y, Y)$, i.e.,

$$g(y) = \nabla_Y \Phi(y, Y)\Big|_{Y=y} = \Big((I-A)^T\lambda - p(x) - B^T v; c(\lambda) - (I-A)x; Bx - r(v)\Big).$$

The following lemma establishes sufficient condition for pseudo-gradient $g : \Omega \to \mathbb{R}^{2n+m}_+$ to be strongly monotone.

Lemma 1 *If operators p, c and r strongly monotone, i.e., (9)–(11) hold, then the pseudo-gradient $g : \Omega \to \mathbb{R}^{2n+m}$ given by (13) is also strongly monotone, i.e., for $\delta = \min\{\alpha, \beta, \gamma\} > 0$, and any y_1 and y_2 from Ω, we have*

$$\langle g(y_1) - g(y_2), y_1 - y_2 \rangle \leq -\delta \left\| y_1 - y_2 \right\|^2. \tag{26}$$

Proof Let $y_1 = (x_1, \lambda_1, v_1)$ and $y_2 = (x_2, \lambda_2, v_2)$ be vectors from Ω, then

$$\begin{aligned}
&\langle g(y_1) - g(y_2), y_1 - y_2 \rangle \\
&\quad = \langle -p(x_1) + (I - A)^T \lambda_1, -B^T v_1 + p(x_2) - (I - A)^T \lambda_2 + B^T v_2, x_1 - x_2 \rangle + \\
&\qquad \langle c(\lambda_1) - (I - A)x_1 - c(\lambda_2) + (I - A)x_2, \lambda_1 - \lambda_2 \rangle + \\
&\qquad \langle Bx_1 - r(v_1) - Bx_2 + r(v_2), v_1 - v_2 \rangle \\
&\quad = -\langle p(x_1) - p(x_2), x_1 - x_2 \rangle + \\
&\qquad \langle (I - A)^T (\lambda_1 - \lambda_2), x_1 - x_2 \rangle - \langle B^T (v_1 - v_2), x_1 - x_2 \rangle \\
&\qquad + \langle c(\lambda_1) - c(\lambda_2), \lambda_1 - \lambda_2 \rangle - \langle (I - A)(x_1 - x_2), \lambda_1 - \lambda_2 \rangle \\
&\qquad + \langle B(x_1 - x_2), v_1 - v_2 \rangle - \langle r(v_1) - r(v_2), v_1 - v_2 \rangle \\
&\quad = -\langle p(x_1) - p(x_2), x_1 - x_2 \rangle + \langle c(\lambda_1) - c(\lambda_2), \lambda_1 - \lambda_2 \rangle - \langle r(v_1) - r(v_2), v_1 - v_2 \rangle.
\end{aligned}$$

From (9)–(11) follows

$$\langle g(y_1) - g(y_2), y_1 - y_2 \rangle \leq -\alpha \left\| x_1 - x_2 \right\|^2 - \beta \left\| \lambda_1 - \lambda_2 \right\|^2 - \gamma \left\| v_1 - v_2 \right\|^2.$$

Therefore, for $\delta = \min\{\alpha, \beta, \gamma\} > 0$, we obtain (26). □

Theorem 2 *If p, c and r are continuous and strongly monotone operators, that is (9)–(11) hold, then the NPCE $y^* = (x^*, \lambda^*, v^*)$ exists and it is unique.*

Proof Let consider $y_0 \in \mathbb{R}^{2n+m}_{++}$ such that $\|y_0\| \leq 1$ and large enough number $M > 0$. We replace Ω by $\Omega_M = \{y \in \Omega : \|y - y_0\| \leq M\}$. Then, Ω_M is a bounded, convex and closed set and so is the set

$$\omega_M(y) = \text{Argmax}\,\{\Phi\,(y, Y)\,|Y \in \Omega_M\}, \tag{27}$$

because for any given $y \in \Omega$ function $\Phi\,(y, Y)$ is linear in Y. Moreover, due to the continuity of p, c, and r, the map $y \to \omega_M(y)$ is upper semi-continuous. Therefore, from Kakutani's theorem follows existence of $y^*_M \in \Omega_M : y^*_M \in \omega\left(y^*_M\right)$ (see, for example, Lancaster [13]).

It turns out, the constraint $\|y - y_0\| \leq M$ is irrelevant in (27). In fact, from (26) with $y_1 = y_0$ and $y_2 = y^*_M$ follows

$$\begin{aligned}
\delta \left\| y_0 - y^*_M \right\|^2 &\leq \langle g\left(y^*_M\right) - g\left(y_0\right), y_0 - y^*_M \rangle \\
&= \langle g\left(y^*_M\right), y_0 - y^*_M \rangle + \langle g\left(y_0\right), y^*_M - y_0 \rangle
\end{aligned} \tag{28}$$

From (27) and $y^*_M \in \omega\left(y^*_M\right)$ follows $\left(g\left(y^*_M\right), y_0 - y^*_M\right) \leq 0$, therefore from (28), we obtain

$$\delta \left\| y_0 - y_M^* \right\|^2 \leq \|g(y_0)\| \left\| y_M^* - y_0 \right\|,$$

or

$$\left\| y_0 - y_M^* \right\| \leq \delta^{-1} \|g(y_0)\|.$$

Therefore, $\left\| y_M^* \right\| \leq \|y_0\| + \left\| y_0 - y_M^* \right\| \leq 1 + \delta^{-1} \|g(y_0)\|$. Thus, for large enough $M > 0$, the constraint $\|y_0 - y\| \leq M$ can't be active in (27); therefore, it is irrelevant in (27) and there is $y^* \in \Omega : y^* \in \omega(y^*)$. Uniqueness follows from (26). In fact, assuming that there are two vectors y^* and y^{**}, which solve VI (12), we obtain

$$\langle g(y^*), y^{**} - y^* \rangle \leq 0 \text{ and } \langle g(y^{**}), y^* - y^{**} \rangle \leq 0,$$

therefore

$$\langle g\left(y^*\right) - g\left(y^{**}\right), y^{**} - y^* \rangle \leq 0$$

or

$$\langle g\left(y^*\right) - g\left(y^{**}\right), y^* - y^{**} \rangle \geq 0,$$

which contradicts (26) with $y_1 = y^*$, $y_2 = y^{**}$. Thus, assumptions (9)–(11) are sufficient for existence and uniqueness NPCE $y^* = (x^*, \lambda^*, v^*)$. □

In what is following we replace the global strong monotonicity assumptions (9)–(11) for strong monotonicity only at the NPCE $y^* = (x^*, \lambda^*, v^*)$, i.e., we assume existence of $\alpha > 0$, $\beta > 0$ and $\gamma > 0$, that

$$\langle p(x) - p(x^*), x - x^* \rangle \geq \alpha \left\| x - x^* \right\|^2, \quad \forall x \in \mathbb{R}_+^n \tag{29}$$

$$\langle c(\lambda) - c(\lambda^*), \lambda - \lambda^* \rangle \leq -\beta \left\| \lambda - \lambda^* \right\|^2, \quad \forall \lambda \in \mathbb{R}_+^n \tag{30}$$

$$\langle r(v) - r(v^*), v - v^* \rangle \geq \gamma \left\| v - v^* \right\|^2, \quad \forall v \in \mathbb{R}_+^n \tag{31}$$

hold. In the next section, we also assume that operator p, c, and r satisfy local Lipschitz condition, i.e., there exist $L_p > 0$, $L_c > 0$ and $L_r > 0$ that

$$\left\| p(x) - p(x^*) \right\| \leq L_p \left\| x - x^* \right\|, \quad \forall x \in \mathbb{R}_+^n \tag{32}$$

$$\left\| c(\lambda) - c(\lambda^*) \right\| \leq L_c \left\| \lambda - \lambda^* \right\|, \quad \forall \lambda \in \mathbb{R}_+^n \tag{33}$$

$$\left\| r(v) - r(v^*) \right\| \leq L_r \left\| v - v^* \right\|, \quad \forall v \in \mathbb{R}_+^n \tag{34}$$

hold. We will say that operator p, c and r are well defined if conditions (29)–(34) are satisfied. Assumption (29) means that production operator p is sensitive to the change of the production output only at x^*. Assumption (30) means that consumption operator c is sensitive to the consumption price change only at λ^*. Finally, from (31) follows that operator r is sensitive to the factor price change only at v^*. Conditions (32)–(34) mean that production, consumption, and factors availability are under control in the neighborhood of equilibrium.

3 Pseudo-gradient Projection Method for Finding NPCE

The projection method for solving optimization problems was introduced in the 60s (see [7, 15, 20]). Later projection methods were used for solving VI problems (see, for example, [1–4, 9, 17, 18] and references therein). Often projection methods have only theoretical value because projection on a convex set in $\mathbb{R}^n$ is numerically costly operation. In case of simple feasible set, however, projection methods can be efficient (see, for example, Polyak [17, 18]). From Theorem 1 follows: for finding NPCE y^* it is sufficient to solve VI (12) with operator g given by (13). The feasible set Ω a very simple and projection on it is a low-cost operation. Therefore, for finding $y^* = (x^*, \lambda^*, v^*)$, we use two projection methods. The projection of $u \in \mathbb{R}^q$ on $\mathbb{R}^q_+$ is given by the following formula:

$$v = P_{\mathbb{R}^q_+}(u) = [u]_+ = \left([u_1]_+, \dots, \left[u_q\right]_+\right),$$

where, for any $1 \le i \le q$, we have

$$\left[u_j\right]_+ = \begin{cases} u_i, & u_i \ge 0 \\ 0, & u_i < 0 \end{cases}.$$

The projection operator $P_\Omega : \Omega \to \Omega$ is given by formula

$$P_\Omega = \left([x]_+, [\lambda]_+, [v]_+\right),$$

where $y = (x, \lambda, v) \in \mathbb{R}^{2n+m}$. The following well-known properties of operator P_Ω will be used later. First, the operator P_Ω is not expansive, i.e., for any u_1 and u_2 from Ω, we have

$$\|P_\Omega(u_1) - P_\Omega(u_2)\| \le \|u_1 - u_2\|. \tag{35}$$

Second, vector y^* solves VI (12) iff for any $t > 0$ vector y^* is a fixed point of a map $P_\Omega(I + tg) : \Omega \to \Omega$, i.e.,

$$y^* = P_\Omega\left(y^* + tg\left(y^*\right)\right) \tag{36}$$

for any $t > 0$. We are ready to describe the Pseudo-gradient Projection (PGP) method for finding NPCE y^*. Let $y_0 = (x_0, \lambda_0, v_0) \in \Omega$ be the starting point and let $y_s = (x_s, \lambda_s, v_s) \in \Omega$ has been found already. The PGP method finds the next approximation by the following formula:

$$y_{s+1} = P_\Omega\left(y_s + tg(y_s)\right). \tag{37}$$

The step-length $t > 0$ we will specify later. The PGP method (37) updates the primal and dual variables by the following formula:

$$x_{s+1,j} = \left[x_{s,j} + t\left((I - A)^T\lambda_s - p(x_s) - B^T v_s\right)_j\right]_+ \quad 1 \le j \le n \tag{38}$$

$$\lambda_{s+1,j} = \left[\lambda_{s,j} + t\left(c(\lambda_s) - (I - A)x_s\right)_j\right]_+ \quad 1 \le j \le n \tag{39}$$

$$v_{s+1,i} = \left[v_{s,i} + t\left(Bx_s - r(v_s)\right)_i\right]_+ \quad 1 \le i \le m. \tag{40}$$

In fact, the PGP is a decomposition method, which computes the primal x_s and the dual $(\lambda_s; v_s)$ vectors independently. On the other hand, Formulas (38)–(40) is a pricing mechanism for establishing NPCE y^*. From (38) follows: if for any product $1 \le j \le n$, the market value of one unit $\left[(I - A)^T \lambda_s\right]_j$ is greater than the sum of the production and factor cost $\left(p(x_s) + \left(B^T v_s\right)\right)_j$, then production $(x_s)_j$ has to be increased, while if $\left[(I - A)^T \lambda_s\right]_j < \left[p(x_s) + \left(B^T v_s\right)\right]_j$, then production $(x_s)_j$ has to be reduced. From (39) follows: if for any product $1 \le j \le n$ the current consumption $c_j(\lambda_s)$ is greater than the production output $[(I - A)\,x_s]_j$, then the price $(\lambda_s)_j$ has to be increased, while if $c_j(\lambda_s) < [(I - A)\,x_s]_j$, then the price $(\lambda_s)_j$ has to be reduced. Finally, from (40) follows: if for any factor $1 \le i \le m$ the consumption $(Bx_s)_i$ is greater than the availability $r_i(v_s)$ of this factor under price v_s, then the price $(v_s)_i$ has to be increased, while if $(Bx_s)_i < r_i(v_s)$, then the price $(v_s)_i$ has to be reduced. The PGP method (37) can be viewed as explicit Euler's projection method for solving the following system of ordinary differential equations:

$$\begin{aligned}
\frac{dx}{dt} &= (I - A)^T\lambda - p(x) - B^T v \\
\frac{d\lambda}{dt} &= c(\lambda) - (I - A)x \\
\frac{dv}{dt} &= Bx - r(v)
\end{aligned} \tag{41}$$

with $x(0) = x_0$, $\lambda(0) = \lambda_0$ and $v(0) = v_0$. For convergence analysis of the PGP method (37), we need the following Lemma similar to Lemma 1.

Lemma 2 *If operators p, c and r are strongly monotone at the equilibrium $y^* = (x^*, \lambda^*, v^*)$, i.e., (29)–(31) hold, then operator $g : \Omega \to \mathbb{R}^{2n+m}_+$ given by (13) is strongly monotone at y^*, i.e., for $\delta = \min\{\alpha, \beta, \gamma\} > 0$ the following bound*

$$\langle g(y) - g(y^*), y - y^*\rangle \le -\delta \left\|y - y^*\right\|^2, \quad \forall y \in \Omega \tag{42}$$

holds.

The proof is similar to proof of Lemma 1.

Lemma 3 *If operators p, c and r satisfy local Lipschits condition (32)–(34), then there is $L > 0$ such that the bound*

$$\|g(y) - g(y^*)\| \leq L \|y - y^*\| \tag{43}$$

holds.

The proof of the Lemma 3 and the upper bound for L can be established using technique similar to the one used in [17].

Remark 1 We assume that for given $x \in \mathbb{R}^n_+$ and $\lambda \in \mathbb{R}^n_+$ computing $p(x)$ and $c(\lambda)$ does not requires more than $O(n^2)$ operations and for a given $v \in \mathbb{R}^m_+$ computing $r(v)$ does not requires more than $O(m^2)$ operations. It is true, for example, if $p(x) = \nabla\left(\frac{1}{2}x^T Px + q^T x\right)$, where P is symmetric positive semi-defined matrix, $c(\lambda) = \nabla\left(\frac{1}{2}\lambda^T C\lambda + d^T\lambda\right)$, where C is symmetric negative semi-defined matrix in $\mathbb{R}^n_+$ and $r(v) = \nabla\left(\frac{1}{2}v^T Rv + s^T v\right)$, where R is symmetric positive semi-defined matrix in $\mathbb{R}^m_+$. In what is following we assume that for any given $x \in \mathbb{R}^n$, $\lambda \in \mathbb{R}^n$ and $v \in \mathbb{R}^m$ computing $p(x), c(\lambda), r(v)$ does not require more than $O(n^2)$ operations, $n > m$. Then, each PGP step (37) does not need more than $O(n^2)$ operations as well, because Formulas (38)–(40) require few matrix by vector multiplication. Let $\varkappa = \gamma L^{-1}$ be the condition number of the VI operator g. The following Theorem establishes global Q-linear convergence rate of the PGP method (37). The proof is similar to the proof of Theorem 3 in [17]. We sketch the proof for completeness.

Theorem 3 *If operators p, c and r are well defined, i.e., conditions (29)–(34) holds then:*

1. *for any $0 < t < 2\delta L^{-2}$ the PGP method (37) globally converges to NPCE $y^* = (x^*, \lambda^*, v^*)$ with Q-linear rate with a ratio $0 < q(t) = \left(1 - 2t\delta + t^2L^2\right)^{1/2} < 1$, i.e.,*

$$\|y_{s+1} - y^*\| \leq q(t) \|y_s - y^*\|; \tag{44}$$

2. *for $t = \delta L^{-2} = \text{Argmin}\{q(t)|t > 0\}$ the following bound holds*

$$\|y_{s+1} - y^*\| \leq (1 - \varkappa^2)^{1/2} \|y_s - y^*\| = q \|y - y^*\|; \tag{45}$$

3. *the PGP complexity is*

$$Comp\,(PGP) = O\left(n^2\varkappa^{-2} \ln \epsilon^{-1}\right), \tag{46}$$

where $\epsilon > 0$ is the required accuracy.

Proof From (35)–(37) follows

$$\left\|y_{s+1}-y^*\right\|^2=\left\|P_\Omega\left(y_s+tg(y_s)\right)-P_\Omega\left(y^*+tg(y^*)\right)\right\|^2 \tag{47}$$

$$\le\left\|y_s+tg(y_s)-y^*-tg(y^*)\right\|^2 \tag{48}$$

$$\le\left\|y_s-y^*\right\|^2+2t\left(y_s-y^*,g(y_s)-g(y^*)\right)+t^2\left\|g(y_s)-g(y^*)\right\|^2. \tag{49}$$

From Lemmas 2, 3 and (47) follows

$$\left\|y_{s+1}-y^*\right\|\le\left\|y_s-y^*\right\|\left(1-2t\delta+t^2L^2\right)^{1/2},$$

therefore, for $0<t<2\delta L^{-2}$, we have $0<q(t)=\left(1-2t\delta+t^2L^2\right)^{1/2}<1$. Let $0<\epsilon\ll 1$ is the required accuracy. In view of (45), it takes $O\left((\ln q)^{-1}\ln\epsilon\right)$ steps to find ϵ-approximation for NPCE y^*. Each step requires $O(n^2)$ operations. Therefore, it takes

$$N=O\left(n^2\frac{\ln\epsilon^{-1}}{\ln q^{-1}}\right) \tag{50}$$

operations for finding an ϵ-approximation for NPCE y^*. From (45) follows

$$\left(\ln q^{-1}\right)^{-1}=\left(-\frac{1}{2}\ln(1-\varkappa^2)\right)^{-1},$$

then, from $\ln(1+x)\le x,\ \forall x>-1$ follows $\left(\ln q^{-1}\right)\le 2\varkappa^{-2}$, which together with (50) leads to the bound (46). □

If $\delta=\min\{\alpha,\beta,\gamma\}=0$, then Theorem 3 does not guarantee even convergence of the PGP method. Therefore, we need another tool for finding **NPCE** y^*. In the following section, we apply Extra Pseudo-gradient (EPG) method for solving VI (12). The Extra Gradient (EG) method for finding saddle points was introduced by G. Korpelevich in the 70s (see [11]). Since then, the EG method became a popular tool for solving VI (see [1, 4, 9, 17, 18] and references therein). We assume that operators p, c, and r are monotone and satisfy Lipschitz condition, so is the operator g, i.e., for any y_1 and y_2 from Ω, we have

$$\langle g(y_1)-g(y_2),y_1-y_2\rangle\le 0 \tag{51}$$

and there exists $L>0$, that

$$\|g(y_1)-g(y_2)\|\le L\|y_1-y_2\|. \tag{52}$$

4 Extra Pseudo-gradient Method for Finding NPCE

Application of G.Korpelevich's extra-gradient method [11] for solving VI (12) leads to two phases extra pseudo-gradient (EPG) method for finding NPCE. At the prediction phase, we predict the production vector $\hat{x}$, the consumption price vector $\hat{\lambda}$, and the factor price vector $\hat{v}$. Then we find the predicted production cost per unit vector $p(\hat{x})$, predicted consumption vector $c(\hat{\lambda})$ and predicted availability factor vector $r(\hat{v})$. At the correction phase, we use $p(\hat{x})$, $c(\hat{\lambda})$ and $r(\hat{v})$ to find the new production, new consumption, and new factor prices. Let $y_0 = (x_0, \lambda_0, v_0) \in \Omega$ be the starting point and approximation $y_s = (x_s, \lambda_s, v_s)$ has been found already. At the prediction phase, one finds

$$\hat{y}_s = P_\Omega (y_s + tg(y_s)) = [y_s + tg(y_s)]_+ . \tag{53}$$

At the correction phase one finds the new approximation

$$y_{s+1} = P_\Omega \left(y_s + tg(\hat{y}_s)\right) = \left[y_s + tg(\hat{y}_s)\right]_+ . \tag{54}$$

In other words, the EPG method first predicts the production vector

$$\hat{x}_s = \left[x_s + t\left((I - A)^T \lambda_s - p(x_s) - B^T v_s\right)\right]_+ , \tag{55}$$

the consumption price vector

$$\hat{\lambda}_s = [\lambda_s + t\,(c(\lambda_s) - (I - A)x_s)]_+ , \tag{56}$$

and the factor price vector

$$\hat{v}_s = [v_s + t\,(Bx_s - r(v_s))]_+ . \tag{57}$$

Then, EPG method finds new approximation for the production vector

$$x_{s+1} = \left[x_s + t\left((I - A)^T \hat{\lambda}_s - p(\hat{x}_s) - B^T \hat{\lambda}_s\right)\right], \tag{58}$$

as well as, new approximation for the price vector

$$\lambda_{s+1} = \left[\lambda_s + t\left(c(\hat{\lambda}_s) - (I - A)\hat{x}_s\right)\right]_+ , \tag{59}$$

and the factor price vector

$$v_{s+1} = \left[v_s + t\left(B\hat{x}_s - r(\hat{v}_s)\right)\right]_+ . \tag{60}$$

The meaning of Formulas (55)–(57) and (58)–(60) is similar to the meaning of Formulas (38)–(40). The step-length $t > 0$ we specify in the course of proving the following theorem. The EPG method (55)–(60) is a pricing mechanism for finding NPCE, which guarantee convergence under minimum assumptions in the input data.

Theorem 4 *If p, c and r be monotone operators, which satisfy Lipschitz condition, i.e., (51) and (52) holds, then, for any $0 < t < \left(\sqrt{2}L\right)^{-1}$, the EPG method generates a converging sequence $\{y_s\}_{s=0}^{\infty}$ and $\lim_{s\to\infty} y_s = y^*$.*

Proof Form (53) and (54), non-expansiveness of the operator P_Ω and Lipschitz condition (52) follows

$$\begin{aligned}\left\|y_{s+1} - \hat{y}_s\right\| &= \left\|P_\Omega(y_s + tg\left(\hat{y}_s\right)) - P_\Omega(y_s + tg\left(y_s\right))\right\| \\ &\le t\left\|g(\hat{y}_s) - g(y_s)\right\| \le tL\left\|\hat{y}_s - y_s\right\|.\end{aligned} \tag{61}$$

From (53) we have

$$\langle tg(y_s) + (y_s - \hat{y}_s), y - \hat{y}_s)\rangle \le 0, \quad \forall y \in \Omega.$$

Therefore, for $y = y_{s+1}$, we obtain

$$\langle tg(y_s) + (y_s - \hat{y}_s), y_{s+1} - \hat{y}_s)\rangle \le 0,$$

so

$$\langle y_s - \hat{y}_s, y_{s+1} - \hat{y}_s\rangle + t\langle g(\hat{y}_s), y_{s+1} - \hat{y}_s\rangle - t\langle g(\hat{y}_s) - g(y_s), y_{s+1} - y_s\rangle \le 0. \tag{62}$$

Then, from (61) and Lipschitz condition (52), we obtain

$$t\langle g(\hat{y}_s) - g(y_s), y_{s+1} - \hat{y}_s\rangle \le t\left\|g(\hat{y}_s) - g(y_s)\right\|\left\|y_{s+1} - \hat{y}_s\right\| \le (tL)^2\left\|\hat{y}_s - y_s\right\|. \tag{63}$$

From (62) and (63) follows

$$\langle y_s - \hat{y}_s, y_{s+1} - \hat{y}_s\rangle + t\langle g(\hat{y}_s), y_{s+1} - \hat{y}_s\rangle - (tL)^2\left\|\hat{y}_s - y_s\right\|^2 \le 0. \tag{64}$$

From (54) follows

$$\langle tg(\hat{y}_s) + y_s - y_{s+1}, y - y_{s+1}\rangle \le 0, \quad \forall y \in \Omega. \tag{65}$$

Hence, for $y = y^*$ we obtain

$$\langle y_s - y_{s+1} + tg(\hat{y}_s), y^* - y_{s+1}\rangle \le 0. \tag{66}$$

From $(g(y^*), y - y^*) \le 0$, $\forall y \in \Omega$ follows

$$\langle g(y^*), \hat{y}_s - y^*\rangle \le 0,$$

so, for any $t > 0$, we have

$$t\langle -g(y^*), y^* - \hat{y}_s\rangle \leq 0. \tag{67}$$

By adding (64), (66), and (67) and using monotonisity of operator g, i.e.,

$$\langle g(\hat{y}_s) - g(y^*), y^* - \hat{y}_s\rangle \geq 0$$

we obtain

$$2\langle y_s - y_{s+1}, y^* - y_{s+1}\rangle + 2\langle y_s - \hat{y}_s, y_{s+1} - \hat{y}_s\rangle - 2\,(tL)^2 \left\|\hat{y}_s - y_s\right\|^2 \leq 0. \tag{68}$$

Form the following identity

$$2\langle u - v, w - v\rangle = \|u - v\|^2 + \|v - w\|^2 - \|u - w\|^2 \tag{69}$$

with $u = y_s$, $v = y_{s+1}$ and $w = y^*$ follows

$$2\langle y_s - y_{s+1}, y^* - y_{s+1}\rangle = \|y_s - y_{s+1}\|^2 + \left\|y_{s+1} - y^*\right\|^2 - \left\|y_s - y^*\right\|^2. \tag{70}$$

Using identity (69) with $u = y_s$, $v = \hat{y}_s$ and $w = y_{s+1}$, we obtain

$$2\langle y_s - \hat{y}_s, y_{s+1} - \hat{y}_s\rangle = \left\|y_s - \hat{y}_s\right\|^2 + \left\|\hat{y}_s - y_{s+1}\right\|^2 - \|y_s - y_{s+1}\|^2. \tag{71}$$

Then, from (68), (70) and (71) follows

$$\left\|y_{s+1} - y^*\right\|^2 + (1 - 2(tL)^2)\left\|y_s - \hat{y}\right\|^2 + \left\|\hat{y}_s - y_{s+1}\right\|^2 \leq \left\|y_s - y^*\right\|^2. \tag{72}$$

Summing up (72) from $s = 0$ to $s = N$, we obtain

$$\left\|y_{N+1} - y^*\right\|^2 + (1 - 2(tL)^2)\sum_{s=0}^{N}\left\|y_s - \hat{y}_s\right\|^2 + \sum_{s=0}^{N}\left\|\hat{y}_s - y_{s+1}\right\|^2 \leq \left\|y_0 - y^*\right\|^2. \tag{73}$$

From (73) and $0 < t < \left(\sqrt{2}L\right)^{-1}$, we obtain

$$\sum_{s=0}^{\infty}\left\|y_s - \hat{y}_s\right\|^2 < \infty \quad \text{and} \quad \sum_{s=0}^{\infty}\left\|\hat{y}_s - y_{s+1}\right\|^2 < \infty. \tag{74}$$

Therefore,

$$\text{(a)}\ \lim_{s\to\infty}\left\|y_s - \hat{y}_s\right\| = 0,\ \ \text{(b)}\ \lim_{s\to\infty}\left\|\hat{y}_s - y_{s+1}\right\| = 0. \tag{75}$$

Also from (72) and $0 < t < \left(\sqrt{2}L\right)^{-1}$ follows

$$\left\| y_{s+1} - y^* \right\| \le \left\| y_s - y^* \right\|. \tag{76}$$

Hence, $\{y_s\}_{s=0}^{\infty}$ is a bounded sequence, therefore there is a converging subsequence $\{y_{s_k}\}_{k=0}^{\infty}$: $\lim_{k\to\infty} y_{s_k} = \bar{y}$. Then, from (75(a)), we have $\lim_{k\to\infty} \hat{y}_{s_k} = \bar{y}$ and, from (75(b)), we obtain $\lim_{k\to\infty} y_{s_k+1} = \bar{y}$.

Keeping in mind continuity g, we obtain

$$\bar{y} = \lim_{k\to\infty} y_{s_k+1} = \lim_{k\to\infty} \left[y_{s_k} + tg\left(\hat{y}_{s_k}\right)\right]_+ = [\bar{y} + tg\left(\bar{y}\right)]_+ .$$

From (36) follows $\bar{y} = y^*$, which together with (76) leads to $\lim_{s\to\infty} y_s = y^*$. □

4.1 Convergence Rate of the EPG Method

At this point, we replace the assumptions (9)–(11) by less restrictive assumptions (29)–(31). From (29)–(31) and Lemma 2 follows

$$\langle g(y) - g(y^*), y - y^* \rangle \le -\delta \left\| y - y^* \right\|^2 \tag{77}$$

with $\delta = \min\{\alpha, \beta, \gamma\} > 0$.

Therefore,

$$\langle g(y), y - y^* \rangle - \langle g(y^*), y - y^* \rangle \le -\delta \left\| y - y^* \right\|^2, \quad \forall y \in \Omega.$$

Then, $(g(y^*), y - y^*)) \le 0, \quad \forall y \in \Omega$, therefore

$$\langle g(y), y - y^* \rangle \le -\delta \left\| y - y^* \right\|^2, \quad \forall y \in \Omega. \tag{78}$$

Theorem 5 *If assumptions (29)–(31) are satisfied and Lipschitz condition (52) holds, then, for* $v(t) = 1 + 2\delta t - 2(tL)^2$ *and ratio* $q(t) = 1 - 2\delta t + 4(\delta t)^2(v(t))^{-1}$, *the following statements are taking place:*

1. $\|y_{s+1} - y^*\| \le \sqrt{q(t)}\, \|y_s - y^*\|, \quad 0 < q(t) < 1, \ \forall t \in \left(0, \left(\sqrt{2}L\right)^{-1}\right);$
2. *for* $t = (2L)^{-1}$, *we have*

$$q\left((2L)^{-1}\right) = (1 + \varkappa)(1 + 2\varkappa)^{-1};$$

3. *for any* $\varkappa \in [0, 0.5]$, *we obtain*

$$\left\| y_{s+1} - y^* \right\| \le \sqrt{1 - 0.5\varkappa}\, \left\| y_s - y^* \right\|; \tag{79}$$

4.

$$comp(EPG) \leq O\left(n^2 \varkappa^{-1} \ln \epsilon^{-1}\right), \tag{80}$$

where $\epsilon > 0$ is the required accuracy.

Proof 1. From (53) to (54), non-expansiveness of operator P_Ω and Lipschitz condition (52) follows

$$\left\|\hat{y}_s - y_s\right\| = \left\|P_\Omega(y_s + tg(y_s)) - P_\Omega(y_s + tg(\hat{y}_s))\right\| \tag{81}$$

$$\leq t\left\|g(y_s) - g(\hat{y}_s)\right\| \leq tL\left\|y_s - \hat{y}_s\right\|. \tag{82}$$

Using argument from the proof of Theorem 4, we obtain

$$\langle y_s - y_{s+1}, y^* - y_{s+1}\rangle + \langle y_s - \hat{y}_s, y_{s+1} - \hat{y}_s\rangle \tag{83}$$

$$+ t\left(g(\hat{y}_s), y^* - \hat{y}_s\right) - (Lt)^2\left\|\hat{y}_s - y_s\right\|^2 \leq 0. \tag{84}$$

From (78) with $y = \hat{y}_s$, we obtain

$$\langle g(\hat{y}_s), y^* - \hat{y}_s\rangle \geq \delta\left\|\hat{y}_s - y^*\right\|^2.$$

Therefore, from (83) follows

$$2\langle y_s - y_{s+1}, y^* - y_{s+1}\rangle + 2\langle y_s - \hat{y}_s, y_{s+1} - \hat{y}_s\rangle \tag{85}$$

$$+ 2\delta t\left\|\hat{y}_s - y^*\right\|^2 - 2(Lt)^2\left\|\hat{y}_s - y_s\right\|^2 \leq 0. \tag{86}$$

By applying identity (69) to the scalar products in (85) first with $u = y_s$, $v = y_{s+1}$, $w = y^*$ and then with $u = y_s$, $v = \hat{y}_s$, $w = y_{s+1}$, we obtain

$$\|y_s - y_{s+1}\|^2 + \left\|y_{s+1} - y^*\right\|^2 - \left\|y_s - y^*\right\|^2 + \left\|y_s - \hat{y}_s\right\|^2$$
$$+ \left\|\hat{y}_s - y_{s+1}\right\|^2 - \|y_s - y_{s+1}\|^2 + 2\delta t\left\|\hat{y}_s - y^*\right\|^2 - 2(Lt)^2\left\|y_s - \hat{y}_s\right\|^2 \leq 0$$

or

$$\left\|y_{s+1} - y^*\right\|^2 + \left\|\hat{y}_s - y_{s+1}\right\|^2 + \left(1 - 2(Lt)^2\right)\left\|y_s - \hat{y}_s\right\|^2 \tag{87}$$
$$+ 2\delta t\left\|\hat{y}_s - y^*\right\|^2 \leq \left\|y_s - y^*\right\|^2.$$

From

$$\left\|\hat{y}_s - y^*\right\|^2 = \langle\hat{y}_s - y_s + y_s - y^*, \hat{y}_s - y_s + y_s - y^*\rangle \tag{88}$$
$$= \left\|\hat{y}_s - y_s\right\|^2 + 2\langle\hat{y}_s - y_s, y_s - y^*\rangle + \left\|y_s - y^*\right\|^2$$

and (87) follows

$$\|y_{s+1} - y^*\|^2 + \|\hat{y}_s - y_{s+1}\|^2 + \left(1 + 2\delta t - 2(Lt)^2\right) \|\hat{y}_s - y_s\|^2 \qquad (89)$$
$$+4\delta t \langle \hat{y}_s - y_s, y_s - y^* \rangle \le (1 - 2\delta t) \|y_s - y^*\|^2 .$$

By introducing $\nu(t) = 1 + 2\delta t - 2(Lt)^2$, we can rewrite the third and fourth terms of left-hand side of (89) as follows:

$$\left\| \sqrt{\nu(t)}(\hat{y}_s - y_s) + 2(y_s - y^*)\frac{\delta t}{\sqrt{\nu(t)}} \right\|^2 - \frac{4(\delta t)^2 \|y_s - y^*\|^2}{\nu(t)}.$$

Then, from (89) follows

$$\|y_{s+1} - y^*\|^2 + \|\hat{y}_s - y_{s+1}\|^2 + \left\| \sqrt{\nu(t)}(\hat{y}_s - y_s) + 2(y_s - y^*)\frac{\delta t}{\sqrt{\nu(t)}} \right\|^2$$
$$\le \left(1 - 2\delta t + \frac{4(\delta t)^2}{\nu(t)}\right) \|y_s - y^*\|^2 . \qquad (90)$$

Hence, for $q(t) = 1 - 2\delta t + 4(\delta t)^2(\nu(t))^{-1}$ from (90) follows

$$\|y_{s+1} - y^*\|^2 \le q(t) \|y_s - y^*\|^2 . \qquad (91)$$

and $0 < q(t) < 1, \ \forall t \in \left(0, \left(\sqrt{2}L\right)^{-1}\right)$.

2. For $t = (2L)^{-1}$ and $\varkappa = \delta L^{-1}$, we obtain

$$q\left((2L)^{-1}\right) = 1 - \varkappa + \varkappa^2(0.5 + \varkappa)^{-1} = (1 + \varkappa)(1 + 2\varkappa)^{-1}. \qquad (92)$$

For any $\varkappa \in [0, 0.5]$ we have $(1 + \varkappa)(1 + 2\varkappa)^{-1} \le 1 - 0.5\varkappa$, therefore from (91) and (92) follows (79).

3. From (79), $q = \sqrt{1 - 0.5\varkappa}$, $\varkappa \in [0, 0.5]$ and given accuracy $0 < \epsilon \ll 1$ follows that EPG method requires

$$s = O\left(\ln \epsilon^{-1} \left(\ln q^{-1}\right)^{-1}\right)$$

steps to find an approximation with accuracy ϵ. Then,

$$\left(\ln q^{-1}\right)^{-1} = \left(-\frac{1}{2}\ln(1 - 0.5\varkappa)\right)^{-1}. \qquad (93)$$

From $\ln(1 + x) \le x \ \ \forall x > -1$ follows $\ln(1 - 0.5\varkappa) \le -0.5\varkappa$, therefore from (93), we obtain $\left(\ln q^{-1}\right)^{-1} \le 4\varkappa^{-1}$. Keeping in mind that each step requires $O(n^2)$ operations, we obtain complexity bound (80).

□

Remark 2 For small $\varkappa > 0$, complexity bound (80) is much better than bound (46). Therefore, although EPG requires two projection on Ω instead of one for PGP, the EPG is still more efficient. For very large NPCE problems the PGP can compete with the EPG.

5 Concluding Remarks

The NPCE is a fundamental departure from both LP and IO models, when it comes to optimal distribution limited resources. It became possible by replacing the fixed production cost, fixed consumption, and fixed factors availability with production, consumption, and factor operators. It leads to data "symmetrization" and allows to replace finding NPCE by solving a VI with a very simple feasible set Ω, projection on which is a low-cost operation. Therefore, for solving VI, we use two projection methods PGP and EPG, which do not require solving a QP or a system of linear equations at each step. Instead, they require at each step few matrix by vector multiplication. Moreover, the primal and dual variables one computes independently, in parallel, that is both methods decompose the problem, which allows solving large-scale NPCE. On the other hand, both method are nothing, but pricing mechanisms for finding NPCE. From (5) follows that at NPCE the total consumption is equal to the sum of production and factors costs. From (6)–(8) follows that at the NPCE the market clearing takes place. From (46) and (80) follows that only three parameters are responsible for complexity of any given NPCE problem: the condition number of the VI operator, the size of the problem, and the given accuracy. It follows from the complexity bounds that, in some instances, finding NPCE can be easier than solving LP of the same size. The numerical results obtained so far support the theoretical findings. The practical use of NPCE, however, will require much more work for developing the input data, numerical realization of the PGP and EPG methods, conducting numerical experiments as well as careful analyzing the results obtained.

References

1. Antipin, A.: The gradient and exstragradient approaches in bilinear equilibrium programming. A. Dorodnizin Computing Center RAS (in Russian) (2002)
2. Auslender, A., Teboulle, M.: Interior projection-like methods for monotone variational inequalities. Math. Program. **104**, 39–68 (2005)
3. Bakushinskiy, A., Polyak, B.: On the solution of variational inequalities. Sov. Math Doklady **14**, 1705–1710 (1974)
4. Censor, Y., Gibali, A., Reich, S.: Extensions of Korpelevich's extragradient method for the variational inequality problem in Euclidean space. Optimization **61**, 1119–1132 (2012)
5. Dantzig, G.: Linear Programming and Extensions. Princeton University, Princeton (1963)
6. Gale, D.: The Theory of Linear Economic Models. McGraw-Hill Book Companym Inc. (1960)
7. Goldstein, A.: Convex programing in Hilbert space. Bull. Am. Math. Soc. **70**, 709–710 (1964)

8. Kantorovich, L.: The Best Use of Economic Resources, vol. 349. Oxford: Pergamon Press (1965)
9. Konnov, I.: Equilibrium Models and Variational Inequalities, vol. 210. Elsevier, Amsterdam (2001)
10. Koopmans, T.: Efficient allocation of resources. Econometrica. **19**, 455–465 (1957)
11. Korpelevich, G.: An extragradient method for finding saddle points and for other problems. Matecon **12**, 747–756 (1976)
12. Kuhn, H.: On a theorem of Wald. In: Linear Inequalities and Related System. Annals of Mathematics Studies, pp. 265–273. Princeton University Press (1956)
13. Lancaster, K.: Mathematical Economics. Macmillan (1968)
14. Leontief, W. Input–Output Economics. Oxford University Press. (1966)
15. Levitin, E., Polyak, B.: Minimaliztion methods in the presents of counstraince. Z.Vycisl. Mat and Mat Fiz. **6**, 787–823 (1966)
16. Morishima, M.: Equilibrium, Stability and Growth: A Multi-sectoral Analysis. Oxford University (1964)
17. Polyak, R.: Nonlinear equilibrium for resource allocation problems. Contemp. Math. **638**, 181–197. AMS (2015)
18. Polyak, R.: Nonlinear input–output equlibrium. Contemp. Math. **659**, 246–262. AMS (2016)
19. Polyak, R., Primak, M.: Methods of controlling sequences for solving equilibrium problems, part 2. Cybernetics **4**, 56–62 (1977)
20. Rosen, J.: The gradient projection method for nonlinear programming. Soc. Indus. Appl. Math. **8**, 181–218 (1960)

Systemic Risk in Financial Systems: Properties of Equilibria

John Stachurski

Abstract Eisenberg and Noe (Manag Sci 47:236–249, 2001) analyze systemic risk for financial institutions linked by a network of liabilities. They show that the solution to their model is unique when the financial system satisfies a regularity condition involving risk orbits. We show that this condition is not needed: under a mild convention use to handle edge cases, a unique solution always exists.

Keywords Systemic risk · Uniqueness

1 Introduction

Eisenberg and Noe [3] analyzed systemic risk for institutions linked by a complex network of liabilities. The fundamental importance of their work was brought to the fore by the financial crisis of 2007–2008, where the solvency of individual firms became unclear due to interlocking financial obligations. In such settings, defaults can trigger negative feedback loops, where restructuring compromises the balance sheets of other firms in the network, restricting their ability to make payments to debt holders, and forces more rounds of restructuring.

Because of these cyclical interactions in the networks, obtaining a system of payments that clears the market is a fixed point problem. While existence of a fixed point is easy to show, uniqueness is challenging. Eisenberg and Noe [3] show that uniqueness holds when the financial system is "regular." This condition has become common in the literature (see, e.g., [1] or [4]). We prove it is essentially unnecessary: under a mild condition that covers edge cases, the limited liability and priority conditions in [3] are themselves enough to uniquely identify an equilibrium clearing vector.

The author thanks Shu Hu for research assistance. Financial support from ARC grant FT160100423 is gratefully acknowledged.

J. Stachurski (✉)
Australian National University, Canberra, Australia
e-mail: john.stachurski@anu.edu.au

M. A. Khan et al. (eds.), *Matching, Dynamics and Games for the Allocation of Resources*, Monographs in Mathematical Economics 7,
https://doi.org/10.1007/978-981-96-6661-4_4

2 Results

If $a = (a(i))_{i=1}^n$ and $b = (b(i))_{i=1}^n$ are vectors in $\mathbb{R}^n$, then $a \leq b$ means $a(i) \leq b(i)$ for all i. We write $a \ll b$ if $a(i) < b(i)$ for all i. The symbol $[a, b]$ represents the order interval $\{x \in \mathbb{R}^n : a \leq x \leq b\}$. A Markov matrix is a nonnegative square matrix with unit row sums. For a given Markov matrix Π, we call j *accessible* from i under Π if either $j = i$ or $\Pi^k(i, j) > 0$ for some $k \in \mathbb{N}$.

As in [3], a *financial system* is a set of nodes $I := \{1, 2, \ldots, n\}$, an $n \times n$ matrix $\Pi = (\Pi(i, j))$ of relative liabilities, a vector $\bar{p} = (\bar{p}(i)) \in \mathbb{R}^n_+$ of nominal obligations ($\bar{p}(i)$ is the sum of all nominal obligations held by node i) and a vector $e = (e(i)) \in \mathbb{R}^n_+$ of external cash balances. The matrix of relative liabilities is a Markov matrix. Following [3], we assume that $\bar{p} \gg 0$, so that, for any given node, the total sum of liabilities to other nodes is not zero.[1]

A *clearing vector* for a financial system $(I, \Pi, \bar{p}, e)$ is a vector of payments $p \in [0, \bar{p}]$ satisfying the limited liability restriction:

$$p(j) \leq \sum_{i \in I} p(i)\Pi(i, j) + e(j) \tag{1}$$

and the absolute priority condition:

$$p(j) = \sum_{i \in I} p(i)\Pi(i, j) + e(j) \quad \text{or} \quad p(j) = \bar{p}(j) \tag{2}$$

for all $j \in I$. Combining these two restrictions and writing them in vector form (cf. Eisenberg and Noe [3, p. 240]), the set of clearing payment vectors is seen to coincide with the set of fixed points of the mapping $p \mapsto \Phi p$ on $[0, \bar{p}]$ defined by

$$\Phi p := (p\Pi + e) \wedge \bar{p}. \tag{3}$$

In (3) and below, all n-vectors are treated as row vectors.

We now consider uniqueness of clearing vectors, which we also refer to as equilibria.

Remark 2.1 Limited liability and absolute priority cannot by themselves pin down equilibrium for the extreme case where every firm in the network has zero operating cash.[2] Hence, in what follows, we adopt the convention $p = 0$ when $e = 0$. This is natural because a payment sequence cannot be initiated without outside capital.

In [3], the *risk orbit* of a given node i is the set of all nodes that are accessible from i. The financial system is called *regular* if the risk orbit of every node contains

---

[1] This is required for Π to have unit row sums. See Eisenberg and Noe [3, p. 239].

[2] For example, if ψ is a stationary distribution for Π, λ is a constant in $[0, 1]$, $p = \lambda\psi$, and $\bar{p} = \psi$, then $e = 0$ implies $\Phi p = (\lambda\psi\Pi + e) \wedge \bar{p} = (\lambda\psi) \wedge \psi = \lambda\psi = p$. Since λ was arbitrary in $[0, 1]$, there is a continuum of equilibria.

Fig. 1 Failure of regularity

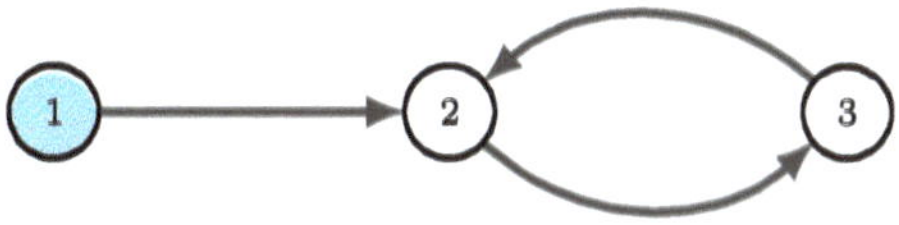

at least one j with $e(j) > 0$. A system where regularity fails is shown below, in Fig. 1. Suppose $e(1) = 1$ and $e(2) = e(3) = 0$. Arrows represent nonzero liabilities. The risk orbit from node 2 is $\{2, 3\}$. Regularity fails because $e(2) = e(3) = 0$.

Fortunately, regularity is not needed for uniqueness, as the next theorem shows.

Theorem 2.1 *Every financial system has exactly one clearing payment vector.*

Theorem 2.1 is proved in Sect. 3. Prior to that we note that the convention discussed in Remark 2.1 is significantly weaker than regularity. For example, regularity fails in the financial network in Fig. 1, as discussed above, so the regularity-based uniqueness result does not apply. At the same time, the network has positive aggregate cash balances. In particular, Theorem 2.1 applies without the need to invoke the convention discussed in Remark 2.1.

3 Proof of Uniqueness

To prove Theorem 2.1, we begin with a lemma. In the lemma, we say that node j in a financial system $(I, \Pi, \bar{p}, e)$, is *cash accessible* if there exists an $i \in I$ such that $e(i) > 0$ and j is accessible from i.

Lemma 3.1 *If every node in $S = (I, \Pi, \bar{p}, e)$ is cash accessible, then S has a unique clearing vector p^*. Moreover, $p^* \gg 0$ and $\Phi^k p \to p^*$ as $k \to \infty$ when $0 \le p \le \bar{p}$.*

Proof Let S be as described. By the fixed point theorem in the appendix (Theorem 1.1), it suffices to show that Φ is an increasing concave self-map on $[0, \bar{p}]$ with $\Phi^k 0 \gg 0$ for some $k \in \mathbb{N}$. As confirmed in [3], Φ is increasing and concave, so only the last statement needs to be verified. To this end, we set

$$\delta := \frac{1}{n^2} \cdot \min \{\{\bar{p}(i) : i \in I\} \cup \{e(i) : i \in I \text{ s.t. } e(i) > 0\}\}.$$

Let $\hat{e}$ be defined by $\hat{e}(i) = 1$ if $e(i) > 0$ and zero otherwise. We claim that, for all $m \le n$,

$$\Phi^m 0 \ge \delta(\hat{e} + \hat{e}\Pi + \cdots + \hat{e}\Pi^{m-1}). \tag{4}$$

This holds at $m = 1$ because $\Phi 0 = e \wedge \bar{p} \ge \delta \hat{e}$. Now suppose (4) holds at some $m \le n - 1$. Then, since Φ is increasing, we obtain

$$\Phi^{m+1} 0 \ge (\delta(\hat{e} + \hat{e}\Pi + \cdots + \hat{e}\Pi^{m-1})\Pi + e) \wedge \bar{p}$$

$$\geq (\delta(\hat{e} + \hat{e}\Pi + \cdots + \hat{e}\Pi^m)) \wedge \bar{p}.$$

Since $\hat{e} + \hat{e}\Pi + \cdots + \hat{e}\Pi^m \leq n^2\mathbb{1}$, where $\mathbb{1}$ is a vector of ones, and since $(\delta n^2\mathbb{1}) \leq \bar{p}$ by the definition of δ, we have $\Phi^{m+1}0 \geq \delta(\hat{e} + \hat{e}\Pi + \cdots + \hat{e}\Pi^m)$. This argument confirms that (4) holds for all $m \leq n$.

We now claim that $\Phi^n 0 \gg 0$. In view of (4), it suffices to show that, for any $j \in I$, there exists a $k < n$ with $(\hat{e}\Pi^k)(j) = \sum_{i \in I} \hat{e}(i)\Pi^k(i, j) > 0$. Since every node in S is cash accessible, we know there exists an $i \in I$ with $e(i) > 0$ and j is accessible from i. For this i we can choose $k \in \mathbb{N}$ with $k < n$ and $\Pi^k(i, j) = \hat{e}(i)\Pi^k(i, j) > 0$. We conclude that $\Phi^n 0 \gg 0$, as claimed. □

In what follows, a subset $\mathbb{J}$ of I is called *absorbing* if no element of its complement $\mathbb{J}^c := \{i \in I : i \notin \mathbb{J}\}$ is accessible from $\mathbb{J}$. Also, for a given vector v on I and some $\mathbb{J} \subset I$, we write $v \mid \mathbb{J}$ for the restriction of v to $\mathbb{J}$. For matrix M on $I \times I$ we write $M \mid \mathbb{J}$ for the restriction of M to $\mathbb{J} \times \mathbb{J}$.

Proof of Theorem 2.1 As pointed out by [3], the operator Φ is increasing and concave on $[0, \bar{p}]$. It follows from the increasing property and Tarski's fixed point theorem that at least one clearing vector always exists.[3] The remainder of the proof focuses on uniqueness.

Let P be the set of all nodes in I that are cash accessible. Let A be all i in P^c such that P is accessible from i. Let N be all i in P^c such that P is not accessible from i. Note that $I = P \cup A \cup N$ and that these sets are disjoint.

The set N is an absorbing set, since, by definition, P is not accessible from N, and A cannot be accessible because otherwise P would also be accessible. The set P is also absorbing because, if $j \in P^c$ is accessible from some $i \in P$, then j is cash accessible. But then $j \in P$, which is a contradiction.

Iterating k times on (1) gives

$$p \leq e + e\Pi + e\Pi^2 + \cdots + e\Pi^{k-1} + p\Pi^k. \tag{5}$$

If j is in A, then j is not cash accessible, so $e(j) = 0$ and $(e\Pi^m)(j) = 0$ for all m. Hence (5) reduces to

$$p(j) \leq (p\Pi^k)(j) \leq (\bar{p}\Pi^k)(j) = \sum_{i \in I} \bar{p}(i)\Pi^k(i, j) = \sum_{i \in A} \bar{p}(i)\Pi^k(i, j). \tag{6}$$

The last equality uses the fact that both P and N are absorbing sets. Since P is accessible from every element of A, when $i, j \in A$ we have $\Pi^k(i, j) \to 0$ as $k \to \infty$.[4] By taking k large in (6), we see that $p(j) = 0$ for all $j \in A$.

---

[3] In fact Φ is a continuous self-map on the convex compact set $[0, \bar{p}]$, so Brouwer's fixed point theorem gives the same conclusion.

[4] A Markov chain started at $i \in A$ leaves for P with $\epsilon > 0$ probability every $n = |I|$ periods. Since P is absorbing it never returns. Hence the probability that the chain hits $j \in A$ after k periods converges to zero with k.

Since N is absorbing, $\Pi \mid N$ is a Markov matrix and $(N, \Pi \mid N, \bar{p} \mid N, e \mid N)$ is itself an independent financial system.[5] Moreover, P contains all cash accessible nodes so no element of N is cash accessible and, in particular, $e \mid N = 0$. Hence $p \mid N = 0$ by the convention in Remark 2.1.

It remains only to treat nodes in P. Since P is absorbing, $\Pi \mid P$ is a Markov matrix and $(P, \Pi \mid P, \bar{p} \mid P, e \mid P)$ is also an independent financial system.[6] Moreover, every node in P is cash accessible, so, by Lemma 3.1, a unique clearing vector $p^* \gg 0$ exists on P. After extending to all nodes by setting $p^*(i) = 0$ for all $i \notin P$, we have a unique clearing vector. The proof is now complete. □

Appendix 1: Remaining Proofs

In what follows, a self-map F from a subset Y of $\mathbb{R}^d$ to itself is called *globally stable* if F has a unique fixed point $\bar{y}$ in Y and $F^m y \to \bar{y}$ as $m \to \infty$ for all $y \in Y$. We use the following fixed point theorem, which slightly modifies Theorem 3.1 of [2]. (See also Corollary 2.1.1. of [5].)

Theorem 1.1 *Let A be an increasing self-map on $[0, b] \subset \mathbb{R}^d$. If A is concave and there exists an $n \in \mathbb{N}$ such that $A^n 0 \gg 0$, then A is globally stable on $[0, b]$.*

Proof Theorem 3.1 of [2] implies that any increasing concave self-map F on $[0, b]$ satisfying $F0 \gg 0$ is globally stable. Since compositions of increasing concave operators are increasing and concave, this implies that A^n is globally stable on $[0, b]$. Denote its fixed point by $\bar{v}$. Since $\{A^m 0\}_{m \in \mathbb{N}}$ is monotone increasing and since the subsequence $\{A^{mn} 0\}_{m \in \mathbb{N}}$ converges up to $\bar{v}$ as $m \to \infty$, we must have $A^m 0 \to \bar{v}$. A similar argument gives $A^m b \to \bar{v}$. For any $v \in [0, b]$ we have $A^m 0 \leq A^m v \leq A^m b$, so $A^m v \to \bar{v}$ as $m \to \infty$.

The last step is to show that $\bar{v}$ is the unique fixed point of A. From Tarski's fixed point theorem we know that at least one fixed point exists. Now suppose $v \in [0, b]$ is such a point. Then $v = A^m v$ for all m. At the same time, $A^m v \to \bar{v}$ by the results just established. Hence $v = \bar{v}$. □

References

1. Cifuentes, R., Ferrucci, G., Shin, H.S.: Liquidity risk and contagion. J. Eur. Econ. Assoc. **3**, 556–566 (2005)
2. Du, Y.: Fixed points of increasing operators in ordered banach spaces and applications. Appl. Anal. **38**, 1–20 (1990)

---

[5] Although nodes in N may have inbound links from A, we have just shown that payments from A are zero. At the same time, there are no inbound links from P, since P is absorbing.

[6] Again, while nodes in P may have inbound links from A, we have just shown that payments from A are zero.

3. Eisenberg, L., Noe, T.H.: Systemic risk in financial systems. Manag. Sci. **47**, 236–249 (2001)
4. Feinstein, Z., Pang, W., Rudloff, B., Schaanning, E., Sturm, S., Wildman, M.: Sensitivity of the Eisenberg-Noe clearing vector to individual interbank liabilities. SIAM J. Finan. Math. **9**, 1286–1325 (2018)
5. Zhang, Z.: Variational, Topological, and Partial Order Methods with Their Applications, vol. 29. Springer Science, Berlin (2012)

The Coalitional TU Extensions of Gale–Mas-Colell Economy

Nengfa Wang and Zhe Yang

Abstract In this paper, we provide some coalitional transferable utility extensions of Gale–Mas-Colell economy. By the allocation rule of wealth, we introduce the notion of coalitional TU equilibria and prove the existence theorem under the regular conditions. Furthermore, we give a fuzzy extension of coalition TU equilibria.

Keywords Coalitional TU (fuzzy) equilibria · Existence · Gale–Mas-Colell economy

1 Introduction

It is well known that Arrow and Debreu [2] first proved the existence of competitive equilibria for a production economy. Later, Mas-Colell [17] improved the result by providing an equilibrium existence theorem without complete or transitive preferences. Following Mas-Colell [17], Gale and Mas-Colell [11] proved the existence of equilibria for a general economy with nonordered preferences. Moreover, Shafer and Sonnenschein [23] established the equilibrium existence theorem of generalized games with nonordered preferences.

Note that the noncooperative idea is the principle for above games and market theory. Moreover, the cooperative solution is another concept in the game theory. Bondareva [8] and Shapley [24] first proved the existence of the core for a transferable utility (TU) cooperative game. Later, Scarf [21] established an NTU core existence

---

Appreciation on the 100th Birthday of David Gale.

N. Wang
School of Mathematics and Statistics, Guizhou University of Finance and Economics, Guiyang, China
e-mail: nengfa_wang@163.com

Guizhou Key Laboratory of Big Data Statistical Analysis, Guiyang, China

Z. Yang (✉)
School of Economics, Shanghai University of Finance and Economics, Shanghai, China
e-mail: zheyang211@163.com

Key Laboratory of Mathematical Economics (SUFE), Ministry of Education, Shanghai, China

M. A. Khan et al. (eds.), *Matching, Dynamics and Games for the Allocation of Resources*, Monographs in Mathematical Economics 7,
https://doi.org/10.1007/978-981-96-6661-4_5

theorem for an NTU cooperative game. Inspired by cooperative games, Scarf [22] first proved the existence of the α-core for strategy-form games, which was extended by Border [9] and Kajii [13]. Moreover, Zhao [35, 36] provided TU α-core and TU β-core existence theorems for normal-form games by following the concept of TU core in the cooperative game. Recently, the cooperative equilibria of strategy-form games were studied in games with incomplete information [4, 6, 18–20], games with infinitely many players [3, 5, 29–34], and games with discontinuous payoffs [27].

The classical cooperative equilibria were studied in the model with the crisp coalition. As an extension, Aubin [7] first introduced the notion of fuzzy coalitions. Later, the fuzzy core existence theorem of economic models was given by Florenzano [10], Allouch and Predtetchinski [1]. Liu and Liu [14] first provided a necessary and sufficient condition for an NTU fuzzy game to have a nonempty fuzzy core. Furthermore, the fuzzy KKMS theorem and fuzzy coalitional production economy were studied by Liu and Tian [16], Liu and Liu [15]. On the other hand, van Gulick and Norde [28] obtained the necessary and sufficient condition for a TU fuzzy game to have a nonempty fuzzy core.

Observe that the work of TU cooperative solutions was initiated by Bondareva [8] and Shapley [24]. Later, the relation between TU game and market becomes an important research field. Shapley and Shubik [25] analyzed the TU market game generated from an exchange economy with TU utilities. Sun et al. [26] showed that the c-core of TU games is equivalent to the set of competitive outcomes of correspondence market. Inoue [12] gave a representation of TU games by coalition production economies. Moreover, the TU cooperative solution was first extended to strategy-form games by Zhao [35, 36]. The existence theorems of TU α-core and TU β-core were provided by Zhao [35, 36].

The motivation of our paper is given as follows. First, we follow the model of Gale and Mas-Colell [11] and assume that there exists the cooperative behavior and transferable utilities in the model. By following the TU cooperative solutions of an economy without externalities, we propose the notion of coalitional TU equilibria, and prove its existence theorem. Note that our coalitional TU equilibrium concept is a development of TU core for an economy without externalities. The coalitional TU equilibrium is consisted of wealth allocation, utility allocation, trading strategy, and price. Second, inspired by the work on fuzzy core, we establish a fuzzy coalitional TU Gale–Mas-Colell economy and prove the existence of fuzzy coalitional TU equilibria. In a word, the work of Gale and Mas-Colell [11] is our fundamental model. We follow the idea of TU cooperative solution of an economy without externalities. Inspired by above work, we shall add the cooperative idea to the Gale–Mas-Colell economy. By the allocation rule of wealth, we introduce the notion of coalitional TU equilibria in the Gale–Mas-Colell economy. Our main result is the coalitional TU equilibrium existence theorem. Furthermore, we provide a fuzzy generalization of coalitional TU equilibrium existence theorem.

The rest of our paper is organized as follows. Section 2 recalls some models and results. Section 3 provides a coalitional TU extension of Gale–Mas-Colell economy. In Sect. 4, we establish a coalitional TU fuzzy equilibrium existence theorem, and Sect. 5 is the conclusion.

2 Classical Models and Results

2.1 The Gale–Mas-Colell Economy

Gale and Mas-Colell [11] considered a market

$$\mathcal{E} = \Big(N, \Delta, (X_i, P_i, \alpha_i)_{i\in N}, Y\Big),$$

where $N = \{1, \dots, n\}$ is the set of traders;

$$\Delta = \left\{p \in R^L_+ | \sum_{i=1}^{L} p_i = 1\right\}$$

is the set of prices; for any $i \in N$, $X_i \subseteq R^L_+$ is the trading set of trader i, denote

$$X = \prod_{i\in N} X_i,\ X_B = \prod_{i\in B} X_i, \forall B \subseteq N;$$

$P_i : X_i \rightrightarrows X_i$ is the preference correspondence of trader i; $\alpha_i : \Delta \longrightarrow R_+$ is the income function of trader i, that is, $\alpha_i(p)$ represents the wealth of trader i at the price p; $Y \subset R^L$ is the technology of the economy.

An equilibrium of $\mathcal{E}$ is a point $(x^*, p^*) \in X \times \Delta$ such that

(i) for any $i \in N$, $p^* \cdot x_i^* = \alpha_i(p^*)$;
(ii) for any $i \in N$, if $z_i \in P_i(x_i^*)$, then $p^* \cdot z_i > p^* \cdot x_i^*$;
(iii) $\sum_{i\in N} x_i^* \in Y$.

Remark 2.1 In the model of Gale and Mas-Colell [11], the technology of the economy can be represented by the sum of subtechnologies of firms in a classical production economy. Moreover, the income $\alpha_i(p)$ of trader i also covers the Arrow–Debreu income function, and the preferences of all traders were extended to nonordered cases.

2.2 TU Core and TU Fuzzy Core

Let $N = \{1, \dots, n\}$ be the set of players, and $\mathcal{N}$ denote the set of all nonempty subsets of N. Denote by $\mathcal{F}^N = [0, 1]^N \setminus \{0\}$ the set of fuzzy coalitions. A point $s = (s_1, \dots, s_n) \in \mathcal{F}^N$ is a fuzzy coalition, where s_i is the participation level of player i to the coalition $car(s) = \{i \in N | s_i > 0\}$. For any $B \in \mathcal{N}$, define $e^B \in \mathcal{F}^N$ by $e_i^B = 1, i \in B$; $e_i^B = 0, i \notin B$.

A family β of $\mathcal{N}$ is balanced if there exist $\{\lambda_B > 0 | B \in \beta\}$ such that $\sum_{B\in\beta, B\ni i} \lambda_B = 1, \forall i \in N$.

A finite family β of $\mathcal{F}^N$ is balanced if there exist $\{\lambda_s > 0 | s \in \beta\}$ such that $\sum_{s\in\beta} \lambda_s s = e^N$.

A transferable utility (TU) game is defined by $v : \mathcal{N} \longrightarrow R$, which is balanced if $\sum_{B\in\beta} \lambda_B v(B) \leq v(N)$ for any balanced family β with balancing weights $\{\lambda_B > 0 | B \in \beta\}$.

A TU fuzzy game is defined by $v : \mathcal{F}^N \longrightarrow R$, which is balanced if $\sum_{s\in\beta} \lambda_s v(s) \leq v(e^N)$ for any balanced family β of $\mathcal{F}^N$ with balancing weights $\{\lambda_s > 0 | s \in \beta\}$.

The TU core of a TU game v is defined by

$$C(v) = \left\{ x \in R^n \middle| \sum_{i\in N} x_i = v(N), \sum_{i\in B} x_i \geq v(B), \forall B \in \mathcal{N} \right\}.$$

The TU fuzzy core of a TU fuzzy game v is defined by

$$C_F(v) = \left\{ x \in R^n \middle| \sum_{i\in N} x_i = v(e^N), \sum_{i\in car(s)} s_i x_i \geq v(s), \forall s \in \mathcal{F}^N \right\}.$$

Theorem 2.1 ([8, 21]) *A TU game has a nonempty TU core if and only if it is balanced.*

Theorem 2.2 ([28]) *A TU fuzzy game has a nonempty TU fuzzy core if and only if it is balanced.*

3 Coalitional TU Equilibria of Gale–Mas-Colell Economy

We now consider a coalitional TU extension of Gale–Mas-Colell economy

$$\mathcal{E} = \left(N, \Delta, (X_i, u_i)_{i\in N}, (\alpha_B)_{B\in\mathcal{N}}, Y \right),$$

where N, Δ, $(X_i)_{i\in N}$, and Y are defined as in Sect. 2.1; for any $i \in N$, $u_i : X_i \longrightarrow R$ is the transferable utility function of trader i; for any $B \in \mathcal{N}$, $\alpha_B : \Delta \longrightarrow R_+$ is the wealth function of the coalition B, and for any $p \in \Delta$,

$$\alpha_N(p) = \sup_{y\in Y} \left\{ p \cdot y \right\} = \Pi(p).$$

For above model, we assume that there exist the wealth allocations of all players such that every coalition cannot block the allocations. Furthermore, we assume that the players maximize the sum of their utilities, which cannot be blocked by any coalition.

Basing above idea, we say that a point

$$(\eta^*, v^*, x^*, p^*) \in R^n \times R^n \times X \times \Delta$$

is a coalitional TU equilibrium of $\mathcal{E}$ if

(A.1) $\sum_{i \in N} \eta_i^* = \alpha_N(p^*), \sum_{i \in B} \eta_i^* \geq \alpha_B(p^*), \forall B \in \mathcal{N}$;

(A.2) $\sum_{i \in N} v_i^* = \sum_{i \in N} u_i(x_i^*) = \max \left\{ \sum_{i \in N} u_i(y_i) | y \in G_N(p^*, \eta^*) \right\}$;

(A.3) for any $B \in \mathcal{N}$, there exists no $(y_i)_{i \in B} \in G_B(p^*, \eta^*)$ such that $\sum_{i \in B} u_i(y_i) > \sum_{i \in B} v_i^*$;

(A.4) $\sum_{i \in N} x_i^* \in Y$, where for any $(p, \eta) \in \Delta \times R^n$,

$$G_B(p, \eta) = \left\{ (y_i)_{i \in B} \in X_B \middle| p \cdot \sum_{i \in B} y_i \leq \sum_{i \in B} \eta_i \right\}.$$

Remark 3.1 The following points are worth noting for the definition of coalitional TU equilibria.

(1) The condition (A.1) represents the allocation of wealth that cannot be blocked by any coalition.
(2) The condition (A.2) means that (v^*, x^*) is the best choice of the grand coalition N. Moreover, the condition (A.3) represents that the allocation of utilities cannot be blocked by any coalition.
(3) The feasible allocation set of the coalition B is defined by $G_B(p, \eta)$ at the price p and the allocation η of wealth.
(4) The condition (A.4) means the feasibility of supply and demand.

Remark 3.2 Note that an exchange economy and a private ownership production economy are special cases of coalitional TU Gale–Mas-Colell economy.

(i) An exchange economy is defined by $\widehat{\mathcal{E}} = \left(N, \Delta, (X_i, u_i, w_i)_{i \in N}\right)$, where $N, \Delta, (X_i, u_i)_{i \in N}$ are defined as in our model, and $w_i \in X_i$ is the initial endowment of trader i. By the exchange economy $\widehat{\mathcal{E}}$, we can construct a coalitional TU Gale–Mas-Colell economy $\mathcal{E} = \left(N, \Delta, (X_i, u_i)_{i \in N}, (\alpha_B)_{B \in \mathcal{N}}, Y\right)$ as follows. For any $B \in \mathcal{N}$,

$$\alpha_B(p) = p \cdot \sum_{i \in B} w_i, \forall p \in \Delta,$$

and $Y = \left\{\sum_{i\in N} w_i\right\}$.

(ii) Following Arrow and Debreu [2], a private ownership production economy is defined by $\widehat{\mathcal{E}} = \left(N, \Delta, (X_i, u_i, w_i)_{i\in N}, M, (Y_j)_{j\in M}, (\theta_{ij})_{i\in N, j\in M}\right)$, where $N, \Delta, (X_i, u_i, w_i)_{i\in N}$ are defined as in the part (i), $M = \{1, \ldots, m\}$ is the set of firms, $Y_j \subset R^L$ is the production set of firm j, and $\theta_{ij} \in [0, 1]$ is the trader $i's$ share of firm j with $\sum_{i=1}^{n} \theta_{ij} = 1, \forall j \in M$. By the private ownership production economy $\widehat{\mathcal{E}}$, we can construct a coalitional TU Gale–Mas-Colell economy $\mathcal{E} = \left(N, \Delta, (X_i, u_i)_{i\in N}, (\alpha_B)_{B\in\mathcal{N}}, Y\right)$ as follows. For any $B \in \mathcal{N}$,

$$\alpha_B(p) = p \cdot \sum_{i\in B} w_i + \sum_{i\in B}\sum_{j\in M} \theta_{ij} \sup\left\{p \cdot y | y \in Y_j\right\}, \forall p \in \Delta,$$

and $Y = \sum_{i\in N} w_i + \sum_{j\in M} Y_j$.

Remark 3.3 Our coalitional TU equilibrium concept is more related to the TU core of economies without externalities. Note that Zhao [35, 36] provided TU α-core and TU β-core concepts for normal-form games with externalities. When there exists no externality, TU α-core, TU β-core, and TU core overlap.

We now prove the existence of coalitional TU equilibria in this model.

Theorem 3.1 *Suppose that an economy*

$$\mathcal{E} = \left(N, \Delta, (X_i, u_i)_{i\in N}, (\alpha_B)_{B\in\mathcal{N}}, Y\right)$$

satisfies the following conditions:

(1) for any $i \in N$, $X_i \subseteq R^L_+$ is nonempty convex compact;
(2) for any $i \in N$, u_i is concave and continuous on X_i;
(3) for any $B \in \mathcal{N}$, α_B is continuous on Δ, and for any $p \in \Delta$, $\alpha_B(p) > 0$; α_N is convex on Δ;
(4) $\{\alpha_B(p) | B \in \mathcal{N}\}$ is balanced for any $p \in \Delta$, that is, for any balanced family β of $\mathcal{N}$ with balancing weights $\{\lambda_B > 0 | B \in \beta\}$, we have $\sum_{B\in\beta} \lambda_B \alpha_B(p) \leq \alpha_N(p)$;
(5) if $y \notin Y$, then $p \cdot y > \Pi(p)$ for some $p \in \Delta$;
(6) Y is nonempty convex compact.

Then there exists a coalitional TU equilibrium.

Proof We shall organize the proof by the following steps.

Step 1. The allocation of wealth.

Since α_N is continuous on Δ, there exists $M > 0$ such that

$$M = \max\left\{\alpha_N(p) | p \in \Delta\right\}.$$

We define a correspondence $T_1 : \Delta \rightrightarrows [0, M]^n$ by

$$T_1(p) = \left\{\eta \in [0, M]^n \middle| \sum_{i \in N} \eta_i = \alpha_N(p), \sum_{i \in B} \eta_i \geq \alpha_B(p), \forall B \in \mathcal{N}\right\}.$$

By Theorem 2.1 and the condition (4), we have that $T_1(p)$ is nonempty for any $p \in \Delta$. For any given $p \in \Delta$, any $\eta^1, \eta^2 \in T_1(p)$ and any $t \in [0, 1]$, we have that

$$\sum_{i \in N}\left[t\eta_i^1 + (1-t)\eta_i^2\right] = t\sum_{i \in N}\eta_i^1 + (1-t)\sum_{i \in N}\eta_i^2 = \alpha_N(p),$$

and for any $B \in \mathcal{N}$,

$$\sum_{i \in B}\left[t\eta_i^1 + (1-t)\eta_i^2\right] = t\sum_{i \in B}\eta_i^1 + (1-t)\sum_{i \in B}\eta_i^2 \geq \alpha_B(p).$$

Thus, $t\eta^1 + (1-t)\eta^2 \in T_1(p)$, i.e., $T_1(p)$ is convex.

Furthermore, for any $(p^m, \eta^m) \in \Delta \times [0, M]^n$ with $\eta^m \in T_1(p^m)$ and $(p^m, \eta^m) \longrightarrow (p, \eta) \in \Delta \times [0, M]^n$, we have

$$\sum_{i \in N}\eta_i^m = \alpha_N(p^m), \sum_{i \in B}\eta_i^m \geq \alpha_B(p^m), \forall B \in \mathcal{N},$$

implying that when $m \longrightarrow +\infty$,

$$\sum_{i \in N}\eta_i = \alpha_N(p), \sum_{i \in B}\eta_i \geq \alpha_B(p), \forall B \in \mathcal{N}.$$

Thus, T_1 is upper semicontinuous with nonempty convex compact values.

Step 2. The maximal choice.

For any $B \in \mathcal{N}$, it is easy to verify that G_B is continuous with nonempty convex compact values.

We next define a correspondence $T_2 : \Delta \times [0, M]^n \rightrightarrows X$ by

$$T_2(p, \eta) = \left\{x \in G_N(p, \eta) \middle| \sum_{i \in N} u_i(x_i) = \max_{z \in G_N(p,\eta)} \sum_{i \in N} u_i(z_i)\right\}.$$

We easily verify that T_2 is upper semicontinuous with nonempty convex compact values.

Step 3. The feasibility of supply and demand.

We define a correspondence $T_3 : X \rightrightarrows \Delta$ by

$$T_3(x) = \left\{ p \in \Delta \middle| p \cdot \sum_{i \in N} x_i - \alpha_N(p) = \max_{q \in \Delta} \left[q \cdot \sum_{i \in N} x_i - \alpha_N(q) \right] \right\}.$$

It is easy to verify that T_3 is also upper semicontinuous with nonempty convex compact values.

Step 4. Existence (I).

By Steps 1–3, $T = T_1 \times T_2 \times T_3 : [0, M]^n \times X \times \Delta \rightrightarrows [0, M]^n \times X \times \Delta$ is upper semicontinuous with nonempty convex compact values. Therefore, there exists $(\eta^*, x^*, p^*) \in [0, M]^n \times X \times \Delta$ such that

$$\sum_{i \in N} \eta_i^* = \alpha_N(p^*), \ \sum_{i \in B} \eta_i^* \geq \alpha_B(p^*), \forall B \in \mathcal{N},$$
$$x^* \in G_N(p^*, \eta^*), \ \sum_{i \in N} u_i(x_i^*) = \max_{z \in G_N^*(p^*, \eta^*)} \sum_{i \in N} u_i(z_i),$$
$$p^* \cdot \sum_{i \in N} x_i^* - \alpha_N(p^*) = \max_{q \in \Delta} \left[q \cdot \sum_{i \in N} x_i^* - \alpha_N(q) \right].$$

From $x^* \in G_N(p^*, \eta^*)$, we have that for any $q \in \Delta$,

$$q \cdot \sum_{i \in N} x_i^* - \alpha_N(q) \leq p^* \cdot \sum_{i \in N} x_i^* - \alpha_N(p^*) = p^* \cdot \sum_{i \in N} x_i^* - \sum_{i \in N} \eta_i^* \leq 0,$$

implying that

$$q \cdot \sum_{i \in N} x_i^* \leq \Pi(q), \forall q \in \Delta.$$

We claim that $\sum_{i \in N} x_i^* \in Y$. Otherwise, if $\sum_{i \in N} x_i^* \notin Y$, then, by the condition (5), there exists $q \in \Delta$ such that

$$q \cdot \sum_{i \in N} x_i^* > \Pi(q) = \alpha_N(q).$$

We get a contradiction.

Step 5. Existence (II).

We construct a TU game $W : \mathcal{N} \longrightarrow R$ by

$$W(B) = \max \left\{ \sum_{i \in B} u_i((y_B)_i) \middle| y_B \in G_B(p^*, \eta^*) \right\}.$$

We next prove that

$$\sum_{B\in\beta}\lambda_B W(B) \le W(N)$$

for any balanced family β of $\mathcal{N}$ with balancing weights $\{\lambda_B > 0 | B \in \beta\}$.

For any balanced family β of $\mathcal{N}$ with balancing weights $\{\lambda_B > 0 | B \in \beta\}$, we have that there exist $y_B \in G_B(p^*, \eta^*), \forall B \in \beta$ such that

$$W(B) = \sum_{i\in B} u_i((y_B)_i), \forall B \in \beta.$$

Since $y_B \in G_B(p^*, \eta^*), \forall B \in \beta$, it follows that

$$p^* \cdot \sum_{i\in B}(y_B)_i \le \sum_{i\in B}\eta_i^*, \forall B \in \beta,$$

implying that

$$\begin{aligned}
p^* \cdot \sum_{i\in N} y_i' &= p^* \cdot \sum_{i\in N}\ \sum_{B\in\beta, B\ni i} \lambda_B (y_B)_i \\
&= p^* \cdot \sum_{B\in\beta}\sum_{i\in B}\lambda_B (y_B)_i \\
&= \sum_{B\in\beta}\lambda_B\left(p^* \cdot \sum_{i\in B}(y_B)_i\right) \\
&\le \sum_{B\in\beta}\lambda_B\left(\sum_{i\in B}\eta_i^*\right) \\
&= \sum_{i\in N}\eta_i^*\left(\sum_{B\in\beta, B\ni i}\lambda_B\right) \\
&= \sum_{i\in N}\eta_i^*,
\end{aligned}$$

where

$$y_i' = \sum_{B\in\beta, B\ni i}\lambda_B (y_B)_i, \forall i \in N.$$

Furthermore, since each u_i is concave, we have that

$$\sum_{B\in\beta}\lambda_B W(B) = \sum_{B\in\beta}\sum_{i\in B}\lambda_B u_i((y_B)_i)$$

$$
\begin{aligned}
&= \sum_{i \in N} \sum_{B \in \beta, B \ni i} \lambda_B u_i((y_B)_i) \\
&\leq \sum_{i \in N} u_i \left(\sum_{B \in \beta, B \ni i} \lambda_B (y_B)_i \right) \\
&= u_i(y_i') \\
&\leq \max \left\{ \sum_{i \in N} u_i(z_i) | z \in G_N(p^*, \eta^*) \right\} \\
&= W(N).
\end{aligned}
$$

Therefore, W is balanced. By Theorem 2.1, there exists $v^* \in R^n$ such that

$$
\begin{aligned}
&\sum_{i \in N} v_i^* = W(N) = \max \left\{ \sum_{i \in N} u_i(z_i) \middle| z \in G_N(p^*, \eta^*) \right\}, \\
&\sum_{i \in B} v_i^* \geq W(B) = \max \left\{ \sum_{i \in B} u_i(z_i) \middle| (z_i)_{i \in B} \in G_B(p^*, \eta^*) \right\}, \forall B \in \mathcal{N}.
\end{aligned}
$$

By Steps 1–5, we obtain that (η^*, v^*, x^*, p^*) is a coalitional TU equilibrium of $\mathcal{E}$. The proof is completed.

As a special case of Gale–Mas-Colell economy, an exchange economy $\widehat{\mathcal{E}} = \left(N, \Delta, (X_i, u_i, w_i)_{i \in N} \right)$ is defined as in Remark 3.2(i). For any $B \in \mathcal{N}$, the feasible allocation set of the coalition B is defined by

$$
F_B = \left\{ (y_i)_{i \in B} \in X_B \middle| \sum_{i \in B} y_i \leq \sum_{i \in B} w_i \right\}.
$$

For any $p \in \Delta$, any $i \in N$, and any $B \in \mathcal{N}$, the budget sets of player i and the coalition B are defined by

$$
\begin{aligned}
D_i(p) &= \{ y_i \in X_i | p \cdot y_i \leq p \cdot w_i \}, \\
D_B(p) &= \left\{ (y_i)_{i \in B} \in X_B \middle| p \cdot \sum_{i \in B} y_i \leq p \cdot \sum_{i \in B} w_i \right\}.
\end{aligned}
$$

We say that a point $(v^*, x^*) \in R^n \times X$ is in the TU core of $\widehat{\mathcal{E}}$ if

(a1) $\sum_{i \in N} v_i^* = \sum_{i \in N} u_i(x_i^*) = \max \left\{ \sum_{i \in N} u_i(y_i) \middle| y \in F_N \right\}$,

(a2) $\sum_{i \in B} v_i^* \geq \max \left\{ \sum_{i \in B} u_i(y_i) \middle| (y_i)_{i \in B} \in F_B \right\}, \forall B \in \mathcal{N}$.

We say that a point $(x^*, p^*) \in X \times \Delta$ is a Walrasian equilibrium of $\widehat{\mathcal{E}}$ if

(b1) for any $i \in N$, $x_i^* \in D_i(p^*)$ and $u_i(x_i^*) = \max_{y_i \in D_i(p^*)} u_i(y_i)$,

(b2) $\sum_{i \in N} x_i^* \leq \sum_{i \in N} w_i$.

We say that a point (η^*, v^*, x^*, p^*) is a coalitional TU equilibrium of $\widehat{\mathcal{E}}$ if

(c1) $\sum_{i \in N} \eta_i^* = p^* \cdot \sum_{i \in N} w_i, \sum_{i \in B} \eta_i^* \geq p^* \cdot \sum_{i \in B} w_i, \forall B \in \mathcal{N}$,

(c2) $\sum_{i \in N} v_i^* = \sum_{i \in N} u_i(x_i^*) = \max \left\{ \sum_{i \in N} u_i(y_i) \middle| y \in G_N(p^*, \eta^*) \right\}$,

(c3) $\sum_{i \in B} v_i^* \geq \max \left\{ \sum_{i \in B} u_i(y_i) \middle| (y_i)_{i \in B} \in G_B(p^*, \eta^*) \right\}, \forall B \in \mathcal{N}$,

(c4) $\sum_{i \in N} x_i^* \leq \sum_{i \in N} w_i$.

Remark 3.4 (i) Since the components of TU core, coalitional TU equilibrium, and Walrasian equilibrium are different, they are three different solution concepts.

(ii) Suppose that (η^*, v^*, x^*, p^*) is a coalitional TU equilibrium of $\widehat{\mathcal{E}}$. By (c1) and (c4), we have $x^* \in F_N$ and

$$p^* \cdot \sum_{i \in N} x_i^* \leq p^* \cdot \sum_{i \in N} w_i = \sum_{i \in N} \eta_i^*.$$

By (c2), we get that

$$\begin{aligned} \max \left\{ \sum_{i \in N} u_i(y_i) \middle| y \in F_N \right\} &\geq \sum_{i \in N} u_i(x_i^*) = \sum_{i \in N} v_i^* \\ &\geq \max \left\{ \sum_{i \in N} u_i(y_i) \middle| y \in G_N(p^*, \eta^*) \right\} \\ &\geq \max \left\{ \sum_{i \in N} u_i(y_i) \middle| y \in F_N \right\}, \end{aligned}$$

implying that

$$\sum_{i \in N} v_i^* = \sum_{i \in N} u_i(x_i^*) = \max \left\{ \sum_{i \in N} u_i(y_i) \middle| y \in F_N \right\}.$$

Thus, the condition (a1) is satisfied.

Since for any $B \in \mathcal{N}$, $F_B \subset D_B(p^*) \subset G_B(p^*, \eta^*)$, it from (c3) follows that

$$\sum_{i \in B} v_i^* \geq \max \left\{ \sum_{i \in B} u_i(y_i) \middle| (y_i)_{i \in B} \in G_B(p^*, \eta^*) \right\}$$

$$\geq \max\left\{\sum_{i\in B} u_i(y_i)\middle|(y_i)_{i\in B}\in F_B\right\}.$$

Thus, the condition (a2) is satisfied. Then, (v^*, x^*) is in the TU core of $\widehat{\mathcal{E}}$.

(iii) The Walrasian equilibrium is defined by the noncooperative idea. Every player maximizes own utility according to the individual budget set. However, the coalitional TU equilibrium is a cooperative solution concept. The trading strategy profile of coalitional TU equilibrium is a maximum solution of the sum of all players' utilities. It differs from the individual optimal solution of every player for the Walrasian equilibrium.

4 A Fuzzy Extension

Observe that the results of Sect. 3 are obtained in the model with the crisp coalition. Inspired by the work on fuzzy core, we provide a fuzzy extension of Sect. 3.

We consider a fuzzy coalitional TU Gale–Mas-Colell economy

$$\mathcal{E} = \Big(N, \Delta, (X_i, u_i)_{i\in N}, (\alpha_s)_{s\in\mathcal{F}^N}, Y\Big),$$

where N, Δ, $(X_i, u_i)_{i\in N}$, and Y are defined as in Sect. 3; $\mathcal{F}^N$ is the set of fuzzy coalitions; for any $s\in\mathcal{F}^N$, $\alpha_s : \Delta \longrightarrow R_+$ is the wealth function of the fuzzy coalition s, where for any $(p, \eta)\in\Delta\times R^n$,

$$G_s(p,\eta) = \left\{(y_i)_{i\in car(s)}\in X_{car(s)}\middle| p\cdot\sum_{i\in car(s)} s_i y_i \leq \sum_{i\in car(s)} s_i\eta_i\right\},$$

and for any $p\in\Delta$,

$$\alpha_{e^N}(p) = \sup_{y\in Y}\left\{p\cdot y\right\} = \Pi(p).$$

We say that a point (η^*, v^*, x^*, p^*) is a fuzzy coalitional TU equilibrium of $\mathcal{E}$ if

(B.1) $\sum_{i\in N}\eta_i^* = \alpha_{e^N}(p^*)$, $\sum_{i\in car(s)} s_i\eta_i^* \geq \alpha_s(p^*)$, $\forall s\in\mathcal{F}^N$;

(B.2) $\sum_{i\in N} v_i^* = \sum_{i\in N} u_i(x_i^*) = \max\left\{\sum_{i\in N} u_i(z_i)|z\in G_{e^N}(p^*,\eta^*)\right\}$;

(B.3) for any $s\in\mathcal{F}^N$, there exists no $(z_i)_{i\in car(s)}\in G_s(p^*,\eta^*)$ such that $\sum_{i\in car(s)} s_i u_i(z_i) > \sum_{i\in car(s)} s_i v_i^*$;

(B.4) $\sum_{i\in N} x_i^*\in Y$.

Remark 4.1 Note that our (B.1)–(B.4) are some fuzzy extensions of (A.1)–(A.4), where the crisp coalition is replaced by fuzzy coalition.

We next prove the existence of fuzzy coalitional TU equilibria in the economy.

Theorem 4.1 *Suppose that an economy*

$$\mathcal{E} = \left(N, \Delta, (X_i, u_i)_{i\in N}, (\alpha_s)_{s\in\mathcal{F}^N}, Y\right)$$

satisfies the following conditions:

(1) for any $i \in N$, $X_i \subseteq R_+^L$ is nonempty convex compact;
(2) for any $i \in N$, u_i is concave and continuous on X_i;
(3) for any $s \in \mathcal{F}^N$, α_s is continuous on Δ, and for any $p \in \Delta$, $\alpha_s(p) > 0$; α_{e^N} is convex on Δ;
(4) $\{\alpha_s(p) | s \in \mathcal{F}^N\}$ is balanced for any $p \in \Delta$, that is, for any balanced family β of $\mathcal{F}^N$ with balancing weights $\{\lambda_s > 0 | s \in \beta\}$, we have $\sum_{s\in\beta} \lambda_s \alpha_s(p) \le \alpha_{e^N}(p)$;
(5) if $y \notin Y$, then $p \cdot y > \Pi(p)$ for some $p \in \Delta$;
(6) Y is nonempty convex compact.

Then there exists a fuzzy coalitional TU equilibrium.

Proof We shall use the similar proof technique of Theorem 3.1. For the sake of completeness, we only give the proof frame. We shall define:

$$M = \max\left\{\alpha_{e^N}(p) | p \in \Delta\right\},$$

$$T_1(p) = \left\{\eta \in [0, M]^n \,\middle|\, \sum_{i\in N} \eta_i = \alpha_{e^N}(p),\ \sum_{i\in car(s)} s_i \eta_i \ge \alpha_s(p), \forall s \in \mathcal{F}^N\right\}, \forall p \in \Delta,$$

$$T_2(p, \eta) = \left\{x \in X \,\middle|\, \sum_{i\in N} u_i(x_i) = \max_{z\in G_{e^N}(p,\eta)} \sum_{i\in N} u_i(z_i)\right\}, \forall (p, \eta) \in \Delta \times [0, M]^n,$$

$$T_3(x) = \left\{p \in \Delta \,\middle|\, p \cdot \sum_{i\in N} x_i - \alpha_{e^N}(p) = \max_{q\in\Delta}\left[q \cdot \sum_{i\in N} x_i - \alpha_{e^N}(q)\right]\right\}, \forall x \in X,$$

$$T(\eta, x, p) = T_1(p) \times T_2(p, \eta) \times T_3(x), \forall (\eta, x, p) \in [0, M]^n \times X \times \Delta.$$

We have that T is upper semicontinuous with nonempty convex compact values. By the similar method, there exists $(\eta^*, x^*, p^*) \in [0, M]^n \times X \times \Delta$ such that

$$\sum_{i\in N} \eta_i^* = \alpha_{e^N}(p^*),\ \sum_{i\in car(s)} s_i \eta_i^* \ge \alpha_s(p^*), \forall s \in \mathcal{F}^N,$$

$$\sum_{i\in N} u_i(x_i^*) = \max_{z\in G_{e^N}(p^*,\eta^*)} \sum_{i\in N} u_i(z_i);$$

$$\sum_{i\in N} x_i^* \in Y.$$

Furthermore, we define a fuzzy TU game $W : \mathcal{F}^N \longrightarrow R$ by

$$W(s) = \max\left\{\sum_{i\in car(s)} s_i u_i(z_i)\middle|(z_i)_{i\in car(s)} \in G_s(p^*, \eta^*)\right\}, \forall s \in \mathcal{F}^N.$$

For any balanced family β of $\mathcal{F}^N$ with balancing weights $\{\lambda_s > 0 | s \in \beta\}$, we have that for any $s \in \beta$, there exists $z_s \in G_s(p^*, \eta^*)$ such that

$$W(s) = \sum_{i\in car(s)} s_i u_i((z_s)_i).$$

We define $z' \in X$ by

$$z'_i = \sum_{s\in\beta, car(s)\ni i} \lambda_s s_i (z_s)_i, \forall i \in N.$$

Since $z_s \in G_s(p^*, \eta^*)$, $\forall s \in \beta$, it follows that

$$p^* \cdot \sum_{i\in car(s)} s_i (z_s)_i \le \sum_{i\in car(s)} s_i \eta_i^*, \forall s \in \beta,$$

implying that

$$\begin{aligned}
p^* \cdot \sum_{i\in N} z'_i &= p^* \cdot \sum_{i\in N} \sum_{s\in\beta, car(s)\ni i} \lambda_s s_i (z_s)_i \\
&= \sum_{s\in\beta} \lambda_s \left(p^* \cdot \sum_{i\in car(s)} s_i (z_s)_i\right) \\
&\le \sum_{s\in\beta} \lambda_s \left(\sum_{i\in car(s)} s_i \eta_i^*\right) \\
&= \sum_{i\in N} \eta_i^* \left(\sum_{s\in\beta, car(s)\ni i} \lambda_s s_i\right) \\
&= \sum_{i\in N} \eta_i^*.
\end{aligned}$$

That is, $z' \in G_{e^N}(p^*, \eta^*)$.

Furthermore, since each u_i is concave, we have that

$$\sum_{s\in\beta} \lambda_s W(s) = \sum_{s\in\beta} \sum_{i\in car(s)} \lambda_s s_i u_i((z_s)_i)$$

$$
\begin{aligned}
&= \sum_{i \in N} \sum_{s \in \beta, car(s) \ni i} \lambda_s s_i u_i((z_s)_i) \\
&\leq \sum_{i \in N} u_i \left(\sum_{s \in \beta, car(s) \ni i} \lambda_s s_i (z_s)_i \right) \\
&= \sum_{i \in N} u_i(z_i') \\
&\leq \max \left\{ \sum_{i \in N} u_i(z_i) \middle| z \in G_{e^N}(p^*, \eta^*) \right\} \\
&= W(e^N).
\end{aligned}
$$

By Theorem 2.2, there exists $v^* \in R^n$ such that

$$
\sum_{i \in N} v_i^* = W(e^N) = \max \left\{ \sum_{i \in N} u_i(z_i) | z \in G_{e^N}(p^*, \eta^*) \right\} = \sum_{i \in N} u_i(x_i^*),
$$

$$
\sum_{i \in car(s)} s_i v_i^* \geq W(s) = \max \left\{ \sum_{i \in car(s)} s_i u_i((z_s)_i) | z_s \in G_s(p^*, \eta^*) \right\}, \forall s \in \mathcal{F}^N.
$$

Thus, we obtain that (η^*, v^*, x^*, p^*) is a fuzzy coalitional TU equilibrium of $\mathcal{E}$. The proof is completed.

5 Conclusion

By following the Gale–Mas-Colell economy, we add the cooperative behavior in the model. By following the coalitional TU equilibrium of economies without externalities, we define the coalitional TU equilibrium in our model. The coalitional TU equilibrium existence theorem is our main contribution. Inspired by the work on fuzzy core, we further prove the existence of fuzzy coalitional TU equilibria in the Gale–Mas-Colell economy with fuzzy coalitions.

References

1. Allouch, N., Predtetchinski, A.: On the nonemptiness of the fuzzy core. Int. J. Game Theory **37**, 203–210 (2008)
2. Arrow, K.J., Debreu, G.: Existence of an equilibrium for a competitive economy. Econometrica **22**, 265–290 (1954)
3. Askoura, Y.: The weak-core of a game in normal form with a continuum of players. J. Math. Econ. **47**, 43–47 (2011)

4. Askoura, Y.: An interim core for normal form games and exchange economies with incomplete information. J. Math. Econ. **58**, 38–45 (2015)
5. Askoura, Y.: On the core of normal form games with a continuum of players. Math. Soc. Sci. **89**, 32–42 (2017)
6. Askoura, Y., Sbihi, M., Tikobaini, H.: The ex-ante α-core for normal form games with uncertainty. J. Math. Econ. **49**, 157–162 (2013)
7. Aubin, J.P.: Cooperative fuzzy games. Math. Oper. Res. **6**, 1–13 (1981)
8. Bondareva, O.N.: Some applications of linear programming methods to the theory of cooperative games. Problemi Kibemitiki **10**, 119–139 (1963)
9. Border, K.C.: A core existence theorem for games without ordered preferences. Econometrica **52**, 1537–1542 (1984)
10. Florenzano, M.: Edgeworth equilibria, fuzzy core and equilibria of a production economy without order preferences. J. Math. Anal. Appl. **153**, 18–36 (1990)
11. Gale, D., Mas-Colell, A.: An equilibrium existence theorem for a general model without ordered preferences. J. Math. Econ. **2**, 9–15 (1975)
12. Inoue, T.: Representation of transferable utility games by coalition production economies. J. Math. Econ. **48**, 143–147 (2012)
13. Kajii, A.: A generalization of Scarf's theorem: an α-core existence theorem without transitivity or completeness. J. Econ. Theory **56**, 194–205 (1992)
14. Liu, J., Liu, X.: A necessary and sufficient condition for an NTU fuzzy game to have a nonempty fuzzy core. J. Math. Econ. **49**, 150–156 (2013)
15. Liu, J., Liu, X.: Existence of Edgeworth and competitive equilibria and fuzzy cores in coalition production economies. Int. J. Game Theory **43**, 975–990 (2014)
16. Liu, J., Tian, H.Y.: Existence of fuzzy cores and generalizations of the K-K-M-S theorem. J. Math. Econ. **52**, 148–152 (2014)
17. Mas-Colell, A.: An equilibrium existence theorem without complete or transitive preferences. J. Math. Econ. **1**, 237–246 (1974)
18. Noguchi, M.: Cooperative equilibria of finite games with incomplete information. J. Math. Econ. **55**, 4–10 (2014)
19. Noguchi, M.: Alpha cores of games with nonatomic asymmetric information. J. Math. Econ. **75**, 1–12 (2018)
20. Noguchi, M.: Essential stability of the alpha cores of finite games with incomplete information. Math. Soc. Sci. **110**, 34–43 (2021)
21. Scarf, H.E.: The core of an N-person game. Econometrica **35**, 50–69 (1967)
22. Scarf, H.E.: On the existence of a cooperative solution for a general class of n-person games. J. Econ. Theory **3**, 169–181 (1971)
23. Shafer, W., Sonnenschein, H.: Equilibrium in abstract economies without ordered preferences. J. Math. Econ. **2**, 345–348 (1975)
24. Shapley, L.S.: On balanced sets and cores. Naval Reseach Logistics Quarterly **14**, 453–460 (1967)
25. Shapley, L.S., Shubik, M.: On market games. J. Econ. Theory **1**, 9–25 (1969)
26. Sun, N., Trockel, W., Yang, Z.: Competitive outcomes and endogenous coalition formation in an n-person game. J. Math. Econ. **44**, 853–860 (2008)
27. Uyanik, M.: On the nonemptiness of the α-core of discontinuous games: Transferable and nontransferable utilities. J. Econom. Theory **158**, 213–231 (2015)
28. van Gulick, G., Norde, H.: Fuzzy cores and fuzzy balancedness. Math. Methods Oper. Res. **77**, 131–146 (2013)
29. Yang, Z.: Some infinite-player generalizations of Scarf's theorem: finite-coalition α-cores and weak α-cores. J. Math. Econ. **73**, 81–85 (2017)
30. Yang, Z.: Some generalizations of Kajii's theorem to games with infinitely many players. J. Math. Econ. **76**, 131–135 (2018)
31. Yang, Z.: The weak α-core of exchange economies with a continuum of players and pseudo-utilities. J. Math. Econom. **91**, 43–50 (2020)

32. Yang, Z., Song, Q.P.: A weak α-core existence theorem of generalized games with infinitely many players and pseudo-utilities. Math. Soc. Sci. **116**, 40–46 (2022)
33. Yang, Z., Yuan, G.X.: Some generalizations of Zhao's theorem: Hybrid solutions and weak hybrid solutions for games with nonordered preferences. J. Math. Econom. **84**, 94–100 (2019)
34. Yang, Z., Zhang, X.: A weak α-core existence theorem of games with nonordered preferences and a continuum of agents. J. Math. Econom. **94**, 102464 (2021)
35. Zhao, J.: The existence of TU α-core in normal form games. Int. J. Game Theory **28**(1), 25–34 (1999)
36. Zhao, J.: A β-core existence result and its application to oligopoly markets. Games Econ. Behav. **27**, 153–168 (1999)

Simple Visibility Design in Network Games

Gabrielle Demange

Abstract Individuals interact through social media networks by posting contributions, comments, and the like. These actions entail complementarities and generate spillovers. Social media carefully designs their platform, often with the objective of increasing users' activity. Here I study visibility strategies, which promote some individuals by making their contributions more visible than others'. Under some specifications, optimal strategies are simple, making visible the individuals whose total impact (total number of followers for example) is maximal. Comparisons with other targeting strategies are discussed.

Keywords Networks · Bilateral interactions · Strategic complements · Targeting

1 Introduction

The tremendous growth in the social networks industry has triggered a variety of research. Statisticians and empiricists develop tools to analyze large networks and exploit the access to a huge amount of data, game theorists model how individuals form links and interact on a social network, marketing, and computer scientists design targeting strategies taking advantage of detailed information on users. The growth of social networks is mainly conducted by companies, called social media platforms, say, Facebook, Twitter, or Instagram to name a few. Companies not only facilitate interactions but also shape them through the platform's design encapsulated in recommendation algorithms. The design matters for users, as illustrated by Instagram project in May 2022 to change its algorithm. In the current setup, the users see mostly pictures of friends or people they follow; the proposed algorithm would have recommended ads and videos from creators they do not follow. Given the users' protests,[1] the proposal is not yet implemented (as of December 2022). The

[1] *Stop trying to be TikTok: how video-centric Instagram sparked a revolt* (Guardian 31 July 2022).

G. Demange (✉)
Paris School of Economics-EHESS, Paris, France
e-mail: demange@pse.ens.fr

M. A. Khan et al. (eds.), *Matching, Dynamics and Games for the Allocation of Resources*, Monographs in Mathematical Economics 7,
https://doi.org/10.1007/978-981-96-6661-4_6

design chosen by a platform, whose objective is to maximize its revenues, may create tensions with the users and lower their incentives to participate. I propose a simple analysis of one aspect of this design, namely *visibility*, by considering strategies that promote some individuals' actions.[2]

The analysis is based on a simple game in which individuals embedded in a social network interact directly with their neighbors. For example, in a friendship network, an individual exerts an effort level to help his/her friends and benefits from their effort to help; on a social media network, a user posts contributions accessible to his/her followers and reads the contributions of those he/she follows (his/her influencers). Individuals' payoffs positively depend on a weighted combination of their neighbors' actions, called exposure. As a result, the game exhibits strategic complementarities, meaning that the incentives for an individual to increase one's action rise with increases in the others' actions. Nash equilibria describe the full extent of these incentives, in particular, they embody not only the influence of neighbors but also that of neighbors of neighbors, and, by extension, the indirect influence channeled through the network of relationships. This paper analyzes the equilibria for specific payoffs for which the best responses are linear in exposure. Equilibria are unique and easy to compute when they exist. Uniqueness allows for an analysis of design issues.

On an Internet platform, a variety of tools can be used for discriminating actions. Contributions can be presented in a biased order or displayed during different periods of time, making them more or less visible. How does a platform design visibility when its objective is to maximize aggregate activity (the sum of the actions)? I examine this question by modeling visibility in a simple way by *weights*, interpreted, for example, as the relative time to display contributions. Importantly, the platform, which knows the network structure, correctly anticipates how its strategies affect equilibria.

Empirical studies conducted by computer scientists provide a guide for modeling social networks, especially social media networks. They analyze the topology of social networks by computing various statistics such as the distributions of in- and out-degrees, the proportion of the directed links, or the clustering coefficients. For example, Kwak, Lee, Park, and Moon [13] analyze the topology of the follower graph on Twitter. They show important differences with other social networks; in particular, the graph is directed due to the lack of reciprocity of links. An important insight of these studies is that influence—measured by the number of retweets—is related but not aligned with the number of followers. Furthermore, statistics such as centrality indices are rather bad predictors of behaviors and real interactions. This justifies a strategic analysis that distinguishes the network structure from the behaviors of agents embedded in the network.

The problem of a designer (planner, firm) trying to influence individuals' actions in situations with complementarities has been investigated in the economics literature. Studies differ along several dimensions, in particular, in the interaction model, the designer 's objective, and the tools of intervention. A first class of interaction models builds on Ballester, Calvo Armengol, and Zenou [1] who introduced games on a

[2] To avoid confusion, I use the word "action" for the individuals and "strategy" for the designer.

network with a linear best responses as in this paper. They consider the problem of removing a well-chosen agent—the key player- so as to minimize aggregate activity. It has been followed by an extensive literature studying interventions aiming at modifying aggregate activity, as referred to in Demange [4], where I study an extension of the game with increasing but non-linear best responses in exposure. Galeotti, Golub, and Goyal [11] consider the maximization of welfare. These interventions modify the marginal incentives of individuals to contribute but do not change the importance of the links. In contrast, a visibility strategy as considered here (partially) changes the network structure; in particular, all the links toward an agent whose visibility weight is null are withdrawn. The problem of changing the network structure is challenging and results are limited (Ballester, Calvo Armengol, and Zenou [1] already cited and Belhaj, Bervoets, Deroian [2]). In a second class of interaction models, actions are discrete, most often binary. They describe, for instance, the choice of a location, the adoption of a new technology, or the health status, see, e.g., Schelling [18] in a behavioral model, Morris [16] in a strategic framework, or Domingos and Richardson [6] in a marketing context. Analytical results on equilibria and optimal interventions are difficult to obtain in these models; simulations suggest that some characteristics of the behavioral process (say the probability of adoption as a function of neighbors and history) play an important role (Dodds and Watts [5], Kempe, Kleinberg and Tardos [15]).

The paper relies on linear analysis (especially on positive matrices) and networks, domains to which David Gale made important contributions. He wrote a famous book on linear models [8], and proved an important result on the feasibility of flows in networks [7]. Gale and Politof [9] extend to general network's Shapley's results on maximal network flow [19] and complements or substitutes in assignment problems [20].

The plan is as follows. Section 2 introduces the model and describes equilibria. Section 3 examines aggregate activity and visibility strategies.

2 The Basic Interaction Model

There is a set of n agents, $N = \{1, \ldots, n\}$, who are in a situation of interaction. Interactions are bilateral, specified by a weighted network. Section 2.1 describes the game and provides two illustrations. Section 2.2 defines and solves the equilibria of the game.

2.1 The Game

Each agent chooses an action represented by a nonnegative real number, x_i for i. The payoff to i, denoted u_i, depends not only on i's own action but also on others' actions. Specifically, denoting $\boldsymbol{x} = (x_i)_{i \in N}$ the vector of all actions[3] u_i is given by:

$$u_i(\boldsymbol{x}) = x_i(\underline{x}_i + \sum_j X_{ij}x_j) - \frac{x_i^2}{2} \text{ for each } i \tag{1}$$

where $\underline{x}_i$ is positive, X_{ij} is nonnegative, and $X_{ii} = 0$. u_i is made of benefit (the first term) and cost (the second term). Cost is quadratic in the action level. i's benefit is linear in i's action with a return that depends on j's action if $X_{ij} > 0$. X_{ij} measures the strength of this dependence and is called the *impact* of j on i. Given $\boldsymbol{x}$, i's *exposure* is defined as the linear combination of others' actions weighted by their impact, $\sum_j X_{ij}x_j$. The return to x_i is thus equal to $\underline{x}_i$ augmented of i's exposure. $\underline{x}_i$ is the action that i would take if i's exposure was null. Call $\underline{x}_i$ i's *standalone action*. An agent i who has no interaction with others, that is, X_{ij} and X_{ji} are null for each j, optimally plays the standalone action whatever other players are doing.

The game is thus characterized by the n-vector of positive standalone actions $\underline{\boldsymbol{x}} = (\underline{x}_i)$ and the $n \times n$ impact matrix $\boldsymbol{X} = (X_{ij})$ with nonnegative elements and null diagonal.

Examples

In the following two examples, the impacts are derived from an observed network of links between agents. The network has N as the set of nodes and is described by its adjacency matrix $\boldsymbol{G}$: $G_{ij} = 1$ if (i, j) is a link and $G_{ij} = 0$ if (i, j) is not a link. $\boldsymbol{G}$ is not necessarily symmetric (Example 1) and, even if symmetric, the impact matrix $\boldsymbol{X}$ derived from $\boldsymbol{G}$ is not necessarily symmetric (Example 2).

Example 1 *unweighted network* Let $\boldsymbol{G}$ represent the follower graph of a platform such as Twitter. Let $G_{ij} = 1$ if i "follows" j, meaning that i receives the tweets posted by j. j thus influences or impacts i. But j does not have to reciprocate and follow i: as emphasized by Kwak et al. [13], the Twitter follower graph is not symmetric, in contrast with social media such as Facebook where links are reciprocated. Individuals post contributions and x_i represents the number or quality of i's contributions. Suppose that there is no congestion between followers nor influencers in the access to contributions, meaning that the impact of j on i is independent of the number of j's followers and the number of i's influencers. In that case, i's exposure is equal to $\delta \sum_j G_{ij}x_j$, the sum of the contributions of i's influencers up to a discount

---

[3] In what follows, vectors and matrices are denoted in bold letters. $\mathbb{1}$ denotes the n-vector of ones and $\mathbb{I}_n$ the n-identity matrix. Given two n-vectors $\boldsymbol{v}$ and $\boldsymbol{v}'$, $\boldsymbol{v} \geq \boldsymbol{v}'$ if $v_i \geq v_i'$ for each i. Similarly, for two matrices $\boldsymbol{A}$ and $\boldsymbol{A}'$ of the same dimension, $\boldsymbol{A} \geq \boldsymbol{A}'$ if $A_{ij} \geq B_{ij}$ for each (i, j). $\widetilde{\boldsymbol{v}}$ denotes the transpose of vector $\boldsymbol{v}$ and similarly for a matrix. Finally, given vector $\boldsymbol{v}$, $\boldsymbol{dg}(\boldsymbol{v})$ denotes the diagonal matrix with v_j as the j-th element on the diagonal.

factor δ that measures the effect of contributions relative to the standalone actions and cost. Hence $\boldsymbol{X} = \delta \boldsymbol{G}$.

Example 2 Let $\boldsymbol{G}$ represent a network between relatives. $G_{ij} = 1$ if j cares for i. In a friendship network, $\boldsymbol{G}$ is symmetric. Let x_j represent the total effort (care or overall time) spent by j with the relatives he/she cares of. Assuming effort to be split equally among them, j's effective effort on each relative is equal to $\frac{x_j}{G_{+j}}$ where $G_{+j} = \sum_i G_{ij}$ is j's in-degree. Define i's exposure as the sum of the effective attention i receives, up to a discount factor δ: $\delta \sum_j G_{ij} \frac{x_j}{G_{+j}}$. This defines the impact matrix $\boldsymbol{X}$ by $X_{ij} = \frac{G_{ij}}{G_{j+}}$. Defining j's impact total by $X_{+j} = \sum_i X_{ij}$, total impacts are equal across individuals.

2.2 Equilibrium

The analysis of interactions relies on the notion of (Nash) equilibrium, which requires that no player observing others' actions could increase his/her payoff by changing his/her own action. Formally:

Definition 1 Let $\boldsymbol{x}_{-i} = (x_j)_{j \neq i}$ denote the actions of agents other than i.

i's *best responses* to $\boldsymbol{x}_{-i}$ are i's actions that maximize i's payoff taking $(\boldsymbol{x}_{-i})$ as given, i.e., maximize: $x_i \rightarrow u_i(x_i, \boldsymbol{x}_{-i})$.

$\boldsymbol{x}$ is an *equilibrium* if for each i, x_i is a best response to others' actions $\boldsymbol{x}_{-i}$.

In our game, payoffs are concave and playing a null action is never optimal because the standalone actions are assumed to be positive. As a result, best responses are surely positive, characterized by first-order conditions. Specifically, i's best response to $\boldsymbol{x}_{-i}$ is $x_i = \underline{x}_i + \sum_j X_{ij} x_j$, which is a linear and increasing function of i's exposure. The game therefore exhibits *strategic complementarities*, meaning that players have incentives to (weakly) increase their own action when others' actions increase. In Example 1, strategic complementarities mean that the more numerous the influencers' contributions, the more a follower wants to contribute. In Example 2, strategic complementarities mean that the larger the relatives' effective effort toward an individual, the more effort the individual is ready to exert.

The problems of existence and computation of equilibria are not easy to solve in general since equilibria are fixed points of the collection of best responses. However, in our game, the problems are very easy to solve due to the linearity of best responses, as stated in Property 1:

Property 1 Let ρ denote the dominant eigenvalue of the nonnegative matrix $\boldsymbol{X}$. If $\rho < 1$, then an equilibrium exists and is unique given by

$$\boldsymbol{x} = (\mathbb{I}_n - \boldsymbol{X})^{-1} \underline{\boldsymbol{x}}. \tag{2}$$

If $\rho \geq 1$ there is no equilibrium.

Formula (2) is standard and first appears in Ballester et al. [1], based on an explicit computation of the fixed points of the best responses. I give the proof for completeness. I also provide an alternative proof based on strategic complementarities, which allows for a better understanding of the condition for existence; furthermore, it extends to games with non-linear increasing best responses without invoking a topological fixed point theorem.[4] As shown by Topkis [21], equilibria (if they exist) can be obtained by iterating best responses. Here, starting from standalone actions, iterated responses reinforce each other and increase. As a result, they converge to an equilibrium, except if the spillover effects due to complementarities are strong enough to make the sequence of best responses unbounded. The dominant eigenvalue of the impact matrix measures the strength of these spillover effects: when the eigenvalue is equal or greater than 1, iterated best responses become infinitely large and no equilibrium exists.

Proof To prove (2), recall that $\boldsymbol{x}$ is an equilibrium if each i plays a best response to others' actions: $x_i = \underline{x}_i + \sum_j X_{ij} x_j$ or in matrix form $(\mathbb{I}_n - \boldsymbol{X})\boldsymbol{x} = \underline{\boldsymbol{x}}$. According to Perron–Frobenius theorem on positive matrices (see footnote 5), $\mathbb{I}_n - \boldsymbol{X}$ is invertible and its inverse is positive if the dominant eigenvalue of $\boldsymbol{X}$ is less than 1: (2) follows.

Consider now the alternative proof based on Topkis's process of iterated best responses. Start by assuming that players exert their standalone actions. Denote $\boldsymbol{x}(0) = \underline{\boldsymbol{x}}$. Standalone actions are optimal against a null exposure, so are not optimal against $\boldsymbol{x}(0)$ for players who follow someone. Their best responses, denoted by $\boldsymbol{x}(1)$, are larger: $x_i(1) = x_i(0) + \sum_j X_{ij} x_j(0)$ for each i, or $\boldsymbol{x}(1) = \boldsymbol{x}(0) + \boldsymbol{X}\boldsymbol{x}(0)$. But now, $x_i(1)$ is not optimal for i against $\boldsymbol{x}(1)$; let players adjust $\boldsymbol{x}(1)$ to their best responses $\boldsymbol{x}(2) = \boldsymbol{x}(0) + \boldsymbol{X}\boldsymbol{x}(1)$. Iterating best responses, actions $\boldsymbol{x}(t)$ at step t follow the process $\boldsymbol{x}(t) = \boldsymbol{x}(0) + \boldsymbol{X}\boldsymbol{x}(t-1)$. Successive adjustments thus satisfy $\boldsymbol{x}(t) - \boldsymbol{x}(t-1) = \boldsymbol{X}(\boldsymbol{x}(t-1) - \boldsymbol{x}(t-2))$ for $t \geq 1$, which implies $\boldsymbol{x}(t) - \boldsymbol{x}(t-1) = \boldsymbol{X}^{t-1}(\boldsymbol{x}(1) - \boldsymbol{x}(0))$. Since $\boldsymbol{x}(1) - \boldsymbol{x}(0) = \boldsymbol{X}\boldsymbol{x}(0)$, we obtain that the adjustment at t is equal to $\boldsymbol{X}^t\boldsymbol{x}(0)$. Writing $\boldsymbol{x}(t)$ as the sum of the successive adjustments starting from $\boldsymbol{x}(0) = \underline{\boldsymbol{x}}$, we obtain

$$\boldsymbol{x}(t) = [\mathbb{I}_n + \boldsymbol{X} + \boldsymbol{X}^2 + ... + \boldsymbol{X}^t]\underline{\boldsymbol{x}}. \tag{3}$$

When t increases, there are two cases.

Case 1: $\rho < 1$. Perron–Frobenius theorem[5] implies that the sums $\sum_{k=0}^{t} \boldsymbol{X}^k$ converge and their limit is the inverse of $\mathbb{I}_n - \boldsymbol{X}$:

$$(\mathbb{I}_n - \boldsymbol{X})^{-1} = \sum_{k=0}^{\infty} \boldsymbol{X}^k. \tag{4}$$

[4] In Demange [4], I consider best responses that are increasing in exposure. When they are not linear, the conditions for the existence of an equilibrium differ from those for their uniqueness.

[5] According to the theorem, ρ is at least as large as the absolute value of any other eigenvalue. This implies that for $\rho < 1$, the sums $\sum_{k=0}^{t} \boldsymbol{X}^k$ converge when t increases; their limit is the inverse of $\mathbb{I}_n - \boldsymbol{X}$ by taking the limit in the identity $(\mathbb{I}_n - \boldsymbol{X})(\sum_{k=0}^{t} \boldsymbol{X}^k) = \mathbb{I}_n - \boldsymbol{X}^{t+1}$.

Taking the limit in (3) proves (2).

Case 2: $\rho \geq 1$. The iterated best responses $x(t)$ become increasingly large because $X^t \underline{x}$ does not tend to zero when t increases. To show this, consider the transpose of X, denoted by $\widetilde{X}$. Matrix $\widetilde{X}$ is a nonnegative matrix with the same eigenvalues as X. Hence, $\widetilde{X}$ has a nonnegative dominant eigenvector y associated to ρ: $\widetilde{y}X = \rho\widetilde{y}$. Iterating implies $\widetilde{y}X^t\underline{x} = \rho^t\widetilde{y}\underline{x}$. Since $\underline{x}$ is positive, $\widetilde{y}\underline{x}$ is positive. This implies that $\widetilde{y}X^t\underline{x}$ either stays constant and positive (for $\rho = 1$) or increases indefinitely (for $\rho > 1$).[6] Whatever case, $X^t\underline{x}$ does not go to zero and iterated best responses become increasingly large when t increases. □

Remark For general games, an equilibrium is quite demanding in terms of computational point of view (as fixed points are) and players' rationality. As the argument is only based on best responses, the analysis applies also to a behavioral setting, in which agents' behaviors are described by exogenous reaction functions that are linear and increasing in others' actions.

3 Aggregate Activity and Visibility

Aggregate activity in our model is measured by the sum of the actions $\sum_i x_i$. Enhancing aggregate activity[7] is the objective for a social media platform, which gives free access to users and draws its revenues from ads. These revenues are linked to the time users spend on the platform, proxied by aggregate activity. Maximizing revenues is thus achieved by maximizing aggregate activity. This section assumes that the designer can change the visibility of users' actions. Section 3.1 introduces how visibility is modeled and presents a result that illustrates the spillover effects it generates (Proposition 1). Section 3.2 considers the maximization problem of aggregate activity by a designer who can change the visibility of agents' actions and who anticipates the effect of its interventions on the equilibria.

3.1 Visibility Strategy and Supermodularity of Aggregate Activity

A visibility strategy is described by individuals' *multipliers* that multiply their impact. Proposition 1 shows that the increase in aggregate action from increasing the multiplier on a particular individual is larger the larger the others' multipliers are, a property called supermodularity.

---

[6] The result holds even if the sequence X^t/ρ^t does not converge, which may happen when X is not regular.

[7] Enhancing aggregate activity is an objective for a designer in other contexts. For example, a health agency aims to maximize individuals' precautionary efforts, possibly building on peer networks.

Formally, a visibility strategy is described by a nonnegative vector $\boldsymbol{v} = (v_i)$; v_j is called j's multiplier as it has the same effect as multiplying j's impact on each other player by v_j, i.e., each x_{ij} is modified into $x_{ij}v_j$. Equivalently, for each i, i's payoffs are:

$$u_i(\boldsymbol{x}, \boldsymbol{v}) = x_i(\underline{x}_i + \sum_j X_{ij}v_j x_j) - \frac{x_i^2}{2}.$$

Setting a null multiplier on j eliminates j's impact on anyone but j is still impacted by influencers' actions. In an unweighted network (Example 1), a null multiplier on j is akin to delete all links to j but not from j.

Consider equilibria associated to the payoffs transformed by $\boldsymbol{v}$. Since $\boldsymbol{v}$ has the same effect as changing the matrix $\boldsymbol{X}$ into $\boldsymbol{Xdg}(\boldsymbol{v})$, Property 1 applies and yields the following results. Denote by $\rho(\boldsymbol{v})$ the dominant eigenvalue of $\boldsymbol{Xdg}(\boldsymbol{v})$. An equilibrium exists if (and only if) $\rho(\boldsymbol{v}) < 1$, in which case the equilibrium is $\boldsymbol{x}(\boldsymbol{v}) = (\mathbb{I}_n - \boldsymbol{Xdg}(\boldsymbol{v}))^{-1}\underline{\boldsymbol{x}}$. Let V measure aggregate activity given $\boldsymbol{v}$: $V(\boldsymbol{v}) = \sum_i x_i(\boldsymbol{v})$. Observe that lowering the multipliers, the matrix $\boldsymbol{Xdg}(\boldsymbol{v})$ is (elementwise) decreased hence its dominant eigenvalue is decreased as well. Thus $\rho(\boldsymbol{v}) < 1$ implies that, for any $\boldsymbol{v}' \leq \boldsymbol{v}$, an equilibrium exists and V is well defined. The next proposition states the supermodularity of V.

Proposition 1 *Let the dominant eigenvalue $\rho(\boldsymbol{v})$ of $\boldsymbol{Xdg}(\boldsymbol{v})$ be strictly less than 1. $\boldsymbol{v} \geq \boldsymbol{v}'$ implies for each i*

$$V(\boldsymbol{v}_{-i}, v_i) - V(\boldsymbol{v}_{-i}, v_i') \geq V(\boldsymbol{v}'_{-i}, v_i) - V(\boldsymbol{v}'_{-i}, v_i'). \quad (5)$$

Proof Let us first prove the following expression for the incremental value:

$$V(\boldsymbol{v}_{-i}, v_i) - V(\boldsymbol{v}_{-i}, v_i') = \widetilde{\mathbb{1}}(\mathbb{I}_n - \boldsymbol{Xdg}(\boldsymbol{v}_{-i}, v_i))^{-1}\boldsymbol{Xdg}(\boldsymbol{0}_{-i}, v_i - v_i')(\mathbb{I}_n - \boldsymbol{Xdg}(\boldsymbol{v}_{-i}, v_i'))^{-1}\underline{\boldsymbol{x}}, \quad (6)$$

where $\mathbf{0_{-i}}$ is the null ob $N - i$. Writing $V(\boldsymbol{v}) = \widetilde{\mathbb{1}}(\mathbb{I}_n - \boldsymbol{Xdg}(\boldsymbol{v}))^{-1}\underline{\boldsymbol{x}}$, we have

$$V(\boldsymbol{v}_{-i}, v_i) - V(\boldsymbol{v}_{-i}, v_i') = \widetilde{\mathbb{1}}\big[(\mathbb{I}_n - \boldsymbol{Xdg}(\boldsymbol{v}_{-i}, v_i))^{-1} - (\mathbb{I}_n - \boldsymbol{Xdg}(\boldsymbol{v}_{-i}, v_i'))^{-1}\big]\underline{\boldsymbol{x}}. \quad (7)$$

Rewrite the matrix inside the square brackets by using the following computation. Letting $\boldsymbol{A} = \boldsymbol{Xdg}(\boldsymbol{v}_{-i}, v_i)$ and $\boldsymbol{A}' = \boldsymbol{Xdg}(\boldsymbol{v}_{-i}, v_i')$:

$$\begin{aligned}(\mathbb{I}_n - \boldsymbol{A})^{-1} - (\mathbb{I}_n - \boldsymbol{A}')^{-1} &= (\mathbb{I}_n - \boldsymbol{A})^{-1}[(\mathbb{I}_n - \boldsymbol{A}') - (\mathbb{I}_n - \boldsymbol{A})](\mathbb{I}_n - \boldsymbol{A}')^{-1} \\ &= (\mathbb{I}_n - \boldsymbol{A})^{-1}(\boldsymbol{A} - \boldsymbol{A}')(\mathbb{I}_n - \boldsymbol{A}')^{-1}.\end{aligned}$$

We obtain

$$(\mathbb{I}_n - \boldsymbol{X}\boldsymbol{dg}(\boldsymbol{v}_{-i}, v_i))^{-1} - (\mathbb{I}_n - \boldsymbol{X}\boldsymbol{dg}(\boldsymbol{v}_{-i}, v_i'))^{-1}$$
$$= (\mathbb{I}_n - \boldsymbol{X}\boldsymbol{dg}(\boldsymbol{v}_{-i}, v_i))^{-1}\boldsymbol{X}(\boldsymbol{dg}(\boldsymbol{v}_{-i}, v_i) - \boldsymbol{dg}(\boldsymbol{v}_{-i}, v_i'))(\mathbb{I}_n - \boldsymbol{X}\boldsymbol{dg}(\boldsymbol{v}_{-i}, v_i'))^{-1}.$$

Now $\boldsymbol{dg}(\boldsymbol{v}_{-i}, v_i) - \boldsymbol{dg}(\boldsymbol{v}_{-i}, v_i')$ is the diagonal matrix whose elements are all null except the i-th diagonal term which is positive equal to the $v_i - v_i'$. Using (7), this proves (6).

To end the proof, observe that the incremental value $V(\boldsymbol{v}_{-i}, v_i) - V(\boldsymbol{v}_{-i}, v_i')$ is obtained by multiplying the product of the three matrices in (6) by the positive vectors $\mathbb{1}$ (on the left) and $\underline{\boldsymbol{x}}$ (on the right). For $v_i \geq v_i'$ the matrix $X\boldsymbol{dg}(\boldsymbol{0}, v_i - v_i')$ has all its elements nonnegative. The two matrices on the left and the right in (6) are nonnegative. To prove the inequality (5), it thus suffices to show that the two positive matrices on the left and the right in (6) are (weakly) decreased elementwise when $\boldsymbol{v}_{-i}$ is decreased to $\boldsymbol{v}'_{-i}$. This holds true since $(\mathbb{I}_n - \boldsymbol{X}\boldsymbol{dg}(\boldsymbol{v}_{-i}, v_i))^{-1} \geq (\mathbb{I}_n - \boldsymbol{X}\boldsymbol{dg}(\boldsymbol{v}'_{-i}, v_i))^{-1}$: Matrix $\boldsymbol{X}\boldsymbol{dg}(\boldsymbol{v}_{-i}, v_i)$ is obtained by multiplying the elements in j's column of $\boldsymbol{X}$ by v_j. Hence $\boldsymbol{v}_{-i} \geq \boldsymbol{v}'_{-i}$ implies $\boldsymbol{X}\boldsymbol{dg}(\boldsymbol{v}_{-i}, v_i) \geq \boldsymbol{X}\boldsymbol{dg}(\boldsymbol{v}'_{-i}, v_i)$. Formula (4) implies the desired inequality:

$(\mathbb{I}_n - \boldsymbol{X}\boldsymbol{dg}(\boldsymbol{v}_{-i}, v_i'))^{-1} \geq (\mathbb{I}_n - \boldsymbol{X}\boldsymbol{dg}(\boldsymbol{v}'_{-i}, v_i'))^{-1}$: same proof. □

3.2 Manager's Visibility Strategies

This section investigates strategies that modify the visibility of individuals' actions with the objective of maximizing aggregate activity. Of course, without constraints, activity can be made infinitely large. I assume here a constraint on the overall time to display actions, as I describe now. For that purpose, it is more convenient to normalize the multipliers and introduce a display factor.

A *visibility* strategy is of the form $(\boldsymbol{w}, \tau)$ where $\boldsymbol{w} = (w_i)$ are nonnegative and sum to n and τ is a positive scalar. Call $\boldsymbol{w}$ the *weights* and τ the *display factor*. $\boldsymbol{w}$ determines the relative visibility of individuals' actions: i's actions are presented w_i/w_j times more than j's ones. Given the display factor τ, j's actions have the same impact on others as if there were multiplied by τw_j. Setting $v_i = \tau w_i$ for each i, strategy $(\boldsymbol{w}, \tau)$ is equivalent to the multipliers $\boldsymbol{v}$ as defined in the previous section. The chosen representation makes the interpretation easier.

The overall time to display actions must be unchanged by the strategies. Specifically, let one unit of action be displayed per unit of time (a normalization). The overall time to display actions $\boldsymbol{x}$ is $\sum_i x_i$. Since strategy $(\boldsymbol{w}, \tau)$ transforms this overall time into $\tau \sum_i w_i x_i$, the time constraint writes

$$\tau \sum_i w_i x_i \leq \sum_i x_i. \tag{8}$$

Basically, given the weights, the overall time to display $\boldsymbol{x}$ is reallocated across individuals.

As explained in the previous section, strategy $(\boldsymbol{w}, \tau)$ transforms the impact matrix $\boldsymbol{X}$ into $\tau \boldsymbol{X}\boldsymbol{dg}(\boldsymbol{w})$, which has dominant eigenvalue $\tau\rho(\boldsymbol{w})$. An equilibrium exists if (and only if) $\tau\rho(\boldsymbol{w}) < 1$, in which case the equilibrium is unique given by $\boldsymbol{x}(\boldsymbol{w}, \tau) = (\mathbb{I}_n - \tau \boldsymbol{X}\boldsymbol{dg}(\boldsymbol{w}))^{-1}\underline{\boldsymbol{x}}$.

The designer objective is to find a visibility strategy that maximizes aggregate activity over all strategies that satisfy the time constraint. The designer has full information on the game (i.e., on $\boldsymbol{X}$ and $\underline{\boldsymbol{x}}$) and anticipates how interventions affect the equilibria. This leads to the following definitions of feasible and optimal strategies.

Definition 2 A visibility strategy $(\boldsymbol{w}, \tau)$ is said to be *feasible* if an equilibrium exists, i.e., if $\tau\rho(\boldsymbol{w}) < 1$, and the equilibrium $\boldsymbol{x}(\boldsymbol{w}, \tau)$ satisfies the following constraint:

$$\tau \sum_i w_i x_i(\boldsymbol{w}, \tau) \leq \sum_i x_i(\boldsymbol{w}, \tau). \tag{9}$$

The feasible strategy $(\boldsymbol{w}, \tau)$ is *optimal* if it maximizes aggregate activity $\sum_i x_i(\boldsymbol{w}, \tau)$ over all the feasible strategies.

Strategy $(\mathbb{1}, 1)$ does not modify the impacts. It is therefore feasible if $\rho(\boldsymbol{X}) < 1$ but typically not optimal, as the next proposition shows. j's impact total is defined as the sum of j's impact values: $X_{+j} = \sum_i X_{ij}$ (it is equal to in-degree in a graph)

Proposition 2 *Let σ denote the maximal impact total over N: $\sigma = \max_j X_{+j}$.*

If $\sigma < 1$, then optimal visibility strategies exist. An optimal strategy $(\boldsymbol{w}, \tau)$ sets positive weights on the individuals whose impact total is maximal, i.e., $w_j > 0$ only if $X_{+j} = \sigma$, and sets the factor τ to the maximal level so that constraint (9) is binding. The aggregate activity is equal to $\frac{\sum_i \underline{x}_i}{1-\sigma}$.

If $\sigma \geq 1$, then aggregate activity can be made arbitrarily large by feasible visibility strategies.

Proof Let us first prove three claims. □

Claim 1 Let $\boldsymbol{w}$ be a weight vector. Assume $\rho(\boldsymbol{w}) > 0$. Then

$$\lim_{\tau \overset{<}{\to} \frac{1}{\rho(\boldsymbol{w})}} \sum_j w_j x_j(\boldsymbol{w}, \tau) = \infty. \tag{10}$$

Proof To prove (10), let $\boldsymbol{y}$ be a nonnegative left eigenvector of $\boldsymbol{X}\boldsymbol{dg}(\boldsymbol{w})$: $\widetilde{\boldsymbol{y}}\boldsymbol{X}\boldsymbol{dg}(\boldsymbol{w}) = \rho(\boldsymbol{w})\widetilde{\boldsymbol{y}}$. Applying (4) and multiplying by $\widetilde{\boldsymbol{y}}$ yields $\widetilde{\boldsymbol{y}}(\mathbb{I}_n - \tau \boldsymbol{X}\boldsymbol{dg}(\boldsymbol{w}))^{-1} = \frac{1}{1-\tau\rho(\boldsymbol{w})}\widetilde{\boldsymbol{y}}$. Multiplying by $\underline{\boldsymbol{x}}$ yields: $\widetilde{\boldsymbol{y}}\boldsymbol{x}(\boldsymbol{w}, \tau) = \frac{1}{1-\tau\rho(\boldsymbol{w})}\widetilde{\boldsymbol{y}}\underline{\boldsymbol{x}}$. As $\widetilde{\boldsymbol{y}}\underline{\boldsymbol{x}} > 0$, this implies $\lim_{\tau \overset{<}{\to} \frac{1}{\rho(\boldsymbol{w})}} \sum_j y_j x_j(\boldsymbol{w}, \tau) = \infty$. Since $\widetilde{\boldsymbol{y}}\boldsymbol{X}\boldsymbol{dg}(\boldsymbol{w}) = \rho(\boldsymbol{w})\widetilde{\boldsymbol{y}}$, $\boldsymbol{y}$ has all its components null outside the support of $\boldsymbol{w}$. Thus $\boldsymbol{y} \leq k\boldsymbol{w}$ for some positive constant k. Equation (10) follows. □

Equation (10) is not necessarily true if $\rho(\boldsymbol{w}) = 0$. For example, set all the weights null except for one agent, say agent 1, the dominant eigenvalue is null and 1's action is equal to the standalone value whatever τ so that (10) fails. Claim 1 extends as follows.

Claim 2 : Let $\boldsymbol{w}$ be a weight vector. By convention set $1/\rho(\boldsymbol{w}) = \infty$ if $\rho(\boldsymbol{w}) = 0$. Then

$$\lim_{\tau \overset{<}{\to} \frac{1}{\rho(\boldsymbol{w})}} \tau \sum_j w_j x_j(\boldsymbol{w}, \tau) = \infty. \tag{11}$$

Proof If $\rho(\boldsymbol{w}) > 0$, (11) follows directly from Claim 1. If $\rho(\boldsymbol{w}) = 0$, it suffices to observe that $x_j(\boldsymbol{w}, \tau) \geq \underline{x}_j$ for each j and standalone actions are positive. □

Claim 3 Let $\boldsymbol{w}$ be a weight vector. The following inequality holds at equilibrium:

$$\sum_i x_i(\boldsymbol{w}, \tau) \leq \sum_i \underline{x}_i + \tau\sigma \sum_j w_j x_j(\boldsymbol{w}, \tau) \tag{12}$$

with an equality for $\boldsymbol{w}$ with support included in $I^{\max}$.

Proof At equilibrium, $\boldsymbol{x}(\boldsymbol{w}, \tau) - \tau \boldsymbol{X} \boldsymbol{dg}(\boldsymbol{w})\boldsymbol{x}(\boldsymbol{w}, \tau) = \underline{\boldsymbol{x}}$ implies:

$$\sum_i x_i(\boldsymbol{w}, \tau) = \textstyle\sum_i \underline{x}_i + \tau \sum_i (\sum_j X_{ij} w_j x_j)$$

$$= \textstyle\sum_i \underline{x}_i + \tau \sum_j [\sum_i X_{ij}] w_j x_j.$$

By definition $\sum_i X_{ij} \leqq \sigma$ for any j with an equality for j in $I^{\max}$: this proves (12). □

To proceed we distinguish two cases according to the value of σ.

Case 1: $\sigma \geq 1$. Let us show that aggregate activity is unbounded at feasible strategies. Consider $\boldsymbol{w}$ with support in $I^{\max}$ and values of τ in $[0, 1/\rho(\boldsymbol{w})[$. (12) is satisfied as an equality. Thus $\sum_i x_i(\boldsymbol{w}, \tau) \geq \tau\sigma(\sum_j w_j x_j(\boldsymbol{w}, \tau))$. Since $\sigma \geq 1$, this implies that strategy $(\boldsymbol{w}, \tau)$ is feasible for any τ in $[0, 1/\rho(\boldsymbol{w})[$. As $\lim_{\tau\to\infty} \tau \sum_j w_j x_j(\boldsymbol{w}, \tau) = \infty$ from Claim 2, it follows that aggregate activity is unbounded at feasible strategies.

Case 2: $\sigma < 1$. Let us first show that given $\boldsymbol{w}$ there is a maximal value of τ for which the strategy $(\boldsymbol{w}, \tau)$ is feasible. The feasibility constraint writes as $\phi(\tau) \geq 0$ where the function ϕ is defined for τ on $[0, 1/\rho(\boldsymbol{w})[$ by: $\phi(\tau) = \sum_i x_i(\boldsymbol{w}, \tau) - \tau \sum w_i x_i(\boldsymbol{w}, \tau)$. When τ is null, the equilibrium is equal to $\underline{\boldsymbol{x}}$ and $\phi(0) = \sum_i \underline{x}_i$ is positive. Let us show $\phi(\tau) < 0$ for large enough τ. By (12), $\phi(\tau) \leq \sum_i \underline{x}_i + (\sigma - 1)\tau \sum_j w_j x_j(\boldsymbol{w}, \tau)$. By Claim 2, $\lim_{\tau \overset{<}{\to} \frac{1}{\rho(\boldsymbol{w})}} \tau \sum_j w_j x_j(\boldsymbol{w}, \tau) = \infty$. Hence $\sigma < 1$ implies $\phi(\tau) < 0$ for τ large enough in $[0, 1/\rho(\boldsymbol{w})[$. Since $\phi(0) > 0$ and ϕ is continuous, this proves that the set of τ in $[0, 1/\rho(\boldsymbol{w})[$ for which $\phi(\tau) \geq 0$ has a maximum τ^* and that $\phi(\tau^*) = 0$, i.e., the feasibility constraint (9) is binding at τ^*.

To conclude the proof, let us show that aggregate activity is maximized at a feasible strategy with support on $I^{\max}$. At a feasible strategy, the display constraint (9) together with (12) implies

$$\sum_i x_i(\boldsymbol{w}, \tau) \le \sum_i \underline{x}_i + \sigma(\sum_j x_j(\boldsymbol{w}, \tau)), \tag{13}$$

with an equality only if both (9) and (12) are satisfied as equality. It follows that aggregate activity is at most equal to $\frac{\sum_i \underline{x}_i}{1-\sigma}$ and strictly less than this upper bound for $\boldsymbol{w}$ with support not included in $I^{\max}$. For $\boldsymbol{w}$ with support on $I^{\max}$, choosing the value τ^* for which (9) is satisfied as an equality, (13) is satisfied as an equality: at $(\boldsymbol{w}, \tau^*)$, aggregate activity is equal to its upper bound $\frac{\sum_i \underline{x}_i}{1-\sigma}$. □

Interestingly, optimal visibility strategies are simple and easy to compute. Information on the standalone actions are not necessary. In an unweighted network (Example 1), it suffices to observe the in-degrees. An optimal strategy makes visible only the actions of individuals with maximal impact total. In Example 2 where each individual shares effort among those impacted, impact totals are all equal, hence interventions have all the same effect on aggregate activity and are useless.[8]

In general, σ is at least as large as the dominant eigenvalue ρ. But the optimality of a strategy is not based on raising the dominant eigenvalue of the transformed matrix to σ. In fact, the dominant eigenvalue is even null for a strategy making a single agent in I^{max} visible, as is surely the case if I^{max} is a singleton. Let this agent be 1. Then $w_1 = n$ and all other weights are null. All the eigenvalues of matrix $\boldsymbol{Xdg}(\boldsymbol{w})$ are null since $(\boldsymbol{Xdg}(\boldsymbol{w}))^2$ is null (recall that X_{11} is null). The equilibrium is easy to compute given τ: player 1 is not impacted by anyone and plays 1's standalone action $x_1 = \underline{x}_1$. The other players are impacted by 1 only: $x_i = \underline{x}_i + n\tau X_{i1}\underline{x}_1$. The maximal value of τ satisfies $\tau n\underline{x}_1 = \sum_i \underline{x}_i + n\sigma\underline{x}_1$. The increase in aggregate activity is due to the display time, not to the spillovers between actions since all the cycles are cut. If there are several agents in I^{max}, other strategies are optimal, making several agents visible and generating spillover between them (if their mutual impacts are positive). The equilibrium follows a simple "hierarchical" structure: compute the equilibrium between agents in I^{max} and derive others' actions by their exposure to these equilibrium actions. The additional activity generated by making several agents interacting between each other and impacting other because of their visibility is factored out by lowering the weights of each and the display unit time.

3.3 Discussion

Different visibility strategies could be considered. For example, let the designer change the relative individual impacts without display time consideration. A strategy is simply described by weights $\boldsymbol{w}$, as τ is fixed equal to 1. Optimal strategies

[8] With equal total sums, $\sigma = \rho$ and $\tilde{\mathbb{1}}\boldsymbol{X} = \sigma\mathbb{1}$. It follows that aggregate activity without intervention is already equal to $\frac{\sum_i \underline{x}_i}{1-\sigma}$.

are difficult to characterize in that setting. The difficulty comes from the fact that aggregate activity is neither concave nor convex in the weights. The supermodularity of V (Proposition 1) implies its convexity in each individual weight but the overall convexity may not hold due to cross derivatives. We cannot use a first-order approach nor conclude that a single agent is made visible (as is the case for a convex V).

Interventions that change standalone actions are much easier to analyze than visibility strategies. The reason is that aggregate activity is linear in $\underline{x}$, given by $\widetilde{\mathbb{1}}(\mathbb{I}_n - X)^{-1}\underline{x}$. In Demange [4], I consider a designer who is endowed with an amount that can be allocated to agents to increase their standalone actions, i.e., their marginal return to actions. An optimal strategy is characterized by centrality indices computed in the direction of influence, i.e., on $\widetilde{X}$, the transpose of X. Specifically, consider the power/prestige indices introduced in sociology by Katz [14] and Bonacich [3] defined by $(\mathbb{I}_n - \widetilde{X})^{-1}\mathbb{1}$.[9] The amount is optimally allocated to the agents whose index is maximal. Optimal strategies are thus very easy to describe and depend only of the positions of the agents in the network. When best responses are not linear in exposure, however, optimal strategies are not so easy to describe as aggregate activity is no longer linear in standalone actions. In that case, optimal strategies depend on the concavity/convexity of the best response functions and the amount to be allocated.

Acknowledgements This is written for the volume "David Gale, Mathematical Economist: Essays in Appreciation on his 100th Birthday" edited by Ali Khan, Nobusumi Sagara, and Alexander Zaslavski. I am grateful to them for this initiative and to the referee for useful comments. I thank the participants of the workshop "Digital Social Networks" at the Fondation des Treilles in Tourtour (France) where I presented part of this work.

References

1. Ballester, C., Calvo Armengol, A., Zenou, Y.: Who's who in networks. wanted: the key player. Econometrica **74**(5), 1403–1417 (2006)
2. Belhaj, M., Bervoets, S., Deroian, F.: Efficient networks in games with local complementarities. Theor. Econ. **11**(1), 357–380 (2016)
3. Bonacich, P.: Power and centrality: a family of measures. Am. J. Sociol. **92**(5), 1170–1182 (1987)
4. Demange, G.: Optimal targeting strategies in a network under complementarities. Games Econom. Behav. **105**, 84–103 (2017)
5. Dodds, P.S., Watts, D.J.: Universal behavior in a generalized model of contagion. Phys. Rev. Lett. **92**(21), 218701 (2004)
6. Domingos, P., Richardson, M.: Mining the network value of customers. In: Proceedings of the Seventh ACM SIGKDD International Conference on Knowledge Discovery and Data Mining, pp. 57–66 (2001)
7. Gale, D.: A theorem on flows in networks. Pac. J. Math. **7**(2), 1073–1082 (1957)
8. Gale, D.: The Theory of Linear Economic Models, 1st edn. McGraw-Hill Book Company, London 1960 (1960). Reprinted University of Chicago Press (1989)

[9] When standalone actions are equal across individuals, actions are proportional to similar indices but computed on X, i.e., in the direction of following.

9. Gale, D., Politof, T.: Substitutes and complements in network flow problems. Discret. Appl. Math. **3**(3), 175–186 (1981)
10. Eckles, D., Kizilcec, R.F., Bakshy, E.: Estimating peer effects in networks with peer encouragement designs. Proc. Natl. Acad. Sci. **113**(27), 7316–7322 (2016)
11. Galeotti, A., Golub, B., Goyal, S.: Targeting interventions in networks. Econometrica **88**(6), 2445–2471 (2020)
12. Halberstam, Y., Knight, B.: Homophily, group size, and the diffusion of political information in social networks: evidence from Twitter. J. Public Econ. **143**, 73–88 (2016)
13. Kwak, H., Lee, C., Park, H., Moon, S.: What is Twitter, a social network or a news media?. In: Proceedings of the 19th International Conference on World wide web, pp. 591–600 (2010)
14. Katz, L.: A new status index derived from sociometric analysis. Psychometrika **18**(1), 39–43 (1953)
15. Kempe, D., Kleinberg, J., Tardos, E.: Maximizing the spread of influence through a social network. In: Proceedings of the Ninth ACM SIGKDD International Conference on Knowledge Discovery and Data Mining, pp. 137–146 (2003)
16. Morris, S.: Contagion. Rev. Econ. Stud. **67**(1), 57–78 (2000)
17. Richardson, M., Domingos, P.: Mining knowledge-sharing sites for viral marketing. In: Proceedings of the Eighth ACM SIGKDD International Conference on Knowledge Discovery and Data Mining, pp. 61–70 (2002)
18. Schelling, T.C.: Models of segregation. Am. Econ. Rev. **59**(2), 488–493 (1969)
19. Shapley, L.S.: On network flow functions. Naval Res. Logist. Quart. **8**(2), 151–158 (1961)
20. Shapley, L.S.: Complements and substitutes in the optimal assignment problem. Naval Res. Logist. Quart. **9**(1), 45–48 (1962)
21. Topkis, D.M.: Equilibrium points in nonzero-sum n-person submodular games. SIAM J. Control. Optim. **17**(6), 773–787 (1979)

Repeated Games with Tail-Measurable Payoffs

János Flesch and Eilon Solan

Abstract We study multiplayer Blackwell games, which are repeated games where the payoff of each player is a bounded and Borel-measurable function of the infinite stream of actions played by the players during the game. These games are an extension of the two-player perfect-information games studied by David Gale and Frank Stewart [14]. Recently, various new ideas have been discovered to study Blackwell games. In this paper, we give an overview of these ideas by proving, in four different ways, that Blackwell games with a finite number of players, finite action sets, and tail-measurable payoffs admit an ε-equilibrium, for all $\varepsilon > 0$.

Keywords Repeated game · Equilibrium · Tail-measurable payoff

1 Introduction

In a seminal paper, David Gale and Frank Stewart [14] studied two-player win-lose games with perfect information and infinite horizon, where player 1 wins if the realized play lies in a given set of plays, and player 2 wins otherwise. They proved that if the winning set of player 1 is open or closed in the product topology, then the game is determined: one of the players has a winning strategy. Along the years this result has been extended to larger classes of winning sets, until Martin [19] proved that the game is determined as soon as the winning set is Borel-measurable. When the winning set is not Borel-measurable, the game is not always determined.

We thank an anonymous referee for his/her valuable comments. Solan acknowledges the support of the Israel Science Foundation, grants #217/17 and #211/22.

J. Flesch
Department of Quantitative Economics, Maastricht University, Maastricht, The Netherlands
e-mail: j.flesch@maastrichtuniversity.nl

E. Solan (✉)
School of Mathematical Sciences, Tel-Aviv University, Tel-Aviv, Israel
e-mail: eilons@tauex.tau.ac.il

M. A. Khan et al. (eds.), *Matching, Dynamics and Games for the Allocation of Resources*, Monographs in Mathematical Economics 7,
https://doi.org/10.1007/978-981-96-6661-4_7

Blackwell [4] studied analogous games where the two players move simultaneously: in every stage both players simultaneously choose actions, and player 1 wins if and only if the play is in some given winning set of plays. These games are termed *Blackwell games*, and in the language of game theory they are repeated games with general payoff functions.[1] Blackwell [4] proved that if the action sets are finite and the winning set is G_δ in the product topology, that is, a countable intersection of open sets, then the game has a value. The restriction to finite action sets is natural, since even in one-shot games with countable action sets, the value need not exist. Along the years, the value existence result has been extended to larger classes of winning sets, until Martin [20] extended it to all Borel-measurable sets, by reducing the existence problem to the determinacy of two-player win-lose alternating-move games. In fact, Martin's [20] result does not require that the outcome is determined by a winning set: as soon as the outcome is a bounded and Borel-measurable function of the play, the value exists, and as a consequence, each player has an ε-optimal strategy for any error-term $\varepsilon > 0$.

As the determinacy of two-player zero-sum games has been established, it is natural to go beyond zero-sum games and ask whether an ε-equilibrium exists in multiplayer Blackwell games with bounded and Borel-measurable payoff functions, for every $\varepsilon > 0$. This is a challenging and largely uncharted question,[2] and only some special classes have been shown to have an ε-equilibrium. These include multiplayer Blackwell games with perfect information (Mertens and Neyman, see Mertens [22]),[3] with discounted payoffs (see, e.g., the survey by Jaśkiewicz and Nowak [16]), with two players and the long-run average payoff (Vieille [27, 28]), and games with classical computer science objectives (see, e.g., Secchi and Sudderth [24], Chatterjee et al. [10], Chatterjee [8], and Bouyer et al. [6]).

Recently, various new ideas have been discovered to study multiplayer Blackwell games when the payoff functions of the players are tail-measurable. The condition of tail-measurability amounts to requiring that the payoffs are independent of the actions chosen in any finite number of stages. In this paper, we give an overview of these ideas, and we do so by proving, in four different ways, that multiplayer Blackwell games with finite action sets, and bounded, Borel-measurable, and tail-measurable payoffs admit an ε-equilibrium, for all $\varepsilon > 0$. The first proof applies to Blackwell games with two players, and the other proofs apply to Blackwell games with finitely many players. The third proof can even be extended to Blackwell games with countably many players. While the first proof is standard, and the second and third have been published, see Ashkenazi-Golan et al. [2, 3], the fourth proof is new.

[1] The qualification "general" refers to the fact that the payoff is not necessarily the discounted sum of stage payoffs or the long-run average of stage payoffs.

[2] Note that multiplayer Blackwell games include stochastic games with deterministic transitions: in these games, the play is in one of the states in each stage, and the next state is a deterministic function of the current state and the action combination chosen in the current stage.

[3] See also Le Roux and Pauly [17]. For the concept of subgame-perfect ε-equilibrium in perfect-information games, we refer to Flesch and Predtetchinski [12], Bruyère et al. [7], and the references therein.

We hope that the variety of approaches taken to prove the result will attract researchers to study the colorful and fascinating area of Blackwell games, and help advance the research in this area.

2 Model and Main Result

Throughout the paper, the set of natural numbers is denoted by $\mathbb{N} = \{1, 2, \ldots\}$.

2.1 The Model

Definition 1 A *Blackwell game* is a triplet $\Gamma = (I, (A_i)_{i\in I}, (f_i)_{i\in I})$, where

- I is a nonempty and finite set of *players*.
- A_i is a nonempty and finite *action set* for player i, for each $i \in I$. Let $A = \prod_{i\in I} A_i$ denote the set of action profiles, and $A^{\mathbb{N}}$ denote the set of *plays*, i.e., infinite streams of action profiles. We endow the set $A^{\mathbb{N}}$ of plays with the product topology $\mathcal{F}(A^{\mathbb{N}})$, where the action set A_i of each player $i \in I$ has its natural discrete topology.
- $f_i : A^{\mathbb{N}} \to \mathbb{R}$ is a bounded and Borel-measurable *payoff function* for player i, for each $i \in I$.

The game is played in discrete time as follows. In each stage $t \in \mathbb{N}$, each player $i \in I$ selects an action $a_i^t \in A_i$, simultaneously with the other players. This induces an action profile $a^t = (a_i^t)_{i\in I}$ in stage t, which is observed by all players. Given the realized play $p = (a^1, a^2, \ldots)$, the payoff to each player $i \in I$ is $f_i(p)$.
Histories. A *history* is a finite sequence of elements of A, including the empty sequence $\emptyset$. The set of histories is denoted by $H = \bigcup_{t=0}^{\infty} A^t$, where A^t is the t-fold product of A. For a play $p \in A^{\mathbb{N}}$ write $p^{\le t}$ to denote the prefix of p of length t.

For two histories $h, h' \in H$ and a play $p \in A^{\mathbb{N}}$, we denote the concatenation of h and h' by hh' and that of h and p by hp.

Each history induces a subgame of Γ. Given $h \in H$, the *subgame* that starts at h is the game $\Gamma^h = (I, (A_i)_{i\in I}, (f_{i,h})_{i\in I})$, where $f_{i,h} = f_i \circ s_h$ and $s_h : A^{\mathbb{N}} \to A^{\mathbb{N}}$ is given by $p \mapsto hp$.

Strategies. A *mixed action* for player i is a probability measure x_i on A_i. The set of mixed actions for player i is denoted by X_i. A *mixed action profile* is a vector of mixed actions $x = (x_i)_{i\in I}$, one for each player. The set of mixed action profiles is denoted by X.

A (behavior) *strategy* of player i is a function $\sigma_i : H \to X_i$. The interpretation is that when player i adopts the strategy σ_i, at each history $h \in H$ player i chooses an action according to the probability measure $\sigma_i(h)$. We denote by Σ_i the set of strategies of player i.

A *strategy profile* is a vector of strategies $\sigma = (\sigma_i)_{i \in I}$, one for each player. We denote by $\Sigma = \prod_{i \in I} \Sigma_i$ the set of strategy profiles.

Definitions for the opponents of a player. For $i \in I$, let $-i$ denote the set of player i's opponents, $I \setminus \{i\}$. When restricting attention to players in $-i$, we can similarly define mixed action profiles $x_{-i} = (x_j)_{j \neq i}$ and strategy profiles $\sigma_{-i} = (\sigma_j)_{j \neq i}$. The corresponding sets are denoted by X_{-i} and Σ_{-i}, respectively.

Expected payoffs. Using Kolmogorov's extension theorem, each strategy profile σ induces a unique probability measure $\mathbf{P}_\sigma$ on $(A^{\mathbb{N}}, \mathcal{F}(A^{\mathbb{N}}))$. Player i's expected payoff under σ is

$$\mathbf{E}_\sigma[f_i] = \int_{A^{\mathbb{N}}} f_i(p)\mathbf{P}_\sigma(dp),$$

and similarly, player i's expected payoff under σ in the subgame starting at a history h is

$$\mathbf{E}_\sigma[f_i \mid h] = \int_{A^{\mathbb{N}}} f_{i,h}(p)\, \mathbf{P}_{\sigma_h}(dp),$$

where $\sigma_h = (\sigma_{j,h})_{j \in I}$ is the strategy profile defined by $\sigma_{j,h}(h') = \sigma_j(hh')$ for each history h' and player j.

Equilibrium. Let $\varepsilon \geq 0$. A strategy profile σ^* is called an ε *-equilibrium*, if for every player $i \in I$ and every strategy $\sigma_i \in \Sigma_i$, we have $\mathbf{E}_{\sigma^*}[f_i] \geq \mathbf{E}_{\sigma_i,\sigma^*_{-i}}[f_i] - \varepsilon$. This means that no player can gain more than ε by a unilateral deviation.

Minmax value and maxmin value. The minmax value (also called the upper value) is the highest payoff that a player can defend against any strategy profile of her opponents. Formally, player i's *minmax value* in the game is defined as

$$\overline{v}_i = \inf_{\sigma_{-i} \in \Sigma_{-i}} \sup_{\sigma_i \in \Sigma_i} \mathbf{E}_{\sigma_i,\sigma_{-i}}[f_i],$$

and similarly, player i's minmax value in the subgame starting at a history h is

$$\overline{v}_i(h) = \inf_{\sigma_{-i} \in \Sigma_{-i}} \sup_{\sigma_i \in \Sigma_i} \mathbf{E}_{\sigma_i,\sigma_{-i}}[f_i \mid h].$$

Note that if σ^* is an ε-equilibrium, then each player i's payoff is at least her minmax value minus ε, i.e., $\mathbf{E}_{\sigma^*}[f_i] \geq \overline{v}_i - \varepsilon$.

The maxmin value (also called the lower value) is the highest payoff that a player can guarantee to receive. Formally, player i's *maxmin value* in the game is defined as

$$\underline{v}_i = \sup_{\sigma_i \in \Sigma_i} \inf_{\sigma_{-i} \in \Sigma_{-i}} \mathbf{E}_{\sigma_i,\sigma_{-i}}[f_i],$$

and similarly, player i's maxmin value in the subgame starting at a history h is

$$\underline{v}_i(h) = \sup_{\sigma_i \in \Sigma_i} \inf_{\sigma_{-i} \in \Sigma_{-i}} \mathbf{E}_{\sigma_i,\sigma_{-i}}[f_i \mid h].$$

It follows by the definitions that $\overline{v}_i \geq \underline{v}_i$ and $\overline{v}_i(h) \geq \underline{v}_i(h)$ for each $h \in H$, and it is known that these inequalities can be strict. In the special case of two-player Blackwell games, by Martin [20], the minmax and the maxmin values are equal: for each player $i \in I = \{1, 2\}$ we have $\overline{v}_i = \underline{v}_i$ and $\overline{v}_i(h) = \underline{v}_i(h)$ for each $h \in H$. In particular, if the two-player Blackwell game is zero sum, i.e., $f_1 = -f_2$, then $\overline{v}_1 = \underline{v}_1 = -\overline{v}_2 = -\underline{v}_2$, and this common quantity is called the *value* of the game.

Tail-measurability. A function $g : A^{\mathbb{N}} \to \mathbb{R}$ is called *tail-measurable*, if whenever two plays $p = (a^1, a^2, \ldots)$ and $p' = (a'^1, a'^2, \ldots)$ satisfy $a^t = a'^t$ for every t sufficiently large, then $g(p) = g(p')$. In other words, g is tail-measurable if changing finitely many coordinates in the play does not influence the value of g. Not every Borel-measurable function is tail-measurable, and not every tail-measurable function is Borel-measurable, see Rosenthal [23] and Blackwell and Diaconis [5]. Various important evaluation functions in the literature are tail-measurable, for example, the limsup and the liminf of the average stage payoffs (e.g., Sorin [26]), and the limsup and the liminf payoffs (e.g., Maitra and Sudderth [18]). Various classical winning conditions in the computer science literature, such as the Büchi, co-Büchi, parity, Streett, and Müller (e.g., Gimbert and Horn [15], and Chatterjee and Henzinger [9]) are also tail-measurable. The discounted payoff (e.g., Shapley [25]) is not tail-measurable.

One important consequence of tail-measurability is that the minmax value and the maxmin value are independent of the history.

Lemma 1 *Consider a Blackwell game Γ and a player $i \in I$. If player i's payoff function f_i is tail-measurable, then $\overline{v}_i = \overline{v}_i(h)$ and $\underline{v}_i = \underline{v}_i(h)$ for every history $h \in H$.*

2.2 Main Theorem

We now present the main theorem of the paper.

Theorem 1 *Let $\Gamma = (I, (A_i)_{i \in I}, (f_i)_{i \in I})$ be a Blackwell game, where the set I of players is finite, the action set A_i of each player $i \in I$ is finite, and the payoff function f_i of each player $i \in I$ is bounded, Borel-measurable, and tail-measurable. Then, the Blackwell game Γ admits an ε-equilibrium, for every $\varepsilon > 0$.*

In the following sections, we will provide four different proofs for Theorem 1. The first proof is standard and it applies to Blackwell games with two players, the second and fourth proofs apply to Blackwell games with finitely many players, while the third proof even applies to Blackwell games with countably many players.

Fix a Blackwell game Γ that satisfies the conditions of Theorem 1, and fix $\varepsilon > 0$ for the rest of the paper.

The following lemma presents a useful sufficient condition for the existence of an ε-equilibrium. This condition is well known, but for completeness we provide a proof.

Lemma 2 *Assume that there is a play $p^* \in A^{\mathbb{N}}$ such that $f_i(p^*) > \overline{v}_i - \varepsilon$ for each player $i \in I$. Then, Γ admits an ε-equilibrium.*

Proof Choose $\delta > 0$ such that for each player $i \in I$,

$$\delta \leq f_i(p^*) - (\overline{v}_i - \varepsilon). \tag{1}$$

By Lemma 1, $\overline{v}_i = \overline{v}_i(h)$ for each history $h \in H$ and each player $i \in I$. Thus, by the definition of the minmax value, for each history $h \in H$ and each player $i \in I$, there exists a strategy profile $\sigma_{-i}^{h,i} \in \Sigma_{-i}$ for player i's opponents such that for each strategy σ_i of player i we have

$$\mathbf{E}_{\sigma_i, \sigma_{-i}^{h,i}}[f_i \mid h] \leq \overline{v}_i + \delta. \tag{2}$$

Intuitively, the strategy profile $\sigma_{-i}^{h,i}$ can be seen as a punishment against player i.

Denote $p^* = (a^{*1}, a^{*2}, \ldots)$. For each player $i \in I$, define a strategy σ_i^* as follows. For any history h at any stage $t \in \mathbb{N}$:

- If, along h, each player played consistently with p^*, that is, $h = (a^{*1}, \ldots, a^{*t-1})$, then $\sigma_i^*(h) := a_i^{*t}$.
- Otherwise, let $t' < t$ be the first stage at which a player, say player j, deviated from p^* along h.[4] Let $h' = (a^{*1}, \ldots, a^{*t'})$, and set $\sigma_i^*(h) := \sigma_i^{h',j}(h)$ for each $i \neq j$, while $\sigma_j^*(h)$ is arbitrary.

Intuitively, the players are supposed to follow the play p^*. If, however, a player deviates from p^*, all other players switch to a strategy profile that ensures that the deviator's expected payoff is at most her minmax value plus δ.

It is straightforward to verify that the strategy profile $\sigma^* = (\sigma_i^*)_{i \in I}$ is an ε-equilibrium. Indeed, σ^* induces the play p^*, and hence under σ^* each player i's expected payoff satisfies

$$\mathbf{E}_{\sigma^*}[f_i] = f_i(p^*) \geq \overline{v}_i - \varepsilon + \delta. \tag{3}$$

Consider now an arbitrary strategy σ_i for some player $i \in I$. Let μ be the probability that the realized play under $(\sigma_i, \sigma_{-i}^*)$ is p^*:

$$\mu := \mathbf{P}_{\sigma_i, \sigma_{-i}^*}(\{p^*\}).$$

Due to the punishment strategies (cf. Eq. (2)), we have

[4] If more than one player deviated from p^* at stage t', then j is the deviator with the smallest index.

$$\mathbf{E}_{\sigma_i,\sigma^*_{-i}}[f_i] \leq \mu \cdot f_i(p^*) + (1-\mu) \cdot (\overline{v}_i + \delta). \tag{4}$$

By Eqs. (1), (3), and (4),

$$\mathbf{E}_{\sigma_i,\sigma^*_{-i}}[f_i] - \mathbf{E}_{\sigma^*}[f_i] \leq (1-\mu) \cdot (\overline{v}_i + \delta - f_i(p^*)) \leq \varepsilon,$$

which completes the proof. □

One implication of Lemma 2 is the following: if there exists a strategy profile that, in each subgame, gives each player a payoff of at least her minmax value minus some $\delta \in [0, \varepsilon)$, then there exists an ε-equilibrium.

Corollary 1 *Assume that there exist a strategy profile σ and some $\delta \in [0, \varepsilon)$ such that $\mathbf{E}_\sigma[f_i \mid h] \geq \overline{v}_i - \delta$ for each player $i \in I$ and each history $h \in H$. Then, Γ admits an ε-equilibrium.*

Proof By Lévy's zero-one law, we have $\mathbf{P}_\sigma(f_i \geq \overline{v}_i - \delta) = 1$ for each player $i \in I$. Since I is finite, there is a play p^* with $f_i(p^*) \geq \overline{v}_i - \delta > \overline{v}_i - \varepsilon$ for each player $i \in I$. Hence, Lemma 2 implies that Γ admits an ε-equilibrium. □

3 First Proof: Subgame-Perfect Maxmin Strategies

In this section, we prove Theorem 1 when there are only two players: $I = \{1, 2\}$. The proof uses strategies, termed *subgame-perfect δ-maxmin strategies*, that ensure that this player's expected payoff is at least her maxmin value minus δ in all subgames.

Because there are only two players, as mentioned earlier, the minmax and the maxmin values are equal for each player: $\overline{v}_i(h) = \underline{v}_i(h)$ for each player $i \in I$ and each history $h \in H$.

Definition 2 Consider a Blackwell game. Let $i \in I$ and $\delta > 0$. A strategy σ_i^δ for player i is *subgame-perfect δ-maxmin* if for every history $h \in H$ and every strategy profile $\sigma_{-i} \in \Sigma_{-i}$ of player i's opponents:

$$\mathbf{E}_{\sigma_i^\delta,\sigma_{-i}}[f_i \mid h] \geq \underline{v}_i(h) - \delta.$$

The following theorem is shown in Mashiah-Yaakovi [21]; see also Flesch et al. [11] and Ashkenazi-Golan et al. [2].

Theorem 2 (Mashiah-Yaakovi [21]) *Consider a Blackwell game. For every $i \in I$ and $\delta > 0$, player i has a subgame-perfect δ-maxmin strategy.*

We can now turn to the proof of Theorem 1 when $|I| = 2$.

Proof (*Proof of Theorem* 1 *when* $|I| = 2$) For each $i \in I = \{1, 2\}$, let σ_i be a subgame-perfect $\frac{\varepsilon}{2}$-maxmin strategy of player i. Then, by Lemma 1, for each player $i \in \{1, 2\}$ and each history $h \in H$

$$\mathbf{E}_{\sigma_1,\sigma_2}[f_i \mid h] \geq \underline{v}_i - \tfrac{\varepsilon}{2}.$$

As $\underline{v}_i = \overline{v}_i$ for each player $i \in \{1, 2\}$, by Corollary 1, the game Γ has an ε-equilibrium. □

Remark 1 By using subgame-perfect δ-maxmin strategies, Flesch and Solan [13] proved a result that is strongly related to Theorem 1 when there are only two players: each two-player stochastic game with finite state and action spaces and bounded, Borel-measurable, and shift-invariant payoffs admits an ε-equilibrium, for all $\varepsilon > 0$. This result allows to have finitely many states, but uses shift-invariance of the payoffs, which is a stronger property than tail-measurability.

Remark 2 The reader may wonder whether the proof works in the presence of more than two players. Unfortunately, the answer is negative. The reason is that when $|I| > 2$, as mentioned earlier, the minmax value of a player can be strictly higher than her maxmin value. Consequently, when a player adopts a subgame-perfect δ-maxmin strategy, she is no longer guaranteed to obtain a payoff of at least her minmax value minus δ. In particular, when all players follow subgame-perfect δ-maxmin strategies, we cannot apply Corollary 1 as in the proof above.

4 Second Proof: Regularity of the Minmax Value

In this section, we prove Theorem 1 using regularity of the minmax value.

Definition 3 Let Y be a topological space, and let $\mathcal{F}(Y)$ denote the Borel sigma-algebra of Y. A function $g : \mathcal{F}(Y) \to \mathbb{R}$ is *inner-regular* if for each $Z \in \mathcal{F}(Y)$ we have

$$g(Z) = \sup\{g(C) : C \subseteq Z,\ C \text{ compact}\}.$$

It is well known that every probability measure defined on the Borel sets of a compact metric space is inner-regular. In particular, for each strategy profile σ, the probability measure $\mathbf{P}_\sigma$ is inner-regular on the set $A^{\mathbb{N}}$ of plays.

It follows from Martin [20] that, given two nonempty and finite action sets A_1 and A_2, the function that assigns to each Borel-measurable set $W \in \mathcal{F}(A^{\mathbb{N}})$ the value of the two-player zero-sum Blackwell game $(\{1, 2\}, (A_1, A_2), (\mathbf{1}_W, -\mathbf{1}_W))$ is inner-regular, where $\mathbf{1}_W$ denotes the characteristic function of the set W. Ashkenazi-Golan et al. [3] extended this result for the minmax value (and for the maxmin value) of multiplayer Blackwell games, and this is stated next.

Theorem 3 (Ashkenazi-Golan et al. [3]) *Let I and $(A_i)_{i\in I}$ be nonempty and finite sets. For each $i \in I$ and $W_i \in \mathcal{F}(A^{\mathbb{N}})$, let $\overline{v}_i(W_i)$ denote the minmax value of player i in the Blackwell game with I as the set of players, A_j as the action set of player j, for each $j \in I$, and $\mathbf{1}_{W_i}$ as the payoff function of player i. Then,*

$$\overline{v}_i(W_i) = \sup\{\overline{v}_i(C_i) \colon C_i \subseteq W_i,\ C_i \text{ compact}\}.$$

We can now turn to the proof of Theorem 1 using regularity of the minmax value. The following proof is given in Ashkenazi-Golan et al. [3].

Proof (*Proof of Theorem* 1 with regularity) In view of Lemma 2, it suffices to find a play p^* such that $f_i(p^*) \geq \overline{v}_i - \frac{\varepsilon}{2}$ for each player $i \in I$.

Let $\Gamma_\varepsilon = (I, (A_i)_{i\in I}, (\mathbf{1}_{W_i})_{i\in I})$ be the auxiliary Blackwell game, where

$$W_i := \{p \in A^{\mathbb{N}} \colon f_i(p) \geq \overline{v}_i - \tfrac{\varepsilon}{2}\}, \qquad \forall i \in I.$$

For every player $i \in I$, we have $\overline{v}_i(W_i) = 1$. Indeed, consider a player $i \in I$, and a strategy profile σ_{-i} of her opponents. It follows from Theorem 2 that, in the game Γ, player i has a strategy σ_i that is an $\frac{\varepsilon}{2}$-best response to σ_{-i} in each subgame. Hence, in particular, we have $\mathbf{E}_{\sigma_i,\sigma_{-i}}[f_i \mid h] \geq \overline{v}_i - \frac{\varepsilon}{2}$ for each history h. By Lévy's zero-one law, the sequence of random variables $(\mathbf{E}_{\sigma_i,\sigma_{-i}}[f_i \mid p^{\leq t}])_{t\in\mathbb{N}}$ converges $\mathbf{P}_{\sigma_i,\sigma_{-i}}$-a.s. to f_i, and hence $\mathbf{P}_{\sigma_i,\sigma_{-i}}(W_i) = 1$. Consequently, $\overline{v}_i(W_i) = 1$ as claimed.

Fix $\delta \in (0, \frac{1}{|I|})$. By Theorem 3, for each player $i \in I$, there is a compact set $C_i \subseteq W_i$ such that $\overline{v}_i(C_i) \geq 1 - \delta$. For every stage $t \in \mathbb{N}$, let D_i^t be the set of plays that agree with some play in C_i in the first t stages:

$$D_i^t = \{p \in A^{\mathbb{N}} \colon p^{\leq t} = q^{\leq t} \text{ for some } q \in C_i\}.$$

Since $D_i^t \supseteq C_i$, we have $\overline{v}_i(D_i^t) \geq 1 - \delta$.

Consider now the auxiliary Blackwell game $\Gamma_{\varepsilon,t} = (I, (A_i)_{i\in I}, (\mathbf{1}_{D_i^t})_{i\in I})$, which is essentially a t-stage game, because the actions beyond stage t do not influence the payoffs. As the game $\Gamma_{\varepsilon,t}$ has finite horizon and the action sets of the players are finite, $\Gamma_{\varepsilon,t}$ has a 0-equilibrium $\sigma^t = (\sigma_i^t)_{i\in I}$. Then, for each player $i \in I$ we have

$$\mathbf{P}_{\sigma^t}(D_i^t) \geq \overline{v}_i(D_i^t) \geq 1 - \delta.$$

Because $\delta < \frac{1}{|I|}$, we necessarily have $\bigcap_{i\in I} D_i^t \neq \emptyset$. Thus, for each $t \in \mathbb{N}$, there is a play $p_{(t)} \in \bigcap_{i\in I} D_i^t$. Because $A^{\mathbb{N}}$ is compact, the sequence $(p_{(t)})_{t=1}^{\infty}$ has an accumulation point p^*, and as the sets C_i are compact, $p^* \in \bigcap_{i\in I} C_i$. Since $C_i \subseteq W_i$ for each player $i \in I$, we have $f_i(p^*) \geq \overline{v}_i - \frac{\varepsilon}{2}$ for each player $i \in I$, as desired. □

5 Third Proof: Decomposition of the Blackwell Game Into Auxiliary One-Shot Games

A deep result of Martin [20] shows that, in each two-player zero-sum Blackwell game, one can associate with each history a certain auxiliary one-shot game, which has the same action spaces as the Blackwell game, such that a player can play well in the Blackwell game by playing well in the one-shot game at each history. In this section, we prove Theorem 1 by using an extension of Martin's result, which was proven in Ashkenazi-Golan et al. [3].

For a function $d : H \to \mathbb{R}$, a player $i \in I$, and a history $h \in H$, consider the following one-shot game $G_i^{d,h}$: Each player $j \in I$ chooses an action $a_j \in A_j$, simultaneously with the other players, which induces an action profile $a = (a_j)_{j \in I}$. Then, player i receives the payoff $d(ha)$, where ha is the extension of the history h with the action profile a at the next stage. The payoff of the other players in $G_i^{d,h}$ is irrelevant. Let $u_i^{d,h}(x)$ denote player i's expected payoff in this one-shot game under a mixed action profile $x \in X$:

$$u_i^{d,h}(x) := \sum_{a \in A} \left[\prod_{j \in I} x_j(a_j) \cdot d(ha) \right],$$

and let $\overline{\mathrm{val}}_i^{d,h}$ denote player i's minmax value in this one-shot game:

$$\overline{\mathrm{val}}_i^{d,h} := \min_{x_{-i} \in X_{-i}} \max_{x_i \in X_i} u_i^{d,h}(x_i, x_{-i}).$$

The following result is an extension of Martin [20].

Theorem 4 (Ashkenazi-Golan et al. [3]) *Consider a Blackwell game. Let $\eta > 0$, and let $i \in I$ be a player. There exists a function $d : H \to \mathbb{R}$ with the following properties:*

1. $\overline{val}_i^{d,h} \geq \overline{v}_i(h) - \eta$, *for every $h \in H$.*
2. *For every strategy profile σ that satisfies*

$$u_i^{d,h}(\sigma(h)) \geq \overline{val}_i^{d,h}, \qquad \forall h \in H,$$

we have

$$\boldsymbol{E}_\sigma[f_i \mid h] \geq \overline{val}_i^{d,h}, \qquad \forall h \in H.$$

In words, the first property means that player i's minmax value $\overline{\mathrm{val}}_i^{d,h}$ in the one-shot game $G_i^{d,h}$, at any history h, is at least her minmax value $\overline{v}_i(h)$ in the Blackwell game minus η. The second property says that if a strategy profile σ, at each history h, gives player i a payoff of at least her minmax value $\overline{\text{val}}_i^{d,h}$ in the one-shot game $G_i^{d,h}$, then this strategy profile σ, in the subgame of the Blackwell game at any history h, gives player i a payoff of at least $\overline{\text{val}}_i^{d,h}$.

Corollary 2 (Ashkenazi-Golan et al. [3]) *In every Blackwell game, for every $\delta > 0$, there exists a strategy profile σ^* such that for every history $h \in H$ and every player $i \in I$,*

$$\boldsymbol{E}_{\sigma^*}[f_i \mid h] \geq \overline{v}_i(h) - \delta.$$

Proof Let $\eta \in (0, \delta)$, and for each player $i \in I$ let $d_i : H \to \mathbb{R}$ be as in Theorem 4.

At each history h consider the following one-shot game: Each player $i \in I$ chooses an action $a_i \in A_i$, simultaneously with the other players, which induces an action profile $a = (a_i)_{i \in I}$. Then, each player $i \in I$ receives the payoff $d_i(ha)$. Since the action sets are finite, this game has an equilibrium $x^h \in X$. Note that x^h gives each player $i \in I$ an expected payoff of at least her minmax value in this one-shot game, i.e., $u_i^{d_i,h}(x^h) \geq \overline{\text{val}}_i^{d_i,h}$ for each player $i \in I$.

Define the strategy profile σ^* by letting $\sigma^*(h) = x^h$ for each history h. By Theorem 4, for each player $i \in I$ we have

$$\mathbf{E}_{\sigma^*}[f_i \mid h] \geq \overline{\text{val}}_i^{d,h} \geq \overline{v}_i(h) - \eta \geq \overline{v}_i(h) - \delta, \qquad \forall h \in H, \tag{5}$$

which completes the proof. □

Proof (*Proof of Theorem* 1 *with decomposing the Blackwell game into auxiliary one-shot games*) In the Blackwell game Γ, let σ^* be a strategy profile as in Corollary 2 for $\delta = \frac{\varepsilon}{2}$. As the payoff functions of Γ are tail-measurable, Lemma 1 implies that $\overline{v}_i = \overline{v}_i(h)$ for each player $i \in I$ and each history $h \in H$. Hence, Corollary 1 implies that Γ has an ε-equilibrium. □

Remark 3 The proof method above can be slightly adjusted to show that Theorem 1 also holds when the number of players is countably infinite, see Ashkenazi-Golan et al. [3]. Since this involves more technical details on the measure-theoretic foundation of the model and the results, we only briefly discuss how the method can be applied.

Theorem 4 extends to the case of countably many players, namely, $I = \mathbb{N}$, when instead of considering the minmax value, we consider the so-called *finitistic minmax value*, where to punish a player, her opponents are restricted to strategies in which at every history only finitely many players play a mixed action. For each player $i \in I$ let d_i be a function as given by Theorem 4 for the finitistic minmax value. Also, for each player $i \in I$ fix an arbitrary action $\widehat{a}_i \in A_i$.

Consider a strategy profile that prescribes to play at each stage $t \in \mathbb{N}$ as follows: Each player $k > t$ plays the action $\widehat{a}_k$. Given these actions, the remaining finitely many players, i.e., players $1, \ldots, t$, play an equilibrium in the one-shot game given by the functions $d_1, \ldots, d_t$.

Thus, each player i plays the action $\widehat{a}_i$ at stages $1, \ldots, i - 1$, and plays according to equilibria in the auxiliary one-shot games at stages $i, i + 1, \ldots$. Since the payoffs are tail-measurable, the first $i-1$ stages do not matter for player i's payoff, and therefore this strategy profile guarantees that player i's expected payoff is at least $\overline{v}_i - \delta$, similar to Eq. (5).

6 Fourth Proof: Detection of Deviations

All the ε-equilibria we constructed so far had a similar structure: the players follow some fixed play that yields all players a payoff at least their minmax value minus ε, and a deviation triggers an indefinite punishment. In this section, we construct an ε-equilibrium that involves mixtures on the equilibrium path.

Alon et al. [1] studied a model where there is a certain *goal*, defined by a set $D \subseteq A^{\mathbb{N}}$, and the players are supposed to follow a given strategy profile σ^* such that $\mathbf{P}_{\sigma^*}(D)$ is close to 1. Suppose that the realized play turns out to be outside D. Assuming that at most one player deviated, the question is whether an outside observer who observes the play can correctly identify the deviator with high probability. As the following result states, the answer is positive.

For every history $h \in H$, denote by $C(h)$ the set of plays that extend h.

Theorem 5 (Alon et al. [1]) *Consider a Blackwell game. Let $\sigma^* \in \Sigma$ be a strategy profile, let $\delta > 0$, and let $D \subseteq A^{\mathbb{N}}$ satisfy $\mathbf{P}_{\sigma^*}(D) > 1 - \delta$. There exists a function $g : D^c \to I$ such that for every $i \in I$ and every $\sigma_i \in \Sigma_i$,*

$$\mathbf{P}_{\sigma_i,\sigma^*_{-i}}(D^c \text{ and } \{g \neq i\}) < 2\sqrt{(|I|-1)\delta}.$$

Moreover, g can be chosen to be constant over $C(h)$, for every $h \in H$ such that $C(h) \subseteq D^c$.

Proof (*Proof of Theorem* 1 *with detection of deviations*) Fix $\delta > 0$ such that

$$\varepsilon > 4\delta + 4 \cdot \left(\sqrt{(|I|-1)\cdot 2\delta} + 2\delta\right) \cdot \max_{i \in I} \|f_i\|_\infty. \tag{6}$$

By Corollary 2, there is a strategy profile $\widehat{\sigma}$ such that

$$\mathbf{E}_{\widehat{\sigma}}[f_i \mid h] \geq \overline{v}_i(h) - \delta = \overline{v}_i - \delta, \qquad \forall i \in I,\ \forall h \in H.$$

Since the payoff functions are bounded, there is $c \in \mathbb{R}^I$ such that $c_i \geq \overline{v}_i - \delta$ for every $i \in I$ and

$$\mathbf{P}_{\widehat{\sigma}}(|f_i - c_i| < \delta, \forall i \in I) > 0,$$

where $f = (f_i)_{i \in I}$. Since f is Borel-measurable, there is a history $h^* \in A^{t^*}$, for some stage $t^* \in \mathbb{N}$, such that

$$\mathbf{P}_{\widehat{\sigma}_{h^*}}(|f_i - c_i| < \delta, \forall i \in I) > 1 - \delta.$$

Denote

$$D^* := \left\{p \in A^{\mathbb{N}} : |f_i(p) - c_i| < \delta, \forall i \in I\right\}.$$

Since every probability measure on $A^{\mathbb{N}}$ is inner-regular, there is a compact subset $D \subseteq D^*$ such that

$$\mathbf{P}_{\widehat{\sigma}_{h^*}}(D) \geq \mathbf{P}_{\widehat{\sigma}_{h^*}}(D^*) - \delta > 1 - 2\delta. \tag{7}$$

Since D is compact, its complement D^c is open, hence it is the union of basic open sets: there is a minimal subset $H^* \subseteq H$ such that

$$D^c = \bigcup\{C(h) : h \in H^*\}.$$

By Theorem 5, there exists a function $g : D^c \to I$ such that for every $i \in I$ and every $\sigma_i \in \Sigma_i$:

$$\mathbf{P}_{\sigma_{i,h^*}, \widehat{\sigma}_{-i,h^*}}(D^c \text{ and } \{g \neq i\}) < \widehat{\varepsilon} := 2\sqrt{(|I| - 1) \cdot 2\delta}. \tag{8}$$

Consider the following strategy profile σ^*:

- In the first t^* stages, the players follow h^*. Any deviation from this history triggers an indefinite punishment at the minmax level; that is, if player i deviated from h^*, then the other players switch to the strategy profile $\sigma_{-i}^{h,i}$ (cf. Eq. (2)), which ensures that player i's payoff is at most $\overline{v}_i + \delta$, where h is the minimal history that is not a prefix of h^*.
- From stage $t^* + 1$ and on, the players follow $\widehat{\sigma}_{h^*}$, i.e., the continuation of the strategy profile $\widehat{\sigma}$ after h^*.
- If the realized history at some stage coincides with some history $h \in H^*$, the players switch to a punishment strategy against player $g(h)$; that is, the players $I \setminus \{g(h)\}$ switch to the strategy profile $\sigma_{-g(h)}^{h,g(h)}$, which ensures that player $g(h)$'s payoff is at most $\overline{v}_{g(h)} + \delta$.

We argue that the strategy profile σ^* is an ε-equilibrium. Indeed, by Eq. (7), under σ^*, a history in H^* is reached with probability at most 2δ. Since the payoff at all plays in D is δ-close to c, we obtain for each player $i \in I$ that

$$\begin{aligned}
\mathbf{E}_{\sigma^*}[f_i] &\geq (1 - 2\delta)(c_i - \delta) - 2\delta\|f_i\|_\infty \\
&= c_i - \delta - 2\delta \cdot (c_i - \delta + \|f_i\|_\infty) \\
&\geq c_i - 2\delta(1 + 2\|f_i\|_\infty).
\end{aligned}$$

Fix now a player $i \in I$ and a strategy $\sigma_i \in \Sigma_i$. Denote by μ the probability that under $(\sigma_i, \sigma_{-i}^*)$ the realized history at some stage lies in H^*. Since once the realized history h lies in H^* player $g(h)$ is punished, since the probability that in this case $g(h) \neq i$ is at most $\widehat{\varepsilon}$ (see Eq. (8)), and since $c_i \geq \overline{v}_i - \delta$, we have

$$\begin{aligned}\mathbf{E}_{\sigma_i,\sigma^*}[f_i] &\leq (1-\mu)(c_i+\delta)+\widehat{\varepsilon}\|f_i\|_\infty+\max\{0,\mu-\widehat{\varepsilon}\}\cdot(\overline{v}_i+\delta)\\ &\leq c_i+2\delta+2\widehat{\varepsilon}\|f_i\|_\infty.\end{aligned}$$

Hence, by Eq. (6) we have $\mathbf{E}_{\sigma_i,\sigma^*}[f_i]-\mathbf{E}_{\sigma^*}[f_i]\leq\varepsilon$, which proves σ^* is an ε-equilibrium.

7 Concluding Remarks

We studied the existence of ε-equilibrium in Blackwell games under the condition that the payoffs are tail-measurable. When the number of players is finite, this condition was only used to obtain that the minmax values of the players are independent of the history (cf. Lemma 1). For countably infinitely many players, however, our proof technique does require tail-measurability in its full generality (cf. Remark 3).

There are many interesting directions for further research. For example, to find conditions under which a Blackwell game with tail-measurable payoffs admits a 0-equilibrium or a subgame-perfect ε-equilibrium, for every $\varepsilon > 0$, or to investigate the existence of ε-equilibrium in Blackwell games without requiring that the payoffs are tail-measurable. A challenging but rather open direction is to investigate Blackwell games when there is an underlying state space (cf. Remark 1).

We hope that our paper will attract researchers to the study of the fascinating area of Blackwell games.

References

1. Alon N., Gunby B., He X., Shmaya E., Solan E.: Identifying the deviator. Ann. Appl. Probab. **34**(5), 4694–4708 (2024)
2. Ashkenazi-Golan, G., Flesch, J., Predtetchinski, A., Solan, E.: Existence of equilibria in repeated games with long-run payoffs. Proc. Natl. Acad. Sci. **119**(11), e2105867119 (2022)
3. Ashkenazi-Golan G., Flesch J., Predtetchinski A., Solan E.: Regularity of the minmax value and equilibria in multiplayer Blackwell games. Isr. J. Math. **266**, 25–67 (2025)
4. Blackwell, D.: Infinite G_δ-games with imperfect information. Appl. Math. **10**, 99–101 (1969)
5. Blackwell, D., Diaconis, P.: A Non-measurable Tail Set. Lecture Notes - Monograph Series vol. 30, pp. 1–5 (1996)
6. Bouyer, P., Brenguier, R., Markey, N., Ummels, M.: Pure Nash equilibria in concurrent deterministic games. Log. Methods Comput. Sci. **11**, 1–72 (2015)
7. Bruyère, V., Le Roux, S., Pauly, A., Raskin, J.F.: On the existence of weak subgame perfect equilibria. Inf. Comput. **276**, 104553 (2021)
8. Chatterjee K.: Two-player nonzero-sum ω-regular games. In: International Conference on Concurrency Theory, pp. 413–427. Berlin, Heidelberg (2005)
9. Chatterjee, K., Henzinger, T.A.: A survey of stochastic ω-regular games. J. Comput. Syst. Sci. **78**, 394–413 (2012)
10. Chatterjee K., Majumdar R., Jurdzinski M.: On Nash equilibria in stochastic games. In: CSL'04, 26–40. LNCS, vol. 3210. Berlin Heidelberg (2004)

11. Flesch, J., Herings, P., Maes, J., Predtetchinski, A.: Subgame maxmin strategies in zero-sum stochastic games with tolerance levels. Dyn. Games Appl. **11**, 704–737 (2021)
12. Flesch, J., Predtetchinski, A.: A characterization of subgame-perfect equilibrium plays in Borel games of perfect information. Math. Oper. Res. **42**, 1162–1179 (2017)
13. Flesch, J., Solan, E.: Equilibrium in two-player stochastic games with shift-invariant payoffs. J. des Mathématiques Pures et Appliquées **179**, 68–122 (2023)
14. Gale, D., Stewart, F.M.: Infinite games with perfect information. Contrib. Theory of Games **2**, 245–266 (1953)
15. Gimbert, H., Horn, F.: Optimal strategies in perfect-information stochastic games with tail winning conditions (2008). arXiv:0811.3978
16. Jaśkiewicz, A., Nowak, A.S.: Non-zero-sum stochastic games. Handbook of Dynamic Game Theory, 1–64 (2018)
17. Le Roux, S., Pauly, A.: Infinite sequential games with real-valued payoffs. In: Proceedings of the Joint Meeting of the Twenty-third EACSL Annual Conference on Computer Science Logic (CSL) and the Twenty-ninth Annual ACM/IEEE Symposium on Logic in Computer Science (LICS), pp. 1–10 (2014)
18. Maitra, A., Sudderth, W.: Borel stochastic games with lim sup payoff. Ann. Prob. 861–885 (1993)
19. Martin, D.A.: Borel determinacy. Ann. Math. **102**, 363–371 (1975)
20. Martin, D.A.: The determinacy of Blackwell games. J. Symb. Logic **63**, 1565–1581 (1998)
21. Mashiah-Yaakovi, A.: Correlated equilibria in stochastic games with Borel measurable payoffs. Dyn. Games Appl. **5**, 120–135 (2015)
22. Mertens, J.F.: Repeated games. In: Proceedings of the International Congress of Mathematicians, 1986 Providence. American Mathematical Society (1987)
23. Rosenthal, J.: Nonmeasurable invariant sets. Am. Math. Mon. **82**(5), 488–491 (1975)
24. Secchi, P., Sudderth, W.D.: Stay-in-a-set games. Int. J. Game Theory **30**, 479–490 (2002)
25. Shapley, L.S.: Stochastic games. Proc. Natl. Acad. Sci. **39**, 1095–1100 (1953)
26. Sorin, S.: Repeated games with complete information. Handbook of Game Theory with Econ. Appl. **1**, 71–107 (1992)
27. Vieille, N.: Two-player stochastic games I: a reduction. Israel J. Math. **119**, 55–91 (2000)
28. Vieille, N.: Two-player stochastic games II: the case of recursive games. Israel J. Math. **119**, 93–126 (2000)

Calibrated Forecasts: The Minimax Proof

Sergiu Hart

Abstract We provide a formal write-up of the simple proof (1995) of the existence of calibrated forecasts by the minimax theorem, which moreover shows that N^3 periods suffice to guarantee a calibration error of at most $1/N$.

Keywords Calibration · Forecasting · Minimax theorem · Simple proof · Calibrated forecasts · Two-person zero-sum game

Consider a weather forecaster who announces each day a probability p that there will be rain tomorrow. The forecaster is said to be *calibrated* if, for each forecast p that is used, the relative frequency of rainy days out of those days in which the forecast was p is equal to p in the long run.

The surprising result of Foster and Vohra [3] is that calibration can be *guaranteed*, no matter what the weather will be. There are various proofs of this result, and there is a large literature on calibration and its uses; see the survey of Olszewski [5] and the more recent paper of Foster and Hart [1].

A simple proof of the existence of calibrated forecasts, based on the *minimax theorem*, was provided by the author in 1995.[1] The basic argument is as follows (see below for details). If the forecaster knew the strategy of the "rainmaker" (which could well be a mixed, i.e., probabilistic, strategy), then the forecaster could clearly get calibrated forecasts by announcing in every period the corresponding known probability of rain. Incorporating this into a finite game (by using a finite grid of forecasts and a finite horizon) yields, by von Neumann's [7] minimax theorem for

[1] At a lecture given by Dean Foster at the Center for Rationality of the Hebrew University of Jerusalem; see Sect. 4, "An argument of Sergiu Hart," in Foster and Vohra [3].

Dedicated with great admiration to David Gale, in commemoration of his 100th birthday. Previous versions: August 2018; October 2021. The author thanks Jérôme Renault for asking about the relation between the calibration error and the number of periods, and Benjy Weiss for providing inequality (2).

S. Hart (✉)
Einstein Institute of Mathematics, Bogen Department of Economics, and Federmann Center for the Study of Rationality, The Hebrew University of Jerusalem, Jerusalem, Israel
e-mail: hart@huji.ac.il

M. A. Khan et al. (eds.), *Matching, Dynamics and Games for the Allocation of Resources*, Monographs in Mathematical Economics 7,
https://doi.org/10.1007/978-981-96-6661-4_8

two-person zero-sum finite games, the existence of a strategy of the forecaster that guarantees calibration against *any* strategy of the rainmaker. This is a striking use of the minimax theorem, since the fact that there is a calibrated *reply* to any given strategy of the rainmaker is clear, whereas the consequence that there is a *single* strategy that is calibrated against *all* strategies of the rainmaker comes as a big surprise.[2]

More formally, consider a two-person finite game where player 1 has m strategies, player 2 has n strategies, and u_{ij} is the payoff when player 1 plays his i-th strategy and player 2 plays his j-th strategy.[3] A mixed strategy x of player 1 is a probability distribution over his pure strategies $\{1, ..., m\}$, i.e., $x = (x_1, ..., x_m)$, where $x_i \geq 0$ for every $i = 1, ..., m$ and $\sum_{i=1}^{m} x_i = 1$; similarly, a mixed strategy y of player 2 is a probability distribution over his pure strategies $\{1, ..., n\}$, i.e., $y = (y_1, ..., y_n)$, where $y_j \geq 0$ for every $j = 1, ..., n$ and $\sum_{j=1}^{n} y_j = 1$. When the two players play the mixed strategies x and y, respectively, the (expected) payoff is $U(x, y) := \sum_{i=1}^{m} \sum_{j=1}^{n} x_i y_j u_{ij}$.

We now state the minimax theorem, formulated in a useful but perhaps less standard way.

Theorem 1 (Minimax) *Assume that the real number v satisfies the following:*
(i) for every mixed strategy y of player 2 there is a mixed strategy $x \equiv x(y)$ of player 1 such that the payoff is at least v (i.e., $U(x(y), y) \geq v$).
Then
(ii) there is a mixed strategy x^* of player 1 that guarantees that the payoff is at least v (i.e., $U(x^*, y) \geq v$ for every mixed strategy y of player 2).

Indeed, the premise (i) says that $\max_x U(x, y) \geq v$ for every y, i.e., $\min_y \max_x U(x, y) \geq v$; since $\max_x \min_y U(x, y) = \min_y \max_x U(x, y)$ by von Neumann's [7] minimax theorem, we get $\max_x \min_y U(x, y) \geq v$, and so, taking x^* to be a maximizer there, $\min_y U(x^*, y) \geq v$, which is the conclusion (ii).[4]

Stated this way, the minimax theorem may look surprising, since from a premise of "for every y there is an x" it gets a conclusion of "there is an x such that for every y," a false logical argument in general (while "every child has a mother" is true, "there is a mother of all children" is not). Nevertheless, the result is correct (and far from trivial) under the assumptions that, first, there are finitely many pure strategies, and second, one uses mixed strategies (the result is easily seen to be false if either one of these assumptions fails[5]).

---

[2] Indeed, Foster and Vohra had a hard time getting their paper published: they got many desk rejections saying that the result "cannot be true" (the technical report came out in 1991, and the published paper only 7 years later).

[3] It does not matter who gets this "payoff" (it could be, say, player 1's payoff); also, the game need not be a zero-sum game, as only one payoff function is considered.

[4] Since (ii) trivially implies (i), the two conditions (i) and (ii) are in fact equivalent. Also, the premise (i) is easily seen to be equivalent to "for every mixed strategy y of player 2 there is a *pure* strategy $i \equiv i(y)$ of player 1 such that $U(i(y), y) \geq v$."

[5] Consider the "choose the higher integer" infinite game, and the "matching pennies" game with pure strategies only.

In the above calibration setup, the premise (i) is that for every strategy of the rainmaker there is a strategy of the forecaster that yields a small calibration score,[6] and the conclusion (ii) is that there is a strategy of the forecaster that yields a small calibration score for every strategy of the rainmaker (apply the minimax theorem, taking as payoff the negative of the calibration score). Let us show how to get a calibration score of, say, 10%. To see that the premise (i) holds, assume that the strategy of the rainmaker is given. We will round each forecast to a multiple of 10% (the finite grid of forecasts is thus 0%, 10%, ..., 100%); therefore, the forecast of, say, 70% is announced when the probability of rain is between 65% and 75%. Assume that this has occurred on a large number of days so that by the law of large numbers (i.e., Chebyshev's inequality) the expected error between expectation and realization is at most 5%; the relative frequency of rain out of these days will then be between 65% − 5% = 60% and 75% + 5% = 80%—i.e., with a calibration error of at most 10%. Since the same holds for every forecast (that is used nonnegligibly often), taking an appropriately large horizon proves the premise (i)—and thus the conclusion (ii).

We now provide a formal write-up of this proof, which moreover shows that an expected calibration error of size ε is guaranteed after $1/\varepsilon^3$ periods.

For each period (day) $t = 1, 2, ...,$ let $a_t \in \{0, 1\}$ be the *weather*, with 1 for rain and 0 for no rain, and let $c_t \in [0, 1]$ be the *forecast*. For convenience, we will let our forecasts lie in the grid $D := \{1/(2N), 3/(2N), ..., (2N-1)/(2N)\}$ for some positive integer N; thus, each point in $[0, 1]$ is within a distance of at most $1/(2N)$ from a point in D (see Remark (e) for the standard $1/N$-grid).

The *calibration score* K_T at time T is computed as follows. For each $d \in D$ let[7]

$$n(d) \equiv n_T(d) := \sum_{t=1}^{T} \mathbf{1}_{c_t=d}$$

be the number of periods in which the forecast was d, and let

$$\overline{a}(d) \equiv \overline{a}_T(d) := \frac{1}{n(d)} \sum_{t=1}^{T} \mathbf{1}_{c_t=d}\, a_t$$

be the (relative) frequency of rain in those $n(d)$ periods; the calibration score K_T is then the average distance between forecasts and rain frequencies, namely,[8]

---

[6] The "calibration score" will be formally defined below, as the average distance between forecasts and relative frequencies (and so being calibrated means that the calibration score is equal to zero).

[7] We write $\mathbf{1}_X$ for the indicator of the event X; thus, $\mathbf{1}_{c_t=d}$ equals 1 if $c_t = d$ and 0 otherwise.

[8] An alternative score averages the squared errors: $\mathcal{K}_T := \sum_{d\in D}(n(d)/T)(\overline{a}(d) - d)^2$. The two scores are essentially equivalent, because $(K_T)^2 \leq \mathcal{K}_T \leq K_T$ (the first inequality is by Jensen's inequality, and the second is by $|\overline{a}(d) - d| \leq 1$, since $\overline{a}(d)$ and d are both in $[0, 1]$).

$$K_T := \sum_{d \in D} \left(\frac{n(d)}{T} \right) |\overline{a}(d) - d| .$$

This setup can be viewed as a finite T-period game in which in every period $t = 1, ..., T$ the rainmaker chooses the weather $a_t \in \{0, 1\}$ and the forecaster chooses the forecast $c_t \in D$, and the payoff is the calibration score K_T. Both players are assumed to have perfect recall of past weather and forecasts (thus allowing for an "adversarial" rainmaker); since the number of periods T and the sets of choices of the players, $\{0, 1\}$ and D, are all finite, the game is a finite game (i.e., each player has finitely many pure strategies).

Theorem 2 (Calibration) *Let $T \geq N^3$. Then there exists a mixed strategy of the forecaster that guarantees that*[9] $\mathbb{E}[K_T] \leq 1/N$ *against any mixed strategy of the rainmaker.*

This follows from the proposition below by applying the minimax theorem to the payoff function $-K_T$.

Proposition 3 *Let $T \geq N^3$. Then for every mixed strategy of the rainmaker there is a strategy of the forecaster such that* $\mathbb{E}[K_T] \leq 1/N$.

Proof Let τ be a mixed strategy of the rainmaker. For every $t \geq 1$ and history $h_{t-1} = (a_1, c_1, ..., a_{t-1}, c_{t-1}) \in (\{0, 1\} \times D)^{t-1}$ of rain and forecasts before time t, let $p_t := \mathbb{P}\left[a_t = 1 | h_{t-1}\right] = \mathbb{E}\left[a_t | h_{t-1}\right]$ be the probability of rain induced by the rainmaker's strategy τ. We then let the forecast c_t after the history h_{t-1} be the rounding of p_t to the grid D, with a fixed tie-breaking rule when p_t is the midpoint of two consecutive points in D; this makes c_t a deterministic function of the history—i.e., c_t is h_{t-1} -measurable—and we always have $|c_t - p_t| \leq 1/(2N)$.

The calibration score K_T can be expressed as

$$K_T = \frac{1}{T} \sum_{d \in D} |G(d)| ,$$

where[10]

$$G(d) := n(d)(\overline{a}(d) - d) = \sum_{t=1}^{T} \mathbf{1}_{c_t = d}(a_t - d) = \sum_{t=1}^{T} \mathbf{1}_{c_t = d}(a_t - c_t)$$

for every $d \in D$. Replacing each c_t with p_t yields the scores

[9] The expectation is over the random choices of the two players.

[10] $G(d)$ is the difference between the actual number of rainy days, $n(d)\overline{a}(d)$, and the predicted number of rainy days, $n(d)d$, in the $n(d)$ days in which the forecast was d; it is referred to as the (total) "gap" in Foster and Hart [1].

$$\widetilde{G}(d) := \sum_{t=1}^{T} \mathbf{1}_{c_t=d}(a_t - p_t) \text{ and}$$
$$\widetilde{K}_T := \frac{1}{T}\sum_{d\in D} \left|\widetilde{G}(d)\right|;$$

since $|c_t - p_t| \le 1/(2N)$ it follows that $|G(d) - \widetilde{G}(d)| \le n(d)/(2N)$ and

$$\left|K_T - \widetilde{K}_T\right| \le \frac{1}{T}\sum_{d\in D}\frac{n(d)}{2N} = \frac{1}{2N} \tag{1}$$

(because $\sum_d n(d) = T$).

We claim that[11]

$$\mathbb{E}\left[\widetilde{G}(d)^2\right] \le \frac{1}{4}\mathbb{E}\left[n(d)\right] \tag{2}$$

for each $d \in D$. Indeed, $\widetilde{G}(d) = \sum_{t=1}^{T} \mathbf{1}_{c_t=d} Z_t$ where $Z_t := a_t - p_t$, for which we have $\mathbb{E}\left[Z_t|h_{t-1}\right] = 0$ (because $p_t = \mathbb{E}\left[a_t|h_{t-1}\right]$) and $\mathbb{E}\left[Z_t^2|h_{t-1}\right] \le 1/4$ (because this is the variance of a Bernoulli random variable, namely, $a_t|h_{t-1}$). Then, for $s < t$ we get

$$\begin{aligned}\mathbb{E}\left[\left(\mathbf{1}_{c_s=d}\, Z_s\right)\cdot\left(\mathbf{1}_{c_t=d}\, Z_t\right)\right] &= \mathbb{E}\left[\mathbb{E}\left[(\mathbf{1}_{c_s=d}\, Z_s)\cdot(\mathbf{1}_{c_t=d}\, Z_t)|h_{t-1}\right]\right]\\ &= \mathbb{E}\left[\mathbf{1}_{c_s=d}\, Z_s \mathbf{1}_{c_t=d}\,\mathbb{E}\left[Z_t|h_{t-1}\right]\right] = 0\end{aligned}$$

(because the random variables c_s, Z_s, and c_t are h_{t-1}-measurable), and for $s = t$ we get

$$\begin{aligned}\mathbb{E}\left[\left(\mathbf{1}_{c_t=d}\, Z_t\right)^2\right] &= \mathbb{E}\left[\mathbb{E}\left[\left(\mathbf{1}_{c_t=d}\, Z_t\right)^2 |h_{t-1}\right]\right]\\ &= \mathbb{E}\left[\mathbf{1}_{c_t=d}\,\mathbb{E}\left[Z_t^2|h_{t-1}\right]\right] \le \frac{1}{4}\mathbb{E}\left[\mathbf{1}_{c_t=d}\right];\end{aligned}$$

summing all these terms yields $\mathbb{E}\left[\widetilde{G}(d)^2\right] \le (1/4)\sum_{t=1}^{T}\mathbb{E}\left[\mathbf{1}_{c_t=d}\right] = (1/4)\mathbb{E}\left[n(d)\right]$, which is (2).

Therefore,

$$\begin{aligned}\mathbb{E}\left[\widetilde{K}_T\right] &= \frac{1}{T}\sum_{d\in D}\mathbb{E}\left[\left|\widetilde{G}(d)\right|\right] \le \frac{1}{T}\frac{1}{2}\sum_{d\in D}\left(\mathbb{E}\left[n(d)\right]\right)^{1/2}\\ &\le \frac{1}{T}\frac{1}{2}\left(N\sum_{d\in D}\mathbb{E}\left[n(d)\right]\right)^{1/2} = \frac{1}{2}\left(\frac{N}{T}\right)^{1/2},\end{aligned} \tag{3}$$

[11] If one does not care about the bound N^3 on T one may use at this point various simpler Chebyshev or law-of-large-number inequalities (see also Remarks (c) and (d) below).

where we have used $\mathbb{E}\left[|\widetilde{G}(d)|\right] \le \left(\mathbb{E}\left[\widetilde{G}(d)^2\right]\right)^{1/2}$ and (2) for the first inequality, the Cauchy–Schwarz inequality for the second one, and finally $\sum_d \mathbb{E}\left[n(d)\right] = T$. When $T \ge N^3$ this gives $\mathbb{E}\left[\widetilde{K}_T\right] \le 1/(2N)$, and hence $\mathbb{E}\left[K_T\right] \le 1/(2N) + 1/(2N) = 1/N$ by (1).

Remarks. *(a)* Since the game between the rainmaker and the forecaster is a game of perfect recall, by Kuhn's [4] theorem one can replace mixed strategies with their equivalent *behavior* strategies. A behavior strategy of the forecaster, which is referred to as a *forecasting procedure*, consists of a separate randomization after each history, i.e., it is a mapping from the set of histories to the set of probability distributions on D.

(b) N^3 is the right order of magnitude for the horizon T that guarantees a calibration error of $1/N$ when the forecaster rounds the rain probabilities p_t to the grid D, because if the rainmaker chooses p_t to be uniform on [0, 1] then each one of the N forecasts d in D is used about T/N times, and so in order to get an error of $1/N$ one needs T/N to be of the order of N^2.

(c) A tighter estimation in the proof of Proposition 3 uses $\mathbb{E}\left[Z_t^2|h_{t-1}\right] = p_t(1 - p_t)$, which is close to $d(1-d)$, instead of $\mathbb{E}\left[Z_t^2|h_{t-1}\right] \le 1/4$ (recall that $\mathbb{E}\left[Z_t^2|h_{t-1}\right]$ is the variance of a *Bernoulli*(p_t) random variable); this yields $\mathbb{E}\left[K_T\right] \le 1/N$ for T starting approximately at $(2/3)N^3$. More precisely: let $f(d) := d'(1-d')$ where $d' = d + 1/(2N)$ for $d < 1/2$, $d' = d$ for $d = 1/2$, and $d' = d - 1/(2N)$ for $d > 1/2$; then $|p_t - d| \le 1/(2N)$ implies $p_t(1-p_t) \le f(d)$ (because $x(1-x)$ increases for $x < 1/2$ and decreases for $x > 1/2$), and then the coefficient 1/4 in inequality (2) may be replaced with $f(d)$. This yields

$$\mathbb{E}\left[\widetilde{K}_T\right] \le \frac{1}{T}\sum_{d\in D}\left(f(d)\mathbb{E}\left[n(d)\right]\right)^{1/2} \le \frac{1}{T}\left(\sum_{d\in D} f(d)\right)^{1/2}\left(\sum_{d\in D}\mathbb{E}\left[n(d)\right]\right)^{1/2}$$

$$= \frac{1}{T^{1/2}}\left(\sum_{d\in D} f(d)\right)^{1/2}.$$

Now it is a straightforward computation to see that $\sum_{d\in D} f(d) = N/6 + 1/4 - 1/(6N)$, and so for all $T \ge (2/3)N^3 + N^2 - (2/3)N$ we have $\mathbb{E}\left[\widetilde{K}_T\right] \le 1/(2N)$, and thus $\mathbb{E}\left[K_T\right] \le 1/N$.

(d) A looser but slightly simpler estimation in the proof of Proposition 3 that uses $n(d) \le T$ for each d instead of $\sum_d n(d) = T$ yields $\mathbb{E}\left[\widetilde{K}_T\right] \le (1/T)(1/2)NT^{1/2}$, and so $\mathbb{E}\left[K_T\right] \le 1/N$ for $T \ge N^4$.

(e) If instead of D we were to use the standard $1/N$-grid $D' = \{0, 1/N, 2/N, ..., 1\}$ we would need to replace N (the size of D) with $N+1$ (the size of D') in (3), which would yield $\mathbb{E}\left[K_T\right] \le 1/N$ for $T \ge (N+1)N^2 = N^3 + N^2$.

(f) A *lower* bound on the guaranteed calibration error as a function of the number of periods T has recently been obtained by Qiao and Valiant [6]; it is of the order

of $T^{-0.472}$ (improving on the trivial lower bound of the order of $T^{-1/2}$, which is obtained when the rain is an i.i.d. *Bernoulli*$(1/2)$ process; note that what we have shown here is an upper bound of $T^{-1/3}$).

(g) The minimax approach can be further used to obtain calibrated forecasts that are "calibeating," a concept introduced by Foster and Hart [2]: they are guaranteed to beat the Brier score of any other forecast by that forecast's calibration score. See Appendix A.2 of the `arxiv` version of Foster and Hart [2].

(h) The minimax proof does not construct a calibrated procedure; it only shows its existence. There are various specific such constructions in the literature, the simplest being the one in Sect. V of Foster and Hart [1].

References

1. Foster, D.P., Hart, S.: Forecast hedging and calibration. J. Polit. Econ. **129**, 3447–3490 (2021)
2. Foster, D.P., Hart, S.: 'Calibeating': beating forecasters at their own game. Theor. Econ. **18**, 1441–1474 (2023)
3. Foster, D.P., Vohra, R.V.: Asymptotic calibration. Biometrika **85**, 379–390 (1998)
4. Kuhn, H.W.: Extensive games and the problem of information. In: Kuhn, H.W., Tucker, A.W. (eds.) Contributions to the Theory of Games, Vol. II Annals of Mathematics Studies, vol. 28, Princeton, NJ: Princeton University Press, pp. 193–216 (1953)
5. Olszewski, W.: Calibration and expert testing. In: Young H.P., Zamir, S. (eds.) Handbook of Game Theory, vol. 4, Berlin: Springer, pp. 949–984 (2015)
6. Qiao, M., Valiant, G.: Stronger calibration lower bounds via sidestepping. In: Proceedings of the 53rd Annual ACM SIGACT Symposium on Theory of Computing (STOC '21) (2021). https://doi.org/10.1145/3406325.3451050
7. von Neumann, J.: Zur Theorie der Gesellschaftsspiele. Mathematische Annalen **100**, 295–320 (1928)

Getting Permission When Options Are Partially Ordered

Peicong Hu and Joel Sobel

Abstract A manager has access to expert advisers. The manager selects at most one project and can implement it only if one expert provides support. The game in which the manager consults experts simultaneously typically has multiple equilibria including one in which at least one expert supports the manager's favorite project. We describe the set of outcomes that survive iterative deletion of weakly dominated strategies. These outcomes typically exclude the manager's most preferred equilibrium outcome. We introduce sequential procedures and compare their performance to the simultaneous game. In general, sequential consultation may be superior or inferior to simultaneous consultation.

Keywords Expert advice · Sequential decision making · Persuasion

1 Introduction

Hu and Sobel [4] study a problem in which a manager needs permission in order to carry out a plan. Permission can come from one of a finite number of experts. The earlier paper imposed strong assumptions on the set of projects available and on the ways in which the manager can consult the experts. This paper extends the analysis in Hu and Sobel [4] to allow more general sets of projects and more general ways to organize consultation.

We assume that there is a finite set of projects, a finite set of (at least two) experts, and a manager. The experts and manager have preferences over projects. The manager wishes to carry out a project, but can do so only if an expert supports it. Without support, the outcome is a status quo project. We are interested in the relationship

P. Hu (✉)
Faculty of Business and Economics, The University of Hong Kong, Pok Fu Lam, Hong Kong
e-mail: kerihu@hku.hk

J. Sobel
Department of Economics, University of California, San Diego, CA, USA
e-mail: jsobel@ucsd.edu

M. A. Khan et al. (eds.), *Matching, Dynamics and Games for the Allocation of Resources*, Monographs in Mathematical Economics 7,
https://doi.org/10.1007/978-981-96-6661-4_9

between how the manager requests support and the project selected. In one organization, the manager asks experts to report a project that they will support. If no expert supports any project, then the outcome is the status quo. Otherwise, the manager implements the best project consistent with the experts' approvals. If projects are linearly ordered (the case studied in Hu and Sobel [4]) the best project consistent with the agents' approvals is the maximal project supported by an expert. In the more general case we study in this paper, there may be no project that is maximal in the set of approved projects. We assume that the set of projects available is a complete lattice and if two projects are approved then the manager can implement the maximum of these projects. Under these conditions, the best project consistent with the experts' approvals is well defined.

The simultaneous game typically has multiple equilibria, one of which permits the manager to implement his favorite project. To see this, notice that if the first expert supports all projects, then it is a best response for all experts to do so. We view this outcome as implausible and impose a refinement. The refinement, iterative deletion of weakly dominated strategies, works to impose the constraint that experts use strategies that perform well when their recommendation turns out to be pivotal. When projects are linearly ordered and preferences satisfy a generic regularity condition, Hu and Sobel [4] show that there is a unique outcome that satisfies the refinement and that this outcome is the manager's least preferred equilibrium outcome. We obtain weaker results when the regularity condition fails or when projects are not linearly ordered.

We want to know how the manager should organize consultation to maximize his payoff. The result that the simultaneous-move game includes an equilibrium that supports the manager's favorite outcome suggests that the manager achieves his best possible outcome by consulting simultaneously. We believe that the manager-preferred equilibrium is an implausible prediction in many cases, however. We ask whether an alternative organization of consultation could lead to outcomes that the manager prefers to the refined equilibria of the simultaneous-move game. We describe a **canonical sequential protocol** in Sect. 5 and demonstrate a sense in which it performs at least as well as any other sequential protocol. We compare the performance of this protocol to simultaneous consultation. The optimal sequential protocol has a simple form. The manager approaches the first expert and asks her to approve his favorite project. If she does, the process stops. If not, he gives the second expert an opportunity to approve his favorite option. Consultation stops as soon as an expert approves the project. If all experts decline to approve the project, the manager returns to the first expert and asks her to approve his next-best project. If no expert approves any project, the outcome is the status quo. An essential feature of this procedure is that the manager must be able to consult experts more than once.

Hu and Sobel [4] show that properly designed sequential organization does at least as well as, but not better than, simultaneous consultation when projects are linearly ordered and experts have strict preferences over distinct projects. In the multi-dimensional setting where the manager can combine approvals of x and x' to carry out a hybrid job that he strictly prefers to both x and x', the canonical sequential protocol may perform strictly better than the simultaneous-move game.

In the multi-dimensional case, it is also possible that there is an equilibrium of the simultaneous-move game satisfying our refinement that the manager prefers to any equilibrium outcome of a sequential procedure. However the outcome provided by the canonical sequential protocol is at least as attractive to the manager as the worst equilibrium outcome of the simultaneous-move game.

Simultaneous protocols may benefit the manager if he can use the possibility of coordination failure to induce different experts to support different projects and then can combine their support to carry out a hybrid project that he strictly prefers to either of the individual projects. In this kind of situation, the manager needs more than one expert to carry out the equilibrium project. Section 6 discusses how one-dimensional environments differ from higher-dimensional ones.

The canonical sequential protocol permits the manager to restrict the actions of the experts. In Sect. 7, we permit the manager to prevent experts from approving certain projects. We show that limiting the options of the experts is strictly beneficial for the manager in both sequential and simultaneous consultations. With commitment, we show that there is a sequential procedure that does as well as (but no better) than the (refined) equilibrium of the simultaneous-move game.

We organize the remainder of the paper as follows. Section 2 describes the basic model. Section 3 describes an application. Section 4 contains the analysis of the simultaneous-move game. Section 5 contains the analysis of the sequential game. Section 6 compares simultaneous to sequential institutions. Section 7 describes what happens when the manager has the power to restrict the set of projects that experts can support. Section 8 contains concluding remarks. The appendices contain proofs that are not in the main text and supplementary examples.

2 Underlying Strategic Environment

There is a finite set of players, who we call experts. I denotes the finite player set.[1] We assume that there is a finite set $X \subset \mathbb{R}^N$ available to each player. We call elements of $X \subset \mathbb{R}^N$ projects. We assume that X is ordered by the usual $\geq$ relation on $\mathbb{R}^N$. We assume that $(X, \geq)$ is a lattice so that $\min\{z \in X : z \geq x, x'\}$ and $\max\{z \in X : z \leq x, x'\}$ exist for all x and x'. We let $x \vee x' = \min\{z \in X : z \geq x, x'\}$ and $x \wedge x' = \max\{z \in X : z \leq x, x'\}$.

For $\mathbf{x} = (x_1, \ldots, x_L)$,[2] $x_i \in X$ let $M(\mathbf{x}) = x_1 \vee \cdots \vee x_L$.[3] Each expert i has a payoff function $\tilde{u}_i : X^I \to \mathbb{R}$.[4] For each i, we assume that there exists $u_i : X \to \mathbb{R}$

[1] In an abuse of notation, we also let I denote the cardinality of the player set.

[2] We use boldface to denote profiles that consist of actions or strategies of multiple players.

[3] We define M on lists that contain L elements; typically $L = I$, but sometimes we apply the function to the strategies of a proper subset of the players, so $L \leq I$ is possible.

[4] Formally, the payoff function should be defined on strategy profiles. For the simultaneous game we study in Sect. 4, X^I is the set of strategy profiles. When we study sequential games in Sect. 5, the strategy sets are more general, but we still denote payoff functions by $\tilde{u}_i$.

such that $\tilde{u}_i : X^I \to \mathbb{R}$ is defined as $\tilde{u}_i(\mathbf{x}) \equiv u_i(M(\mathbf{x}))$. We denote the minimum element of X by $\underline{x}$ and the maximum element by $\bar{x}$.

We assume each $u_i(\cdot)$ is quasi-supermodular.[5] Quasi-supermodularity is a complementarity assumption that implies, roughly, that increasing one dimension of a project makes increases in another dimension more attractive. It will hold if utility is separable across components, but it will fail if one dimension substitutes for another dimension (and experts strictly prefer intermediate projects). In addition to the experts, for many applications we assume that there is a (non-strategic) manager. The manager has preferences $\succ$[6] defined over X. Assume that if $x \neq x'$, then either $x \succ x'$ or $x' \succ x$. Consequently, $\succeq$ is complete. Further assume that if $x, x' \in X$ and $x \geq x'$, then $x \succeq x'$. When $N > 1$, $\succeq$ does not agree with the usual order $\geq$ on $\mathbb{R}^N$ ($\succeq$ is complete; $\geq$ is not). We interpret $\underline{x}$ as the status quo project. The manager prefers any other project to $\underline{x}$.

We study strategic interactions between the experts in this basic strategic environment. The environment is abstract. There are several ways to interpret the environment. We describe one application in Sect. 3 and others in Hu and Sobel [4].

To gain an understanding of the strategic environment, assume that $X \subset \mathbb{R}_+$ so that $\succ$ is the usual order ("greater than"), and that $\underline{x} = 0$. The manager cannot implement a project different from the status quo without the assistance of at least one expert. We will study strategic environments in which experts announce which projects they support. When offered a variety of projects, the manager will select the maximum (his most preferred project from the set). For this reason, we assume that experts report only a single project and that preferences over profiles $\mathbf{x} \in X^I$ depend only on the maximum of $\mathbf{x}$. That is, we study a reduced form of a game in which the manager is a strategic player who selects his favorite project among those offered by experts. In this environment, there are no natural restrictions on the experts' preferences over X. For example, let $I = 2$ and $X = \{0, .1, \ldots, .9, 1\}$, where x describes a project that generates total surplus x. If the manager cares about total surplus, then he prefers x to x' if and only if $x > x'$. But different projects may distribute the share of the surplus across experts differently. This example suggests how transfers could be compatible with our framework as long as they are included in the description of elements of X. (If all divisions of the surplus were feasible, then they would need to be added to X as additional projects. Our formulation does not permit an environment in which projects are associated with the surplus that they generate, but that once supported, the manager can freely distribute the surplus a project generates. Instead, we require that actual division of surplus also requires approval.)

We close this section with a comment on the interpretation of $\succ$. We assume that the manager has strict preferences over projects and the order represents these preferences.[7] This suggests that if the manager has complete preferences, we can

---

[5] The function $f : X \to \mathbb{R}$ is supermodular if $f(v \vee w) + f(v \wedge w) \geq f(v) + f(w)$ and quasi-supermodular if $f(v) \geq (>) f(v \wedge w)$ implies $f(v \vee w) \geq (>) f(w)$.

[6] For $x, x' \in X$, we write $x \succeq x'$ if $x \succ x'$ or $x \sim x'$. Relations $\prec$, $\nsucc$, and $\nsucceq$ are defined in the standard way.

[7] Allowing the manager to be indifferent between projects adds complexity without insight.

take X to be a completely ordered set. Specifically, we can assume $X \subset \mathbb{R}$ and (by reordering elements of X) that $\succ$ agrees with the usual order on $\mathbb{R}$ ($>$). We wish to extend the analysis to allow for the possibility that $M(\mathbf{x}) > x_i$ for all i. When this happens, the manager can combine support from different experts to carry out a project he prefers to any single project authorized by an expert. Consequently, we want to examine situations in which X is not completely ordered by $\geq$. When we do so, we maintain the following assumptions: $X \subset \mathbb{R}^N$, and the manager's complete preference ordering is monotonic in the sense that $x > x'$ implies that the manager prefers x to x' ($x \succ x'$).

3 Interpretation: Project Choice

We present an example that illustrates the value of studying partially ordered projects. In September 2017, Amazon announced that it was planning to build new headquarters for the company to supplement the main operations center in Seattle. The company claimed that the project would bring investment and employment to the host region and requested bids from different locations. Two-hundred and thirty-eight locations across North America responded to the announcement. Although Amazon originally claimed to be looking to find a single location for its new headquarters, in November 2018 it announced that it would make substantial investments in two areas, Northern Virginia and New York City (Stevens et al. [7]).[8]

Location choices of this kind are common. Firms decide where to place new operations. Movie producers consider several locations for filming. Sports teams periodically move to new cities. Organizations must decide where to hold conferences. These bargaining problems share features of our formal model. The firm looking for a location plays the role of the manager in our model. The experts represent the locations themselves. The firm seeks to leverage competition between locations to obtain a better deal (tax subsidies or reduced regulation). It is intuitive that expanding the number of potential locations is beneficial to the firm. It is not clear how the firm should structure the bidding process. The Amazon example suggests the relevance of a multi-dimensional formulation. Amazon attempted to use competition between locations to combine elements of offers from two different locations.

Location i can only agree to proposals that involve placing the new operation in Location i. For example, Virginia cannot make an agreement that involves placing Amazon's Headquarters in San Francisco. Although our formal model assumes that all locations can agree to any proposal and that it takes only one agreement to carry out a proposal, our model applies with a simple modification. We assume that all locations can propose anything, but that agreements that involve placing the operation at any other location are dominated. This modification (combined with the use of

[8] In February 2019, probably in response to local opposition, Amazon announced that it would not move forward with plans to build new headquarters in New York City (see Goodman [2]).

an equilibrium refinement that restricts the use of dominated strategies) permits our general formulation to apply to location-choice problems.

Quite similar to the location-choice problem is a contracting problem in which a decision-maker seeks to find someone to perform a job. As in the location-choice problem, in order to describe the situation accurately within our framework we must assume that one contractor will not commit other contractors to perform the job. It is also sensible to assume that the decision-maker can divide the job over several contractors, suggesting that a multi-dimensional model is relevant.

In practice, decision-makers structure the negotiation process in a way that combines features of simultaneous and sequential games that we analyze. The decision-maker may ask for locations to submit bids simultaneously, but after receiving initial bids, the decision-maker may refine agreements. Our model suggests that the second round of negotiations need not benefit the decision-maker. We can think of at least two reasons not included in our model why it may be beneficial to go beyond simultaneous bargaining. First, there could be incomplete information. Information revealed in an initial stage may make coordination easier. Second, specifying a complete contract may be costly. There may be efficiency gains to institutions that identify a small set of locations. Once these locations have been identified, they can specify the details of offers more completely.

4 Simultaneous Moves

In this section, we study the game in which each expert simultaneously selects an element in X. If $\mathbf{x} = (x_1, \ldots, x_I)$ is the profile of projects, then Expert i's payoff is $\tilde{u}_i(\mathbf{x}) = u_i(M(\mathbf{x}))$. We interpret the minimum element of X, $\underline{x}$, to be a status quo. So when Expert i wishes to support no project, she uses the strategy $x_i = \underline{x}$.

Section 4.1 points out basic properties of the Nash equilibria of this game. Section 4.2 describes the equilibrium refinement. Section 4.3 states our main characterization result for the simultaneous-move game. Appendix 2 contains examples that further illustrate the result and demonstrates the tightness of our characterization theorem.

4.1 Basic Properties

A profile $\mathbf{x}^* = (x_1^*, \ldots, x_I^*)$ with the property that $u_i(M(\mathbf{x}^*)) \geq u_i(M(\mathbf{x}))$ for all $\mathbf{x}$ such that $M(\mathbf{x}) \geq M(\mathbf{x}^*)$ and all i is a Nash equilibrium profile. If $\mathbf{x}^*$ is a Nash equilibrium, we refer to $M(\mathbf{x}^*)$ as an equilibrium outcome. For any equilibrium profile $\mathbf{x}^*$, a strategy profile $\mathbf{x}$ that satisfies $x_i \leq x_i^*$ and at least two $x_j = M(\mathbf{x}^*)$ is a Nash equilibrium. The manager obtains his most preferred outcome when $M(\mathbf{x}^*)$ is equal to the maximum element in X, $\bar{x}$. Task $\bar{x}$ is always an equilibrium outcome. It can be supported by a strategy profile in which at least two experts play $\bar{x}$. Typically there are other Nash equilibria.

We claim that the pure-strategy Nash equilibria are Pareto ranked from the perspective of the experts when $\geq$ is a complete order. More generally, if x^* and x^{**} are equilibrium outcomes and $x^{**} > x^*$, then all experts prefer x^* to x^{**}. To see this, observe that if any expert preferred the outcome x^{**} to x^*, then she could deviate by using the strategy x^{**} instead of the strategy she uses in the equilibrium that leads to the outcome x^*. Consequently if $\geq$ is complete, the equilibria are Pareto ranked. In this case, the manager's preferences $\succ$ coincide with $>$, so that the manager's preferences (restricted to equilibria) are completely opposed to the (common) preferences of the experts. The equilibria are not necessarily Pareto ranked from the perspective of the experts if projects are partially ordered because in this case it is possible that $x^{**} \succ x^*$ but not $x^{**} > x^*$. Even when $\succ$ does not coincide with $>$, if $\mathbf{x}^*$ and $\mathbf{x}^{**}$ are both Nash equilibria and $M(\mathbf{x}^{**}) \geq M(\mathbf{x}^*)$, then $\tilde{u}_i(\mathbf{x}^*) \geq \tilde{u}_i(\mathbf{x}^{**})$ for all i.

4.2 Weak Dominance

The possibility of multiple equilibria leads us to consider a more restrictive solution concept.

Definition 1 Given subsets $X_i' \subset X$, with $X' = \prod_{i \in I} X_i'$, Expert i's strategy $x_i^* \in X_i'$ is a best response to $\mathbf{x}_{-i} \in X_{-i}'$ relative to X_i if $\tilde{u}_i(x_i^*, \mathbf{x}_{-i}) \geq \tilde{u}_i(x_i, \mathbf{x}_{-i})$ for all $x_i \in X_i$. Expert i's strategy $x_i \in X_i'$ is weakly dominated relative to X' if there exists $x_i' \in X_i'$ such that $\tilde{u}_i(x_i, \mathbf{x}_{-i}) \leq \tilde{u}_i(x_i', \mathbf{x}_{-i})$ for all $\mathbf{x}_{-i} \in X_{-i}'$, with strict inequality for at least one $\mathbf{x}_{-i} \in X_{-i}'$.

Definition 2 The set $S = S_1 \times \cdots \times S_I \subset X$ survives iterated deletion of weakly dominated strategies (IDWDS) if for $m = 0, 1, 2, \ldots$, there are sets $S^m = S_1^m \times \cdots \times S_I^m$, such that $S^0 = X$, $S^m \subset S^{m-1}$ for $m > 0$; S_i^m is obtained by (possibly) removing strategies in S_i^{m-1} that are weakly dominated relative to S^{m-1}; $S^m = S^{m-1}$ if and only if for each i no strategy in S_i^{m-1} is weakly dominated relative to S^{m-1}; and $S_i = \bigcap_{m=1}^{\infty} S_i^m$ for each i.[9]

For finite games, it must be the case that there exists an m such that $S^r = S^m$ for all $r > m$. There are typically many different procedures that are consistent with Definition 2. These procedures may lead to different sets that survive the process. We discuss properties that are common to all sets that survive and give conditions under which all sets that survive lead to the same maximum project. Appendix 2 contains examples that illustrate the possibility that different projects may survive depending on the order of elimination of weakly dominated strategies.

IDWDS is a powerful concept that makes strong demands on the rationality of agents. It is also delicate—the order of deletion matters and it is sometimes poorly

[9] Our notation follows these rules: superscripts denote steps in an iterative process; subscripts denote players; arguments are components.

behaved in games with continuous strategy spaces.[10] Nevertheless, this concept appears appropriate in contexts such as ours where an individual agent's decision is relevant to her own payoff in a limited number of circumstances. Just as in voting models one wants to condition behavior on the event that a voter is pivotal, in our model, one wants to focus attention on circumstances when an expert's strategy is pivotal. Weak dominance arguments capture these strategic circumstances.[11]

We analyze the implications of applying iterated deletion of weakly dominated strategies. Sobel [6] introduces a class of games called WID-supermodular games and describes general properties of strategies that survive the process of iteratively deleting weakly dominated strategies in these games. He shows that if $X \subset \mathbb{R}$, then the simultaneous-move game is a WID-supermodular game. It is straightforward to show that when $u_i(\cdot)$ is quasi-supermodular, the simultaneous-move game is a WID-supermodular game even when X is multi-dimensional.

4.3 Description of Equilibria

This section describes the outcomes that survive IDWDS in the simultaneous-move game. We begin with some general properties of the equilibrium set that follow from quasi-supermodularity. Topkis [8, Theorem 2.72] reports the following fact.

Fact For any sublattice $X' \subset X$, $\arg\max_{z \in X'} u_i(z)$ is a sublattice of X. □

Hence the set of best replies to any pure-strategy forms a sublattice and the smallest best response exists.

Lemma 1 *If π' and π'' are equilibrium outcomes, then $\pi' \wedge \pi''$ is an equilibrium outcome.*

Lemma 1 is obvious when X is completely ordered because in that case $\pi' \wedge \pi''$ is equal to the minimum of π' and π''. In general, the result is a consequence of quasi-supermodularity, which guarantees that if there exists a profitable deviation from $\pi' \wedge \pi''$, then there exists a profitable deviation from π' or π''.

We have observed that the maximal project is an equilibrium outcome. Lemma 1 implies that there is a minimum equilibrium outcome. We next show that the manager prefers every project that survives iterated deletion of weakly dominated strategies, whether it is an equilibrium outcome or not, to the minimum equilibrium outcome. Before we describe the result, we define two quantities π^* and $\tilde{\pi}^*$.

---

[10] In particular, in large games there is no guarantee that there exists a Nash equilibrium in strategies that are not weakly dominated. It is for this reason that we limit attention to finite strategy spaces.

[11] It is possible that alternative solution concepts lead to the same selection as IDWDS. A natural candidate, trembling-hand perfection, is not sufficient. It is not difficult to construct generic examples with multiple trembling-hand perfect equilibria.

Definition 3 The smallest strictly preferred equilibrium outcome is

$$\pi^* = \min\{\pi : u_i(\pi) > u_i(x_i) \text{ for all } x_i > \pi \text{ and all } i\}.$$

Definition 4 The smallest equilibrium outcome is

$$\tilde{\pi}^* = \min\{\pi : u_i(\pi) \geq u_i(x_i) \text{ for all } x_i > \pi \text{ and all } i\}.$$

Lemma 1 implies the existence of $\tilde{\pi}^*$. A small variation of the proof of Lemma 1 (stated and proved in Appendix 1) implies the existence of π^*.

These outcomes are Pareto efficient (from perspective of the experts) in the set of Nash equilibrium payoffs. "Strictly preferred" in the definition of π^* refers to the preferences of the experts (and not those of the manager).

We note several consequences of these definitions. Clearly, $\pi^* \geq \tilde{\pi}^*$. If $u_i(\cdot)$ is one-to-one for each player, then $\pi^* = \tilde{\pi}^*$. We emphasize that π^* and $\tilde{\pi}^*$ differ only in non-generic cases.[12] Any strategy profile $\mathbf{x}$ that satisfies $x_i \leq \pi$ and at least two $x_j = \pi$ is a Nash equilibrium for $\pi = \pi^*$ or $\tilde{\pi}^*$. We will show that $\tilde{\pi}^*$ is a lower bound of the set of outcomes that survive IDWDS, i.e., all outcomes that survive IDWDS are greater than or equal to $\tilde{\pi}^*$.

We need a bit more terminology and notation to state our main result. Recall that $\bar{x}$ is the maximal element of X so that $\bar{x} = (\bar{x}(1), \ldots, \bar{x}(N))$. Let $\bar{x}(-k)$ be the collection of largest components in dimensions other than k.

Definition 5 The bounding project is $\bar{\pi}^* = (\bar{\pi}^*(1), \ldots, \bar{\pi}^*(N))$ where, for each $k = 1, \ldots, N$,

$$\bar{\pi}^*(k) = \min\{\pi(k) : \text{ for all } i, u_i(\pi(k), \bar{x}(-k)) > u_i(x_i(k), \bar{x}(-k)) \text{ whenever } x_i(k) > \pi(k)\}.$$

Lemma 3 (Appendix 1) demonstrates that $\bar{\pi}^*$ is an equilibrium outcome such that $u_i(\bar{\pi}^*) > u_i(x_i)$ for all $x_i > \bar{\pi}^*$ and all i. If X is completely ordered, then $\pi^* = \bar{\pi}^*$. Hence in the generic, one-dimensional case $\tilde{\pi}^* = \pi^* = \bar{\pi}^*$. When X is multi-dimensional, $\bar{\pi}^*$ generally differs from π^*. The outcome π^* has the property that no expert would prefer a strictly higher project. The outcome $\bar{\pi}^*$ imposes the requirement, which Lemma 3 demonstrates is stronger, that expert would prefer to not increase any component of the project whenever the other components are at their highest level. The new definition takes into account the possibility that increasing a component of π may be attractive to an expert if the expert fears that another expert will increase another component.

Example 9 (in Appendix 2) illustrates that it is possible that $\pi^* < \bar{\pi}^*$. Example 9 also provides insight into why the bounding projects may be strictly greater than π^* when X is not one dimensional. Each expert must decide if she is providing "too much" opportunity to the manager dimension by dimension. The test imposed in the

[12] We say that a property is generic if it holds for an open set of Lebesgue measure one and the property is non-generic otherwise. In this paper, genericity always refers to the property that utility functions are one-to-one.

definition of bounding project requires that when an expert considers reducing the kth component of her strategy, she assumes that the manager faces no constraints in other dimensions. It is possible to refine the definition of bounding project (to obtain a weakly lower bound) by replacing $\bar{x}$ by the upper bound of strategies that survive deletion of weakly dominated strategies. We omit this discussion because it adds complexity without much insight.

We can now state the main result of this section.

Proposition 1 *If* $\mathbf{x}$ *is a strategy profile that survives IDWDS in the simultaneous-move game, then* $M(\mathbf{x}) \in [\tilde{\pi}^*, \bar{\pi}^*]$. *For any equilibrium strategy profile* $\mathbf{x}$, $\tilde{u}_i(\mathbf{x}) \geq u_i(\bar{\pi}^*)$ *for all* i.

The proposition bounds the set of outcomes that survive iterative deletion of weakly dominated strategies. The lower bound is the lower bound of the set of Nash equilibria. The upper bound is typically lower than the maximal project. We provide examples to demonstrate that the bounds need not survive IDWDS in Appendix 2, but we are not able to describe situations when the bounds are tight.

To prove the proposition, we first show that there is always an outcome less than or equal to π^* that survives IDWDS. This observation follows because if $x_j \leq \pi^*$ for all $j \neq i$, then Expert i must have an undominated best reply to $\mathbf{x}_{-i}$ that is less than or equal to π^*. Next we show that strategy profiles $\mathbf{x}$ with $M(\mathbf{x}) < \tilde{\pi}^*$ must eventually be eliminated. This argument uses the definition of $\tilde{\pi}^*$ and, in particular, the fact that for any $\tilde{\pi} \ngeq \tilde{\pi}^*$, there must exist an expert i and an $x_i' > \tilde{\pi}$ such that $u_i(x_i') > u_i(\tilde{\pi})$ and we can find a strategy that weakly dominates Expert i's strategy in $\mathbf{x}$. Finally, we show how to delete strategies that are not less than or equal to $\bar{\pi}^*$. Proving this is more involved. The argument involves constructing a strategy that dominates the smallest remaining strategy that is not below $\bar{\pi}^*$. Appendix 1 contains the details.

Sobel [6, Proposition 3] proves this result when $X \subset \mathbb{R}$.

Corollary 1 *If* $\bar{\pi}^* = \tilde{\pi}^*$, *then for all* $\mathbf{x}$ *that survive IDWDS,* $M(\mathbf{x}) = \pi^* = \bar{\pi}^* = \tilde{\pi}^*$.

Corollary 1 follows directly from Proposition 1.

Remark 1 When X is one dimensional, the project that survives is unique if $u_i(\cdot)$ is one-to-one for all i (so that $\pi^* = \tilde{\pi}^*$).

As the examples in Appendix 2 demonstrate, we cannot generally say more than that equilibrium projects must be in the interval $[\tilde{\pi}^*, \bar{\pi}^*]$. In particular, the bounds are not attained in every game.

5 Sequential Protocols

The manager's preferred outcome does not always survive iterative deletion of weakly dominated strategies when experts move simultaneously. This leaves open the question of whether the manager could do better by consulting the experts in a different way. This section discusses the issue. We begin with an example that suggests that simple sequential procedures may perform poorly relative to the simultaneous-move game. We then introduce a family of sequential protocols and describe a simple member of the family that (generically) performs at least as well as any other sequential protocol (from the standpoint of the manager) when X is one dimensional and is undominated by other sequential protocols in general. This protocol generates the same outcome as the simultaneous-move game when X is one dimensional. It may perform better or worse than simultaneous consultation when X is multi-dimensional.

Section 5.1 defines general sequential procedures. Section 5.2 introduces two canonical sequential consultation procedures. Section 5.3 provides lower bounds on the outcomes generated by these protocols. Section 5.4 provides general upper bounds to the performance of sequential procedures. Combined with the results in Sect. 5.3 these show why the manager favors the canonical procedures.

5.1 Preliminaries

In this section, we describe general sequential consultation procedures. In each period t, the procedure selects an expert to make a choice and a choice set for that expert. After a choice, the manager either consults with another expert or stops. Formally, let $H_0 = \emptyset$, $H_t = (I_+ \times X)^t$, where X is the (finite) set of possible projects and $I_+ = \{0\} \cup I$ is the union of "0" and the set of players. Let $H = \bigcup_{t=0}^{T} H_t$ be the set of histories, where T is the number of rounds.[13] If $h_t = (h_t^1, \ldots, h_t^t) \in H_t$ and $h_{t'} = (h_{t'}^1, \ldots, h_{t'}^{t'}) \in H_{t'}$, then $h_t h_{t'} \in H_{t+t'}$ is the history obtained by the natural concatenation: $h_t h_{t'} = (h^1, \ldots, h^t, h^{t+1}, \ldots, h^{t+t'})$ where

$$h^m = \begin{cases} h_t^m & \text{if } 1 \le m \le t \\ h_{t'}^{m-t} & \text{if } t < m \le t + t'. \end{cases}$$

If $h' \in H_{t'}$ and $h'' \in H_{t''}$ for $t'' \ge t'$, we say $h' \subset h''$ if there exists $h \in H_{t''-t'}$ such that $h'' = (h', h)$.

Definition 6 A sequential protocol is a mapping $P = (P_I, P_X) : H \to I_+ \times 2^X$ and $T > 0$ such that for all $h, h_t \in H$,

$$P_I(h_T) = 0 \text{ for all } h_T \in H_T, \tag{1}$$

$$P_I(h_t) = 0 \Longrightarrow P_I(h_t h) = 0, \tag{2}$$

$$P_I(h_t) = 0 \Longrightarrow P_X(h_t) = \emptyset, \tag{3}$$

[13] We will restrict attention to procedures in which the number of consultations is uniformly bounded.

and

$$P_I(h_t) \neq 0 \Longrightarrow \underline{x} \in P_X(h_t). \tag{4}$$

The manager observes a history, h_t. He then decides whether to stop the process ($P_I(h_t) = 0$) or to consult Expert i ($P_I(h_t) = i$). Condition (1) expresses the fact that the protocol must stop after T periods. Condition (2) means that once the decision-maker stops the process, he cannot restart it. $P_X(h_t)$ describes the set of choices available to the expert consulted. Condition (3) states that no choices are available when the process stops. Condition (4) states that when the manager consults an expert, the expert always has the option to support the status quo $\underline{x}$. We limit attention to deterministic protocols that end after a finite number of periods. It is possible to view a simultaneous game as a sequential protocol in which the manager consults each expert once without stopping and experts do not learn the recommendations of other experts. Treating sequential and simultaneous procedures separately permit us to identify the extent to which limiting the information that experts have about the behavior of other experts influences outcomes.

A sequential protocol (we shorten this to "protocol") induces a game in which the players are the experts. Player i's strategy specifies a project as a function of h_t for each h_t such that $P_I(h_t) = i$. Given a history of length t, $h_t = (h_t^1, \ldots, h_t^t)$, let $i_t(h_t) = (i_t^1, \ldots, i_t^t)$ be the list of experts consulted and $p_t(h_t) = (p_t^1, \ldots, p_t^t)$ be the list of projects supported (the projection of h_t onto X^t), and let $\mu(h_t) = p_t^1 \vee \cdots \vee p_t^t$. A strategy profile $\mathbf{s} = (s_1, \ldots, s_I)$ determines projects $\bar{p}_t(\mathbf{s}) = (\bar{p}_t^1(\mathbf{s}), \ldots, \bar{p}_t^t(\mathbf{s}))$ and histories $\bar{h}_t(\mathbf{s})$ for $t = 1, \ldots, T$ where $\bar{h}_1(\mathbf{s}) = (P_I(\emptyset), \bar{p}_1(\mathbf{s})) = (P_I(\emptyset), s_{P_I(\emptyset)}(\emptyset))$, $\bar{p}_2(\mathbf{s}) = (\bar{p}_1(\mathbf{s}), s_{P_I(\bar{h}_1(\mathbf{s}))}(\bar{h}_1(\mathbf{s})))$, $\bar{h}_2(\mathbf{s}) = (\bar{h}_1(\mathbf{s}), (P_I(\bar{h}_1(\mathbf{s})), \bar{p}_2^2(\mathbf{s})))$, and, in general, $\bar{p}_k(\mathbf{s}) = (\bar{p}_{k-1}(\mathbf{s}), s_{P_I(\bar{h}_{k-1}(\mathbf{s}))}(\bar{h}_{k-1}(\mathbf{s})))$, $\bar{h}_k(\mathbf{s}) = (\bar{h}_{k-1}(\mathbf{s}), (P_I(\bar{h}_{k-1}(\mathbf{s})), \bar{p}_k^k(\mathbf{s})))$.[14] Expert i's payoff as a function of the strategy profile is $\tilde{u}_i(\mathbf{s}) = u_i(\mu(\bar{h}_T(\mathbf{s})))$. We say that a project π is generated by a sequential protocol if the induced game has a strategy profile that survives IDWDS in which π is implemented. Formally, π is generated by a sequential protocol if there exists a strategy profile $\mathbf{s}$ that survives IDWDS such that $\pi = p_T^1(\mathbf{s}) \vee \cdots \vee p_T^T(\mathbf{s})$. A project π is uniquely generated by a sequential protocol if π is the only project generated by the protocol.

The specification of the game assumes that the manager implements $\mu(\bar{h}_T(\mathbf{s}))$. One can imagine games in which the manager does not do this. If the manager implements $\gamma(h)$ given the history h, then our specification of the game requires setting $\tilde{u}_i(\mathbf{s}) = u_i(\gamma(\bar{h}_T(\mathbf{s})))$ and a project π would be generated by a sequential protocol if there exists a strategy profile that survives IDWDS such that $\pi = \gamma(\bar{h}_T(\mathbf{s}))$. In fact, in Sect. 6, we consider an alternative in which the manager cannot combine supports from different experts and instead implements his most preferred project from $\{p_T^1(\mathbf{s}), \ldots, p_T^T(\mathbf{s})\}$.

The formal definition of sequential protocols defines P on "too many" histories. The protocol determines which expert to consult next and (possibly) restricts the sets

[14] These formula require a specification of s_0 because it is possible that $P_I(h) = 0$. We set $s_0(h) = \mu(h)$.

of projects that an expert can support at any point. Assuming that the manager can commit to the consultation procedure, we can concentrate on behavior defined on a subset of allowable histories.

Definition 7 A history h_t is allowable if $p_t^{t'}(h_t) \in P_X(h_{t'-1})$ and $i_t^{t'}(h_t) \in P_I(h_{t'-1})$ for $t' = 1, \ldots, t$.

A history is allowable if it specifies projects that are elements in available choice sets and in which the experts are as specified by the protocol. We can restrict strategies to be defined on the set of allowable histories.

Definition 8 A protocol is finite if there exists a finite T^* such that $P_I(h_{T^*}) = 0$ for all allowable $h_{T^*} \in H_{T^*}$.

Definition 9 A protocol is constrained if there exists h_t such that $P_I(h_t) \neq 0$ and $P_X(h_t) \neq X$. A protocol that is not constrained is unconstrained.

A protocol is unconstrained if whenever the manager consults an expert, the expert can support any project. We will show that the ability to constrain choices is valuable to the manager. We do not want to limit attention to unconstrained protocols. We would like to point out a restriction that our procedure will satisfy.

Definition 10 A protocol is neutral if for each history h, i, and $x \succ \mu(h)$, there exists $h' \subset h$ such that $i = P_I(h')$ and $x \in P_X(h')$.

Neutrality prevents the manager from "skipping" an option in a protocol. Suppose that the protocol generates the project π. Neutrality requires that at some point in the consultation process leading to π every agent had an opportunity to support any project the manager prefers to π. In order to implement a non-neutral protocol, the manager must be able to commit to resist the temptation to try to get support for a project he prefers to the one he actually implements. We view neutrality as a restriction on the manager's commitment power. Neutrality guarantees that the protocol does not ignore options.

The manager can gain by using a constrained protocol. In the next section, we describe a protocol that is constrained and neutral. This protocol may perform better than simultaneous consultation in multi-dimensional models. We trace this improved performance to the ability to limit the set of projects that experts can support during a stage of the consultation. We will show how to improve performance of simultaneous protocols by giving the manager additional authority. We also provide conditions under which simultaneous consultation does at least as well as sequential consultation using unconstrained protocols.

The ability to create a protocol assumes that the manager has commitment power. Commitment enters into the analysis in at least three ways. First, we assume that the action that the manager takes is always the best available given the strategy of the experts. By making this assumption, we implicitly assume that the manager cannot commit to implementing a project that he likes less than a project he could implement. Second, the manager might find it attractive to consult experts in an order different from what the protocol recommends or continue consultations when the protocol specifies termination. Third, we assume that the manager can restrict the set of projects that an expert can support.[15] An extreme way to do this would be to exclude some elements of X from every choice set. This strong form of commitment is valuable. If the manager is able to rule out approvals of a particular project at any point in a sequential protocol, then he should be able to rule out that project in a simultaneous-move game. Under some conditions, the manager may have this power. In Sect. 7, we demonstrate that if the manager can limit consultations to a proper subset of projects (or, equivalently, can commit to not implementing some projects even if they receive support), then he can do better than the bounds established for simultaneous games or for what is possible with neutral protocols. For example, if a single expert prefers $\bar{x}$ to $\underline{x}$, then the manager can guarantee his favorite option by refusing to permit experts to support intermediate projects (or by refusing to implement these projects when supported). If the manager can never restrict the choice set of an expert, then unconstrained protocols are the appropriate consultation procedures. Neutral protocols give the manager more commitment power than in unconstrained protocols and less commitment power than in constrained protocols. In the next subsection, we present and study the properties of an intuitive unconstrained protocol and a constrained, but neutral, protocol. These protocols generate the same outcome when projects are one dimensional, but the constrained protocol performs better for the manager in general.

One way to restrict commitment ability is to study sequential games in which the manager can make choices during the consultation procedure. For example, one can imagine an environment in which the manager first selects an expert, offers her a set of projects, and, based on her choice, decides whether to terminate the consultation procedure or continue the consultation procedure. If he terminates, he must select a project. If he continues, he must decide who to consult and how to constrain the expert's choices. Standard notions of sequential rationality (subgame perfection or rationalizability) would impose the restriction that the manager selects the maximum of all projects approved (we make this assumption in the definition of sequential protocols). The power to terminate consultations and restrict choices will be valuable and some projects may not arise as equilibria in a game in which the manager is an active player. We do not pursue the analysis of these extended games for two reasons. First, we believe that it is realistic to assume that managers have some commitment power. It is common to have rules governing consultation procedures.[16] Second, and related, we are interested in protocols that perform well even when the manager does not know the preferences of the experts. If we studied a game in which managers were active players, standard solution concepts would require that preferences are common knowledge. Instead, we study the implications of different restrictions on the set of protocols available.

[15] We always require that the protocol includes the status quo as an option. That is, the manager cannot force an expert to support a project.

[16] Robert's Rules of Order (Vries [9]), which establishes rules governing who can speak and what can be discussed in a meeting is a leading example.

5.2 *The Canonical Sequential Protocols*

This subsection introduces two canonical sequential protocols. The next subsection will explain the importance and limits of these protocols.

Assume that the manager has strict preferences over projects. Specifically, suppose that the projects can be ranked $\bar{x} = \pi_K \succ \pi_{K-1} \succ \cdots \succ \pi_1 = \underline{x}$.

Definition 11 The canonical sequential protocol (CSP) has the properties:

- $T = KI$
  and if $t = mI + r$, for $r = 0, \ldots, I-1$ and $m = 0, \ldots, K-1$, then
- $$P_I(h_t) = \begin{cases} 0 & \text{if } \mu(h_t) \succeq \pi_{K-m} \\ r+1 & \text{otherwise} \end{cases}.$$
- $$P_X(h_t) = \begin{cases} \emptyset & \text{if } P_I(h_t) = 0 \\ \{\pi_1, \pi_{K-m}\} & \text{if } P_I(h_t) \neq 0 \end{cases}.$$

Let us describe the behavior of the canonical sequential protocol after allowable histories. The manager approaches experts in sequence. The first time he approaches (round 0), he asks experts to support a project and stops if any of the experts supports his favorite, π_K. If all of the experts decline to support the best project, then the manager goes back to Expert 1 (round 1) and asks for support of his next most preferred project. The manager continues to consult the experts, stopping after his round $m-1$ consultation with an expert if he has received enough support to implement a project at least as good as the mth best project. The canonical protocol is neutral and constrained.

Definition 12 The unconstrained canonical sequential protocol (CP) has the properties

- $T = KI$
  and if $t = mI + r$, for $r = 0, \ldots, I-1$ and $m = 0, \ldots, K-1$, then
- $$P_I(h_t) = \begin{cases} 0 & \text{if } \mu(h_t) \succeq \pi_{K-m} \\ r+1 & \text{otherwise} \end{cases}.$$
- $$P_X(h_t) = \begin{cases} \emptyset & \text{if } P_I(h_t) = 0 \\ X & \text{if } P_I(h_t) \neq 0 \end{cases}.$$

In the unconstrained canonical sequential protocol (CP), the manager follows CSP, but if the manager consults Expert i, then she can support any project. This means, for example, when the manager first consults Expert 1 in the CP, Expert 1 can support π_{K-1}. The protocol then specifies that the manager will continue to consult with the other experts.

CP and CSP have the same stopping rule. If, in consultation t in round m, one of them supports π_{K-m} (or any project π such that $\pi \vee \mu(h_{t-1}) \succeq \pi_{K-m}$), the consultation stops. Otherwise, the manager continues to consult. Observe that the CSP and CP depend directly on the manager's preferences (that is, the order of projects in the protocol depends on $\succ$). The preferences of the manager do play a role in the simultaneous-move game ($M(\mathbf{x})$ determines payoffs and $M(\mathbf{x})$ depends on $\succ$), but the strategies in the simultaneous-move game do not depend on the manager's preferences.

The next example demonstrates that CSP may perform differently than CP.

Example 1

There are six projects, of the form (i, j) for $i = 1, 2, 3$ and $j = 1, 2$, and two experts whose preferences are

$$u_1(1, 1) > u_1(1, 2) > u_1(3, 1) > u_1(2, 1) > u_1(3, 2) > u_1(2, 2)$$

and

$$u_2(2, 1) > u_2(1, 1) > u_2(2, 2) > u_2(3, 1) > u_2(1, 2) > u_2(3, 2).$$

These preferences satisfy quasi-supermodularity. In this case, $\tilde{\pi}^* = \pi^* = \bar{\pi}^* = (3, 1)$. Consequently, the outcome of the simultaneous-move game is $(3, 1)$. Assume that the manager's preferences are

$$(3, 2) \succ (2, 2) \succ (3, 1) \succ (1, 2) \succ (2, 1) \succ (1, 1).$$

One can verify that the outcome of the CSP is $(3, 2)$. (This will be a consequence of Propositions 3.) Why does the sequential protocol yield a better result for the manager? When played simultaneously, $(3, 2)$ is dominated for Expert 2 and $(2, 1)$ cannot be dominated. Furthermore, Expert 2 will eventually delete $(2, 2)$ and $(1, 2)$. As long as Expert 2 thinks that it is possible that Expert 1 will support $(3, 1)$, Expert 2 will avoid using strategies that support "2" in the second dimension. But once Expert 2 deletes strategies of the form $(x(1), 2)$, Expert 1 will have no reason to support "2" in the second dimension. In the CSP, however, if both experts fail to support $(3, 2)$, then the manager will offer the experts a chance to support $(2, 2)$ without danger of $(3, 2)$ being selected. Expert 2 is willing to support $(2, 2)$ because she understands that if she fails to do so, Expert 1 will support $(3, 1)$. But then Expert 1 will be motivated to support $(3, 2)$ in the first round because she knows that if she does not, then

Expert 2 will support (2, 2) before either expert has the opportunity to support (1, 1) or (2, 1).

The example identifies two differences between sequential and simultaneous procedures. On one hand, (3, 1) is the outcome of the CP. Expert 1 can support (3, 1) in the first round. Expert 2 realizes that if she supports (2, 2) in the second round, then the outcome will be (3, 2). Consequently, this protocol generates the outcome (3, 1), because once (3, 1) has been approved, neither expert would support a project that permits the manager to carry out (3, 2). It is for this reason we say that CSP requires commitment ability. The manager would like to be able to decline to implement (3, 1) following an "impossible" history in which an expert supports (3, 1) in the first round of consultation.

On the other hand, one can obtain (3, 2) in the equilibrium of a simultaneous-move game provided that project implemented given the strategy profile (x_1, x_2) is $\max_{\succ}\{x_1, x_2\}$ (and not $x_1 \vee x_2$, which may be strictly preferred to both x_1 and x_2). That is, the ability to commit not to combine projects may be valuable to the manager. In this case, (3, 2) is the only outcome that survives IDWDS in the simultaneous-move game. In particular, (3, 1) cannot be the outcome because if Export 1 supports (3, 1), Expert 2 will support (2, 2) and the result will be the maximum of (3, 1) and (2, 2) with respect to the manager's preference ($(2, 2) \equiv \max_{\succ}\{(3, 1), (2, 2)\}$) and not $(3, 2) = (3, 1) \vee (2, 2)$).[17]

5.3 *Lower Bounds for Canonical Sequential Protocols*

We describe the performance of (CSP) and (CP) in this section

Proposition 2 *If π is a project that survives IDWDS in the game determined by CP, then $\pi \geq \tilde{\pi}^*$.*

Proposition 2 establishes a lower bound on the manager's outcome for sequential games. For every $\pi < \tilde{\pi}^*$, there will be an expert who strictly prefers a higher project. From this observation, it is straightforward to show that a project π such that $\pi < \tilde{\pi}^*$ cannot be generated by CP. To prove the proposition, we must further show that if $\pi \ngeq \tilde{\pi}^*$, then π cannot be the project induced by a strategy profile that survives IDWDS. We construct a dominating strategy by showing that it is possible to increase the action of one expert at her final turn to move. When $\pi \ngeq \tilde{\pi}^*$, it is possible to find such a deviation. The altered strategy will either not change the outcome or immediately terminate the consultation process at an outcome strictly better for the deviating expert.

[17] (3, 2) would also be the outcome of CP if the manager can commit to implementing $\max_{\succ}\{x_1, x_2\}$ rather than $x_1 \vee x_2$.

To describe the performance of CSP we need two new terms. In the definitions, we write "$\min_\succ$" to denote minimization with respect to the manager's preferences $\succ$.

Definition 13

$$\pi_\succ^* = \min_\succ\{\pi : u_i(\pi) > u_i(x_i) \text{ for all } x_i \succ \pi \text{ and all } i\}.$$

Definition 14

$$\tilde{\pi}_\succ^* = \min_\succ\{\pi : u_i(\pi) \geq u_i(x_i) \text{ for all } x_i \succ \pi \text{ and all } i\}.$$

The quantities $\pi_\succ^*$ ($\tilde{\pi}_\succ^*$) and π^* ($\tilde{\pi}^*$) differ because $>$ does not coincide with $\succ$ when X is not one dimensional. The idea of the new bounds is to use the manager's preferences to provide a complete order over projects even when projects are multi-dimensional. The set of x_i such that $x_i > \pi$ is identical to the set of x_i such that $x_i \succ \pi$ in the one-dimensional case because the manager's preferences are monotonic. In general, $x_i > \pi$ implies $x_i \succ \pi$, but not conversely. Consequently, $\pi_\succ^* \geq \pi^*$ and $\tilde{\pi}_\succ^* \geq \tilde{\pi}^*$, with equality in the one-dimensional case. Furthermore, $\pi_\succ^* \geq \tilde{\pi}_\succ^*$. If $u_i(\cdot)$ is one-to-one for each player, then $\pi_\succ^* = \tilde{\pi}_\succ^*$.

We can provide a lower bound to the performance of CSP similar (but generally higher) than the bound for CP established in Proposition 2.

Proposition 3 *If π is a project that survives IDWDS in the game determined by CSP, then $\pi \succeq \tilde{\pi}_\succ^*$.*

5.4 General Bounds on Sequential Consultation

We have focused on the canonical protocols in the previous subsection. This subsection explains what is and is not possible using general protocols.

Proposition 4 *If $x > \pi^*$, then there exists no unconstrained sequential protocol that generates the project x in a pure-strategy, subgame-perfect equilibrium. If $x \succ \pi_\succ^*$, then there exists no neutral sequential protocol that generates the project x in a pure-strategy, subgame-perfect equilibrium.*

Proposition 4 follows from backward induction. If the protocol is unconstrained, then after each history that supports no more than π^*, it is never a best response to support more than π^*. So if the first expert anticipates that the final project supported will be more than π^*, then she can do strictly better by approving π^* and no one else will add more to π^*. Consequently there will never be projects greater than π^* in equilibrium. Proving the second sentence in the proposition is more involved. Appendix 1 contains the proof; here we indicate the idea behind the proof. Because experts can always support only the status quo, there is a terminal history in which the outcome generated is the status quo. If the protocol is neutral, then every expert has an opportunity to support $\pi^*_{\succ}$. We can show that this means that every expert i is guaranteed a payoff of at least $u_i(\pi^*_{\succ})$, which guarantees that $\pi^*_{\succ} \succeq x$.

Example 3 shows that it is possible to generate projects greater than $\pi^*_{\succ}$ (and hence π^*) using a protocol that is not neutral.

Proposition 5 *For any unconstrained protocol, there is always an outcome that survives IDWDS that is not greater than* π^*.

Proposition 5 follows from backward induction and the observation that if π^* has already been supported, no expert wishes to support a project that would lead to a strictly higher outcome. Hence if an expert supports π^* when it is her turn (and π^* is in her choice set), then she knows that there will exist undominated strategies for future experts that involve no other project being supported.

Sequential protocols generate perfect-information games. There is a close connection between subgame-perfect equilibria in perfect-information games and outcomes that survive IDWDS. So we can use Proposition 4 to obtain a characterization for our solution concept.

Proposition 6 *Fix a sequential protocol. If the experts' utility functions are one-to-one, there is a unique outcome that survives IDWDS.*

Corollary 2 *Assume the experts' utility functions are one-to-one. If the protocol is unconstrained, then no outcome that survives IDWDS is greater than* π^*. *If the protocol is neutral, then the manager prefers no outcome that survives IDWDS to* $\pi^*_{\succ}$.

The first assertion in Corollary 2 is a consequence of Proposition 5, which shows existence of a project that satisfies IDWDS no greater than π^*, and Proposition 6, which shows that there is no other project that survives IDWDS. The second assertion follows from Propositions 4 and 6.

When X is one dimensional, our results illustrate what is possible using neutral sequential protocols and provide a sense in which the canonical protocol is optimal for the manager. In the generic, one-dimensional case, the results of this subsection show that every neutral protocol generates only projects less than or equal to π^*. Combined with Propositions 2 and 3, this means that CP and CSP perform at least as well as any neutral sequential protocol from the manager's point of view. When X has more than one dimension, the results are incomplete in two ways. First, there is a difference between "less than or equal to" and "not larger than." The propositions leave open the possibility that there may be an unconstrained protocol that generates an outcome that is not comparable to π^*. Example 2 demonstrates that this possibility really arises. Second, it is possible that a protocol that is not neutral may generate an outcome strictly greater than π^*. Example 3 illustrates this possibility.

Example 2
There are six projects. They take the form (i, j) for $i = 1, 2, 3$ and $j = 1, 2$. There are three experts whose preferences are

$$u_1(1,1) > u_1(1,2) > u_1(3,1) > u_1(3,2) > u_1(2,1) > u_1(2,2),$$

$$u_2(2,1) > u_2(1,1) > u_2(2,2) > u_2(3,1) > u_2(1,2) > u_2(3,2),$$

and

$$u_3(1,1) > u_3(1,2) > u_3(2,1) > u_3(3,1) > u_3(3,2) > u_3(2,2).$$

These preferences satisfy quasi-supermodularity. In this case, $\tilde{\pi}^* = \pi^* = \bar{\pi}^* = (3,1)$. Consider the protocol in which the manager consults all three experts in order. Expert 1 is consulted first. Expert 1's favorite project is $(1,1)$, but if she supports $(1,1)$, then Expert 2 will support $(2,1)$, which will be the final outcome because Expert 3 prefers $(2,1)$ to all higher projects. If Expert 1 supports $(1,2)$, then this will be the final outcome (if Expert 2 supports a higher project, then $(3,2)$ will be the outcome). Hence the protocol generates project $(1,2)$. The project $(1,2)$ is not comparable to $(3,1)$, which is the outcome of the unconstrained canonical sequential protocol. Hence if $(1,2) \succ (3,1)$ it is possible for the manager to do better using an alternative to the unconstrained canonical sequential protocol.

This result is consistent with the findings of this section. The protocol generates a unique outcome and this outcome is not greater than π^*. It demonstrates that it may be possible to generate an outcome that is not comparable to π^*. Hence the manager prefers the protocol of asking Expert 1, then 2, then 3 to the unconstrained canonical sequential protocol (that leads to the project π^*) if he prefers $(1,2)$ to $(3,1)$.

Example 3

There are three projects, $C \succ B \succ A$, with A the status quo. There are two experts whose preferences are

$$u_1(B) > u_1(C) > u_1(A) \text{ and } u_2(A) > u_2(B) > u_2(C).$$

This is a generic, one-dimensional game in which $\tilde{\pi}^* = \pi^* = \pi^*_{\succ} = B$. Consider the protocol in which the manager consults Expert 1 first and offers her the option of supporting A or C. After consulting Expert 1, the protocol specifies that the manager consults Expert 2, who can support any project.

The protocol stops after this consultation. This protocol is constrained, but not neutral. It generates project C, because Expert 1 is willing to support C to avoid the implementation of the status quo. CSP, CP, and the simultaneous-move game generate B.

Example 3 demonstrates that if the manager can design the protocol with knowledge of the experts' preferences, then he may be able to generate outcomes he prefers to those provided by CSP. The protocol has the property that Expert 1 is never given the opportunity to support π^*. This restriction is not necessary. For example, the protocol could specify that the manager consult Expert 1 a second time if some expert supported either B or C earlier. What is essential is that Expert 1 cannot support project B before Expert 2 supports A. Hence the protocol requires that the manager can commit not to ask Expert 1 about project B before "settling" for project A. Neutrality rules out this kind of commitment power. The construction of the protocol requires that the manager knows the preferences of the experts. The protocol would perform poorly if

$$u_1(B) \geqslant u_1(A) > u_1(C) \text{ and } u_2(A) > u_2(B) > u_2(C),$$

because in this case it generates outcome A while CSP and CP generate B.

In Sect. 7, we discuss optimal protocols when the manager has commitment power.

6 Comparing Outcomes in Simultaneous and Sequential Games

This section reviews the connection between the simultaneous and sequential procedures.

When projects are one dimensional and the experts' preferences are one-to-one, our results are clear. CSP and CP provide the best outcome among all sequential protocols and this outcome is identical to the outcome generated by the simultaneous-move game. When projects are completely ordered and experts have non-generic preferences, the results are less clearcut.[18] Example 4 (in Appendix 2) shows that when $\pi^* > \tilde{\pi}^*$, it is possible that the outcome π^* may fail to survive IDWDS in the simultaneous-move game. The same phenomenon is possible in sequential protocols.[19]

[18] We have not constructed examples in which the simultaneous-move game generates an outcome better than the outcomes of a canonical sequential protocol or vice versa.

[19] We can show that if there is an expert who is indifferent between $\tilde{\pi}^*$ and π^*, there is an order of deleting weakly dominated strategies such that project π^* survives in the simultaneous-move game, CSP, and CP.

When projects are multi-dimensional, giving the manager the ability to commit to not implementing hybrid projects eliminates the advantage that CSP has over simultaneous consultation.

Finally, we note that while the simultaneous-move game may have an outcome that the manager prefers to the outcome of the canonical sequential procedure, the sequential procedure always has an equilibrium outcome that is at least as good for the manager as some equilibrium of the simultaneous-move game.

When X is multi-dimensional, CSP may or may not be superior to the simultaneous protocol. One can trace the advantage of CSP to the power that the manager has to limit the projects that an expert can approve. Indeed, it is straightforward to show that the bounds $\bar{\pi}^*$ and $\tilde{\pi}^*$ established in Proposition 1 can be replaced by $\pi^*_{\succ}$ and $\tilde{\pi}^*_{\succ}$ if the strategy profile $\mathbf{x}$ induces the manager's most preferred project from $\{x_1, \ldots, x_I\}$ instead of the (potentially preferred) project $x_1 \vee \cdots \vee x_I$. In this way, the advantage of CSP depends on the ability to rule out hybrid projects that the manager can implement using the support of more than one expert. If the manager could commit not to implement hybrid projects in sequential consultation, then CP would generate the same outcome as CSP for generic payoffs.[20]

We introduced the unconstrained canonical protocol to analyze sequential games in which the manager lacked the power to restrict the experts' approvals. When projects are multi-dimensional, the simultaneous-move game may generate outcomes superior to CSP. Example 9 (Appendix 2) illustrates this possibility. We argue that the manager's most preferred project survives IDWDS in the simultaneous-move game. Neither CSP nor CP can generate this outcome because no expert wants to be the first to support a project other than the status quo ((0, 0)).

One reason the results are incomplete is that the characterization of the set of possible outcomes in simultaneous-move games is looser. The project $\bar{\pi}^*$ may exceed π^* and, although $\bar{\pi}^*$ is an upper bound, it may not be attained. We can describe the outcome of CSP more precisely, but cannot say much about when CSP performs better than the simultaneous-move game.

Permitting multi-dimensional projects adds the possibility that the manager can exploit complementarities to combine different projects supported by different experts into a third project that he likes better than either of the two individual projects. This suggests that the manager's preferred equilibrium in the simultaneous-move game is the result of a coordination problem and that other equilibria exist. Proposition 7 confirms this intuition.

Proposition 7 *There exists an outcome of the simultaneous game that survives IDWDS that is less than or equal to* π^*.

Proposition 7 follows from general properties of the simultaneous-move game. It is straightforward to check that the simultaneous-move game is (WID) supermodular as defined in Sobel [6]. Sobel [6, Theorem 6] demonstrates that in these games there exists a pure-strategy Nash equilibrium equal to the minimum of the set of strategies that survive IDWDS. Call this the minimal IDWDS profile. Lemma 5, which we state

[20] We are not sure if this claim holds for non-generic payoffs.

and prove in Appendix 1, implies that a strategy profile that generates an outcome less than or equal to π^* always survives IDWDS. Consequently the minimal IDWDS profile must generate an outcome no greater than π^*.

Proposition 7 identifies a way in which CSP is superior to the simultaneous game. Proposition 3 guarantees that outcomes generated by CSP are at least $\tilde{\pi}^*_{\succ}$. Because $\pi^*_{\succ} \geq \pi^*$ and $\pi^*_{\succ} = \tilde{\pi}^*_{\succ}$ for generic preferences for the experts, Proposition 7 implies that CSP always generates a project at least as attractive to the manager as some outcome that satisfies IDWDS in the simultaneous-move game.

7 Commitment

Suppose that experts have single-peaked preferences. In both simultaneous and sequential games, the equilibrium outcome permits the manager to implement the maximum of the experts' peaks, which we denote by x^*. What if the manager knows that at least one expert prefers the manager's favorite project $\bar{x}$ to the status quo? If the manager is able to limit the responses of the experts, then he can achieve his most preferred outcome. In a simultaneous-move game, he does so by limiting experts to either supporting $\bar{x}$ or not. In a sequential game, he does so by asking experts, in sequence, whether they support $\bar{x}$ and promising to maintain the status quo if no expert approves $\bar{x}$. That is, the protocol implements the status quo if no expert supports $\bar{x}$. This protocol requires that the manager have the ability to make a commitment not to ask for approval of intermediate projects if no one supports $\bar{x}$. The logic behind these assertions is familiar: If experts must decide between (only) $\bar{x}$ and the status quo, an expert who prefers $\bar{x}$ will support it.

We emphasize that the ability to obtain outcomes greater than x^* depends on commitment power. In the simultaneous-move game, the manager must be able to prevent experts from supporting intermediate projects or resist the temptation to implement an intermediate project if it is the best thing supported by the experts. In the sequential-move game, the manager must be able to commit to implementing the status quo (rather than a project in $(\underline{x}, \bar{x})$) in the event that no expert supports $\bar{x}$.

Hence the characterization of outcomes derived in Proposition 1 does not apply if the manager knows something about the preferences of experts and can make commitments. This observation does not rely on the assumption that preferences are single peaked.

We assume in this section that X is completely ordered and u_i is one-to-one for each i.

Let $X_0 = \{\underline{x}\}$ and define X_k inductively by

$$X_k = \{\pi \in X : \text{ there exists } i \text{ and } \pi_m \in X_m \text{ for } m < k \text{ such that } u_i(\pi) > u_i(\pi_m) \text{ and } \pi > \pi_m\}.$$

Projects in X_1 are those that some expert prefers to the status quo. For $n > 0$, a project π is in X_n if there is a collection of $0 \leq k \leq n$ projects, $\pi_0, \ldots, \pi_k$, such that

$\underline{x} = \pi_0 < \pi_1 < \cdots < \pi_k = \pi$ and some expert prefers π_j to π_{j-1} for $j = 1, \ldots, k$. To construct this **chain of preference**, we argue by induction. If $\pi \in X_1$, then we take $\pi_1 = \pi$. In general, assume that it is possible to construct a chain of preference for all $\pi \in X_m$ for $m < n$. We will show that it is also possible for n. Begin with $\pi \in X_n$, use the definition of X_n to find $m < n$ and $\pi_m \in X_m$ with the properties that $\pi_m < \pi$ and there is an i such that $u_i(\pi) > u_i(\pi_m)$ and then apply the induction hypothesis to find a chain of preference from $\underline{x} = \pi_0$ to π_m.

By definition, $X_{n-1} \subset X_n$. Because X is finite, there exists $n^\dagger$ such that $X_n = X_{n^\dagger}$ for all $n > n^\dagger$. Let $X^\dagger = X_{n^\dagger}$. Let $\pi^\dagger$ be the manager's preferred outcome in $X^\dagger$. We assert that the manager with knowledge of the preferences of the experts and commitment ability can implement $\pi^\dagger$ in either a simultaneous or sequential game. The reason is simple. Project $\pi^\dagger$ is equal to the equilibrium outcome if the manager restricts the projects only to those in a chain of preference.

Call a subset, $X^c \subset X$ **admissible** if $\underline{x} \in X^c$. X^c determines a simultaneous-move game in which experts simultaneously select elements of X^c; the maximum strategy determines the outcome as in the standard simultaneous-move game, and the experts' utility functions are restrictions of $\tilde{u}_i$ to $X^c \times \cdots \times X^c$. Call such a game the simultaneous-move game restricted to X^c.

Assume that the manager can require the experts to support only projects in X^c, how well can he do? If X is completely ordered and u_i is one-to-one for each i, then for each admissible X^c, the simultaneous-move game restricted to X^c has a unique outcome that survives IDWDS. Call this outcome $\tilde{\pi}(X^c)$. So the best that the manager can do by restricting outcomes is $\max\{\tilde{\pi}(X^c) : X^c \text{ is admissible}\}$. The next result states that this value is equal to $\pi^\dagger$. That is, if the manager has commitment power, then the best project he can generate is his most preferred project in $X^\dagger$.

Proposition 8 *Suppose X is completely ordered and u_i is one-to-one for each i. The best outcome that the manager can achieve by restricting strategies in a simultaneous-move game is $\pi^\dagger$. That is,*

$$\pi^\dagger = \max\{\tilde{\pi}(X^c) : X^c \textit{ is admissible}\}.$$

The proposition does not identify the restriction X^c that leads to $\pi^\dagger$. It is tempting to conjecture that the manager can restrict to strategies in $X^\dagger$, but this need not be the case. For example, if for each project $x \neq \underline{x}$ there exists an expert who prefers x to $\underline{x}$, then $X^\dagger = X$. The manager can attain his most preferred outcome $\bar{x}$ by setting $X^c = \{\underline{x}, \bar{x}\}$, but typically he would not attain $\bar{x}$ without some restrictions on X.[21] The fact that $\pi^\dagger \in X^\dagger$ guarantees that there is a minimal collection of projects, starting at $\pi_0 = \underline{x}$ and going to $\pi_K = \pi^\dagger$ such that some expert prefers π_k to π_{k-1} for each k. Restricting to this collection guarantees the outcome $\pi^\dagger$. The harder part of the proof of Proposition 8 is to show that no other restriction can do better for the

[21] Here is an example that demonstrates the claim. Assume that experts have common preferences that rank $\underline{x}$ lowest, but otherwise are monotonically decreasing. That is, for $x' > x'' \neq \underline{x}$, $u_i(x'') > u_i(x') > u_i(\underline{x})$ for all i.

manager than $\pi^{\dagger}$. We prove this by showing that any outcome that survives IDWDS must be in $X^{\dagger}$.

There is a parallel result for sequential protocols. A sequential protocol with the restricted strategy space $X^c \subset X$ is a sequential protocol in which the manager only requests experts to approve projects in X^c. The protocol operates by asking experts, in order, whether they approve the manager's most preferred task in X^c. If they all refuse, then he asks them to approve his next favorite project. The protocol continues until either some expert approves or all projects are rejected and the status quo results.

Proposition 9 *Suppose X is completely ordered and u_i is one-to-one for each i. There exists a sequential game in which $\pi^{\dagger}$ is the unique outcome that survives IDWDS. Furthermore, there is no sequential protocol with restricted strategy spaces that achieves a higher outcome.*

Proposition 9 is a consequence of two simple observations. First, because sequential protocols always give experts the freedom to endorse no project and no expert would support a project not in $X^{\dagger}$ if she could induce a project in $X^{\dagger}$, the outcome of the protocol must be an element of $X^{\dagger}$. Second, if one restricts to a chain of preference ending in $\pi^{\dagger}$, then CSP generates $\pi^{\dagger}$.

When X is completely ordered but u_i are not necessarily one-to-one, our characterization results are not as sharp, but the basic message of the earlier results remains true. We have bounds for the outcomes that can survive IDWDS. Commitment will be beneficial to the manager if commitment leads to a game with smallest equilibrium outcome greater than $\tilde{\pi}^*$. Any restriction that helps the manager in the simultaneous-move game will also lead to a beneficial restriction for sequential consultation. Of course, the exact equivalence between simultaneous and sequential institutions breaks down when ties are possible.

When X is not completely ordered, we can treat projects as completely ordered (by the preferences of the manager). Versions of Propositions 8 and 9 apply in this setting. (If it is not feasible to combine projects—or if the manager can commit to selecting his most preferred single project of those approved rather than forming hybrid projects, then we preserve the equivalence between simultaneous and sequential procedures.)

Finally we note that in addition to commitment power, the manager needs to know the set of preferences of the experts in order to figure out his preferred restriction on reports. The analysis of the simultaneous-move game and the sequential protocol do not require this knowledge. (Preferences must be common knowledge between the players of the game. That is, the experts must know the preferences of other experts.) The constructions in this section do not require that the manager knows which expert has which preference order. If the manager had this information, it is conceivable that he could further improve the outcome by restricting different experts to different subsets of projects.[22]

[22] We do not have an example in which the ability to restrict different experts to different strategies is strictly beneficial, but our arguments do not rule out this possibility.

The commitment model shares features of models of constrained delegation (for example, Holmström [3] and Armstrong and Vickers [1]), in which a principal delegates decision-making authority to an agent provided that the agent's choice remains in a possibly strict subset of feasible choices. Armstrong and Vickers show that the principal might gain by restricting intermediate choices for the agent's choice set, a property that holds in our model.

8 Discussion

One of our goals in this study is to identify features that favors simultaneous versus sequential consultation procedures. Our results suggest that the choice of organization does not matter. Nevertheless, even in our setting, there are differences between the procedures.

A feature of the consultation procedures that we describe is that the manager need not know anything about the preferences of the experts in order to operate them. That is, they are robust institutions because they need not be tailored to individual preferences. Of course, our predictions require that experts know the preferences of other agents (and that they assume the other agents avoid dominated strategies). Section 7 demonstrates how the manager might be able to take advantage of additional information about experts' preferences if he can limit the set of projects.

We have identified a sequential protocol that performs at least as well as simultaneous consultation. Sequential procedures may be superior to the simultaneous-move game in another way if the manager has information about preferences. The simultaneous-move game requires that all experts (or, at least, a subset of experts needed to make the minimum equilibrium outcome for that subset equal to π^*) participate actively. If the manager knows $\pi^*_{\succ}$, however, the sequential protocol that initially asks for support of $\pi^*_{\succ}$ (and then follows the canonical sequential protocol) generates $\pi^*_{\succ}$ with only one consultation. The ability to consult more experts encourages the first expert to support $\pi^*_{\succ}$. The canonical protocol begins by asking experts to support $\bar{x}$, so unless $\bar{x} = \pi^*_{\succ}$ it must consult every expert at least once. Even if the manager does not know $\pi^*_{\succ}$, there is an equilibrium of CSP in which the first expert consulted supports the project generated by the protocol, but unless this project is $\bar{x}$, the manager will not know enough to stop consultations at this point.

There appear to be natural settings in which the manager prefers simultaneous procedures and other settings in which sequential procedures are the norm. On one hand, editors in economics typically consult several reviewers simultaneously to obtain reports on a submission.[23] Committee deliberations, anonymous voting, and obtaining multiple bids for a construction contract have features of simultaneous procedures. On the other hand, it is frequent to consult medical experts in sequence.

There are situations that mix simultaneous and sequential consultations. Here is an example.

---

[23] This institution varies across disciplines and even across editors in economics.

The Amazon negotiations described in Sect. 3 mix features of simultaneous and sequential consultations. A full understanding of the relative merits of simultaneous and sequential procedures requires a richer model. Two directions seem promising. Certainly adding costs (whether direct payment or waiting times) to consultation will change the analysis, presumably in the direction of favoring sequential procedures. In our model experts cannot learn from each other. In a variation of the model in which experts have different strategy sets or where the action of one expert influences the set of actions available to subsequent agents, the manager may favor sequential over simultaneous procedures.

We have studied two ways to organize consultations. It is natural to ask whether there is an organization that is superior to either the simultaneous or sequential procedures that we examined. We lack a complete answer to this question, which differs from standard implementation questions for several reasons. First, we must study something more general than implementation in Nash equilibrium (because otherwise it is straightforward to obtain the manager's favorite outcome). Second, in the standard implementation problem, the decision-maker's set of actions is fixed. We assume that what the manager can do depends on the behavior of the experts (the manager cannot take an action unless an expert supports it). Finally, standard mechanism design gives the designer commitment power, but imposes individual rationality constraints. To study commitment in our model, we wish to prevent the designer from ruling out the status quo if no expert supports an alternative, but this condition is a restriction on the experts' collective response. It does not translate into a lower bound on utility independent of the behavior of other experts.

Acknowledgements This manuscript contains material originally circulated in unpublished versions of Hu and Sobel [4]. Sobel is grateful to the NSF for research funding.

Appendix 1: Proofs

Lemma 1 If π' and π'' are equilibrium outcomes, then $\pi' \wedge \pi''$ is an equilibrium outcome.

Proof If π' and π'' are equilibrium outcomes, then $u_i(\pi'') \geq u_i((\pi' \wedge \hat{\pi}) \vee \pi'')$ and $u_i(\pi') \geq u_i(\pi' \vee \hat{\pi})$ for any $\hat{\pi}$. It follows from quasi-supermodularity that

$$u_i((\pi' \wedge \hat{\pi}) \wedge \pi'') \geq u_i(\pi' \wedge \hat{\pi}) \tag{5}$$

and

$$u_i(\pi' \wedge \hat{\pi}) \geq u_i(\hat{\pi}). \tag{6}$$

□

A small variation on the proof establishes that π^* is well defined.

Lemma 2 *The smallest strictly preferred equilibrium outcome exists.*

Proof We show that if

$$u_i(\pi) > u_i(x_i) \text{ for all } x_i > \pi \text{ and all } i \tag{7}$$

for $\pi = \pi'$ and π'', then Inequality (7) holds for $\pi = \pi' \wedge \pi''$. Suppose that Inequality (7) holds for $\pi = \pi'$ and π''. It follows that, for any $\hat{\pi}$, $u_i(\pi'') \geq u_i((\pi' \wedge \hat{\pi}) \vee \pi'')$ with strict inequality if $\pi' \wedge \hat{\pi} \nleq \pi''$ and $u_i(\pi') \geq u_i(\pi' \vee \hat{\pi})$ with strict inequality if $\hat{\pi} \nleq \pi'$. It follows from quasi-supermodularity that

$$u_i((\pi' \wedge \hat{\pi}) \wedge \pi'') \geq u_i(\pi' \wedge \hat{\pi}) \tag{8}$$

with strict inequality if $\pi' \wedge \hat{\pi} \nleq \pi''$ and

$$u_i(\pi' \wedge \hat{\pi}) \geq u_i(\hat{\pi}) \tag{9}$$

with strict inequality if $\hat{\pi} \nleq \pi'$.

Now assume that $\hat{\pi} > \pi' \wedge \pi''$. Consequently, $(\pi' \wedge \hat{\pi}) \wedge \pi'' = \pi' \wedge \pi''$ so that inequality (8) implies

$$u_i(\pi' \wedge \pi'') \geq u_i(\pi' \wedge \hat{\pi}) \tag{10}$$

with strict inequality if $\pi' \wedge \hat{\pi} \nleq \pi''$. It follows from (9) and (10) that $u_i(\pi' \wedge \pi'') \geq u_i(\hat{\pi})$ with strict inequality if $\pi' \wedge \hat{\pi} \nleq \pi''$ or $\hat{\pi} \nleq \pi'$. Note that if $\pi' \wedge \hat{\pi} \leq \pi''$ or $\hat{\pi} \leq \pi'$, then $\hat{\pi} \leq \pi' \wedge \pi''$. It follows that $\hat{\pi} > \pi' \wedge \pi''$, then $u_i(\pi' \wedge \pi'') > u_i(x_i)$ for all $x_i > \pi' \wedge \pi''$ and all i. □

Lemma 3 *Let π satisfy $u_i(\pi(k), \bar{x}(-k)) > u_i(x_i(k), \bar{x}(-k))$ for all $x_i(k) > \pi(k)$, all dimensions k, and all i. Then π is an equilibrium project such that $u_i(\pi) > u_i(x_i)$ for all $x_i > \pi$ and all i.*

Proof Fix a project $\pi' > \pi$ and an arbitrary expert i. It follows that $\pi'(k) \geq \pi(k)$ in all dimensions and the inequality is strict on at least one dimension. We will show that $u_i(\pi) > u_i(\pi')$. Without loss of generality, suppose that $\pi'(1) > \pi(1)$, where $\pi(1)$ is first component of π and $\pi(-1)$ are the components of π in dimensions other than 1. We will show by induction on r that

$$u_i(\pi) > u_i(\pi'(1), \ldots, \pi'(r), \pi(r+1), \ldots, \pi(n)) \tag{11}$$

for $r = 1, \ldots, n$. The claim then follows from inequality (11) for $r = n$.

Because $u_i(\pi(1), \bar{x}(-1)) > u_i(\pi'(1), \bar{x}(-1))$, it must be that $u_i(\pi) > u_i(\pi'(1), \pi(-1))$ by quasi-supermodularity. Hence inequality (11) holds for $r = 1$. Assume inequality (11) for $r \leq k$. It suffices to show that it holds for $r = k + 1$.

Because $u_i(\pi(k+1), \bar{x}(-(k+1))) \geq u_i(\pi'(k+1), \bar{x}(-(k+1)))$,

$$u_i(\bar{x}(1), \ldots, \bar{x}(k), \pi(k+1), \ldots, \pi(n)) \geq u_i(\bar{x}(1), \ldots, \bar{x}(k), \pi'(k+1), \pi(k+2), \ldots, \pi(n)) \tag{12}$$

by quasi-supermodularity. Inequality (12) and quasi-supermodularity imply that

$$u_i(\pi'(1), \ldots, \pi'(k), \pi(k+1), \ldots, \pi(n)) \geq u_i(\pi'(1), \ldots, \pi'(k+1), \pi(k+2), \ldots, \pi(n)). \tag{13}$$

It follows from inequality (11) for $r = k$ and inequality (13) that inequality (11) holds for $r = k + 1$, which completes the proof. □

Proposition 1 If $\mathbf{x}$ is a strategy profile that survives IDWDS in the simultaneous-move game, then $M(\mathbf{x}) \in [\tilde{\pi}^*, \bar{\pi}^*]$. For any equilibrium strategy profile $\mathbf{x}$, $\tilde{u}_i(\mathbf{x}) \geq u_i(\bar{\pi}^*)$ for all i.

Denote the set of strategies that survive IDWDS by S. We prove the proposition in a series of steps, which we state and prove as lemmas.

Lemma 4 *For all* $\mathbf{x} \in S$ *and every* i*, there exists* $x_i \in S_i$ *that is a best response to* $\mathbf{x}$ *relative to* X.

Proof The result is clear if every best response to $\mathbf{x}$ relative to X has not yet been deleted. If the best response to $\mathbf{x}$ has been deleted, then it was deleted by a strategy that weakly dominates it. This strategy must be a best reply to $\mathbf{x}$. □

Lemma 5 *There exists a strategy profile* $\mathbf{x} \in S$ *such that* $M(\mathbf{x}) \leq \pi^*$.

Proof Suppose that after r iterations, there exists a strategy profile $\mathbf{x}$ satisfying the condition in the claim. In the next iteration, every agent must have a strategy that is a best response to $\mathbf{x}$ by Lemma 4. Suppose that the best response for Expert i to $\mathbf{x}$ is some $x_i' \not\leq \pi^*$. We will argue to a contraction. Because $M(x_i', \mathbf{x}_{-i}) = x_i' \vee M(\mathbf{x}_{-i}) = M(x_i' \vee M(\mathbf{x}_{-i}), \mathbf{x}_{-i})$,

$$\tilde{u}_i(x_i', \mathbf{x}_{-i}) = u_i(M(x_i', \mathbf{x}_{-i})) = u_i(M(x_i' \vee M(\mathbf{x}_{-i}), \mathbf{x}_{-i})) = \tilde{u}_i(x_i' \vee M(\mathbf{x}_{-i}), \mathbf{x}_{-i}). \tag{14}$$

It follows from (14) that $x_i'' \equiv x_i' \vee M(\mathbf{x}_{-i})$ is also a best response to $\mathbf{x}$. Note that because $x_i'' > M(\mathbf{x}_{-i})$,

$$M(x_i'', \mathbf{x}_{-i}) = x_i''. \tag{15}$$

In addition, note that

$$M(x_i'' \wedge \pi^*, \mathbf{x}_{-i}) = (x_i'' \wedge \pi^*) \vee M(\mathbf{x}_{-i}) = x_i'' \wedge \pi^*, \tag{16}$$

where second equality holds because $M(\mathbf{x}_{-i}) < x_i''$ and $M(\mathbf{x}_{-i}) \leq \pi^*$ imply that $M(\mathbf{x}_{-i}) \leq x_i'' \wedge \pi^*$.

Because $x_i' \not\leq \pi^*$ and $x_i'' \geq x_i'$, $x_i'' \not\leq \pi^*$. Therefore, $x_i'' \vee \pi^* > \pi^*$. By definition of π^*, $u_i(\pi^*) > u_i(x_i)$ for all $x_i > \pi^*$, it follows that

$$u_i(\pi^*) > u_i(x_i'' \vee \pi^*). \tag{17}$$

Hence it must be that

$$\tilde{u}_i(x_i'' \wedge \pi^*, \mathbf{x}_{-i}) = u_i(M(x_i'' \wedge \pi^*, \mathbf{x}_{-i})) = u_i(x_i'' \wedge \pi^*) > u_i(x_i'') = u_i(M(x_i'', \mathbf{x}_{-i})) = \tilde{u}_i(x_i'', \mathbf{x}_{-i}) \tag{18}$$

where the second equation follows from (16), the inequality follows by (17) and quasi-supermodularity, and the third equation follows from (15). Expressions (14) and (18) combine to show that x_i' is not a best response to $\mathbf{x}_{-i}$, which is a contradiction. We conclude that no strategy $x_i'' \not\leq \pi^*$ can do at least as well as $x_i'' \wedge \pi^*$ against $\mathbf{x}$. Therefore, all best responses to $\mathbf{x}$ are less than or equal to π^* and a strategy less than or equal to π^* must remain for each agent. □

Next we show that all projects that survive IDWDS are greater than or equal to the smallest equilibrium outcome.

Lemma 6 *If* $\mathbf{x} \in S$, *then* $M(\mathbf{x}) \geq \tilde{\pi}^*$.

Proof Let $\tilde{\pi} \equiv \wedge_{\mathbf{x} \in S} M(\mathbf{x})$ be the meet of all outcomes that survive IDWDS. We wish to show that $\tilde{\pi} \geq \tilde{\pi}^*$. In order to reach a contradiction, assume that $\tilde{\pi} \not\geq \tilde{\pi}^*$. By the definition of $\tilde{\pi}^*$, it must be the case that for some i,

$$\text{there exists } x_i \text{ with } x_i > \tilde{\pi} \text{ such that } u_i(x_i) > u_i(\tilde{\pi}), \tag{19}$$

because otherwise $\tilde{\pi}^*$ would not be the smallest equilibrium outcome. Let

$$x_i' = \min\{\arg\max_{x_i \geq \tilde{\pi}} u_i(x_i)\}$$

be the smallest best response of Expert i to $\tilde{\pi}$, which is well defined by Fact in 4.3. It is apparent that $x_i' > \tilde{\pi}$ and so $x_i'(k) > \tilde{\pi}(k)$ for some k. Select $\tilde{\mathbf{x}} \in S$ such that $M(\tilde{\mathbf{x}})(k) = \tilde{\pi}(k)$. This is possible by the definition of $\tilde{\pi}$. We assert that $x_i' \vee \tilde{x}_i$ weakly dominates $\tilde{x}_i$. The assertion is sufficient to prove the claim because it means that if (19) holds, then we can show an element in S is weakly dominated. This contradicts the definition of S.

It remains to show that $x_i' \vee \tilde{x}_i$ weakly dominates $\tilde{x}_i$. For any $\mathbf{x}_{-i}$ such that $M(\mathbf{x}_{-i}) \geq x_i' \vee \tilde{x}_i$,

$$\tilde{u}_i(x_i' \vee \tilde{x}_i, \mathbf{x}_{-i}) = u_i(M(\mathbf{x}_{-i})) = \tilde{u}_i(\tilde{x}_i, \mathbf{x}_{-i}).$$

For any $\mathbf{x}_{-i}$ such that $M(\mathbf{x}_{-i}) \not\geq x_i' \vee \tilde{x}_i$, i's utility from playing $x_i' \vee \tilde{x}_i$ and $\tilde{x}_i$ are $u_i((x_i' \vee \tilde{x}_i) \vee M(\mathbf{x}_{-i})) = u_i(x_i' \vee (\tilde{x}_i \vee M(\mathbf{x}_{-i})))$ and $u_i(\tilde{x}_i \vee M(\mathbf{x}_{-i}))$, respectively. We now claim that

$$\tilde{u}_i(x_i' \vee \tilde{x}_i, \mathbf{x}_{-i}) = u_i(x_i' \vee (\tilde{x}_i \vee M(\mathbf{x}_{-i}))) \geq u_i(\tilde{x}_i \vee M(\mathbf{x}_{-i})) = \tilde{u}_i(\tilde{x}_i, \mathbf{x}_{-i}). \tag{20}$$

Hence $x_i' \vee \tilde{x}_i$ is weakly better than $\tilde{x}_i$. The equations in (20) follow from the definition of $\tilde{u}_i$. The inequality would hold by quasi-supermodularity if

$$u_i(x_i') \geq u_i(x_i' \wedge (\tilde{x}_i \vee M(\mathbf{x}_{-i}))). \tag{21}$$

Inequality (21) is satisfied. Because $\tilde{x}_i \vee M(\mathbf{x}_{-i}) \geq \tilde{\pi}$ by the definition of $\tilde{\pi}$ and $x_i' > \tilde{\pi}$, $x_i' \wedge (\tilde{x}_i \vee M(\mathbf{x}_{-i})) \geq \tilde{\pi}$. Inequality (21) now follows from the definition of x_i'.

It follows from (19) and the definition of x_i' that $x_i' > \tilde{\pi}$. Also $x_i' > x_i' \wedge M(\tilde{\mathbf{x}}) \geq \tilde{\pi}$, and so the definition of x_i' also implies that $u_i(x_i') > u_i(x_i' \wedge M(\tilde{\mathbf{x}}))$. Consequently, $u_i(x_i' \vee M(\tilde{\mathbf{x}})) > u_i(M(\tilde{\mathbf{x}}))$ by quasi-supermodularity. It follows that

$$\tilde{u}_i(x_i' \vee \tilde{x}_i, \tilde{\mathbf{x}}_{-i}) = u_i(x_i' \vee M(\tilde{\mathbf{x}})) > u_i(M(\tilde{\mathbf{x}})) = \tilde{u}_i(\tilde{\mathbf{x}}) = \tilde{u}_i(\tilde{x}_i, \tilde{\mathbf{x}}_{-i}). \qquad (22)$$

Inequality (22) guarantees that $x_i' \vee \tilde{x}_i$ is strictly better than $\tilde{x}_i$ when $\mathbf{x}_{-i} = \tilde{\mathbf{x}}_{-i}$. Establishing this completes the proof that $x_i' \vee \tilde{x}_i$ weakly dominates $\tilde{x}_i$. □

All projects that survive IDWDS are greater than or equal to $\tilde{\pi}^*$. Next, we show that all projects that survive IDWDS are less than or equal to the bounding project.

Lemma 7 *If* $\mathbf{x} \in S$, *then* $M(\mathbf{x}) \leq \bar{\pi}^*$.

Proof Let S_i^r be the set of strategies remaining for i after r rounds of deleting strategies. For each i let $Q_i^r = \{s_i^r \in S_i^r : s_i^r \nleq \bar{\pi}^*\}$. If there exists r such that $\bigcup_i Q_i^r = \emptyset$, then the proof is complete. Otherwise, there is at least one dimension k such that $s_i^r(k) > \bar{\pi}^*(k)$ for some $s_i^r \in Q_i^r$ and some i. Let $R^r(k) = \bigcup_i \{s_i^r \in Q_i^r : s_i^r(k) > \bar{\pi}^*(k)\}$. Let $z^r \in \arg\min\{s^r(k) : s^r \in R^r(k)\}$. If $\bigcup_i Q_i^r \neq \emptyset$, then z^r exists and $z^r \in Q_j^r$ for some j; we write $z^r = z_j^r$ to indicate that $z_j^r \in Q_j^r$. We claim that z_j^r is weakly dominated by $(\bar{\pi}^*(k), z_j^r(-k))$.[24]

For any $\mathbf{x}$ such that $M(\mathbf{x}_{-j}(k)) \geq z_j^r(k)$,

$$\tilde{u}_j((\bar{\pi}^*(k), z_j^r(-k)), \mathbf{x}_{-j}) = u_j(M(\mathbf{x}_{-j}(k)), M(z_j^r(-k), \mathbf{x}_{-j}(-k))) = \tilde{u}_j(z_j^r, \mathbf{x}_{-j}).$$

This follows because $z_j^r(k) > \bar{\pi}^*(k)$.

For any $\mathbf{x}$ such that $M(\mathbf{x}_{-j}(k)) < z_j^r(k)$, $M(\mathbf{x}_{-j}(k)) \leq \bar{\pi}^*(k)$ by the definition of z_j^r. Hence, Expert j's utility from using z_j^r is $u_j(z_j^r(k), M(z_j^r(-k), \mathbf{x}_{-j}(-k)))$, while j's utility from using $(\bar{\pi}^*(k), z_j^r(-k))$ is $u_j(\bar{\pi}^*(k), M(z_j^r(-k), \mathbf{x}_{-j}(-k)))$. We claim that

$$u_j(\bar{\pi}^*(k), M(z_j^r(-k), \mathbf{x}_{-j}(-k))) > u_j(z_j^r(k), M(z_j^r(-k), \mathbf{x}_{-j}(-k))). \qquad (23)$$

It follows from the definition of $\bar{\pi}^*$ that $u_j(\bar{\pi}^*(k), \bar{\mathbf{x}}(-k)) > u_j(z_j^r(k), \bar{\mathbf{x}}(-k))$, therefore Inequality (23) follows from quasi-supermodularity. That is, j does strictly better using $(\bar{\pi}^*(k), z_j^r(-k))$ than z_j^r whenever the kth dimension of $M(\mathbf{x}_{-j})$ is less than $z_j^r(k)$. Because there always exists a strategy in which $M(\mathbf{x}_{-j}(k))$ is less than $z_j^r(k)$ by Lemma 5 and $\pi^* \leq \bar{\pi}^*$, $(\bar{\pi}^*(k), z_j^r(-k))$ must be strictly better than z_j^r against one strategy profile that survives IDWDS. Consequently, $(\bar{\pi}^*(k), z_j^r(-k))$ weakly dominates z_j^r. Therefore, z_j^r must eventually be deleted. We conclude that there must exist an r^* such that $Q_i^{r^*} = \emptyset$ for all i, which establishes the result. □

---

[24] If $X \subset \mathbb{R}$, $(\bar{\pi}^*(k), z_j^r(-k)) = \bar{\pi}^* = \pi^*$.

Proposition 1 If $\mathbf{x}$ is a strategy profile that survives IDWDS in the simultaneous-move game, then $M(\mathbf{x}) \in [\tilde{\pi}^*, \bar{\pi}^*]$. For any equilibrium strategy profile $\mathbf{x}$, $\tilde{u}_i(\mathbf{x}) \geq u_i(\bar{\pi}^*)$ for all i.

Proof Lemma 6 guarantees that all $\mathbf{x} \in S$ satisfy $M(\mathbf{x}) \geq \tilde{\pi}^*$. Lemma 7 guarantees that all $\mathbf{x} \in S$ satisfy $M(\mathbf{x}) \leq \bar{\pi}^*$. This establishes the first part of the proposition. Given any $\mathbf{x} \in S$, it follows from Lemma 4 that each player has a surviving strategy that is a best response to $\mathbf{x}_{-i}$ relative to the full strategy set. Because $x_i = \bar{\pi}^*$ leads to payoff $u_i(\bar{\pi}^*)$ for Expert i against any surviving strategy by Lemma 7, the second part of the proposition follows. □

Proposition 2 If π is a project that survives IDWDS in the game determined by CP, then $\pi \geq \tilde{\pi}^*$.

Proof Let π be a project that survives IDWDS. Suppose it is generated by the strategy profile $\hat{\mathbf{s}}$. Suppose that $\pi \not\geq \tilde{\pi}^*$. We will show that there exists an expert i such that $\hat{s}_i$ is weakly dominated.

By the definition of $\tilde{\pi}^*$, it must be the case that for some i:

$$\text{there exists } x_i \text{ with } x_i > \pi \text{ such that } u_i(x_i) > u_i(\pi). \tag{24}$$

Find a history $\hat{h}$ consistent with $\hat{\mathbf{s}}$ such that $P_I(\hat{h}) = i$ and if i's play at h is $\hat{s}_i(\hat{h})$ then there is no undominated strategy profile that consults i again. It is possible to find such a history because $\pi \not\geq \tilde{\pi}^*$ implies that the manager must consult every expert at least once and because the protocol never consults an expert more than K times.

Consider an alternative strategy of Expert i, in which

$$s_i'(h) = \begin{cases} x_i' & \text{if } h = \hat{h} \\ \hat{s}_i(h) & \text{otherwise} \end{cases},$$

where x_i' solves $\max u_i(x_i)$ subject to $x_i > \pi$. We know that x_i' exists and satisfies and $u_i(x_i') > u_i(\pi)$ by (24). By the definition of $\hat{h}$, if Expert i supports x_i' after $\hat{h}$, the protocol must stop. (We know that the protocol would stop in the next round with the outcome π, so it must stop immediately when Expert i supports something strictly better for the manager than π.) It follows that s_i' weakly dominates $\hat{s}_i$. The strategy s_i' does exactly as well as $\hat{s}_i$ for any strategy profile that does not induce the history $\hat{h}$. We know that some strategy profile does induce $\hat{h}$ and, by construction, Expert i does strictly better in any such case. Consequently, any outcome $\pi \not\geq \tilde{\pi}^*$ must be generated by a strategy profile in which one player uses a weakly dominated strategy. This establishes the proposition. □

Proposition 3 If π is a project that survives IDWDS in the game determined by CSP, then $\pi \succeq \tilde{\pi}^*_{\succ}$.

Proof Let π be a project that survives IDWDS. Suppose it is generated by the strategy profile $\hat{\mathbf{s}}$. Suppose that $\pi \prec \tilde{\pi}^*_{\succeq}$. We take π to be a minimal such project (that is, no $\pi' \prec \pi$ survives IDWDS). We will show that there exists an expert i such that $\hat{s}_i$ is weakly dominated.

By the definition of $\tilde{\pi}^*_{\succeq}$, it must be the case that for some i:

$$\text{there exists } x_i \text{ with } x_i \succ \pi \text{ such that } u_i(x_i) > u_i(\pi). \tag{25}$$

Using (25), select i and $x_i \in \arg\max_{\{\pi':\pi'\succ\pi\}} u_i(\pi')$ so that $u_i(x_i) > u_i(\pi)$. Find a possible history $\hat{h}$ consistent with $\hat{\mathbf{s}}$ such that $P_I(\hat{h}) = i$ and $x_i \in P_X(\hat{h})$. Because $x_i \succ \pi$, such a history exists and $\hat{s}_i(h)$ does not support x_i.

Consider an alternative strategy of Expert i, in which

$$s_i'(h) = \begin{cases} x_i & \text{if } h = \hat{h} \\ \hat{s}_i(h) & \text{otherwise} \end{cases}.$$

By the definition of $\hat{h}$ and CSP, if Expert i supports x_i after $\hat{h}$, the protocol must stop. It follows that s_i' weakly dominates $\hat{s}_i$. The strategy s_i' does exactly as well as $\hat{s}_i$ for any strategy profile that does not induce the history $\hat{h}$. We know that some strategy profile does induce $\hat{h}$. We claim that Expert i does at least as well in such a case. To see this, note that by construction, Expert i prefers project x_i to any other project $\pi' \succeq \pi$. Furthermore, no project $\pi' \prec \pi$ survives IDWDS by the definition of π. Finally, because $\hat{\mathbf{s}}$ generates π and $u_i(x_i) > u_i(\pi)$, s_i' performs better than $\hat{s}_i$ in one instance and at least as well in all instances.

Consequently, any outcome $\pi \not\succeq \tilde{\pi}^*_{\succeq}$ must be generated by a strategy profile in which one player uses a weakly dominated strategy. This establishes the proposition. □

Proposition 4 If $x > \pi^*$, then there exists no unconstrained sequential protocol that generates the project x in a pure-strategy, subgame-perfect equilibrium. If $x \succ \pi^*_{\succeq}$, then there exists no neutral sequential protocol that generates the project x in a pure-strategy, subgame-perfect equilibrium.

Proof Proposition 4 follows from backward induction. If the protocol is unconstrained, then after each history that supports no more than π^*, it is never a best response to disclose more than π^*. So if the first expert anticipates that the final project supported will be more than π^*, then she can do strictly better by approving π^* and no one else will add more to π^*. Consequently there will never be projects greater than π^* in equilibrium.

Assume that the protocol is neutral. If $\pi^* = \underline{x}$, then all experts prefer $\underline{x}$ to any other project and hence $\underline{x}$ will be the outcome of any protocol. Assume that $\pi^* > \underline{x}$. Protocols must always permit an expert to support the status quo when consulted. Consequently, there exists a terminal history $\underline{h}$ such that $\mu(\underline{h}) = \underline{x}$. Neutrality implies that for each i there is a history $h \subset \underline{h}$ such that $\pi^*_{\succeq} \in P_X(h)$ and $i = P_I(h)$. Furthermore, Expert i can guarantee a payoff of at least $u_i(\pi^*_{\succeq})$ in the subgame that follows h

by supporting $\pi^*_{\succ}$ (after which no other expert would support a larger project). Hence, by the definition of $\pi^*_{\succ}$, the outcome $\pi(h)$ of this subgame must satisfy $\pi^*_{\succ} \succeq \pi(h)$. Let h^* be a minimal subhistory of $\underline{h}$ with the property that the outcome $\pi(h^*)$ of the subgame determined by h^* satisfies $\pi^*_{\succ} \succeq \pi(h^*)$. That is, h^* satisfies $h^* \subset \underline{h}$ and has the property that if $h' \subset h^*$ and the outcome of the subgame determined by h' satisfies $\pi^*_{\succ} \succeq \pi(h')$, then $h' = h^*$.

Either $h^* = \emptyset$, or $\pi(h^*) = \pi^*_{\succ}$, or $\pi^*_{\succ} \succ \pi(h^*)$. We will show that $h^* = \emptyset$. If $\pi(h^*) = \pi^*_{\succ}$, then $h^* = \emptyset$ because otherwise there is an expert i^* who moves immediately before the subgame and Expert i^* can guarantee a payoff of at least $u_{i^*}(\pi^*_{\succ})$ by inducing h^*. Hence her equilibrium action must generate a project that the manager prefers less than $\pi^*_{\succ}$, which contradicts the minimality of h^*. If $\pi^*_{\succ} \succ \pi(h^*)$, then it follows from neutrality that either $h^* = \emptyset$ or there is a history $h' \subset h^*$, $h' \neq h^*$ in which $\pi^*_{\succ} \in P_X(h')$. But the existence of h' would contradict the minimality of h^* because Expert $P_I(h')$ could guarantee the project $\pi^*_{\succ}$ in the subgame starting from h'. Hence Expert $P_I(h')$ must obtain utility at least $u_{P_I(h')}(\pi^*_{\succ})$ and therefore $\pi^*_{\succ} \succeq \pi(h')$. □

Proposition 5 For any unconstrained protocol, there is always an outcome that survives IDWDS that is not greater than π^*.

Proof Fix a protocol. We say that a history generates project π if π is the maximum of projects supported in the history. Because anything greater than π^* is strictly worse than π^* for every expert, supporting π^* is the unique best reply at any history that generates π^* provided that the manager will consult no further experts independent of the choice. Consequently, at least one of her strategies that survive IDWDS must involve supporting π^* at histories that generate π^*. By backward induction, there must exist a strategy profile that survives IDWDS such that experts will choose π^* in any history that generates project π^*, because each expert's unique best response to these histories is to support no further projects when the expert expects future experts to support no new projects. It follows that one of the final projects supported will be π^* if the first expert supports π^*. So if the strategy π^* survives IDWDS for the first expert, then the proof is complete. If the first expert's strategy π^* has been deleted, it is weakly dominated by a strategy that leads to a weakly higher payoff for the first expert against all remaining strategies of the other experts. Consequently at least one project that survives IDWDS does at least as well as π^* for the first expert. By definition of π^*, the first expert strictly prefers π^* to anything greater than π^*. Hence there exists a project that survives IDWDS that is no greater than π^*. □

Proposition 6 Fix a sequential protocol. If the experts' utility functions are one-to-one, there is a unique outcome that survives IDWDS.

Proof If an expert is indifferent between two strategies given any strategy profile of other players, the two strategies must lead to the same project because utility functions are one-to-one. So the other experts are indifferent between the two profiles (differing only in that expert's action) as well. Hence the **transference of decision-maker indifference** (TDI) condition in Marx and Swinkels [5] holds.

Furthermore, it is clear that there is a unique outcome from backward induction in this game (with no ties). Marx and Swinkels [5] show that if an extensive-form game with perfect information that satisfies TDI has a unique backward-induction payoff, then any full reduction by weak dominance also contains only that payoff. Therefore there is a unique project that survives IDWDS, which is the unique backward-induction solution. □

Proposition 8 Suppose X is completely ordered and u_i is one-to-one for each i. The best outcome that the manager can achieve by restricting strategies in a simultaneous-move game is $\pi^\dagger$. That is,

$$\pi^\dagger = \max\{\tilde{\pi}(X^c) : X^c \text{ is admissible}\}.$$

Proof Proposition 1 applies to the simultaneous-move game restricted to X^c. Hence $\tilde{\pi}(X^c)$ is well defined and equal to

$$\min\{\pi : u_i(\pi) > u_i(\pi_i) \text{ for all } \pi_i \in X^c \text{ such that } \pi_i > \pi \text{ and all } i\}$$

(because X is linearly ordered, one can replace $x_i > \pi$ by $x_i \succ \pi$ in this expression).

Because $\pi^\dagger \in X^\dagger$, there exists a finite set of projects $\{\pi_0, \pi_1, \ldots, \pi_K\}$ such that $\underline{x} = \pi_0 < \pi_1 < \cdots < \pi_K = \pi^\dagger$; $\pi_i \in X^\dagger$ for all $i = 0, \ldots, K$ and

$$\text{for each } k = 1, \ldots, K \text{ there exists } i \text{ such that } u_i(\pi_k) > u_i(\pi_{k-1}). \tag{26}$$

Let $\tilde{X}$ denote the set $\{\pi_0, \pi_1, \ldots, \pi_K\}$. $\pi^\dagger$ must be the manager's most preferred element in $\tilde{X}$ because $\tilde{X} \subset X^\dagger$ and $\pi^\dagger$ is the manager's most preferred element of $X^\dagger$. By (26), $\tilde{\pi}(\tilde{X}) = \pi^\dagger$. It remains to show that for all admissible X^c, the manager prefers $\pi^\dagger$ to $\tilde{\pi}(X^c)$.

Recall that $X_0 = \{\underline{x}\}$ and X_k is defined inductively by

$$X_k = \{\pi \in X : \text{ there exists } i \text{ and } \pi_m \in X_m \text{ for } m < k \text{ such that } \pi > \pi_m \text{ and } u_i(\pi) > u_i(\pi_m)\}$$

and that $\tilde{\pi}(X^c)$ is defined to be the project that survives IDWDS when experts are restricted to projects in X^c. We will show that for any admissible X^c for which the manager prefers $\tilde{\pi}(X^c)$ to $\pi^\dagger$, there exists k such that $\tilde{\pi}(X^c) \in X_k$. Hence $\pi^c \equiv \tilde{\pi}(X^c) \in X^\dagger$ and, by definition of $\pi^\dagger$, the manager prefers $\pi^\dagger$ to π^c. This establishes the proposition.

We prove the claim by constructing a sequence $\{\pi_0^c, \ldots, \pi_k^c\}$ such that $\pi_0^c = \underline{x}$, $\pi_{k'}^c > \pi_{k'-1}^c$ for $k' = 1, \ldots, k$, $\pi_{k'}^c \in X_{k'} \cap X^c$, $\pi_k^c = \pi^c$.

Set $\pi_0^c = \underline{x}$. If $\underline{x} = \pi^c$, then we set $k = 0$ and we are done. Otherwise, assume that we have constructed $\{\pi_0^c, \ldots, \pi_m^c\}$ such that $\pi_0^c = \underline{x}$, $\pi_{m'}^c > \pi_{m'-1}^c$ for $m' = 1, \ldots, m$, $\pi_{m'}^c \in X_{m'} \cap X^c$. If $\pi_m^c = \pi^c$, then we set $k = m$ and we are done. Otherwise, observe that $\pi_m^c = \tilde{\pi}(\{\pi_0^c, \ldots, \pi_m^c\})$ because π_m^c is the only element of $\{\pi_0^c, \ldots, \pi_m^c\}$ for which $u_i(\pi_{m'}^c) > u_i(\pi_{m''}^c)$ for all $m'' > m'$. It must be that $\tilde{\pi}(X^c) \geq \tilde{\pi}(\{\pi_0^c, \ldots, \pi_m^c\})$ because $\tilde{\pi}(X^c) \geq \pi^\dagger$ by hypothesis, $\pi^\dagger$ is the largest element of $X^\dagger$, and $\{\pi_0^c, \ldots, \pi_m^c\} \subset X^\dagger$. Therefore, $\pi_m^c \neq \pi^c$ implies $\tilde{\pi}(X^c) > \pi_m^c$.

Consequently, there must be $\pi^c_{m+1} > \pi^c_m$ such that $\pi^c_{m+1} \in X_{m+1} \cap X^c$. Otherwise, the outcome of the game restricted to X^c could be no greater than π^c_m. Hence we can continue to add new elements to the sequence $\{\pi^c_m\}$ until we have included π^c. Because X is finite, the process must eventually include π^c. □

Proposition 9 Suppose X is completely ordered and u_i is one-to-one for each i. There exists a sequential game in which $\pi^\dagger$ is the unique outcome that survives IDWDS. Furthermore, there is no sequential protocol with restricted strategy spaces that achieves a higher outcome.

Proof The manager can obtain $\pi^\dagger$ by restricting the strategy space to a chain of preference ending in $\pi^\dagger$ and using CSP. It remains to show that the manager cannot do better than $\pi^\dagger$. Consider the game generated by a protocol. Because preferences are generic, there will be a unique subgame-perfect equilibrium project. Call a history h nice if the equilibrium outcome induced in the subgame determined by h is in $X^\dagger$. There exist nice histories. For example, a history in which no project has been supported and all actions at this history are terminal is nice. It suffices to show that $\emptyset$ is a nice history. Let $h^* \neq \emptyset$ be a nice history. We claim that the history h that immediately precedes h^* is nice. Because nice histories exist, the claim proves that $\emptyset$ is a nice history and proves the proposition. To establish the claim, note that because h^* is nice, the outcome induced in the subgame determined by h^* and equilibrium strategies is in $X^\dagger$ and that the expert, call her Expert i, consulted initially in the subgame generated by h has an action that generates history h^*. If the equilibrium specifies that Expert i take another action, it must be that Expert i prefers the outcome induced to the outcome induced following history h^*. Consequently, the equilibrium outcome induced in the subgame determined by h is in $X^\dagger$ and therefore h is a nice history. □

Appendix 2: Examples

In this subsection, we present examples that demonstrate that we cannot strengthen the conclusion of Proposition 1 and that, in general, the order of deleting strategies matters. In this case Proposition 1 identifies a unique payoff that survives IDWDS when payoff functions are generic and X is one dimensional. Hence the pathologies when projects are completely ordered are all due to ties in payoff functions. We cannot guarantee that either project π^* or $\tilde{\pi}^*$ will survive IDWDS nor can we guarantee that all payoffs that survive IDWDS are greater than or equal to $u_i(\pi^*)$.

In Examples 4–9 assume that there are three projects, A, B, and C. The manager prefers C to B to A and the projects are completely ordered according to these preferences (so that, for example, $M(A, B) = B$). Project A is the status quo. The specification of expert preferences determines the payoff matrices.

Example 4

Consider the following game:

	A	B	C
A	$2, 0$	$1, 2$	$1, 1$
B	$1, 2$	$1, 2$	$1, 1$
C	$1, 1$	$1, 1$	$1, 1$

Expert preferences are given by: $u_1(A) = 2, u_1(B) = u_1(C) = 1$: and $u_2(A) = 0, u_2(B) = 2, u_2(C) = 1$. We have $\pi^* = C$ and $\tilde{\pi}^* = B$. If we first discard the bottom two strategies of Expert 1 (the row player), (A, B) is the only strategy profile that survives IDWDS; alternatively, discarding Expert 2's A and C, leaves (x_1, B) for $x_1 = A, B, C$ (so either B or C is the project implemented). You cannot delete the profile (A, B). So the set of projects that survive IDWDS always includes $\tilde{\pi}^*$, but may or may not include π^*. Although π^* is a project that survives IDWDS for some order of deletion, this example demonstrates that we cannot guarantee that π^* survives independent of the deletion order.

Example 5

Consider the following game:

	A	B	C
A	$1, 1$	$-1, 0$	$1, 0$
B	$-1, 0$	$-1, 0$	$1, 0$
C	$1, 0$	$1, 0$	$1, 0$

We have $\pi^* = C$ and $\tilde{\pi}^* = A$. If we discard Expert 2's B and C, (A, A) and (C, A) survive IDWDS. Discarding Expert 1's A and B leads to (C, x_2) for $x_2 = A, B, C$ surviving. You cannot delete Expert 1's strategy C. Here the set of projects that survive IDWDS always includes π^*, but may or may not include $\tilde{\pi}^*$.

Taken together, the examples show that you need not select π^* or $\tilde{\pi}^*$. The examples are consistent with the observation that you will select one or the other and that there is a way of deleting weakly dominated strategies that will select both. The second claim is not true, however.

Example 6

Consider the following game:

	A	B	C
A	$0, 0$	$0, -1$	$-1, -1$
B	$0, -1$	$0, -1$	$-1, -1$
C	$-1, -1$	$-1, -1$	$-1, -1$

We have $\pi^* = C$ and $\tilde{\pi}^* = A$. (B, A) and (A, A) are equilibria that survive IDWDS, but it is weakly dominated to enable the project C. Consequently $\tilde{\pi}^*$ is an equilibrium outcome that survives IDWDS; π^* is not an equilibrium outcome that survives IDWDS; and there is an equilibrium outcome that survives IDWDS strictly between $\tilde{\pi}^*$ and π^*.

It is also possible to construct an example in which $\tilde{\pi}^*$ is not an equilibrium outcome that survives IDWDS.

Example 7

Consider the following game:

	A	B	C
A	$2, 1$	$0, 1$	$2, 0$
B	$0, 1$	$0, 1$	$2, 0$
C	$2, 0$	$2, 0$	$2, 0$

We have $\pi^* = C$ and $\tilde{\pi}^* = A$. The only equilibrium outcome that survives IDWDS is π^*.

The next example shows IDWDS need not bound the utility of the experts. That is, the bound on utility given in Proposition 1 requires a restriction to equilibrium strategies.

Example 8

Consider the following game:

	A	B	C
A	$1, 0$	$-1, 0$	$0, 0$
B	$-1, 0$	$-1, 0$	$0, 0$
C	$0, 0$	$0, 0$	$0, 0$

In this example, $\pi^* = C$ and $\bar{\pi}^* = A$. Expert 1's B strategy is weakly dominated, but all other strategies survive. Consequently, profile (A, B) survives IDWDS, although it is not an equilibrium. The project B induced by (A, B) gives Expert 1 a payoff less than $u_1(\pi^*)$.

Now we turn to a situation in which the strategies are not completely ordered.

Example 9

Consider the following game:

	(0, 0)	(1, 0)	(0, 1)	(1, 1)
(0, 0)	$2, 2$	$1, -1$	$-1, 1$	$0, 0$
(1, 0)	$1, -1$	$1, -1$	$0, 0$	$0, 0$
(0, 1)	$-1, 1$	$0, 0$	$-1, 1$	$0, 0$
(1, 1)	$0, 0$	$0, 0$	$0, 0$	$0, 0$

In this example, the set of projects has two dimensions. If one expert supports (1, 0) and the other supports (0, 1), then the manager will implement (1, 1). The experts' preferences are: $\{0, 0\} \succ_1 \{1, 0\} \succ_1 \{1, 1\} \succ_1 \{0, 1\}$ and $\{0, 0\} \succ_2 \{0, 1\} \succ_2 \{1, 1\} \succ_2 \{1, 0\}$. There are 2 pure-strategy equilibrium projects: (0, 0), (1, 1). It is dominated for Expert 1 to support 1 on the second dimension and for Expert 2 to support 1 on the first dimension. So discard (1, 1), (0, 1) for Expert 1 and (1, 1), (1, 0) for Expert 2. No other strategy can be deleted. Every project survives IDWDS. In this example $\pi^* = \bar{\pi}^* = (0, 0)$, but $\bar{\pi}^* > \pi^*$ and there are strategies that survive IDWDS that exceed π^*.

Example 9 illustrates how the manager's most preferred project may survive iterative deletion of weakly dominated strategies in the simultaneous-move game even when the largest strategy is weakly dominated. This cannot happen in the one-dimensional case. In the example, (1, 0) is an unattractive project from the perspective of Expert 2. If this project is a possibility, then Expert 2 will have a justification for using (0, 1). Similarly, Expert 1 would not delete her strategy (1, 0) because she prefers $(1, 1)$ to $(0, 1)$. Although $(0, 0)$ remains an equilibrium outcome that survives IDWDS, the manager's most preferred outcome is more robust than it is in the one-dimensional case.

References

1. Armstrong, M., Vickers, J.: A model of delegated project choice. Econometrica **78**(1), 213–244 (2010)
2. Goodman, J.D.: Amazon pulls out of planned New York City headquarters. The New York Times (2019)
3. Holmström, B.: On the theory of delegation. Bayesian Models in Economic Theory, Studies in Bayesian Econometrics, 5 (1984)
4. Hu, P., Sobel, J.,: Getting permission. Am. Econ. Rev.: Insights **4**(4), 459–472 (2022)
5. Marx, L., Swinkels, J.M.: Order independence for iterated weak dominance. Games Econ. Behav. **18**, 219–245 (1997)
6. Sobel, J.: Iterated weak dominance and interval-dominance supermodular games. Theor. Econ. **14**(1), 71–102 (2019)
7. Stevens, L., Morris, K., Honan, K.: Amazon Picks New York City, Northern Virginia for its HQ2 Locations (2018)
8. Topkis, D.M.: Supermodularity and Complementarity. Princeton University Press (2011)
9. Vries, M.A.D.: The New Robert's Rules of Order. Signet (1998)

Existence of Haar Measures: A Note on Rudin's Fixed-Point Approach

Toru Maruyama

Abstract Rudin [4] proved the existence of Haar measures on compact Hausdorff topological groups by means of the Kakutani fixed-point theorem. In this note, we extend Rudin's idea so as to cover the case of locally compact groups.

Keywords Locally compact group · Haar measure · Kakutani fixed point theorem

1 Introduction

Let G be a locally compact Hausdorff topological group (not necessarily commutative). A positive Radon measure ζ on G is called a left-Haar measure (resp. right-Haar measure) if $\zeta(E) = \zeta(aE)$ (resp. $\zeta(E) = \zeta(Ea)$) for any $E \in \mathscr{B}(G)$ and $a \in G$, where $\mathscr{B}(G)$ is the Borel σ-field on G. The following theorem due to Weil [6] is well known as a basic foundation of harmonic analysis.

Theorem *There exists a left- and a right-Haar measure on G. It is unique up to multiplications by positive constants.*

Some detailed expositions of Weil's proof have been given, say by Hewitt-Ross [2] (Chap. 4, Sect. 15). Rudin [4] (pp. 128–134) gave an alternative proof based on a fixed-point theorem for a group of linear operators[1]:

Kakutani's Fixed-point Theorem *Let $\mathscr{X}$ be a locally convex Hausdorff topological vector space and $K \subset \mathscr{X}$ a nonempty compact convex subset. Suppose that $\mathscr{G}$ is a group of linear operators of G into itself which satisfies*

(i) $\mathscr{G}$ is equi-continuous on K, and

[1] Kakutani [3], Dunford and Schwartz [1] p. 457.

T. Maruyama (✉)
Keio University, Minato, Japan
e-mail: maruyama.a2@keio.jp

M. A. Khan et al. (eds.), *Matching, Dynamics and Games for the Allocation of Resources*, Monographs in Mathematical Economics 7,
https://doi.org/10.1007/978-981-96-6661-4_10

(ii) $T(K) \subset K$ *for all* $T \in \mathscr{G}$.
Then there exists a common fixed point $x^* \in K$ *for* $\mathscr{G}$; *that is,*

$$x^* = T(x^*) \quad \text{for all} \quad T \in \mathscr{G}.$$

However Rudin's proof is restricted to the case G is compact.

In this brief note, a trial to extend Rudin's idea to a general locally compact group will be given.

2 A Lemma

Through this paper, G stands for a locally compact Hausdorff topological group. (We do not assume the commutativity of G.)

Given $x_0 \in G$ and $f \in \mathfrak{C}_\infty(G, \mathbb{R})$, a function $S_{x_0}(f) \in \mathfrak{C}_\infty(G, \mathbb{R})$ is defined by

$$S_{x_0}(f)(x) = f(x_0 x), \quad x \in G. \tag{1}$$

$x \mapsto S_x(f)$ is a mapping of G into $\mathfrak{C}_\infty(G, \mathbb{R})$. $\mathfrak{C}_\infty(G, \mathbb{R})$ is the space of all continuous functions of G into $\mathbb{R}$ which vanish at infinity. It is endowed with the uniform convergence norm $\|\cdot\|_\infty$.

The following lemma is a generalization of Rudin's result.

Lemma *(i) The mapping* $x \mapsto S_x(f)$ *is continuous.*

(ii) The set

$$H(f) = \overline{\text{co}}\{S_x(f) \mid x \in G\}$$

is compact in $\mathfrak{C}_\infty(G, \mathbb{R})$.[2]

Proof (i) 1°. Since $f \in \mathfrak{C}_\infty(G, \mathbb{R})$, there exists, for each $\varepsilon > 0$, a compact set $K \subset G$ such that

$$|f(x)| < \frac{\varepsilon}{2} \quad \text{for} \quad x \in G \backslash K = W_0. \tag{2}$$

For each $a \in K$, there is a neighborhood W_a of the unit e such that

$$|f(x) - f(a)| < \varepsilon \quad \text{for} \quad x \in aW_a. \tag{3}$$

There also exists a neighborhood V_a of e which satisfies

$$V_a V_a^{-1} \subset W_a, \quad a \in K. \tag{4}$$

[2] $\overline{\text{co}}\, A$ signifies the closed convex hull of A.

2°. Since K is compact and $\{aV_a \mid a \in K\}$ is a covering of K, there are finite elements $a_1, a_2, \ldots, a_n \in K$ such that

$$K \subset \bigcup_{i=1}^{n} a_i V_{a_i}. \tag{5}$$

It is obvious that

$$G = \left(\bigcup_{i=1}^{n} a_i V_{a_i} \right) \cup W_0.$$

Let V be a symmetric neighborhood of e which is contained in every V_{a_i}; that is,

$$V \subset \bigcap_{i=1}^{n} V_{a_i}, \tag{6}$$

3°. We now prove that

$$\| S_x(f) - S_y(f) \|_\infty \leqq 2\varepsilon \quad \text{if} \quad x \in yV \ (\Longleftrightarrow y \in xV^{-1}). \tag{7}$$

There are two possible cases for the location of y;
(A) $y \in a_i V_{a_i}$ for some i, (B) $y \in W_0$.
We examine the situation for each case.

In the case of (A),

$$| f(y) - f(a_i) | < \varepsilon \tag{8}$$

holds good by (3) since $y \in a_i V_{a_i} \subset a_i W_{a_i}$. Taking account of the relation

$$\begin{aligned} x \in yV &\subset a_i V_{a_i} V^{-1} \ (V \text{is symmetric}) \\ &\subset a_i V_{a_i} V_{a_i}^{-1} \subset a_i W_{a_i} \quad (\text{by}(4)), \end{aligned}$$

we obtain

$$| f(x) - f(a_i) | < \varepsilon. \tag{9}$$

Combining (8) and (9), we have

$$| f(x) - f(y) | < 2\varepsilon.$$

We next examine the case of (B). If $x \in W_0$, $| f(x) - f(y) | < 2\varepsilon$ holds true obviously. If $x \notin W_0$, on the contrary, $x \in a_i V_{a_i}$ for some i. Hence we must have $x \in a_i V_{a_i} \cap yV$. It follows from $x \in yV$ that

$$y \in xV^{-1} \subset a_i V_{a_i} V^{-1} \subset a_i V_{a_i} V_{a_i}^{-1} \subset a_i W_{a_i} \quad (\text{by}(4)).$$

This implies that

$$|f(y) - f(a_l)| < \varepsilon. \tag{10}$$

By $x \in a_l V_{a_l} \subset a_l W_{a_l}$, we also have

$$|f(x) - f(a_l)| < \varepsilon. \tag{11}$$

It follows again that

$$|f(x) - f(y)| < 2\varepsilon$$

from (10) and (11).

So, regardless of the location of y, we obtain

$$|f(x) - f(y)| < 2\varepsilon \quad \text{provided that} \quad x \in yV. \tag{12}$$

4°. So far, we have consecutively constructed W_a, V_a and then V. By a quite similar procedure, we obtain another symmetric neighborhood V' of e which satisfies[3]

$$|f(x) - f(y)| < 2\varepsilon \quad \text{if} \quad x \in V'y. \tag{12'}$$

Finally let U be a symmetric neighborhood of e such that

$$U \subset V \cap V'.$$

It is clear that

$$|f(x) - f(y)| < 2\varepsilon \quad \text{if either} \quad x \in yU \quad \text{or} \quad x \in Uy. \tag{13}$$

5°. Since $(xu)(yu)^{-1} = xy^{-1}$ for any $u \in G$, $xy^{-1} \in U$ implies that $(xu)(yu)^{-1} \in U$; *i.e.*, $xu \in U(yu)$. Thus it follows, by (13), that

[3] There exists, for each $a \in K$, a neighborhood W'_a of e such that

$$|f(x) - f(y)| < 2\varepsilon \quad \text{if} \quad x \in W'_a a.$$

Let V'_a be a neighborhood of e which satisfies $(V'_a)^{-1}V'_a \subset W'_a$. There exist some finite elements $a'_1, a'_2, \cdots, a'_m \in K$ such that $K \subset \bigcup_{j=1}^{m} V_{a'_j} a'_j$. Then there exists a symmetric neighborhood V' of e contained in every V'_{a_j} ; i.e., $V' \subset \bigcap_{j=1}^{m} V_{a'_j}$. V' satisfies (12′).

The discussion in 4° is required only for the proof of (14).

$$| f(xu) - f(yu) | < 2\varepsilon \quad \text{for any} \quad u \in G$$
$$\text{provided that} \quad x \in yU. \tag{14}$$

Equation (7) is an immediate consequence of (14). This proves (i).

(ii) Fix $x \in G$. Since $(xu)^{-1}(xv) = u^{-1}v$ for any $u, v \in G$:

$$xv \in (xu)\,U \quad \text{if} \quad u^{-1}v \in U.$$

Hence it follows that

$$| f(xu) - f(xv) | < 2\varepsilon \quad \text{if} \quad u^{-1}v \in U$$

by (14), which is equivalent to

$$| S_x(f)(u) - S_x(f)(v) | < 2\varepsilon \quad \text{if} \quad u^{-1}v \in U. \tag{15}$$

This holds true for any $x \in G$, and the choice U is independent of $x \in G$. In other words, $\{S_x(f) \mid x \in G\}$ is an equi-continuous family in $\mathfrak{C}_\infty(G, \mathbb{R})$. It is obvious that $\{S_x(f)(u) \mid x \in G\}$ is bounded in $\mathbb{R}$ for any $u \in G$. Thus $\{S_x(f) \mid x \in G\}$ is relatively compact by the Ascoli–Arzelà theorem (Schwartz [5] pp. 78–82). Applying the Mazur theorem (Dunford and Schwartz [1] pp. 416–417), we conclude that

$$H(f) = \overline{\text{co}}\{S_x(f) \mid x \in G\}$$

is compact. (Q.E.D)

3 Proof of the Theorem

We now proceed to the proof of the theorem. We restrict ourselves to the existence of left-Haar measures. The case of right-Haar measures is similarly proved.

1° Define an linear operator $S_x : \mathfrak{C}_\infty(X, \mathbb{R}) \to \mathfrak{C}_\infty(G, \mathbb{R})$ by

$$S_x : f \mapsto S_x(f) \tag{16}$$

for each $x \in G$. Since S_x is an isometry, $\{S_x \mid x \in G\}$ is an equi-continuous family of linear operators. It is a group with respect to the composition $S_x \circ S_y = S_{xy} (x, y \in G)$. As shown in the lemma (in Sect. 2), $H(f) = \overline{\text{co}}\{S_x(f)\} \mid x \in G\}$ is compact in $\mathfrak{C}_\infty(G, \mathbb{R})$ for each $f \in \mathfrak{C}_\infty(G, \mathbb{R})$. It is easily verified that $S_x(H(f)) = H(f)$.

Fix a $f \in \mathfrak{C}_\infty(G, \mathbb{R})$ and restrict the operator S_x to $H(f)$. Then, by the Kakutani fixed-point theorem:

$$\{S_x : H(f) \to H(f) \mid x \in G\}$$

has a common fixed point φ^* in $H(f)$; i.e.:

$$\varphi^* = S_x(\varphi^*) \quad \text{for all} \quad x \in G. \tag{17}$$

Since $\varphi^*(xe) = \varphi^*(e)$ for all $x \in G$. φ^* is actually a constant function. This φ^* can be uniformly approximated by functions in $\text{co}\{S_x(f) \mid x \in G\}$.

Summing up—there exists, for each $f \in \mathfrak{C}_\infty(G, \mathbb{R})$, a constant function c which can be uniformly approximated by functions in $\text{co}\{S_x(f) \mid x \in G\}$.

2° In this second stage, we show that the above constant function c is determined uniquely for each $f \in \mathfrak{C}_\infty(G, \mathbb{R})$. Suppose that c and c' can be approximated by functions in $\text{co}\{S_x(f) \mid x \in G\}$. Then there exist, for each $\varepsilon > 0$, two finite sets $\{a_i\}$, $\{b_j\}$ in G and two sets $\{\alpha_i\}$, $\{\beta_j\}$ of real coefficients which satisfy

$$\alpha_i \geqq 0, \quad \sum_i \alpha_i = 1,$$

$$\beta_i \geqq 0, \quad \sum_j \beta_i = 1$$

such that

$$|\, c - \sum_i \alpha_i f(a_i x) \,| < \varepsilon \quad \text{for all} \quad x \in G, \tag{18}$$

$$|\, c' - \sum_j \beta_j f(b_j x) \,| < \varepsilon \quad \text{for all} \quad x \in G. \tag{19}$$

Substituting $x = b_j$ in $c - \sum_i \alpha_i f(a_i x)$, multiplying by β_j and adding them over j, we obtain

$$|\, c - \sum_{i,j} \alpha_i \beta_j f(a_i b_j) \,| < \varepsilon. \tag{20}$$

By applying similar procedure to (19),[4] we also have

$$|\, c' - \sum_{i,j} \alpha_i \beta_j f(a_i b_j) \,| < \varepsilon. \tag{21}$$

$c = c'$ immediately follows. Hence c and c' are uniquely determined and they are the same constant.

3° We denote by $I(f)$ the uniquely determined constant c for each f. $I(f)$ satisfies the following properties: for $f, g \in \mathfrak{C}_\infty(G, \mathbb{R})$, $\alpha \in \mathbb{R}$.

(i) $I(f + g) = I(f) + I(g)$,
(ii) $I(\alpha f) = \alpha I(f)$,

[4] Substitute $x = b_j^{-1} a_i b_j$ in $c' - \sum_j \beta_j f(b_j x)$, multiply by α_i and add them up.

(iii) $I(f) \geqq 0$ if $f \geqq 0$,
(iv) $I(S_x(f)) = I(f)$ for $x \in G$.

For the verification of (i)–(iv), see Rudin [4] pp. 131–132.
4° For the Riesz–Markov–Kakutani theorem (cf. Hewitt and Ross [2], for instance), the functional $I(\cdot)$ can be represented as a measure. That is, there exists a regular Borel measure ζ which satisfies

$$I(f) = \int_G f d\zeta = \int_G S_x(f) d\zeta, \quad x \in G. \tag{22}$$

ζ is determined uniquely. Any positive constant multiple ζ' of ζ also satisfies

$$\int_G f d\zeta' = \int_G S_x(f) d\zeta'.$$

The relation (22) is valid for any characteristic function χ_E of $E \in \mathscr{B}(G)$ since $\mathfrak{C}_\infty(G, \mathbb{R})$ is dense in $\mathfrak{L}^1(G, \mathbb{R})$ (with the measure ζ).

This proves that ζ is a left-Haar measure which represents $I(\cdot)$. (Q.E.D)

I expect the fixed-point approach shown above to be helpful for readers familiar with functional analysis, since it seems relatively accessible for those mathematicians in comparison with Weil's ingeniously constructive proof.

Acknowledgements Valuable comments by the anonymous referee are gratefully appreciated.

References

1. Dunford, N.: Schwartz, J.T.: Linear Operators, Part 1. Interscience, New York (1958)
2. Hewitt, E., Ross, K.A.: Abstract Harmonic Analysis, vol. I. Springer, Berlin/Gòttingen/ Heidelberg (1963)
3. Kakutani, S.: Two fixed-point theorems concerning bicompact convex sets. Proc. Imp. Acad. Tokyo **14**, 242–245 (1938)
4. Rudin, W.: Functional Analysis, 2nd edn. McGraw-Hill, New York (1991)
5. Schwartz, L.: Functional Analysis. Courant Institute New York University, New York (1964)
6. Weil, A.: L' integration dans les groupes topologiques et ses applications. Hermann, Paris (1941, 1951)

Economic Dynamics and Growth

Infinite-Horizon Single-Objective and Multiobjective Optimal Control Problems in the Continuous Time Framework

Joël Blot and Hasan Yilmaz

Abstract We build a generalization of the Pontryagin maximum principle in infinite horizon in the framework of the continuous time. We improve a classical result of Halkin by lightening the assumptions. We also provide conditions to ensure that the multiplier associated to the criterion is equal to 1. We extend these results to the multiobjective problems. In this framework, we establish a new sufficient condition of Pareto optimality for multiobjective optimal control problems in infinite horizon.

Keywords Infinite horizon · Pareto optimality · Pontryagin maximum principle

1 Introduction

These last years, we have seen numerous important contributions on the theory of the continuous-time infinite-horizon optimal control notably in the works of H. Halkin, A.J. Zaslavski, S. Pickenhain, S.M. Aseev and V.M. Veliov, P. Michel, L. Montrucchio, and other authors; see, e.g., [1, 5, 6, 9–11, 13] and references therein.

For the infinite-horizon multiobjective optimal control problems, some results on the Pontryagin maximum principle can be found in [12] for the continuous time framework and in [8] for the discrete time framework.

In this paper, we establish an improvement of the Pontryagin maximum principle for single-objective and multiobjective optimal control problems in the continuous time framework. Indeed, we generalize the Pontryagin maximum principle established by Halkin in [6] in infinite horizon for single-objective and multiobjective optimal control problems by using a new Pontryagin maximum principle established in [2] and in [7]. Firstly, the control space is a Hausdorff topological space instead of a closed subset of $\mathbb{R}^r$. Next, for the vector field and the integrand of the criterion, we lighten assumptions of Fréchet differentiability by using assumptions of the Gâteaux differentiability. Besides, we provide qualification conditions to obtain the multiplier associated to the criterion equal to 1 for the single-objective problem and

J. Blot (✉) · H. Yilmaz
Université Paris 1 Panthéon-Sorbonne, Paris, France
e-mail: blot@univ-paris1.fr

M. A. Khan et al. (eds.), *Matching, Dynamics and Games for the Allocation of Resources*, Monographs in Mathematical Economics 7,
https://doi.org/10.1007/978-981-96-6661-4_11

the multipliers associated to the criteria different from zero for the multiobjective problem.
Finally, for the multiobjective optimal control problems, we provide sufficient conditions of Pareto optimality.

We summarize the content of this paper as follows.

In Sect. 2, we recall some definitions, and we introduce our notations.

In Sect. 3, we provide our Pontryagin maximum principle for the single-objective problems. We prove this result by using the reasoning of Halkin. Indeed, we show that the restriction of the solution is a solution of finite-horizon optimal control problem. Next, by using a new Pontryagin maximum principle established in [2] we obtain this result. Finally, we introduce constraint qualifications.

In Sect. 4, we generalize the results established in Sect. 2 for the multiobjective optimal control problems.

In Sect. 5, we provide sufficient conditions of Pareto optimality.

2 Recall and Notations

When X is a Hausdorff topological space, when $a, b \in \mathbb{R}$ s.t. $a < b$, $PC^0([a, b], X)$ denotes the space of the piecewise continuous functions from $[a, b]$ into X.

As in [2], we recall that $x \in PC^0([a, b], X)$ when x is continuous on $[a, b]$ or when there exists a subdivision of $[a, b]$, $a = \tau_0 < \tau_1 < ... < \tau_k < \tau_{k+1} = b$ such that x is continuous at t when $t \notin \{\tau_i : 0 \leq i \leq k+1\}$ and the right-hand limit $x(\tau_i+)$ exists in X and when $i \in \{0, ..., k\}$ and the left-hand limit $x(\tau_i-)$ exists in X when $i \in \{1, ..., k+1\}$.

We define $NPC^0_R([a, b], X)$ as the set of the $x \in PC^0([a, b], X)$ which are right-continuous on $[a, b[$ and left-continuous at b. An element of $NPC^0_R([a, b], X)$ is called a normalized piecewise continuous function, cf. [2].

We define $NPC^0_R(\mathbb{R}_+, X)$ as the set of the functions $x : \mathbb{R}_+ \to X$ s.t. for all $c, d \in \mathbb{R}_+$, $c < d$, $x_{|[c,d]} \in NPC^0_R([c, d], X)$ where $\mathbb{R}_+ := [0, +\infty[$.

When X is included into a real normed vector space, $PC^1([a, b], X)$ denotes the space of the piecewise continuously differentiable functions from $[a, b]$ into X. We recall that $x \in PC^1([a, b], X)$ when x is continuously differentiable on $[a, b]$ or when x is continuous on $[a, b]$ and there exists a subdivision, $a = \tau_0 < \tau_1 < ... < \tau_k < \tau_{k+1} = b$ such that x is continuously differentiable at t if $t \notin \{\tau_i : 0 \leq i \leq k+1\}$ the right derivative $x'_R(\tau_i) = x'(\tau_i+)$ exists when $i \in \{0, ..., k\}$, the left derivative $x'_L(\tau_i) = x'(\tau_i-)$ exists when $i \in \{1, .., k+1\}$, cf. [2].

We define $PC^1(\mathbb{R}_+, X)$ as the set of the functions $x : \mathbb{R}_+ \to X$ s.t. for all $c, d \in \mathbb{R}_+$, $c < d$, $x_{|[c,d]} \in PC^1([c, d], X)$.

As in [2], we consider the extended derivative of $x \in PC^1([a, b], X)$ as follows:

$$\underline{d}x(t) := \begin{cases} x'(t) \text{ if } t \in [a, b] \setminus \{\tau_i : i \in \{0, ..., k+1\}\} \\ x'_R(t) \text{ if } t = \tau_i, i \in \{0, ..., k\} \\ x'_L(t) \text{ if } t = b. \end{cases} \tag{1}$$

Notice that if $x \in PC^1([a,b], X)$ then $\underline{d}x \in NPC^0_R([a,b], X)$. Besides, $\underline{d}$ is a continuous linear operator from $PC^1([a,b], X)$ into $NPC^0_R([a,b], X)$.

When X is normed vector space, let G be an open subset of X. Let $\mathfrak{f} : G \to \mathbb{R}$ be a mapping. The mapping $\mathfrak{f}$ is said to be concave at $x \in G$ when for all $y \in G$, for all $t \in [0,1]$ s.t. $(1-t)x + ty \in G$, $\mathfrak{f}((1-t)x + ty) \geq (1-t)\mathfrak{f}(x) + t\mathfrak{f}(y)$.

Moreover, when X is a finite product of n real normed spaces, $X := \prod_{i=1}^n X_i$, if $i \in \{1, ..., n\}$, $D_{G,i}\mathfrak{f}(x)$ (respectively, $D_{H,i}\mathfrak{f}(x)$, respectively, $D_{F,i}\mathfrak{f}(x)$) denotes the partial Gâteaux (respectively, Hadamard, respectively, Fréchet) differential of $\mathfrak{f}$ at x with respect to the $i-th$ vector variable.

We denote by $\mathcal{B}(\mathbb{R}_+)$ the Borel tribe of $\mathbb{R}_+$, and by $\mathfrak{m}_1$ the canonical positive Borel measure on $\mathcal{B}(\mathbb{R}_+)$.

3 Pontryagin Principle for the Infinite-Horizon Single-Objective Problem

Let Ω be a non-empty subset of $\mathbb{R}^n$, $\xi_0 \in \Omega$, let U be a Hausdorff topological space. We consider the following functions $f : [0,T] \times \Omega \times U \to \mathbb{R}^n$ and $f^0 : [0,T] \times \Omega \times U \to \mathbb{R}$ where $\mathbb{N}_* \equiv \mathbb{N} \setminus \{0\}$ and $n \in \mathbb{N}_*$.

We set $Adm(LI)$ the set of the processes $(x,u) \in PC^1(\mathbb{R}_+, \Omega) \times NPC^0_R(\mathbb{R}_+, U)$ s.t. $x(0) = \xi_0$ and for all $t \in \mathbb{R}_+$ where $x'(t)$ exists, $x'(t) = f(t, x(t), u(t))$.

We denote by $Dom(LI)$ the set of the processes $(x,u) \in Adm(LI)$ such that $\int_0^{+\infty} f^0(t, x(t), u(t))dt$ converges in $\mathbb{R}$.

We consider the function $J : Dom(LI) \to \mathbb{R}$ defined by, for all $(x,u) \in Dom(LI)$, $J(x,u) :\equiv \int_0^{+\infty} f^0(t, x(t), u(t))dt$.

In infinite-horizon optimal control problems, as explained in [5], we can find the following problems:

($\mathcal{P}^n$) Find $(\overline{x}, \overline{u}) \in Dom(LI)$ s.t., for all $(x,u) \in Dom(LI)$, $J(\overline{x}, \overline{u}) \geq J(x,u)$.

($\mathcal{P}^s$) Find $(\overline{x}, \overline{u}) \in Dom(LI)$ s.t., for all $(x,u) \in Adm(LI)$, $J(\overline{x}, \overline{u}) \geq \lim_{T \to +\infty} \sup \int_0^T f^0(t, x(t), u(t))dt$.

($\mathcal{P}^o$) Find $(\overline{x}, \overline{u}) \in Adm(LI)$ s.t., for all $(x,u) \in Adm(LI)$, $\lim_{T \to +\infty} \inf \int_0^T (f^0(t, \overline{x}(t), \overline{u}(t)) - f^0(t, x(t), u(t)))dt \geq 0$.

($\mathcal{P}^w$) Find $(\overline{x}, \overline{u}) \in Adm(LI)$ s.t., for all $(x,u) \in Adm(LI)$, $\lim_{T \to +\infty} \sup \int_0^T (f^0(t, \overline{x}(t), \overline{u}(t)) - f^0(t, x(t), u(t)))dt \geq 0$.

For these problems, the Hamiltonian of Pontryagin is the function $\mathcal{H} : \mathbb{R}_+ \times \Omega \times U \times \mathbb{R}^{n*} \times \mathbb{R} \to \mathbb{R}$ defined by $\mathcal{H}(t, \xi, \zeta, p, \lambda) := \lambda f^0(t, \xi, \zeta) + p \cdot f(t, \xi, \zeta)$ when $(t, \xi, \zeta, p, \lambda) \in \mathbb{R}_+ \times \Omega \times U \times \mathbb{R}^{n*} \times \mathbb{R}$.

Note that a process $(\overline{x}, \overline{u})$ which is a solution of ($\mathcal{P}^n$) is called a natural optimal solution. As in [5], a process $(\overline{x}, \overline{u})$ which is a solution of ($\mathcal{P}^s$) (respectively, ($\mathcal{P}^o$), respectively, ($\mathcal{P}^w$)) is called a strong optimal solution (respectively, overtaking optimal solution, respectively, weak overtaking optimal solution).

As underlined in [5], among the notions of solution introduced, the notion of weak overtaking is the weakest notion of solution. That is why, we will use only this notion of solution in the theorems of this section.

Let $(\overline{x}, \overline{u}) \in Adm(LI)$. Now, we consider the following conditions.

Conditions on the integrand of the criterion.

(AI1) $f^0 \in C^0(\mathbb{R}_+ \times \Omega \times U, \mathbb{R})$, for all $(t, \xi, \zeta) \in \mathbb{R}_+ \times \Omega \times U$, $D_{G,2}f^0(t, \xi, \zeta)$ exists, for all $(t, \zeta) \in \mathbb{R}_+ \times U$, $D_{F,2}f^0(t, \overline{x}(t), \zeta)$ exists and $[(t, \zeta) \mapsto D_{F,2}f^0(t, \overline{x}(t), \zeta)] \in C^0(\mathbb{R}_+ \times U, \mathbb{R}^{n*})$.

(AI2) For all $T \in \mathbb{R}_{+*}$, for all non-empty compact $K \subset \Omega$, for all non-empty compact $M \subset U$, $\sup_{(t,\xi,\zeta)\in[0,T]\times K\times M} \|D_{G,2}f^0(t, \xi, \zeta)\| < +\infty$ where $\mathbb{R}_{+*} :=]0, +\infty[$,

where C^0 means the continuity and $\mathbb{R}^{n*}$ denotes the topological dual space of $\mathbb{R}^n$.

Conditions on the vector field.

(AV1) $f \in C^0(\mathbb{R}_+ \times \Omega \times U, \mathbb{R}^n)$, for all $(t, \xi, \zeta) \in \mathbb{R}_+ \times \Omega \times U$, $D_{G,2}f(t, \xi, \zeta)$ exists, for all $(t, \zeta) \in \mathbb{R}_+ \times U$, $D_{F,2}f(t, \overline{x}(t), \zeta)$ exists and $[(t, \zeta) \mapsto D_{F,2}f(t, \overline{x}(t), \zeta)] \in C^0(\mathbb{R}_+ \times U, \mathcal{L}(\mathbb{R}^n, \mathbb{R}^n))$.

(AV2) For all $T \in \mathbb{R}_{+*}$, for all non-empty compact $K \subset \Omega$, for all non-empty compact $M \subset U$, $\sup_{(t,\xi,\zeta)\in[0,T]\times K\times M} \|D_{G,2}f(t, \xi, \zeta)\| < +\infty$,

$\mathcal{L}(\mathbb{R}^n, \mathbb{R}^n)$ denotes the space of bounded linear mappings from $\mathbb{R}^n$ into $\mathbb{R}^n$.

Now with these elements, for all $T \in \mathbb{R}_{+*}$, for all $\xi \in \Omega$, we consider the following finite-horizon optimal control problem.

$$(\mathcal{L}_{T,\xi}) \begin{cases} \text{Maximize } \int_0^T f^0(t, x(t), u(t))dt \\ \text{subject to } x \in PC^1([0, T], \Omega), u \in NPC^0_R([0, T], U) \\ \qquad\qquad \forall t \in [0, T], \underline{d}x(t) = f(t, x(t), u(t)), \ x(0) = \xi_0 \\ \qquad\qquad x(T) = \xi. \end{cases}$$

Theorem 1 *If $(\overline{x}, \overline{u})$ is a weak overtaking solution then, for all $T \in \mathbb{R}_{+*}$, the process $(\overline{x}_T, \overline{u}_T)$ is a solution of the problem $(\mathcal{L}_{T,\overline{x}(T)})$ where $(\overline{x}_T, \overline{u}_T) := (\overline{x}_{|[0,T]}, \overline{u}_{|[0,T]})$.*

Proof We assume that $(\overline{x}, \overline{u})$ is a weak overtaking solution. Let $T \in \mathbb{R}_{+*}$.

Let (x, u) be an admissible process of $(\mathcal{L}_{T,\overline{x}(T)})$.

Setting, for all $t \in \mathbb{R}_+$, $\tilde{x}(t) := x(t)$ and $\tilde{u}(t) := u(t)$ if $t \leq T$, $\tilde{x}(t) := \overline{x}(t)$ and $\tilde{u}(t) := \overline{u}(t)$ if $t > T$. Note that $(\tilde{x}, \tilde{u}) \in Adm(LI)$.

Since $(\overline{x}, \overline{u})$ is a weak overtaking solution, we have, for all $\varepsilon > 0$, there exists $T_\epsilon > T$ s.t. $\int_0^{T_\epsilon} f^0(t, \overline{x}(t), \overline{u}(t))dt \geq \int_0^{T_\epsilon} f^0(t, \tilde{x}(t), \tilde{u}(t))dt - \varepsilon$.

Consequently, we have, for all $\varepsilon > 0$,
$\int_0^T f^0(t, \overline{x}_T(t), \overline{u}_T(t))dt \geq \int_0^T f^0(t, x(t), u(t))dt - \varepsilon$.

Therefore, we obtain $\int_0^T f^0(t, \overline{x}_T(t), \overline{u}_T(t))dt \geq \int_0^T f^0(t, x(t), u(t))dt$. □

Theorem 2 *When $(\overline{x}, \overline{u})$ is a weak overtaking solution, under (AI1), (AI2), (AV1), and (AV2) there exist a multiplier $\lambda_0 \in \mathbb{R}$ and an adjoint function $p \in PC^1(\mathbb{R}_+, \mathbb{R}^{n*})$ which satisfy the following conditions:*

(NO) $\|(\lambda_0, p(0))\| = 1$.
(Si) $\lambda_0 \geq 0$.
(AE) $p'(t) = -D_{F,2}\mathcal{H}(t, \overline{x}(t), \overline{u}(t), p(t), \lambda_0)$ *for all* $t \in \mathbb{R}_+$ *where* $p'(t)$ *exists.*
(MP) *For all* $t \in \mathbb{R}_+$, *for all* $\zeta \in U$,
$\mathcal{H}(t, \overline{x}(t), \overline{u}(t), p(t), \lambda_0) \geq \mathcal{H}(t, \overline{x}(t), \zeta, p(t), \lambda_0)$.
(CH) $\bar{\mathcal{H}} := [t \mapsto \mathcal{H}(t, \overline{x}(t), \overline{u}(t), p(t), \lambda_0)] \in C^0(\mathbb{R}_+, \mathbb{R})$.

Proof First, we set, for all $x \in \Omega$, for all $i \in \{1, ..., n\}$, $h_i(x) := x_i - \overline{x}_i(T)$. Note that, for all $i \in \{1, ..., n\}$, h_i is Hadamard differentiable at $\overline{x}(T)$ and the Hadamard differential $D_H h_i(\overline{x}(T)) := e_i^*$ where $(e_i^*)_{1 \leq i \leq n}$ is the dual basis of the canonical basis of $\mathbb{R}^n$.

Consequently, by using Theorem 1 and the assumptions (AI1), (AI2), (AV1), and (AV2), for each $T \in \mathbb{R}_{+*}$, the assumptions of Theorem 2.1 and Corollary 2.3, (i), in [2] are verified for the problem $(\mathcal{L}_{T,\overline{x}(T)})$. Therefore, there exist $\lambda_0^T, \mu_1^T, ..., \mu_n^T$ and $p^T \in PC^1([0, T], \mathbb{R}^{n*})$ verifying the following conditions:

(NN^T) $(\lambda_0^T, (\mu_\beta^T)_{1 \leq \beta \leq n})$ is non-zero.
(Si^T) $\lambda_0^T \geq 0$.
(TC^T) $\sum_{\beta=1}^n \mu_\beta^T e_i^* = p^T(T)$.
(AE^T) $\underline{d}p^T(t) = -D_{F,2}\mathcal{H}(t, \overline{x}(t), \overline{u}(t), p^T(t), \lambda_0^T)$ for all $t \in [0, T]$.
(MP^T) For all $t \in [0, T]$, for all $\zeta \in U, \mathcal{H}(t, \overline{x}(t), \overline{u}(t), p^T(t), \lambda_0^T) \geq \mathcal{H}(t, \overline{x}(t), \zeta, p^T(t), \lambda_0^T)$.
(Cor^T) For all $t \in [0, T]$, $(\lambda_0^T, p^T(t)) \neq 0$.

Thanks to (Cor^T), we can take, for each $T \in \mathbb{R}_{+*}$, $\|(\lambda_0^T, p^T(0))\| = 1$. Consequently, there exists $(\lambda_0, p(0)) \in \mathbb{R} \times \mathbb{R}^{n*}$ s.t. $\|(\lambda_0, p(0))\| = 1$, there exists $\varphi : \mathbb{N}_* \to \mathbb{N}_*$ strictly increasing s.t. $\lim_{k \to +\infty} (\lambda_0^{\varphi(k)}, p^{\varphi(k)}(0)) = (\lambda_0, p(0))$.

Since, for all $k \in \mathbb{N}_*$, $\lambda_0^{\varphi(k)} \geq 0$, we have, when $k \to +\infty$, $\lambda_0 \geq 0$. We consider the following Cauchy problem on an inhomogeneous ODE:

$$\left.\begin{array}{l} q'(t) = -D_{F,2}\mathcal{H}(t, \overline{x}(t), \overline{u}(t), q(t), \lambda^0) \\ q(0) = p(0). \end{array}\right\} \tag{2}$$

By using the Theorem 1 in ([4], p. 117), there exists a unique solution $p \in PC^1(\mathbb{R}_+, \mathbb{R}^{n*})$ defined on $\mathbb{R}_+$ all over. We consider the linear ODE

$$q'(t) = -q(t) \circ D_{F,2}f(t, \overline{x}(t), \overline{u}(t)). \tag{3}$$

We denote by $R(\cdot, \cdot)$ the resolvent of (3).

Note that for all $t \in \mathbb{R}_+$, by using the Proposition 3 in ([4], p. 181), we have $p(t) := R(t, 0) \cdot p(0) - \lambda_0 \int_0^t R(t, s) \cdot D_{F,2}f^0(s, \overline{x}(s), \overline{u}(s))ds$.

Let $t \in \mathbb{R}_+$ and $\zeta \in U$. Since φ is strictly increasing, we have there exists $N \in \mathbb{N}_*$, for all $k \geq N$, $\varphi(k) \geq t$. Consequently, for all $k \geq N$, by using (AE$^{\varphi(k)}$), we have $p^{\varphi(k)}(t) := R(t, 0) \cdot p^{\varphi(k)}(0) - \lambda_0^{\varphi(k)} \int_0^t R(t, s) \cdot D_{F,2}f^0(s, \overline{x}(s), \overline{u}(s))ds$.

Consequently, when $k \to +\infty$, we obtain that

$$\left.\begin{aligned}\lim_{k\to+\infty} p^{\varphi(k)}(t) &= R(t,0)\cdot p(0) - \lambda_0 \int_0^t R(t,s)\cdot D_{F,2}f^0(s,\overline{x}(s),\overline{u}(s))ds \\ &= p(t).\end{aligned}\right\} \quad (4)$$

Besides, for all $k \geq N$, by using (MP$^{\varphi(k)}$), we have $\mathcal{H}(t,\overline{x}(t),\overline{u}(t),p^{\varphi(k)}(t),\lambda_0^{\varphi(k)}) \geq \mathcal{H}(t,\overline{x}(t),\zeta,p^{\varphi(k)}(t),\lambda_0^{\varphi(k)})$.

Since $\lim_{k\to+\infty} \lambda_0^{\varphi(k)} = \lambda_0$ and $\lim_{k\to+\infty} p^{\varphi(k)}(t) = p(t)$, when $k \to +\infty$, we have $\mathcal{H}(t,\overline{x}(t),\overline{u}(t),p(t),\lambda_0) \geq \mathcal{H}(t,\overline{x}(t),\zeta,p(t),\lambda_0)$.

It remains to prove (CH). Let $t \in \mathbb{R}_+$. There exists $\overline{T} \in \mathbb{R}_{+*}$ s.t. $\overline{T} > t$. Since $p_{|[0,\overline{T}]} \in C^0([0,\overline{T}],\mathbb{R}^{n*})$, $\overline{x}_{|[0,\overline{T}]} \in C^0([0,\overline{T}],\mathbb{R}^n)$, (AI1) and (AV1), we have $[(t,\zeta) \mapsto p(t)\cdot f(t,\overline{x}(t),\zeta) + \lambda_0 f^0(t,\overline{x}(t),\zeta)] \in C^0([0,\overline{T}],\mathbb{R})$. Consequently, since $\overline{u}_{|[0,\overline{T}]} \in NPC^0_R([0,\overline{T}],U)$, by using (MP) and the Lemma 5.2 in [3], we have $[t \mapsto p(t)\cdot f(t,\overline{x}(t),\overline{u}(t)) + \lambda_0 f^0(t,\overline{x}(t),\overline{u}(t))] \in C^0([0,\overline{T}],\mathbb{R})$. Hence, $\bar{\mathcal{H}}$ is continuous at t.

Therefore, the conclusions (NO), (Si), (AE), (MP), and (CH) are verified. □

We introduce another condition.

(AV3) U is a subset of real normed vector space Y, there exists $\hat{t} \in \mathbb{R}_+$, U is a neighborhood of $\overline{u}(\hat{t})$ in Y, $D_{G,3}f(\hat{t},\overline{x}(\hat{t}),\overline{u}(\hat{t}))$ exists and it is surjective.

Corollary 1 *In the setting and under the assumptions of Theorem 2, if (AV3) is fulfilled, we can take $\lambda_0 = 1$ and we can remove (NO) and (Si).*

Proof First, we want to prove that $\lambda_0 \neq 0$. We proceed by contradiction, we assume that $\lambda_0 = 0$. Since $D_{G,3}f(\hat{t},\overline{x}(\hat{t}),\overline{u}(\hat{t}))$ exists, $D_{G,3}\mathcal{H}(\hat{t},\overline{x}(\hat{t}),\overline{u}(\hat{t}),p(\hat{t}),0)$ exists and $D_{G,3}\mathcal{H}(\hat{t},\overline{x}(\hat{t}),\overline{u}(\hat{t}),p(\hat{t}),0) = p(\hat{t}) \circ D_{G,3}f(\hat{t},\overline{x}(\hat{t}),\overline{u}(\hat{t}))$.

Therefore, by using (MP), we have $p(\hat{t}) \circ D_{G,3}f(\hat{t},\overline{x}(\hat{t}),\overline{u}(\hat{t})) = 0$, and since $D_{G,3}f(\hat{t},\overline{x}(\hat{t}),\overline{u}(\hat{t}))$ is surjective, we have $p(\hat{t}) = 0$. Consequently, since (AE) is linear homogeneous, we have p is equal to zero on $\mathbb{R}_+$. Therefore, we have $\|(0,p(0))\| = 0$ which contradicts (NO).

By setting $p' := \frac{1}{\lambda_0} p$ the conclusions (AE) and (MP) are verified with $\lambda_0 = 1$. □

4 Pontryagin Principle for the Infinite-Horizon Multiobjective Problems

Let Ω be a non-empty subset of $\mathbb{R}^n$, let $\xi_0 \in \Omega$ and let U be a Hausdorff topological space, let $f : [0,T] \times \Omega \times U \to \mathbb{R}^n$ and for all $i \in \{1,...,l\}$, $f_i^0 : [0,T] \times \Omega \times U \to \mathbb{R}$ be functions. We set $Adm(MLI)$ the set of the processes $(x,u) \in PC^1(\mathbb{R}_+,\Omega) \times NPC^0_R(\mathbb{R}_+,U)$ s.t. $x(0) = \xi_0$ and for all $t \in \mathbb{R}_+$ where $x'(t)$ exists, $x'(t) = f(t,x(t),u(t))$.

We denote by $Dom(MLI)$ the set of the processes $(x,u) \in Adm(MLI)$ s.t. for all $i \in \{1,...,l\}$, $\int_0^{+\infty} f_i^0(t,x(t),u(t))dt$ converges in $\mathbb{R}$. We consider the function $J : Dom(MLI) \to \mathbb{R}$ defined by, for all $(x,u) \in Adm(MLI)$, $J(x,u) := (J_1(x,u),...,J_l(x,u)) := (\int_0^{+\infty} f_1^0(t,x(t),u(t))dt, ..., \int_0^{+\infty} f_l^0(t,x(t),u(t))dt)$.

We can consider the following multiobjective optimal control problem in infinite horizon:

$$(\mathcal{P}^n_m) \begin{cases} \text{Maximize } J(x,u) \\ \text{subject to } (x,u) \in Dom(MLI) \end{cases}.$$

Definition 1 A process $(\overline{x}, \overline{u}) \in Dom(MLI)$ is called a natural Pareto optimal solution or a Pareto optimal solution of $(\mathcal{P}^n_m)$ if there does not exist a process $(x,u) \in Dom(MLI)$ s.t. for all $i \in \{1, .., l\}$, $J_i(x,u) \geq J_i(\overline{x}, \overline{u})$ and for some $i_0 \in \{1, .., l\}$ $J_{i_0}(x,u) > J_{i_0}(\overline{x}, \overline{u})$.

Definition 2 A process $(\overline{x}, \overline{u}) \in Dom(MLI)$ is called a weak natural Pareto optimal solution or a weak Pareto optimal solution of $(\mathcal{P}^n_m)$ if there does not exist a process $(x,u) \in Dom(MLI)$ s.t. for all $i \in \{1, .., l\}$, $J_i(x,u) > J_i(\overline{x}, \overline{u})$).

We can also find the following problems.

$(\mathcal{P}^o_m)$ Find $(\overline{x}, \overline{u}) \in Adm(MLI)$ s.t. there does not exist a process $(x,u) \in Adm(MLI)$ s.t. for all $i \in \{1, .., l\}$, $\lim_{T\to+\infty} \sup \int_0^T (f_i^0(t, x(t), u(t))) - f_i^0(t, \overline{x}(t), \overline{u}(t)))dt \geq 0$ and for some $i_0 \in \{1, ..., l\}$, $\lim_{T\to+\infty} \sup \int_0^T (f_{i_0}^0(t, x(t), u(t)) - f_{i_0}^0(t, \overline{x}(t), \overline{u}(t)))dt > 0$.

$(\mathcal{P}^w_m)$ Find $(\overline{x}, \overline{u}) \in Adm(MLI)$ s.t. there does not exist a process $(x,u) \in Adm(MLI)$ s.t. for all $i \in \{1, .., l\}$, $\lim_{T\to+\infty} \inf \int_0^T (f_i^0(t, x(t), u(t))) - f_i^0(t, \overline{x}(t), \overline{u}(t)))dt \geq 0$ and for some $i_0 \in \{1, ..., l\}$, $\lim_{T\to+\infty} \inf \int_0^T (f_{i_0}^0(t, x(t), u(t)) - f_{i_0}^0(t, \overline{x}(t), \overline{u}(t)))dt > 0$.

$(\mathcal{P}'^o_m)$ Find $(\overline{x}, \overline{u}) \in Adm(MLI)$ s.t. there does not exist a process $(x,u) \in Adm(MLI)$ s.t. for all $i \in \{1, .., l\}$, $\lim_{T\to+\infty} \sup \int_0^T (f_i^0(t, x(t), u(t))) - f_i^0(t, \overline{x}(t), \overline{u}(t)))dt > 0$.

$(\mathcal{P}'^w_m)$ Find $(\overline{x}, \overline{u}) \in Adm(MLI)$ s.t. there does not exist a process $(x,u) \in Adm(MLI)$ s.t. for all $i \in \{1, .., l\}$, $\lim_{T\to+\infty} \inf \int_0^T (f_i^0(t, x(t), u(t))) - f_i^0(t, \overline{x}(t), \overline{u}(t)))dt > 0$.

For these problems, the Hamiltonian of Pontryagin is the function $\mathcal{H}^m : \mathbb{R}_+ \times \Omega \times U \times \mathbb{R}^{n*} \times \mathbb{R}^l \to \mathbb{R}$ defined by $\mathcal{H}^m(t, \xi, \zeta, p, (\theta_i)_{1\leq i\leq l}) := \sum_{i=1}^l \theta_i f_i^0(t, \xi, \zeta) + p \cdot f(t, \xi, \zeta)$ when $(t, \xi, \zeta, p, (\theta_i)_{1\leq i\leq l}) \in \mathbb{R}_+ \times \Omega \times U \times \mathbb{R}^{n*} \times \mathbb{R}^l$.

Note that a process $(\overline{x}, \overline{u})$ which is a solution of $(\mathcal{P}^0_m)$ (respectively, $(\mathcal{P}^w_m)$) is called an overtaking Pareto optimal solution (respectively, a weak overtaking Pareto optimal solution).

Besides, a process $(\overline{x}, \overline{u})$ which is a solution of $(\mathcal{P}'^0_m)$ (respectively, $(\mathcal{P}'^w_m)$) is called an overtaking weak Pareto optimal solution (respectively, a weak overtaking weak Pareto optimal solution).

Furthermore, for each problem, the notion of weak Pareto optimal solution is weaker than Pareto optimal solution. As in the single-objective optimal control problem in infinite horizon, we remark that the notion of weak overtaking Pareto optimal solution (respectively, weak overtaking weak Pareto optimal solution) is the weakest notion of Pareto optimal solution (respectively, weak Pareto optimal solution). For this reason, we will use only this notion of Pareto optimal solution in the theorems of this section.

Notice that, for infinite-horizon multiobjective optimal control problems in the discrete time case, we can find the same kind of problems in [8].

Let $(\overline{x}, \overline{u}) \in Adm(MLI)$. Now, we consider the following conditions.

Conditions on the integrands of the criterion.

(AI1$_m$) For all $i \in \{1, ..., l\}$, $f_i^0 \in C^0(\mathbb{R}_+ \times \Omega \times U, \mathbb{R})$, for all $(t, \xi, \zeta) \in \mathbb{R}_+ \times \Omega \times U$, $D_{G,2} f_i^0(t, \xi, \zeta)$ exists, for all $(t, \zeta) \in \mathbb{R}_+ \times U$, $D_{F,2} f_i^0(t, \overline{x}(t), \zeta)$ exists and $[(t, \zeta) \mapsto D_{F,2} f_i^0(t, \overline{x}(t), \zeta)] \in C^0(\mathbb{R}_+ \times U, \mathbb{R}^{n*})$.

(AI2$_m$) For all $i \in \{1, ..., l\}$, for all $T \in \mathbb{R}_{+*}$, for all non-empty compact $K \subset \Omega$, for all non-empty compact $M \subset U$, $\sup_{(t,\xi,\zeta) \in [0,T] \times K \times M} \|D_{G,2} f_i^0(t, \xi, \zeta)\| < +\infty$.

Conditions on the vector field.

(Av1$_m$) $f \in C^0(\mathbb{R}_+ \times \Omega \times U, \mathbb{R}^n)$, for all $(t, \xi, \zeta) \in \mathbb{R}_+ \times \Omega \times U$, $D_{G,2} f(t, \xi, \zeta)$ exists, for all $(t, \zeta) \in \mathbb{R}_+ \times U$, $D_{F,2} f(t, \overline{x}(t), \zeta)$ exists and $[(t, \zeta) \mapsto D_{F,2} f(t, \overline{x}(t), \zeta)] \in C^0(\mathbb{R}_+ \times U, \mathcal{L}(\mathbb{R}^n, \mathbb{R}^n))$.

(Av2$_m$) For all $T \in \mathbb{R}_{+*}$, for all non-empty compact $K \subset \Omega$, for all non-empty compact $M \subset U$, $\sup_{(t,\xi,\zeta) \in [0,T] \times K \times M} \|D_{G,2} f(t, \xi, \zeta)\| < +\infty$.

For all $T \in \mathbb{R}_{+*}$, for all $\xi \in \Omega$, we consider the following finite-horizon multiobjective optimal control problem.

$$(\mathcal{L}^m{}_{T,\xi}) \begin{cases} \text{Maximize } \left(\int_0^T f_1^0(t, x(t), u(t))dt, ..., \int_0^T f_l^0(t, x(t), u(t))dt\right) \\ \text{subject to } x \in PC^1([0, T], \Omega), u \in NPC_R^0([0, T], U) \\ \qquad \forall t \in [0, T],\ \underline{d}x(t) = f(t, x(t), u(t)),\ x(0) = \xi_0 \\ \qquad x(T) = \xi. \end{cases}$$

We denote by $Adm(\mathcal{L}^m{}_{T,\xi})$ the set of the admissible processes of $(\mathcal{L}^m{}_{T,\xi})$.

Theorem 3 *If $(\overline{x}, \overline{u})$ is a weak overtaking Pareto optimal solution then, for all $T \in \mathbb{R}_{+*}$, the process $(\overline{x}_T, \overline{u}_T)$ is a Pareto optimal solution of the problem $(\mathcal{L}^m{}_{T,\overline{x}(T)})$ according to Definition 3.1 in [7] where $(\overline{x}_T, \overline{u}_T) := (\overline{x}_{|[0,T]}, \overline{u}_{|[0,T]})$.*

Proof We assume that $(\overline{x}, \overline{u})$ is a weak overtaking Pareto optimal solution. Let $T \in \mathbb{R}_{+*}$. We proceed by contradiction, we assume that there exists a process $(x, u) \in Adm(\mathcal{L}^m{}_{T,\overline{x}(T)})$ s.t. for all $i \in \{1, ..., l\}$, $\int_0^T f_i^0(t, x(t), u(t))dt \geq \int_0^T f_i^0(t, \overline{x}(t), \overline{u}(t))dt$ and for some $i_0 \in \{1, ..., l\}$, $\int_0^T f_{i_0}^0(t, x(t), u(t))dt > \int_0^T f_{i_0}^0(t, \overline{x}(t), \overline{u}(t))dt$.

Setting, for all $t \in \mathbb{R}_+$, $\tilde{x}(t) := x(t)$ and $\tilde{u}(t) := u(t)$ if $t \leq T$, $\tilde{x}(t) := \overline{x}(t)$ and $\tilde{u}(t) := \overline{u}(t)$ if $t > T$. Notice that $(\tilde{x}, \tilde{u}) \in Adm(MLI)$.

Note that, for all $\overline{T} \geq T$, we have, for all $i \in \{1, ..., l\}$, $\int_0^{\overline{T}} (f_i^0(t, \tilde{x}(t), \tilde{u}(t)) - f_i^0(t, \overline{x}(t), \overline{u}(t)))dt = \int_0^T (f_i^0(t, x(t), u(t)) - f_i^0(t, \overline{x}(t), \overline{u}(t)))dt$.

Consequently, we have, for all $i \in \{1, ..., l\}$, $\underline{\lim}_{\overline{T} \to +\infty} \inf \int_0^{\overline{T}} (f_i^0(t, \tilde{x}(t), \tilde{u}(t)) - f_i^0(t, \overline{x}(t), \overline{u}(t)))dt = \int_0^T (f_i^0(t, \tilde{x}(t), \tilde{u}(t)) - f_i^0(t, \overline{x}(t), \overline{u}(t)))dt \geq 0$.

Moreover, for all $\overline{T} \geq T$, we have $\int_0^{\overline{T}} (f_{i_0}^0(t, \tilde{x}(t), \tilde{u}(t))) - f_{i_0}^0(t, \overline{x}(t), \overline{u}(t)))dt = \int_0^T (f_{i_0}^0(t, x(t), u(t))) - f_{i_0}^0(t, \overline{x}(t), \overline{u}(t)))dt$.

Therefore, we have $\underline{\lim}_{\overline{T} \to +\infty} \inf \int_0^{\overline{T}} (f_{i_0}^0(t, \tilde{x}(t), \tilde{u}(t))) - f_{i_0}^0(t, \overline{x}(t), \overline{u}(t)))dt = \int_0^T (f_{i_0}^0(t, x(t), u(t))) - f_{i_0}^0(t, \overline{x}(t), \overline{u}(t)))dt > 0$. This contradiction with the process $(\overline{x}, \overline{u})$ is a weak overtaking Pareto optimal solution. □

Theorem 4 *When $(\overline{x}, \overline{u})$ is a weak overtaking Pareto optimal solution, under (A*I*1$_m$), (A*I*2$_m$), (A*V*1$_m$), and (A*V*2$_m$) there exist multipliers $(\theta_i)_{1 \leq i \leq l} \in \mathbb{R}^l$ and an adjoint function $p \in PC^1(\mathbb{R}_+, \mathbb{R}^{n*})$ which satisfy the following conditions:*

- *(NO$_m$)* $\|((\theta_i)_{1 \leq i \leq l}, p(0))\| = 1$.
- *(Si$_m$)* *For all $i \in \{1, ..., l\}$, $\theta_i \geq 0$.*
- *(AE$_m$)* $p'(t) = -D_{F,2}\mathcal{H}^m(t, \overline{x}(t), \overline{u}(t), p(t), (\theta_i)_{1 \leq i \leq l})$ *for all $t \in \mathbb{R}_+$ where $p'(t)$ exists.*
- *(MP$_m$)* *For all $t \in \mathbb{R}_+$, for all $\zeta \in U$, $\mathcal{H}^m(t, \overline{x}(t), \overline{u}(t), p(t), (\theta_i)_{1 \leq i \leq l}) \geq \mathcal{H}^m(t, \overline{x}(t), \zeta, p(t), (\theta_i)_{1 \leq i \leq l})$.*
- *(CH$_m$)* $\bar{\mathcal{H}}^m := [t \mapsto \mathcal{H}^m(t, \overline{x}(t), \overline{u}(t), p(t), (\theta_i)_{1 \leq i \leq l})] \in C^0(\mathbb{R}_+, \mathbb{R})$.

Proof First, we set, for all $x \in \Omega$, for all $i \in \{1, ..., n\}$, $h_i(x) := x_i - \overline{x}_i(T)$. Notice that, for all $i \in \{1, ..., n\}$, h_i is Hadamard differentiable at $\overline{x}(T)$ and $D_H h_i(\overline{x}(T)) := e_i^*$.

Therefore, by using Theorem 3 and the assumptions (AI1$_m$), (AI2$_m$), (AV1$_m$), and (AV2$_m$), for each $T \in \mathbb{R}_{+*}$, the assumptions of Theorem 4.9 and Corollary 4.10 in [7] are satisfied for the problem $(\mathcal{L}^m_{T,\overline{x}(T)})$. Therefore, there exist $(\theta_i^T)_{1 \leq i \leq l} \in \mathbb{R}^l$, $(\mu_\beta^T)_{1 \leq \beta \leq n} \in \mathbb{R}^n$ and $p^T \in PC^1([0, T], \mathbb{R}^{n*})$ verifying the following conditions:

- (NN$_m^T$) $((\theta_i^T)_{1 \leq i \leq l}, (\mu_\beta^T)_{1 \leq \beta \leq n})$ is non-zero.
- (Si$_m^T$) For all $i \in \{1, ..., l\}$, $\theta_i^T \geq 0$.
- (TC$_m^T$) $\sum_{\beta=1}^n \mu_\beta^T e_i^* = p^T(T)$.
- (AE$_m^T$) $\underline{d}p^T(t) = -D_{F,2}\mathcal{H}^m(t, \overline{x}(t), \overline{u}(t), p^T(t), (\theta_i^T)_{1 \leq i \leq l})$ for all $t \in [0, T]$.
- (MP$_m^T$) For all $t \in [0, T]$, for all $\zeta \in U$, $\mathcal{H}^m(t, \overline{x}(t), \overline{u}(t), p^T(t), (\theta_i^T)_{1 \leq i \leq l}) \geq \mathcal{H}^m(t, \overline{x}(t), \zeta, p^T(t), (\theta_i^T)_{1 \leq i \leq l})$.
- (Cor$_m^T$) For all $t \in [0, T]$, $((\theta_i^T)_{1 \leq i \leq l}, p^T(t)) \neq 0$.

□

Thanks to (Cor_m^T), we can take also, for each $T \in \mathbb{R}_{+*}$, $\|((\theta_i^T)_{1\leq i\leq l}, p^T(0))\| = 1$. Consequently, there exists $((\theta_i)_{1\leq i\leq l}, p(0)) \in \mathbb{R} \times \mathbb{R}^{n*}$ s.t. $\|((\theta_i)_{1\leq i\leq l}, p(0))\| = 1$, there exists $\varphi : \mathbb{N}_* \to \mathbb{N}_*$ strictly increasing s.t. $\lim_{k\to+\infty}((\theta_i^{\varphi(k)})_{1\leq i\leq l}, p^{\varphi(k)}(0)) = ((\theta_i)_{1\leq i\leq l}, p(0))$.

Since, for all $k \in \mathbb{N}_*$, for all $i \in \{1, ..., l\}$, $\theta_i^{\varphi(k)} \geq 0$, we have, when $k \to +\infty$, $\theta_i \geq 0$. We consider the following Cauchy problem on an inhomogeneous ODE:

$$\left.\begin{array}{l} q'(t) = -D_{F,2}\mathcal{H}^m(t, \overline{x}(t), \overline{u}(t), q(t), (\theta_i)_{1\leq i\leq l}) \\ q(0) = p(0). \end{array}\right\} \tag{5}$$

Thanks to the Theorem 1 in ([4], p. 117), there exists a unique solution $p \in PC^1(\mathbb{R}_+, \mathbb{R}^{n*})$ defined on $\mathbb{R}_+$ all over. We consider the linear ODE

$$q'(t) = -q(t) \circ D_{F,2}f(t, \overline{x}(t), \overline{u}(t)). \tag{6}$$

We denote by $\mathcal{R}(\cdot, \cdot)$ the resolvent of (6). Notice that for all $t \in \mathbb{R}_+$, by using the Proposition 3 in ([4], p. 181), we have $p(t) := \mathcal{R}(t, 0) \cdot p(0) - \sum_{i=1}^l \theta_i \int_0^t \mathcal{R}(t, s) \cdot D_{F,2}f_i^0(s, \overline{x}(s), \overline{u}(s))ds$.

Let $t \in \mathbb{R}_+$ and $\zeta \in U$. Since φ is strictly increasing, we have there exists $N \in \mathbb{N}_*$ s.t. for all $k \geq N$, $\varphi(k) \geq t$. Consequently, for all $k \geq N$, by using ($\text{AE}_m^{\varphi(k)}$), we have $p^{\varphi(k)}(t) := \mathcal{R}(t, 0) \cdot p^{\varphi(k)}(0) - \sum_{i=1}^l \theta_i^{\varphi(k)} \int_0^t \mathcal{R}(t, s) \cdot D_{F,2}f_i^0(s, \overline{x}(s), \overline{u}(s))ds$.

Hence, when $k \to +\infty$, we obtain that

$$\left.\begin{array}{rl} \lim_{k\to+\infty} p^{\varphi(k)}(t) &= \mathcal{R}(t, 0) \cdot p(0) - \sum_{i=1}^l \theta_i \int_0^t \mathcal{R}(t, s) \cdot D_{F,2}f_i^0(s, \overline{x}(s), \overline{u}(s))ds \\ &= p(t). \end{array}\right\} \tag{7}$$

Furthermore, for all $k \geq N$, by using ($\text{MP}_m^{\varphi(k)}$), we have $\mathcal{H}^m(t, \overline{x}(t), \overline{u}(t), p^{\varphi(k)}(t), (\theta_i^{\varphi(k)})_{1\leq i\leq l}) \geq \mathcal{H}^m(t, \overline{x}(t), \zeta, p^{\varphi(k)}(t), (\theta_i^{\varphi(k)})_{1\leq i\leq l})$.

Since $\lim_{k\to+\infty}(\theta_i^{\varphi(k)})_{1\leq i\leq l} = (\theta_i)_{1\leq i\leq l}$ and $\lim_{k\to+\infty} p^{\varphi(k)}(t) = p(t)$, when $k \to +\infty$, we have $\mathcal{H}^m(t, \overline{x}(t), \overline{u}(t), p(t), (\theta_i)_{1\leq i\leq l}) \geq \mathcal{H}^m(t, \overline{x}(t), \zeta, p(t), (\theta_i)_{1\leq i\leq l})$.

It remains to prove (CH_m). Let $t \in \mathbb{R}_+$. There exists $\overline{T} \in \mathbb{R}_{+*}$ s.t. $\overline{T} > t$. Since $p_{|[0,\overline{T}]} \in C^0([0, \overline{T}], \mathbb{R}^{n*})$, $\overline{x}_{|[0,\overline{T}]} \in C^0([0, \overline{T}], \mathbb{R}^n)$, ($\text{AI1}_m$) and ($\text{AV1}_m$), we have $[(t, \zeta) \mapsto p(t) \cdot f(t, \overline{x}(t), \zeta) + \sum_{i=1}^l \theta_i f_i^0(t, \overline{x}(t), \zeta)] \in C^0([0, \overline{T}], \mathbb{R})$. Hence, since $\overline{u}_{|[0,\overline{T}]} \in NPC_R^0([0, \overline{T}], U)$, by using ($\text{MP}_m$) and Lemma 5.2 in [3], we have $[t \mapsto p(t) \cdot f(t, \overline{x}(t), \overline{u}(t)) + \sum_{i=1}^l \theta_i f_i^0(t, \overline{x}(t), \overline{u}(t))] \in C^0([0, \overline{T}], \mathbb{R})$. Therefore, $\bar{\mathcal{H}}^m$ is continuous at t.

Hence, the conclusions (NO_m), (Si_m), (AE_m), (MP_m), and (CH_m) are verified.

We introduce the following condition.

(Av3_m) U is a subset of real normed vector space Y, there exists $\hat{t} \in \mathbb{R}_+$, U is a neighborhood of $\overline{u}(\hat{t})$ in Y, $D_{G,3}f(\hat{t}, \overline{x}(\hat{t}), \overline{u}(\hat{t}))$ exists and it is surjective.

Corollary 2 *In the setting and under the assumptions of Theorem 4, if (Av3_m) is fulfilled, we have* $(\theta_i)_{1\leq i\leq l} \neq 0$.

Proof Firstly, we want to prove that $(\theta_i)_{1\leq i\leq l} \neq 0$.

We proceed by contradiction, we assume that $(\theta_i)_{1\leq i\leq l} \neq 0$. Since $D_{G,3}f(\hat{t}, \overline{x}(\hat{t}), \overline{u}(\hat{t}))$ exists, $D_{G,3}\mathcal{H}^m(\hat{t}, \overline{x}(\hat{t}), \overline{u}(\hat{t}), p(\hat{t}), 0)$ exists and $D_{G,3}\mathcal{H}^m(\hat{t}, \overline{x}(\hat{t}), \overline{u}(\hat{t}), p(\hat{t}), 0) = p(\hat{t}) \circ D_{G,3}f(\hat{t}, \overline{x}(\hat{t}), \overline{u}(\hat{t}))$.

Therefore, by using (MP$_m$), we have $p(\hat{t}) \circ D_{G,3}f(\hat{t}, \overline{x}(\hat{t}), \overline{u}(\hat{t})) = 0$, and since $D_{G,3}f(\hat{t}, \overline{x}(\hat{t}), \overline{u}(\hat{t}))$ is surjective, we have $p(\hat{t}) = 0$. Consequently, since (AE$_m$) is linear homogeneous, we have p is equal to zero on $\mathbb{R}_+$. Therefore, we have $\|(0, p(0))\| = 0$ which is a contradiction with (NO$_m$). □

5 Sufficient Conditions of Pareto Optimality

In the sequel of this section, we replace $\mathbb{R}^n$ by a Banach space E.

Let $(\overline{x}, \overline{u}) \in PC^1(\mathbb{R}_+, \Omega) \times NPC^0_R(\mathbb{R}_+, U)$, we consider the following conditions:

- **(SI1)** For all $i \in \{1, ..., l\}$, $f_i^0 \in C^0(\mathbb{R}_+ \times \Omega \times U, \mathbb{R})$.
- **(SI2)** For all $t \in \mathbb{R}_+$, for all $i \in \{1, ..., l\}$, $D_{F,2}f_i^0(t, \overline{x}(t), \overline{u}(t))$ exists and $[t \mapsto D_{F,2}f_i^0(t, \overline{x}(t), \overline{u}(t))] \in NPC^0_R(\mathbb{R}_+, E^*)$.
- **(SV1)** $f \in C^0(\mathbb{R}_+ \times \Omega \times U, E)$.
- **(SV2)** For all $t \in \mathbb{R}_+$ $D_{F,2}f(t, \overline{x}(t), \overline{u}(t))$ exists and $[t \mapsto D_{F,2}f(t, \overline{x}(t), \overline{u}(t))] \in NPC^0_R(\mathbb{R}_+, \mathcal{L}(E, E))$.

Theorem 5 *When $(\overline{x}, \overline{u}) \in Adm(LMI)$, under (SI1), (SI2) (SV1), (SV2) if there exists $((\theta_i)_{1\leq i\leq l}, p)$ belongs to $\mathbb{R}^l_+ \times PC^1(\mathbb{R}_+, E^*)$ s.t. $\|((\theta_i)_{1\leq i\leq l}, p(0))\| = 1$ and if the following conditions are satisfied:*

- **(SH1)** *for each $(x, u) \in Adm(LMI)$, for all $\mathfrak{m}_1$-a.e. $t \in \mathbb{R}_+$,*

$$\mathcal{H}^m(t, \overline{x}(t), \overline{u}(t), p(t), (\theta_i)_{1\leq i\leq l}) - \mathcal{H}^m(t, x(t), u(t), p(t), (\theta_i)_{1\leq i\leq l}) \geq p'(t) \cdot (x(t) - \overline{x}(t)),$$

- **(SA^{w}1)** *for each $(x, u) \in Adm(LMI)$, $\lim_{T\to+\infty} \inf p(T) \cdot (x(T) - \overline{x}(T)) \geq 0$.*

then we have
if $(\theta_i)_{1\leq i\leq l} \neq 0$, then $(\overline{x}, \overline{u})$ is a weak overtaking weak Pareto optimal solution;
if for all $i \in \{1, ..., l\}$, $\theta_i \neq 0$, then $(\overline{x}, \overline{u})$ is a weak overtaking Pareto optimal solution.

Proof Let $(x, u) \in Adm(LMI)$. Let $T \in \mathbb{R}_{++}$, Note that, for each $t \in [0, T]$, $\sum_{i=1}^l \theta_i f_i^0(t, \overline{x}(t), \overline{u}(t)) = \mathcal{H}^m(t, \overline{x}(t), \overline{u}(t), p(t), (\theta_i)_{1\leq i\leq l}) - p(t) \cdot f(t, \overline{x}(t), \overline{u}(t))$ and $\sum_{i=1}^l \theta_i f_i^0(t, x(t), u(t)) = \mathcal{H}^m(t, x(t), u(t), p(t), (\theta_i)_{1\leq i\leq l}) - p(t) \cdot f(t, x(t), u(t))$.

Consequently, we have, for all $\mathfrak{m}_1 - a.e.\ t \in [0, T]$:

$$\begin{aligned}
&\sum_{i=1}^{l} \theta_i f_i^0(t, \overline{x}(t), \overline{u}(t)) - \sum_{i=1}^{l} \theta_i f_i^0(t, x(t), u(t)) \\
&= \mathcal{H}^m(t, \overline{x}(t), \overline{u}(t), p(t), (\theta_i)_{1\le i\le l}) - p(t)\cdot f(t, \overline{x}(t), \overline{u}(t)) \\
&-\mathcal{H}^m(t, x(t), u(t), p(t), (\theta_i)_{1\le i\le l}) + p(t)\cdot f(t, x(t), u(t)) \\
&= \mathcal{H}^m(t, \overline{x}(t), \overline{u}(t), p(t), (\theta_i)_{1\le i\le l}) - \\
&\mathcal{H}^m(t, x(t), u(t), p(t), (\theta_i)_{1\le i\le l}) + p(t)\cdot \underline{d}x(t) - p(t)\cdot \underline{d}\overline{x}(t) \\
&\ge \underline{d}p(t)\cdot (x(t) - \overline{x}(t)) + p(t)\cdot \underline{d}(x - \overline{x})(t) \text{ thanks to (SH1).}
\end{aligned}$$

Therefore, we have

$$\begin{aligned}
&\int_0^T \sum_{i=1}^{l} \theta_i f_i^0(t, \overline{x}(t), \overline{u}(t))dt - \int_0^T \sum_{i=1}^{l} \theta_i f_i^0(t, x(t), u(t))dt \\
&\ge \int_0^T (\underline{d}p(t)\cdot (x(t) - \overline{x}(t)) + p(t)\cdot \underline{d}(x - \overline{x})(t))dt \\
&= p(T)\cdot (x(T) - \overline{x}(T)).
\end{aligned}$$

Therefore, we obtain that

$$\begin{aligned}
&\lim_{T\to+\infty} \inf \sum_{i=1}^{l} \theta_i \int_0^T (f_i^0(t, \overline{x}(t), \overline{u}(t))dt - f_i^0(t, x(t), u(t)))dt \\
&\ge \lim_{T\to+\infty} \inf p(T)\cdot (x(T) - \overline{x}(T)) \ge 0.
\end{aligned}$$

Now, we assume that $(\theta_i)_{1\le i\le l} \neq 0$.

We want to prove that $(\overline{x}, \overline{u})$ is a weak overtaking weak Pareto optimal solution. We proceed by contradiction, we assume that $(\overline{x}, \overline{u})$ is not a weak overtaking weak Pareto optimal solution, i.e., there exists $(x, u) \in Adm(LMI)$ such that for all $i \in \{1, ..., l\}$, $\lim_{T\to+\infty} \sup \int_0^T (f_i^0(t, x(t), u(t))) - f_i^0(t, \overline{x}(t), \overline{u}(t))dt > 0$.

Consequently, we have $\lim_{T\to+\infty} \sup \sum_{i=1}^{l} \theta_i \int_0^T (f_i^0(t, x(t), u(t))) - f_i^0(t, \overline{x}(t), \overline{u}(t))dt > 0$. Hence, we obtain $\lim_{T\to+\infty} \inf \sum_{i=1}^{l} \theta_i \int_0^T (f_i^0(t, \overline{x}(t), \overline{u}(t))dt - f_i^0(t, x(t), u(t)))dt < 0$. This a contradiction.

Next, we assume that, for all $i \in \{1, ..., l\}$, $\theta_i \neq 0$.

We want to prove that $(\overline{x}, \overline{u})$ is a weak overtaking Pareto optimal solution. We proceed by contradiction, we assume that $(\overline{x}, \overline{u})$ is not a weak overtaking Pareto optimal solution, i.e., there exists $(x, u) \in Adm(LMI)$ such that for all $i \in \{1, ..., l\}$, $\lim_{T\to+\infty} \sup \int_0^T (f_i^0(t, x(t), u(t))) - f_i^0(t, \overline{x}(t), \overline{u}(t))dt \ge 0$ and for some $i_0 \in \{1, ..., l\}$, $\lim_{T\to+\infty} \sup \int_0^T (f_{i_0}^0(t, x(t), u(t))) - f_{i_0}^0(t, \overline{x}(t), \overline{u}(t))dt > 0$. Hence, we obtain that $\lim_{T\to+\infty} \sup \sum_{i=1}^{l} \theta_i \int_0^T (f_i^0(t, x(t), u(t))) - f_i^0(t, \overline{x}(t), \overline{u}(t))dt > 0$. Therefore, we have $\lim_{T\to+\infty} \inf \sum_{i=1}^{l} \theta_i \int_0^T (f_i^0(t, \overline{x}(t), \overline{u}(t))dt - f_i^0(t, x(t), u(t)))dt < 0$. This a contradiction. □

Corollary 3 *When $(\overline{x}, \overline{u}) \in Adm(LMI)$, under (SI1), (SI2), (SV1), (SV2) if there exists $((\theta_i)_{1\le i\le l}, p)$ belongs to $\mathbb{R}^l \times PC^1(\mathbb{R}_+, E^*)$ verifying all the conclusions of Theorem 4 if the condition (SA^{w}1) and the following condition are satisfied:*

(SH2) *for all* $(t,\xi) \in \mathbb{R}_+ \times \Omega$, $\hat{\mathcal{H}}^m(t,\xi,p(t),(\theta_i)_{1\leq i\leq l}) = \max_{\zeta\in U} \mathcal{H}^m(t,\xi,\zeta,p(t),(\theta_i)_{1\leq i\leq l})$ *exists, and for all* $t \in \mathbb{R}_+$, $[\xi \mapsto \hat{\mathcal{H}}^m(t,\xi,p(t),(\theta_i)_{1\leq i\leq l})]$ *is concave at* $\overline{x}(t)$ *and Gâteaux differentiable at* $\overline{x}(t)$,

then we have
if $(\theta_i)_{1\leq i\leq l} \neq 0$, *then* $(\overline{x},\overline{u})$ *is a weak overtaking weak Pareto optimal solution;*
if for all $i \in \{1,...,l\}$, $\theta_i \neq 0$, *then* $(\overline{x},\overline{u})$ *is a weak overtaking Pareto optimal solution.*

Proof For the proof, note that (SH2) implies (SH1). The proof is similar to the Theorem 5.2 in [7]. □

Corollary 4 *When* $(\overline{x},\overline{u}) \in Adm(LMI)$, *under (SI1), (SI2), (SV1), (SV2) if there exists* $((\theta_i)_{1\leq i\leq l}, p)$ *belongs to* $\mathbb{R}^l \times PC^1(\mathbb{R}_+, E^*)$ *verifying all the conclusions of Theorem 4 if the condition* (SA^{w}1) *and the following condition are satisfied:*

(SH3) *U is a subset of a real normed vector space Y s.t. for all* $t \in \mathbb{R}_+$, *U is a neighborhood of* $\overline{u}(t)$, *and for all* $t \in \mathbb{R}_+$, $[(\xi,\zeta) \mapsto \mathcal{H}^m(t,\xi,\zeta,p(t),(\theta_i)_{1\leq i\leq l})$ *is Gâteaux differentiable at* $(\overline{x}(t),\overline{u}(t))$ *and concave at* $(\overline{x}(t),\overline{u}(t))$,

then we have
if $(\theta_i)_{1\leq i\leq l} \neq 0$, *then* $(\overline{x},\overline{u})$ *is a weak overtaking weak Pareto optimal solution;*
if for all $i \in \{1,...,l\}$, $\theta_i \neq 0$, *then* $(\overline{x},\overline{u})$ *is a weak overtaking Pareto optimal solution.*

Proof Notice that (SH3) implies (SH1). The proof is similar to the Theorem 5.3 in [7]. □

Remark 1 In the setting and under the assumptions of each theorem in this section, notice that if $(\overline{x},\overline{u}) \in Dom(LMI)$ and if we replace the assumption (SA^{w}1) by the following assumption:

(SA^{s}1) for each $(x,u) \in Dom(LMI)$, $\lim_{T\to+\infty} p(T)\cdot(x(T)-\overline{x}(T))$ exists in $\mathbb{R}$ and $\lim_{T\to+\infty} p(T)\cdot(x(T)-\overline{x}(T)) \geq 0$,

then we have
if $(\theta_i)_{1\leq i\leq l} \neq 0$, then $(\overline{x},\overline{u})$ is a weak natural Pareto optimal solution;
if for all $i \in \{1,...,l\}$, $\theta_i \neq 0$, then $(\overline{x},\overline{u})$ is a natural Pareto optimal solution.

References

1. Aseev, S.M., Veliov, V.M.: Needle variations in infinite-horizon optimal control, In: Wolenski, G., Zaslavski, A.J (eds.) Variational and Optimal Control Problems on Unbounded Domains, pp. 1–17 (2017). (Contemp. Math. **619**, 2014)
2. Blot, J., Yilmaz, H.: Pontryagin Principle and Envelope Theorem (2022). https://doi.org/10.48550/arXiv.2206.13313
3. Blot, J., Yilmaz, H.: A generalization of Michel's result on the Pontryagin Maximum Principle. J. Opti. Theory Appl. **183**, 792–812 (2019)

4. Bourbaki, N.: Fonctions d'une variable réelle. Hermann, Paris (1976)
5. Carlson, D.A., Haurie, A.B., Leizarowitz, A.: Infinite Horizon Optimal Control, Deterministic and Stochastic Systems. Springer, Berlin (1991)
6. Halkin, H.: Necessary conditions for optimal control problems with infinite horizons. Econometrica **42**, 267–272 (1974)
7. Hayek, N., Yilmaz, H.: Infinite-Dimensional Multiobjective Optimal Control in Continuous Time (2022). https://doi.org/10.48550/arXiv.2207.00358
8. Hayek, N.: Infinite horizon multiobjective optimal control problems in the discrete time case. Optimization **60**(4), 509–529 (2011)
9. Lykina, V., Pickenhain, S., Wagner, M.: On a resource allocation model with infinite horizon. Appl. Math. Comput. **204**(2), 595–601 (2008)
10. Michel, P.: On the transversality condition in infinite horizon. Econometrica **50**(4), 975–985 (1982)
11. Montrucchio, L.: A turnpike theorem for continuous-time optimal control models. J. Econom. Dyna Control **19**, 599–619 (1995)
12. Reddy, P.V, Engwerda, J.C: Necessary and sufficient conditions for Pareto optimality in infinite horizon cooperative differential games. IEEE Trans. Autom. Contr. **59** 2536–2542 (2014)
13. Zaslavski, A.J.: Turnpike Properties in the Calculus of Variations and Optimal Control. Springer, Sciences + Business Media, New York (2006)

Optimal Growth in a Miniature Two-Sector Model With and Without Discounting: A Synthesis

Liuchun Deng, Minako Fujio, and M. Ali Khan

Abstract This essay synthesizes recent findings of the Robinson-Shinkai-Leontief (RSL) model of optimal growth, a miniature two-sector model that has been used as a workhorse in the literature of economic dynamics to demonstrate a variety of optimal outcomes. In terms of a substantive contribution to the subject, the taxonomy that we present elaborates both the importance, and the geometrical shape, of optimal policy correspondences as opposed to functions, a point of view first emphasized in Khan and Mitra (Adv Math Econ 8:349–381, (2006), [28]), Khan and Mitra (Jpn Econ Rev 58(2):191–225, (2007), [30]). Furthermore, our investigation takes an overview of the entire parameter space of the model, and in doing so, addresses several lacunae pertaining to special cases of interest, including the case of an unsustainable technology without discounting. The synthesis also points to the (only) parameter region for which a complete characterization of the optimal policy is still left open. (134 words)

Keywords Optimal growth · Value-loss · Dynamic programming · Golden rule · Cycles · Chaos

> I have tried to convey ... a sort of answer to the question "what is mathematical economics?" For achieving the above purposes the subject of economic development seems a particularly suitable vehicle. The proposition that proper pricing can bring about optimal allocation of

---

The authors acknowledge with gratitude many a stimulating discussion with Mukul Majumdar on positioning the contribution of David Gale to postwar "neoclassical" economics. Liuchun Deng acknowledges the support of the start-up grant from Yale-NUS College. Khan's work towards this chapter was supported by the "2022 JHU Discovery Award Deception and Bad-Faith Communication," and in contract from "Blue Green Future & Rebalance Earth."

L. Deng (✉)
Social Sciences Division, Yale-NUS College, Singapore, Singapore
e-mail: liuchun.deng@dukekunshan.edu.cn

Social Sciences Division, Duke Kunshan University, Kunshan, China

M. Fujio
Department of Economics, Yokohama National University, Yokohama, Japan

M. A. Khan
Department of Economics, The Johns Hopkins University, Baltimore, MD, USA

M. A. Khan et al. (eds.), *Matching, Dynamics and Games for the Allocation of Resources*, Monographs in Mathematical Economics 7,
https://doi.org/10.1007/978-981-96-6661-4_12

> resources may well be *the* central idea in all of economic theory. Modern global optimization theory has only served to emphasize and reenforce this proposition, sometimes in quite spectacular ways.[1]
> Gale (1968)

> David Gale (1966) has treated both the asymptotic convergence of optimal paths defined by the Atsumi-Weizsacker criterion and the existence of such optimal paths in linear production models with a finite set of activities. However, he has avoided some of the problems incident to introducing many goods by using a strictly concave utility function, defined on activity levels rather than on goods.[2]
> McKenzie (1968)

1 Introduction

In a seminal contribution, undoubtedly motivated by von Neumann [52],[3] Gale [18] studies a class of multi-sector models of optimal growth without discounting. Under the overtaking criterion originally proposed by Atsumi [2] and von Weizsäcker [53], he proves the existence of optimal programs through a technique based on the theory of dual variables. The result, and perhaps more importantly the technique, stimulated lasting inquiries into dynamic economic models without discounting.[4] For models with discounting, the optimal policy obtained without discounting served as an important analytical benchmark and helped inform the characterization of the optimal policy with discounting, a point of view consistently espoused by Lionel McKenzie.

In this essay, we remain true to the subject to consider to the theory of optimal growth, with or without discounting, but rather than von Neumann, look towards an aggregate model as originally considered by Ramsey [51], though in a two-sector setting with linear felicity functions and a Leontief technology. The model, which has now come to be referred to as the Robinson-Shinkai-Leontief (RSL) model,[5] was first

---

[1] The author also writes, "These examples will provide useful illustrations of the power of modern optimization theory in attacking economic problems;" see Gale [19], a Russian translation available in *Matematika,* **14** (6), 160-175.

[2] See McKenzie [40], a paper reprinted in Mitra and Nishimura [45]. McKenzie is referring to the working paper date of what became Gale [18]. For a more definite appraisal of the lines of investigation opened up by Gale, McKenzie and their followers, see the Handbook chapter McKenzie [41], the Palgrave entry McKenzie [42], the book McKenzie [43], and the references furnished therein. Malinvaud's classic 1953 paper surely merits a mention as a benchmark in this trajectory.

[3] This paper was translated into English by G. Morgenstern, and a commentary on this article made available in Champernowne [7]. For subsequent appraisals, see Koopmans [38], Chap. XVIII and Appendix C in Hicks [25], and Part II in Morishima [47] titled "The von Neumann revolution." It is interesting that Hicks situates the von Neuamnn model as a contribution to the theory of optimal growth, and writes, "Its [von Neumann 'equilibrium'] proper place is in Optimum Theory, not in the Pure Positive Theory;" see footnote 2, p. 133.

[4] For an elaborate discussion of the contribution of Gale [18], see Mitra [44].

[5] For the naming of this model, see Deng et al. [10]. McKenzie'e criticism in the second epigraph is addressed by the model since it is phrased in terms of a linear felicity function. Indeed, it is the strictly concave case that proves to be difficulty and has so far not been addressed.

brought into mainstream prominence by Nishimura and Yano [48] in their celebrated example of the existence of optimal chaos under arbitrarily weak discounting. In this miniature two-sector model, the economy consists of a consumption-good and an investment-good sector, and the social planner maximizes the sum of future utilities by allocating capital and labor between the two sectors.[6] We provide a complete and synthetic overview of the optimal policy for the RSL model by collecting and classifying results for a variety of cases scattered around in the existing literature. Furthermore, when the production technology cannot sustain any positive capital stock in the long run, the optimal policy has not been obtained without discounting, and we fill the void by providing a complete characterization for this case.

The overview, covering the *entire* parameter space of the model, delivers three main observations. First, there is an intimate connection between the discounted and undiscounted cases of the model. The optimal policy with discounting is identical to that without discounting when the discount factor is above an explicitly identified threshold. Second, the optimal policy is qualitatively different under different capital intensity assumptions. When the discount factor δ is above $1/\theta$, the inverse of the marginal rate of transformation (MRT) between today and tomorrow's capital under full specialization in the investment-good sector, the optimal policy is simple and uniform for the case of a capital-intensive investment-good sector. This is no longer so for the case of a capital-intensive consumption-good sector where it exhibits complex bifurcation structures. When the discount factor is below $1/\theta$, however, this contrasting pattern of the optimal policy is reversed between the two sectors. Third, we identify the last parameter region in which a complete characterization of the optimal policy is still left open. Outside this parameter region, the optimal policy has been fully characterized, and when it is uniquely determined, the optimal dynamics yields a convergence to one of the three possible states: extinction, convergence to the (modified) golden-rule stock, or a two-period cycle.[7] To be sure, some general version of these observations, and the possibilities they illustrate, are all part of the folklore, but it is our hope that their concretization in the specificity of the RSL model that we present here, will be of use for future investigations of the subject of optimal inter-temporal resource allocation.

---

[6] In the context of "equilibrium" rather than optimal growth theory, a comprehensive treatment is available in Part I of Morishima [47] that builds on Morishima [46] and the pioneering treatment of Shinkai. Note however, that unlike the simpler treatment presented here, Morishima's "Walras-type model of match-box size," has two techniques in each of his two sectors. For the distinction between equilibrium and optimal growth, also see the quote from Hicks in footnote 3 above, and Deng et al. [11]. From the viewpoint of this dichotomy, Gale [21] comes closest to what we are referring to as "equilibrium growth theory." If we include Samuelson's overlapping generations model, Gale [23] is an important reference.

[7] Benhabib and Nishimura [4] establish sufficient conditions for optimal cycles in a two-sector model with neoclassical production technologies (instead of the Leontief technology as in the RSL model). Also see Benhabib et al. [3]. It should be noted that both papers obtain optimal cycles for open sets of parameter values. For early examples of cycles and chaos that arise from two-sector models of optimal growth, see Boldrin and Montrucchio [6], Deneckere and Pelikan [8], and Boldrin and Deneckere [5].

The dialectic between the particular and the universal is hardly original to either this essay or to economic theory at large: it is a time-honored theoretical practice to test general conjectural claims by trying them out on specific instances; and alternatively, to try and generalize the results for concrete examples to more general settings. Whereas the work presented here goes from a multi-sectoral direction to a two-sector one, it is important to note that the other direction is also at play. The specific RSL model that we analyze generalizes the setting of Nishimura-Yano, whereby capital lasts for only one period, a setting of circulating capital, to the more general case of durable capital. But this is not the only interesting special case nested in our simplified setting: the RSL model is closely related to the so-called Robinson-Solow-Srinivasan (RSS) model, originally formulated to handle the issue of "choice of technique" in development planning. The RSL model generalizes the one-technique, two-sector case of the RSS model in which only labor, rather than machines and labor, is used to produce machines. As such, the the investment-good sector is by default the labor intensive sector, and the parametric space shrinks only to three parameters: the rate of depreciation and the discount factor, both taking values in the closed unit interval, and the non-negative labor-input, machine output coefficient. It is this that allows for a detailed use of the geometry as an "engine of analysis" rather than simply as "illustration of the analysis."[8] Furthermore, this precludes the possibility of extinction, which arises in the RSL model when the discount factor is below $1/\theta$, an RSL parameter that will be defined below, and has no counterpart in the RSS setting. From a methodological point of view, this essay also synthesizes both the dynamic programming and the value-loss approaches.[9]

The rest of this essay is structured as follows. We present the model setup, its geometry, and some preliminaries in the next section. In Sects. 3 and 4, we present the characterization results of the optimal policy for the case of a capital-intensive investment-good sector and then for the case of a capital-intensive consumption-good sector. We discusses the optimal policy for two special cases in Sect. 5, and conclude in the last Sect. 6 with two remarks. An appendix presents the details of the proofs.

[8] See Khan and Mitra [26, 28] for the full RSS version, and for the two-sector version. By now, a rather rich and mature theoretical treatment is available for this model. For the further development of the geometry, see Khan and Mitra [26, 30, 33]; for a demonstration of the existence of optimal topological chaos, see Khan and Mitra [26]. In parallel with the results for the RSL model summarized in this essay, Khan and Mitra [31, 32, 34] investigate geometrically the optimal dynamics of the RSS model and the optimal policy for various parameter regions. For the turnpike properties arising from the model, see Khan and Zaslavski [36, 37]. For a comprehensive treatment of optimal control problems related to the RSS model, see the monograph by Zaslavski [54].

[9] Khan and Mitra [29] provide an earlier synthesis of those two approaches in the context of the RSS model.

2 The RSL Model of Optimal Growth

2.1 The Model

We consider the two-sector RSL model of optimal growth with and without discounting. The economy consists of two sectors: a consumption-good sector and an investment-good sector. The production technology in both sectors is Leontief. It requires one unit of labor and $a_C > 0$ units of capital to produce one unit of consumption good, and one unit of labor and $a_I \geq 0$ units of capital to produce $b > 0$ units of investment good. Relative factor intensity of a sector depends on the comparison between a_C and a_I. If $a_C > a_I$, the consumption good sector is more capital intensive than the investment good sector. If $a_C < a_I$, the investment good sector is more capital intensive than the consumption good sector. If $a_C = a_I$, the model boils down to its one-sector setting. Note that we assume $a_C > 0$ because otherwise the planner would have no incentives to produce any investment good. However, we do not exclude the possibility of $a_I = 0$ which corresponds to the two-sector RSS setting as in Khan and Mitra [27].

Labor supply is fixed and normalized to be one in each time period t. Denote the capital stock in the current period by x, the capital stock in the next period by x', and the depreciation rate of capital by $d \in (0, 1]$. Capital is said to be *durable* for $d \in (0, 1)$ and said to be *circulating* for $d = 1$. The *transition possibility set* is given by

$$\Omega = \{(x, x') \in \mathbb{R}_+ \times \mathbb{R}_+ : x' - (1-d)x \geq 0, x' - (1-d)x \leq b \min\{1, x/a_I\}\},$$

where $\mathbb{R}_+$ is the set of non-negative real numbers. Denote by y the output of consumption good. For any $(x, x') \in \Omega$, we define a correspondence $\Lambda : \Omega \longrightarrow \mathbb{R}_+$

$$\Lambda(x, x') = \left\{ y \in \mathbb{R}_+ : y \leq \frac{1}{a_C}\left(x - \frac{a_I}{b}(x' - (1-d)x)\right) \text{ and } y \leq 1 - \frac{1}{b}(x' - (1-d)x) \right\}.$$

A felicity function, $w : \mathbb{R}_+ \longrightarrow \mathbb{R}$, is linear and given by $w(y) = y$. The reduced form utility function, $u : \Omega \longrightarrow \mathbb{R}_+$, is defined as

$$u(x, x') = \max\{w(y) : y \in \Lambda(x, x')\}.$$

The future utility is discounted with a discount factor $\delta \in (0, 1]$, with $\delta = 1$ corresponding to the undiscounted case. Define

$$\zeta \equiv \frac{b}{a_C - a_I} - (1-d) \tag{1}$$

to be the marginal rate of transformation (MRT) of capital between today and tomorrow under full utilization of both production factors. Define

Table 1 Definition of parameters in the RSL model

Parameter	Definition
$\delta \in (0, 1]$	Discount factor
$d \in (0, 1]$	Depreciation rate of capital
$a_C > 0$	Capital-labor ratio in the consumption-good sector
$a_I \geq 0$	Capital-labor ratio in the investment-good sector
$b > 0$	Investment-good output by 1 unit of labor and a_I units of capital
$\theta \equiv \frac{b}{a_I} + (1 - d)$	Marginal rate of transformation of capital today into capital tomorrow with zero consumption where there is surplus labor
$\zeta \equiv \frac{b}{a_C - a_I} - (1 - d)$	Marginal rate of transformation of capital today into capital tomorrow with full utilization of capital and labor

$$\theta \equiv \frac{b}{a_I} + (1 - d) \tag{2}$$

to be the MRT when the economy fully specializes in investment-good production with zero consumption good when there is surplus labor. Later we will simply refer to θ as the MRT with zero consumption. In what follows, we assume $\theta > 1$ unless otherwise stated. This assumption ensures that when the economy fully specializes in investment-good production, capital stock can grow. In Sect. 5.2, we will consider the special case of $\theta \leq 1$ in which long-term growth cannot be sustained technologically. Table 1 summarizes the definition of all the parameters.

We write explicitly the reduced-form utility function as

$$u(x, x') = \begin{cases} \frac{a_I \theta}{a_C b} x - \frac{a_I}{a_C b} x', & \text{for } (a_C - a_I)x' \leq ((1 - d)(a_C - a_I) - b)x + a_C b \\ \frac{1-d}{b} x - \frac{1}{b} x' + 1, & \text{for } (a_C - a_I)x' \geq ((1 - d)(a_C - a_I) - b)x + a_C b \end{cases} \tag{3}$$

In the reduced-form utility function above, the first line is for the case of full utilization of capital and the second line is for the case of full employment of labor.

An *economy E* consists of a triplet (Ω, u, δ). A *program starting from* $\bar{x} \in \mathbb{R}_+$ is a sequence $\{x_t, y_t\}_{t=0}^{\infty}$ such that $x_0 = \bar{x}$ and for any non-negative integer t, $(x_t, x_{t+1}) \in \Omega$ and $y_t = \max \Lambda(x_t, x_{t+1})$. For the model with discounting $(0 < \delta < 1)$, a program $\{x_t^*, y_t^*\}_{t=0}^{\infty}$ starting from $\bar{x} \in \mathbb{R}_+$ is said to be *optimal* if

$$\sum_{t=0}^{\infty} \delta^t [u(x_t, x_{t+1}) - u(x_t^*, x_{t+1}^*)] \leq 0$$

for every program $\{x_t, y_t\}_{t=0}^{\infty}$ starting from $\bar{x}$. For the model without discounting $(\delta = 1)$, a program $\{x_t^*, y_t^*\}_{t=0}^{\infty}$ starting from $\bar{x} \in \mathbb{R}_+$ is said to be *optimal* if there does not exist any other program $\{x_t, y_t\}_{t=0}^{\infty}$ starting from $\bar{x}$ that *overtakes* it, which is to say that there does not exist any $\varepsilon > 0$ and a time period t_ε such that

$$\sum_{t=1}^{T}[u(x_t, x_{t+1}) - u(x_t^*, x_{t+1}^*)] \geq \varepsilon \text{ for any } T \geq t_\varepsilon.$$

An optimal program starting from $\bar{x} \in \mathbb{R}_+$ is said to be *a stationary optimal program* if $x_t = \bar{x}$ for any time period t. *A stationary optimal stock* $\bar{x}$ is a non-negative real number such that there is a stationary optimal program from $\bar{x}$. A stationary optimal stock $\bar{x}$ is said to be *non-trivial* if $u(\bar{x}, \bar{x}) > u(0, 0)$.

2.2 Basic Geometry

Before we turn to the formal discussion of the optimal policy, we describe in this subsection the basic geometry of the RSL model illustrated by Fig. 1 (for the case of $a_C < a_I$).[10] Capital stock available in the current period, x, is measured on the horizontal axis and capital stock in the next period, x', is measured on the vertical axis. The 45° line helps track a program from the current period to the next period. The set enclosed by the lines OV, VL, and OD represents the transition possibility set Ω.

The OD line with a slope of $(1-d)$ corresponds to full specialization of the economy in the consumption-good sector. Since we assume that investment is irreversible, the next period capital stock is simply given by $x' = (1-d)x$. The OV and VL lines correspond to full specialization of the economy in the investment-good sector. The slope of the OV line is the MRT with zero consumption, θ. Assuming that the investment good sector is productive, i.e., $\theta > 1$, the OV line is above the 45-degree line.

The MV line corresponds to the case of full utilization of labor and capital. The absolute value of the slope of MV is precisely the MRT under full utilization of labor and capital, ζ. When the investment-good sector is capital intensive as it is in Fig. 1, if a production plan is above the MV line, capital is fully utilized whereas there is surplus labor, and if a production plan is below the MV line, labor is fully employed whereas there is excess capacity of capital.

The indifference curves for per-period utility are kinked on MV line and piece-wise linear illustrated as IC_1, IC_2, and IC_3 in orange, with lower indifference curves being associated with higher utility. The geometry for the case of a capital-intensive consumption-good sector ($a_C > a_I$) is similar. The main difference is that when $a_C > a_I$, the MV line can be downward sloping (for $\zeta > 0$).

[10] It should be noted that all the illustrations in this essay are based on the case of durable capital ($0 < d < 1$). The illustrations for the case of circulating capital ($d = 1$) are qualitatively the same but with horizontal VL and OD lines.

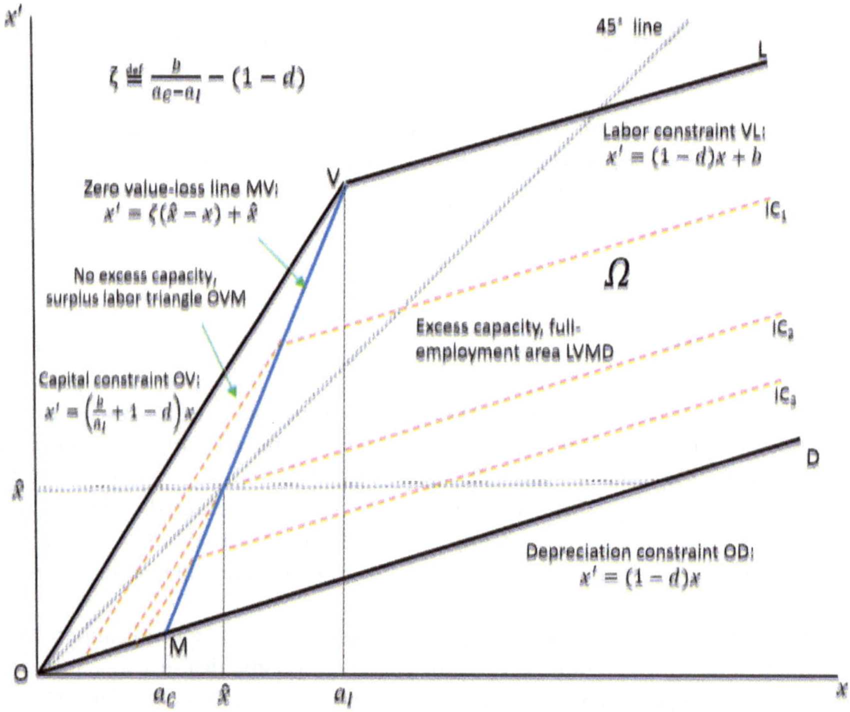

Fig. 1 The basic geometry for $a_C < a_I$

2.3 *Preliminaries*

We first introduce the concept of a golden rule, which plays a central role in the characterization of the optimal policy. In the undiscounted case, a stock $\hat{x}$ is defined as a *golden-rule stock* if $(\hat{x}, \hat{x}) \in \Omega$ and $u(\hat{x}, \hat{x}) \geq u(x, x')$ for all $(x, x') \in \Omega$ with $x \leq x'$. We also define the *golden rule* as a pair $(\hat{x}, \hat{p}) \in \mathbb{R}^2_+$ such that $(\hat{x}, \hat{x}) \in \Omega$ and

$$u(\hat{x}, \hat{x}) \geq u(x, x') + \hat{p}(x' - x) \text{ for all} (x, x') \in \Omega.$$

It should be noted that the price support property of the golden-rule price is originally developed by Gale [18]. In the discounted case, the *modified golden rule* is defined as a pair $(\hat{x}, \hat{p}) \in \mathbb{R}^2_+$ such that $(\hat{x}, \hat{x}) \in \Omega$ and

$$u(\hat{x}, \hat{x}) + (\delta - 1)\hat{p}\hat{x} \geq u(x, x') + \hat{p}(\delta x' - x) \text{ for all } (x, x') \in \Omega.$$

The *modified golden-rule stock* satisfies $u(\hat{x}, \hat{x}) \geq u(x, x')$ for all $(x, x') \in \Omega$ such that $x \leq (1 - \delta)\hat{x} + \delta x'$. Notice that the definition of the modified golden rule accommodates the golden rule for the undiscounted setting ($\delta = 1$) as a special case.

Lemma 1 *If* $1/\theta < \delta \leq 1$, *then there exists a (modified) golden rule given by*

$$(\hat{x}, \hat{p}) = \left(\frac{a_C b}{b + d(a_C - a_I)}, \frac{1}{(a_C - a_I)(1 + \delta\zeta)} \right),$$

and the (modified) golden-rule stock $\hat{x}$ *is the unique non-trivial stationary optimal stock.*

Lemma 1 characterizes the (modified) golden rule for $\delta > 1/\theta$ Fujio [13, 16]; Deng et al. [10]; Fujio et al. [17]. If $\delta > 1/\theta$, the modified golden-rule stock $\hat{x}$ exists and is also the unique non-trivial stationary optimal stock. Notably, $\hat{x}$ is independent of the discount factor and is the same for both discounted and undiscounted cases. Geometrically, it can be found as the intersection between the MV line and the 45-degree line. If $\delta < 1/\theta$, a non-trivial stationary optimal stock does not exist Deng et al. [12].

The Value-Loss Approach: Now we define the *value-loss* associated with a plan $(x, x') \in \Omega$ in terms of the (modified) golden rule stock and price as

$$\ell^{\delta}(x, x') \equiv u(\hat{x}, \hat{x}) + (\delta - 1)\hat{p}\hat{x} - u(x, x') - \hat{p}(\delta x' - x).$$

For the undiscounted case ($\delta = 1$), the value-loss for a plan $(x, x') \in \Omega$ simplifies as

$$\ell(x, x') \equiv u(\hat{x}, \hat{x}) + -u(x, x') - \hat{p}(x' - x).$$

It is straightforward to show that the value-loss comes from either unemployment or excess capacity of capital. Thus, when a plan is on the MV line so that both labor and capital are fully utilized, there is no value-loss.

Lemma 2 *There is no value-loss when labor and capital are fully utilized.*

The MV line is referred to as the *von Neumann facet* in the literature, which is defined as the set of all plans in the transition possibility set with zero value losses at the (modified) golden rule prices Khan and Mitra [26]. Furthermore, we can construct the iso-value-loss lines which are parallel to the MV line. As an iso-value-loss line moves away from MV line, the associated value loss increases.

Appealing to the extension of Brock's theorem formally presented by Khan and Mitra [26, 30] and restated in Deng et al. [10], we can obtain an optimal program by minimizing its aggregate value-loss. When the absolute value of the slope of the von Neumann facet is strictly less than unity, a program that starts from a certain range of the facet will stay on the facet for all periods thereafter and converge to the (modified) golden-rule stock. The aggregate value-loss of such a program is zero and hence minimized. So it is an optimal program. We shall call this particular range of the von Neumann facet *the McKenzie facet*.[11] A policy is said to be the *period-by-period minimum value-loss policy* if it minimizes the value-loss every period. When

[11] This terminology was introduced by Khan and Mitra [30] for the first time. The Mckenzie facet has the property that "any programme starting on it remains on it (p. 209)."

the McKenzie facet is identical to the von Neumann facet, the optimal policy is the period-by-period minimum value-loss policy.

In the undiscounted setting, and also in the discounted setting when the absolute value of the slope of the von Neumann facet is strictly less than unity, we can apply the value loss approach to pin down the optimal policy. Otherwise, we take a dynamic programming approach which we turn to next.

The Dynamic Programming Approach: Let $0 < \delta < 1$. Define the value function $V : \mathbb{R}_+ \to \mathbb{R}$ as

$$V(x) = \sum_{t=0}^{\infty} \delta^t u(x_t, x_{t+1})$$

where $\{x_t, y_t\}_{t=0}^{\infty}$ is an optimal program starting from $x_0 = x$. For each $x \in \mathbb{R}_+$, the Bellman equation

$$V(x) = \max_{x' \in \Gamma(x)} \{u(x, x') + \delta V(x')\}$$

holds where $\Gamma(x) = \{x' : (x, x') \in \Omega\}$. For each $x \in \mathbb{R}_+$, define the *optimal policy correspondence* $h(x) = \arg\max_{x' \in \Gamma(x)} \{u(x, x') + \delta V(x')\}$. If $h(x)$ is a singleton for any $x \in \mathbb{R}_+$, then we define the *optimal policy function* $g : \mathbb{R}_+ \to \mathbb{R}_+$ as $g(x) \in h(x)$ for any $x \in \mathbb{R}_+$. A program $\{x_t, y_t\}_{t=0}^{\infty}$ from x_0 is optimal if and only if it satisfies the equation:

$$V(x_t) = u(x_t, x_{t+1}) + \delta V(x_{t+1}) \text{ for } t \geq 0.$$

To characterize the optimal policy for the discounted case, we repeatedly apply the guess-and-verify method. When a complete characterization cannot be achieved, we investigate the properties of the value function to provide a partial characterization.

3 The Case of a Capital-Intensive Investment-Good Sector

In this section, we consider the case of a capital-intensive investment-good sector: $a_C < a_I$. For $\delta > 1/\theta$, the optimal policy is simple and uniform, whereas for $\delta < 1/\theta$, the optimal policy exhibits intriguing bifurcation with respect to the discount factor. To facilitate the exposition, we define *the straight-down-the-turnpike policy* as

$$g^s(x) = \begin{cases} \theta x & \text{for } x \in (0, \hat{x}\theta) \\ \max\{\hat{x}, (1-d)x\} & \text{for } x \in [\hat{x}/\theta, \infty) \end{cases}.$$

Proposition 1 *Let $a_C < a_I$. If $1/\theta < \delta \leq 1$, then the optimal policy function is the straight-down-the-turnpike policy g^s.*

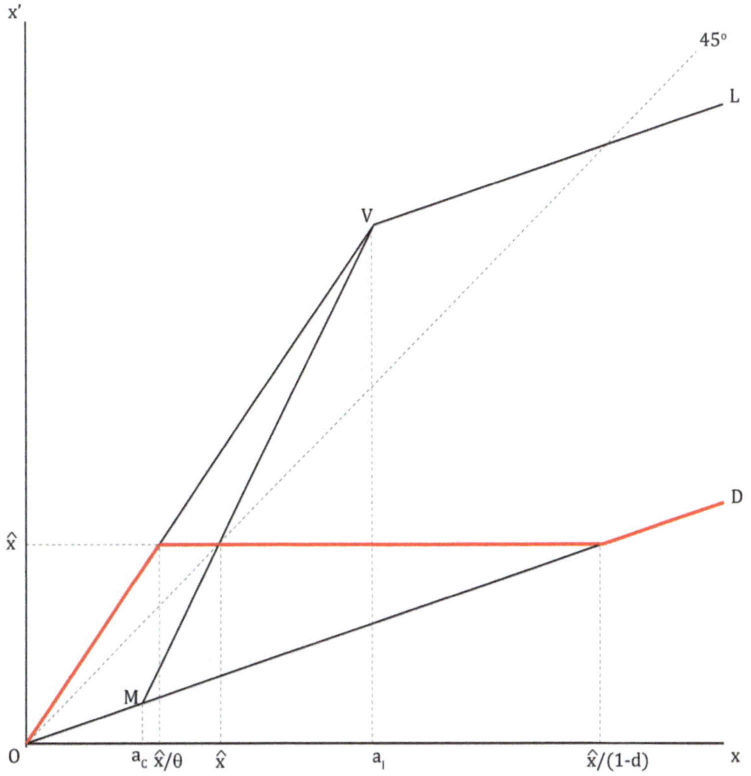

Fig. 2 The optimal policy for $a_C < a_I$ and $1/\theta < \delta \leq 1$

Proposition 1 provides a sharp characterization of the optimal policy for $\delta \in (1/\theta, 1]$ Fujio [13]; Fujio et al. [17].[12] As shown in Fig. 2, when capital stock x is in the intermediate range such that $(x, \hat{x}) \in \Omega$, the economy will jump onto the (modified) golden-rule stock in one period; when capital stock x is sufficiently low such that $\hat{x}$ is not attainable in one period, the economy will fully specialize in the investment-good sector; when capital stock x is sufficiently high such that $(1 - d)x \geq \hat{x}$, the economy will fully specialize in the consumption-good sector. Therefore, the optimal policy yields global convergence to the (modified) golden-rule stock as quickly as possible. That is also why we call the policy g^s the straight-down-the-turnpike policy.

Next we consider the case of $\delta < 1/\theta$ in which the model does not admit a non-trivial stationary optimal stock Deng et al. [12]. To characterize the optimal policy and its bifurcation structure, we define a sequence $\{x_n\}_{n=0}^{\infty}$ as the thresholds for capital stock as follows: $x_0 \equiv a_C$ and for any $n \in \mathbb{N}$,

---

[12] Fujio [13] establishes the optimal policy for the undiscounted case with $0 < d < 1$, and Fujio et al. [17] establish the optimal policy for the discounted case with $0 < d \leq 1$. For the undiscounted case with $d = 1$, we include a proof of the optimal policy in the Appendix, which originally appeared in Fujio [14].

$$x_n = -\frac{1}{\zeta}\left(x_{n-1} - \frac{a_C b}{a_C - a_I}\right).$$

By construction, for any $n \in \mathbb{N}$, (x_n, x_{n-1}) is on the MV line where capital and labor are fully utilized. Geometrically, it is clear that this sequence converges to $\hat{x}$. Formally, we can show that $\{x_n\}_{n=0}^{\infty}$ is strictly increasing and converges to $\hat{x}$. Further, we can construct another sequence $\{\mu_n\}_{n=0}^{\infty}$, which serves as the bifurcation values for the discount factor in the characterization of the optimal policy, with

$$\mu_0 \equiv \frac{1}{\frac{b}{a_C} + (1-d)} < \frac{1}{\frac{b}{a_I} + (1-d)} = \frac{1}{\theta},$$

where the inequality follows from $a_C < a_I$. The sequence $\{\mu_n\}_{n=0}^{\infty}$ is strictly increasing and converges to $1/\theta$. With these two sequences, we are now ready to characterize the optimal policy for $\delta < 1/\theta$.

Proposition 2 *Let $a_C < a_I$. If $\delta < \mu_0$, then the optimal policy function is given by $g(x) = (1-d)x$ for any $x > 0$. If $\mu_{n-1} < \delta < \mu_n$ for $n \in \mathbb{N}$, then the optimal policy function is given by*

$$g(x) = \begin{cases} (1-d)x & \text{for } x \in (0, a_C] \\ -\zeta x + \frac{a_C b}{a_C - a_I} & \text{for } x \in (a_C, x_n] \\ \max\{x_{n-1}, (1-d)x\} & \text{for } x \in (x_n, \infty) \end{cases}.$$

Remark 1 This characterization result also applies to the case of $\theta = 1$.

The characterization result of Proposition 2 is illustrated in the first three panels of Fig. 3.[13] The first panel shows the case of $\delta < \mu_0$. When the discount factor is sufficiently small, then the economy is in the phase of extinction without any investment. As the discount factor increases from below μ_0 to $1/\theta$, as shown in the next two panels, the optimal policy bifurcates with respect to the discount factor at $\delta = \mu_n$ $(n = 0, 1, 2...)$. For $\delta \in (\mu_0, 1/\theta)$, the economy converges globally to extinction in the long run but along the path to extinction, there can be investment and even full utilization of resources. In particular, for $\mu_{n-1} < \delta < \mu_n$, we can show that if the economy starts from $x = x_n$, it will experience exactly n periods of positive investment on the MV line. In our setting, the felicity function is linear, so the motive for intertemporal consumption smoothing is muted, but there can still be positive investment along the path to extinction: it is the technology instead of the preference that creates the incentives for investment.

Proposition 3 *Let $a_C < a_I$. If $\delta = 1/\theta$, then the optimal policy correspondence is given by*

[13] For the details of the construction of the sequence $\{\mu_n\}_{n=0}^{\infty}$, and also the characterization of the optimal policy correspondence for $\delta = \mu_n$, see Deng et al. [12].

$$h(x) = \begin{cases} [\max\{(1-d)x, -\zeta x + \frac{a_C b}{a_C - a_I}\}, \theta x] & for\ x \in (0, \frac{\hat{x}}{\theta}] \\ [\max\{(1-d)x, -\zeta x + \frac{a_C b}{a_C - a_I}\}, \hat{x}] & for\ x \in (\frac{\hat{x}}{\theta}, \hat{x}] \\ \{\max\{\hat{x}, (1-d)x\}\} & for\ x \in (\hat{x}, \infty) \end{cases} .$$

Proposition 3 characterizes the optimal policy for the knife-edge case of $\delta = 1/\theta$ Deng et al. [12]. In this case, the optimal policy becomes a correspondence, as illustrated in the last panel of Fig. 3. The optimal policy correspondence opens up the possibility of a variety of optimal dynamics, giving rise to a continuum of stationary optimal stocks. Any $x \in (0, \hat{x}]$ is a non-trivial stationary optimal stock.

Table 2 summarizes the optimal policy and the dynamics that it yields under different discount factors in the case of a capital-intensive investment-good sector.

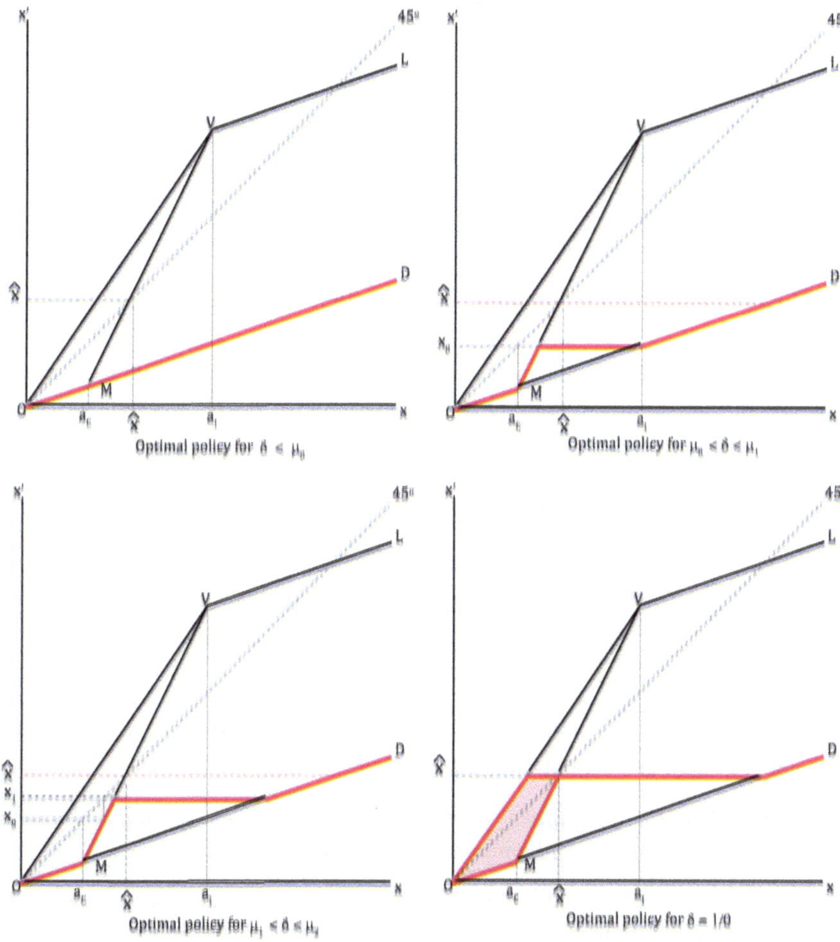

Fig. 3 The optimal policy for $a_C < a_I$ and $\delta \leq 1/\theta$

Table 2 The optimal policy for $a_C < a_I$

Discount factor	Optimal policy	Dynamical properties
$\delta \in (1/\theta, 1]$	Straight-down-the-turnpike policy	Monotonic convergence to golden-rule stock
$\delta = 1/\theta$	Optimal policy correspondence	Optimal program not uniquely determined; a continuum of stationary optimal stocks
$\delta \in [\mu_0, 1/\theta)$	There exit infinitely many cutoffs: $\mu_0 < \mu_1 < \mu_2 \cdots < 1/\theta$. Optimal policy bifurcates with respect to δ at $\delta = \mu_0, \mu_1, \mu_2, \ldots$	Convergence to extinction with positive investment and full utilization of resources along the path to extinction
$\delta \in (0, \mu_0)$	Specialization in consumption-good sector	Convergence to extinction without investment

Except for the cases in which optimal policy correspondence arises, the optimal policy does not yield cycles or complicated dynamics and there is global convergence to either the (modified) golden-rule stock or extinction.

4 The Case of a Capital-Intensive Consumption-Good Sector

In this section, we consider the case of a capital-intensive consumption-good sector: $a_C > a_I$. In sharp contrast with the previous case of $a_C < a_I$, the characterization of the optimal policy under $a_C > a_I$ is rather complex and not fully completed for $\delta > 1/\theta$ but the optimal policy is simple and uniform for $\delta < 1/\theta$. To facilitate the exposition, we define the period-by-period minimum value-loss policy (for $a_I < a_C$) as

$$g^m(x) = \begin{cases} \theta x & \text{for} x \in (0, a_I] \\ -\zeta x + \frac{a_C b}{a_C - a_I} & \text{for } x \in (a_I, a_C) \\ (1-d)x & \text{for } x \in [a_C, \infty) \end{cases}.$$

We start with the case of $\delta > 1/\theta$. In this case, the characterization of the optimal policy hinges on ζ, the MRT of capital between today and tomorrow under full utilization of labor and capital. Since $a_C > a_I$, $\zeta > -1$.

Proposition 4 *Let $a_C > a_I$ and $\zeta \leq 0$. If $1/\theta < \delta \leq 1$, then the optimal policy function is the period-by-period minimum value-loss policy g^m.*

Proposition 4 says the optimal policy for $\zeta \leq 0$ and $\delta > 1/\theta$ is the period-by-period minimum value-loss policy Fujio [15]; Fujio et al. [10, 16]. As shown in Fig. 4, the optimal policy, described by the OV-VM-MD lines, minimizes the value-loss every period and yields a monotonic convergence to the (modified) golden-rule stock $\hat{x}$.

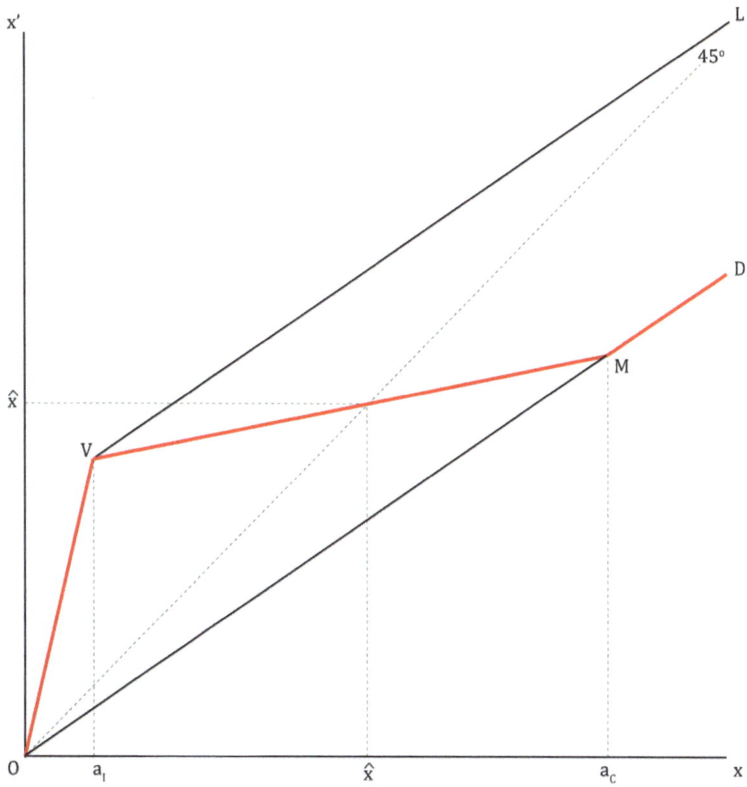

Fig. 4 The optimal policy for $a_C > a_I$, $\zeta \leq 0$, and $1/\theta < \delta \leq 1$

For the special case of $\zeta = 0$, when the MV line is horizontal, the optimal policy is also the straight-down-the-turnpike policy.

Proposition 5 *Let* $a_C > a_I$ *and* $0 < \zeta \leq 1$. *If* $1/\theta < \delta \leq 1$, *then the optimal policy yields global convergence to the (modified) golden-rule stock* $\hat{x}$ *for* $0 < \zeta < 1$ *and global convergence to* $\hat{x}$ *or a two-period cycle for* $\zeta = 1$. *Moreover, if in addition,* $a_C \geq a_I\theta$ *and* $a_C(1-d) \geq a_I$ *are satisfied, then the optimal policy function is the period-by-period minimum value-loss policy* g^m.

Proposition 5 describes the optimal dynamics for the case of $0 < \zeta \leq 1$. For $0 < \zeta < 1$, the optimal policy yields a global convergence to $\hat{x}$, whereas for $\zeta = 1$, the optimal policy yields a global convergence to either $\hat{x}$ or a two-period cycle.[14] For

[14] It should be noted that two-period cycles here arise in a knife-edge case of $\zeta = 1$. However, this result does not rule out the possibility that two-period cycles may still arise for an open set of the parameter values, as in the models with neoclassical technologies for which Benhabib and Nishimura [4] establish the existence of optimal cycles. This is because there is an open set of parameter values for the RSL model under which a complete characterization of the optimal policy

the subcase of $a_C \geq a_I\theta$ and $a_C(1-d) \geq a_I$, the optimal policy remains the period-by-period minimum value-loss policy, which geometrically can again be represented by the OV-VM-MD lines, but with an upward-sloping MV line.[15]

Proposition 6 *Let $a_C > a_I$ and $\zeta > 1$. If $\max\{1/\zeta, 1/\theta\} < \delta \leq 1$, then the optimal policy yields global convergence to the (modified) golden-rule stock $\hat{x}$ in finite periods. Moreover, if in addition, $a_C < a_I\theta$ and $a_C(1-d) \leq a_I$ are satisfied, then the optimal policy function is the straight-down-the-turnpike policy g^s.*

Proposition 6 describes the optimal dynamics for the case of $\zeta > 1$ Fujio [15, 16]; Fujio et al. [9]. When the discount factor is sufficiently close to one, then the optimal policy resembles that in the undiscounted case, and it yields global convergence to $\hat{x}$ in finite periods. For the subcase of $a_C < a_I\theta$ and $a_C(1-d) \leq a_I$, the optimal policy is the straight-down-the-turnpike policy, so the economy converges to $\hat{x}$ as quickly as possible.[16]

Corollary 1 *Let $a_C > a_I$ and $\zeta \geq \theta$. If $1/\theta < \delta \leq 1$, then the optimal policy yields global convergence to the (modified) golden-rule stock $\hat{x}$ in finite periods.*

From Corollary 1, the optimal dynamics yields convergence for $\zeta \geq \theta$. Thus, what remains open is the case of $1 < \zeta < \theta$ and $1/\theta < \delta \leq 1/\zeta$. The following proposition provides a partial characterization for this case.

Proposition 7 *Let $a_C > a_I$ and $1 < \zeta < \theta$. If $1/\theta < \delta \leq 1/\zeta$, then the optimal policy correspondence satisfies*

$$h(x) \subset \begin{cases} \{\theta x\} & \text{for } x \in (0, \hat{x}/\theta] \\ [\hat{x}, \theta x] & \text{for } x \in (\hat{x}/\theta, a_I] \\ [\hat{x}, \zeta(\hat{x}-x)+\hat{x}] & \text{for } x \in (a_I, \hat{x}] \\ [\zeta(\hat{x}-x)+\hat{x}, \hat{x}] & \text{for } x \in (\hat{x}, a_C] \\ [(1-d)x, \max\{\hat{x}, (1-d)x\}] & \text{for } x \in (a_C, \infty). \end{cases}$$

Proposition 7 provides a partial characterization of the optimal policy Deng et al. [10]. Figure 5 illustrates the possible region in which the optimal policy may occur. For $x \leq \hat{x}/\theta$ and $x \geq \hat{x}/(1-d)$ (for $d < 1$), the optimal policy has been fully characterized. For $\hat{x}/\theta < x < \hat{x}/(1-d)$, the optimal policy can appear in the region as illustrated in the two shaded triangles. In the existing literature, it has been demonstrated that chaotic dynamics can arise for this case under rather special conditions of $d = 1$

is yet to be obtained and it is known in the literature optimal cycles and chaos arise in that parameter region (Proposition 7).

[15] A complete characterization for the other subcases is more involved and has been obtained by Fujio [15, 16] for the undiscounted case and by Deng et al. [10] for the discounted case. We refer the interested readers to those two papers for the details.

[16] For the details of the characterization of the other subcases, we refer the interested readers to Fujio [15, 16] and Deng and Fujio [9].

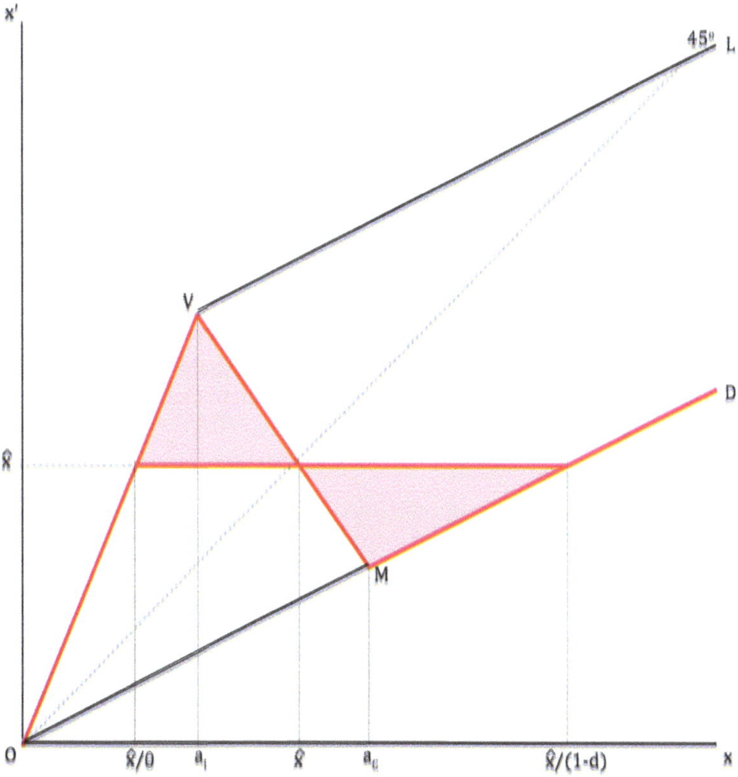

Fig. 5 The optimal policy for $a_C > a_I$, $1 < \zeta < \theta$, and $1/\theta < \delta \leq 1/\zeta$

Nishimura and Yano [48] and $a_I = 0$ Khan and Mitra [27, 32].[17] However, a complete characterization of this case and its implications on cyclical and chaotic dynamics have not been obtained. For us, a satisfactory characterization of the optimal policy for this parameter region is the last main open question for this entire strand of work on the two-sector RSL model of optimal growth.

Proposition 8 *Let $a_C > a_I$. If $\delta < 1/\theta$, then the optimal policy function is given by $g(x) = (1 - d)x$ for any $x > 0$.*

Proposition 8 shows that the optimal policy for $\delta < 1/\theta$ is surprisingly simple Deng et al. [12]. The economy converges globally to extinction ($x = 0$) in the long run and no resources would be allocated to the investment-good sector along the path to the extinction.

Proposition 9 *Let $a_C > a_I$. If $\delta = 1/\theta$, then the optimal policy correspondence is given by*

[17] For the case of circulating capital ($d = 1$), also see Nishimura and Yano [49, 50].

$$h(x) = \begin{cases} [(1-d)x, \min\{a_C, \theta x\}] & for\ x \in (0, a_I] \\ \left[(1-d)x, \min\left\{a_C, -\zeta x + \frac{a_C b}{a_C - a_I}\right\}\right] & for\ x \in (a_I, a_C] \\ \{(1-d)x\} & for\ x \in (a_C, \infty) \end{cases}$$

Proposition 9 characterizes the optimal policy for the knife-edge case of $\delta = 1/\theta$ Deng et al. [12]. Like the case of $a_C < a_I$, the optimal policy is a correspondence, as illustrated in Fig. 6. The optimal policy correspondence again gives rise to a continuum of stationary optimal stocks. Any $x \in (0, \hat{x}]$ is a non-trivial stationary optimal stock.

Table 3 summarizes the optimal policy and the dynamics that it yields for different δ and ζ in the case of a capital-intensive consumption-good sector. We highlight in blue the parameter region under which a complete characterization of the optimal policy is yet to be obtained.

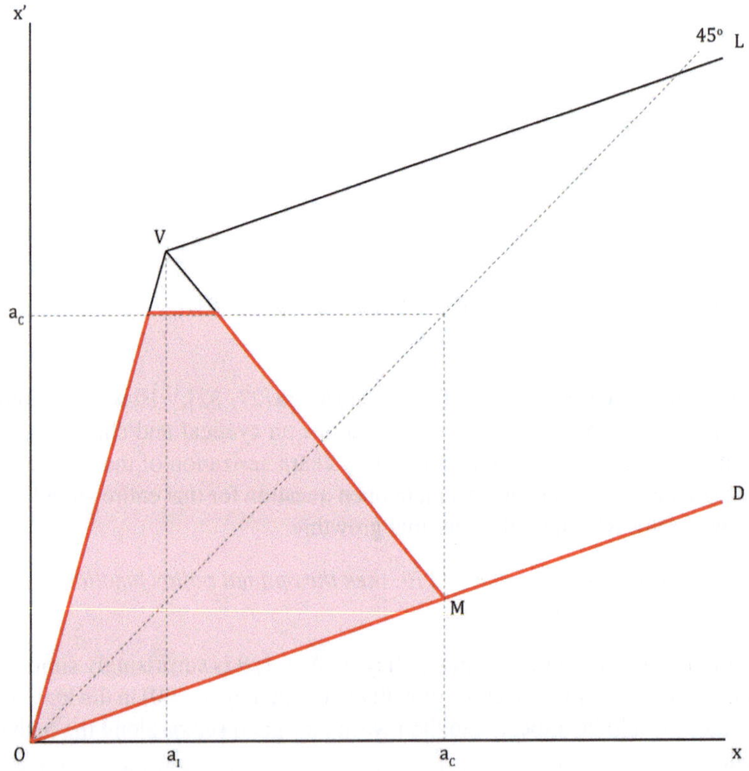

Fig. 6 The optimal policy for $a_C > a_I$ and $\delta = 1/\theta$

Table 3 The optimal policy for $a_C > a_I$

	Optimal policy	Dynamical properties
$1/\theta < \delta \leq 1$		
$\zeta \leq 0$	Period-by-period minimum value-loss policy	Monotonic convergence to golden-rule stock
$\zeta \in (0, 1)$	Period-by-period minimum value-loss policy or its slight variant	Cyclical convergence to golden-rule stock
$\zeta = 1$	Period-by-period minimum value-loss policy or its slight variant	Cyclical convergence to golden- rule stock or two-period cycles
$\zeta \in (1, \theta)$	Straight-down-the-turnpike	Convergence to golden-rule
$\delta \in (1/\zeta, 1]$	policy or its slight variant	stock in finite periods
$\zeta \in (1, \theta)$	A partial characterization; a complete	Possibility to have cycles and
$\delta \in (1/\theta, 1/\zeta]$	characterization only for special cases	Chaos
$\zeta \geq \theta$	Straight-down-the-turnpike policy or its slight variant	Convergence to golden-rule-stock in finite periods
$\delta \leq 1/\theta$		
$\delta = 1/\theta$	Optimal policy correspondence	Optimal program not uniquely determined; a continuum of stationary optimal stocks
$\delta \in (0, 1/\theta)$	Specialization in consumption-good sector	Convergence to extinction without investment

5 Special Cases

5.1 The One-Sector Case

In this subsection, we consider the one-sector case of the model: $a_C = a_I$. The optimal policy for the one-sector model is generally simple. For $\delta > 1/\theta$, the optimal policy resembles that for the case of a capital-intensive investment-good sector, and for $\delta < 1/\theta$, the optimal policy resembles that for the case of a capital-intensive consumption-good sector. The following proposition synthesized from Fujio et al. [17] and Deng et al. [12] provides a complete characterization of the optimal policy for the one-sector case.[18]

Proposition 10 *Let $a_C = a_I$. If $1/\theta < \delta \leq 1$, then the optimal policy function is the straight-down-the-turnpike policy. If $\delta < 1/\theta$, then the optimal policy function*

[18] The one-sector case without discounting has not been characterized in the literature, but though a simple extension of the geometric argument as in Fujio [13], we can show that the optimal policy is also the straight-down-the-turnpike policy.

is given by $g(x) = (1-d)x$ *for any* $x > 0$. *If* $\delta = 1/\theta$, *then the optimal policy correspondence is given by*

$$h(x) = \begin{cases} [(1-d)x, \min\{a_C, \theta x\}] & \text{for } x \in (0, a_C] \\ [(1-d)x, \max\{a_C, (1-d)x\}] & \text{for } x \in (a_C, \infty) \end{cases}$$

5.2 The Case of an Unsustainable Technology

In this subsection, we relax the only parametric assumption that we have implicitly imposed so far: $\theta > 1$. In particular, we now consider instead the case of an unsustainable technology: $\theta \leq 1$. Because $\theta \leq 1$, capital stock cannot grow even when the economy fully specializes in the investment-good sector. In fact, for $\theta < 1$, convergence to extinction is guaranteed in the long run.

Proposition 11 *Let* $a_C \geq a_I$ *and* $\theta \leq 1$. *If* $\delta < 1/\theta$, *then the optimal policy function is given by* $g(x) = (1-d)x$ *for any* $x > 0$.

Proposition 11 shows results in Propositions 8 and 10 carry over to the case of $\theta \leq 1$: the optimal policy yields global convergence to extinction, without any investment along the transition path.[19] Since $\theta \leq 1$, we must have $\delta \leq 1 \leq 1/\theta$. What is left out in the proposition above is then the case of $\delta = 1/\theta$, which holds only if $\delta = \theta = 1$. For this very special case, since $\theta = 1$, we can write the reduced-form utility under $a_C \leq a_I$ as

$$u(x, x') = \frac{a_I}{a_C b}(x - x')$$

for any $x \leq a_C$. Without discounting, for an initial stock below a_C, any program that yields the extinction of capital in the long run would *not* be overtaken by another policy. The optimal policy is therefore not unique, echoing the existence of an optimal policy correspondence for $\delta = 1/\theta$ when $\theta > 1$.

Proposition 12 *Let* $a_C < a_I$ *and* $\theta < 1$. *If* $\mu_0 > 1$, *the optimal policy function is given by* $g(x) = (1-d)x$ *for any* x. *If* $\mu_0 \leq 1$, *there exists* $n_0 \in \mathbb{N}$ *such that* $\mu_{n_0-1} \leq 1 < \mu_{n_0}$. *For* $\delta \leq \mu_{n_0-1}$, *characterization of the optimal policy follows the case of* $\theta \geq 1$. *For* $\mu_{n_0-1} < \delta \leq 1$, *the optimal policy function is given by*

$$g(x) = \begin{cases} (1-d)x & \text{for } x \in (0, a_C] \\ -\zeta x + \frac{a_C b}{a_C - a_I} & \text{for } x \in (a_C, x_{n_0}] \\ \max\{x_{n_0-1}, (1-d)x\} & \text{for } x \in (x_{n_0}, \infty) \end{cases}.$$

[19] The proof is relegated to the Appendix.

Proposition 12 extends the characterization of the optimal policy in Proposition 2 to the case of $\theta < 1$.[20] The optimal policy bifurcates with respect to the discount factor, but unlike the case of $\theta > 1$, as the discount factor rises, there are only finite bifurcations. Since Proposition 2 also applies to the case of $\theta = 1$, what remains to characterize is when $a_C < a_I$ and $\delta = \theta = 1$. Similar to the case of $a_C \geq a_I$, we can show that the optimal policy is no longer unique in such a special case without discounting. For an initial stock below a_I, any program that yields extinction in the long run is optimal under the overtaking criterion.

Remark 2 The optimal policy is not unique for $\delta = \theta = 1$.

6 Concluding Remarks

We conclude this essay with three remarks. First, what remains open is a complete characterization of the optimal policy for the only parameter region in which optimal chaos may arise. More specifically, the optimal policy is partially characterized for the case of a capital-intensive consumption-good sector ($a_C > a_I$) with the discount factor at an intermediate level ($1/\theta < \delta \leq 1/\zeta$) and the MRT under full resource utilization satisfying $1 < \zeta < \theta$. This parameter region is where Nishimura and Yano [49] start their investigation of the two-sector RSL model of optimal growth. In some way, this essay calls for a circling back to the starting point: a complete characterization of the optimal policy for this specific parameter region would enrich our understanding of the sources of optimal chaos and cycles.

Second, specific factor intensity conditions have been identified for cycles and chaos in two-sector models of optimal growth with the neoclassical or Leontief production technology. It is natural to wonder if those factor intensity conditions can be generalized to multi-sector growth models, especially the ones with production linkages between different sectors as in Acemoglu et al. [1]. There will be an interesting interplay between the rich production network structure and potentially complex dynamics along the optimal growth path. We believe this to be an exciting avenue for future analytical research.

Last, as a retrospective observation, we note that this essay synthesizes existing work on the two-sector RSL model of optimal growth, and by providing a taxonomy of the characterization results already available in the existing literature, it also fills lacunae for several special cases of the model without discounting. The characterization results, and their intimate connection with the findings for the model with discounting, underscore the enduring impact of the seminal work of David Gale,[21] and the turn that the subject took under his hands, along with those of Malinvaud,

[20] Deng et al. [12] characterizes the optimal policy for $\theta < 1$ with discounting. Following the same argument as in the proof of Proposition 11, we obtain the optimal policy without discounting as covered in Proposition 12.

[21] See Gale [18, 19] for the multi-sectoral case, and Gale and Sutherland [24] and Gale [20] for its aggregated version. Also see Majumdar [39] for a more recent appraisal.

McFadden, McKenzie, and Radner.[22] Fifty-five years on, it continues to enable new inquiries, and furnish productive windows into postwar economic theory.

A: Appendix

Proof of Proposition 1: As mentioned in footnote 12, the proof of this proposition has been provided in the literature except for the case of circulating capital ($d = 1$) without discounting. Here we provide the characterization of the optimal policy for this special case. One can extend the geometric analysis on the case of durable capital in Fujio [13] to the case of circulating capital. Alternatively, we adopt the approach of undiscounted dynamic programming as in Khan and Mitra [29]. In particular, we define a (modified) value function, $\tilde{V} : \mathbb{R}_+ \to \mathbb{R}$ as

$$\tilde{V}(x) \equiv \sum_{t=0}^{\infty}[u(x_t^*, x_{t+1}^*) - u(\hat{x}, \hat{x})],$$

where $\{x_t^*, y_t^*\}_{t=0}^{\infty}$ is an optimal program starting from $x \in \mathbb{R}_+$. We note that the definition of this value function is different from the standard definition we include in the main text for the discounted case. Without the subtraction of $u(\hat{x}, \hat{x})$, $\tilde{V}(x)$ would be infinity. For the case of a capital-intensive investment-good sector without discounting, we can show that $\tilde{V}$ is well-defined, concave, and non-decreasing on $\mathbb{R}_+$ and continuous on $(0, \infty)$, satisfying $\tilde{V}(\hat{x}) = 0$. The program $\{x_t^*, y_t^*\}_{t=0}^{\infty}$ is optimal if and only if for any non-negative integer t, $\tilde{V}(x_t^*) = [u(x_t^*, x_{t+1}^*) - u(\hat{x}, \hat{x})] + \tilde{V}(x_{t+1}^*)$. Moreover, $\tilde{V}$ satisfies the functional equation of dynamic programming $\tilde{V}(x) = \max_{x' \in \Gamma(x)}\{[u(x, x') - u(\hat{x}, \hat{x})] + \tilde{V}(x')\}$. We apply the guess-and-verify approach to obtain the optimal policy

For $d = 1$ and $a_C < a_I$, the reduced form utility is simplified as

$$u(x, x') = \begin{cases} \frac{a_I \theta}{a_C b} x - \frac{a_I}{a_C b} x', & \text{for } x' \geq \zeta(\hat{x} - x) + \hat{x} \\ \\ -\frac{1}{b} x' + 1, & \text{for } x' < \zeta(\hat{x} - x) + \hat{x} \end{cases}$$

Corresponding to the straight-down-the-turnpike policy, we postulate the value function of the following form:

[22] McFadden is one of the five references in Gale [19], and Malinvaud one of the four in Gale [22]. In their preface, Khan and Zaslavski [35], note that the noted 1967 symposium of *The Review of Economic Studies* contains articles of McFadden, Radner and (two articles) of Gale. As to what we are seeing as a "turn," we refer the reader to a comparison of Hicks [25] with Morishima [47], a comparison that is especially germane since the latter is explicit and aware of Hicks' influence on him. In terms of a summary statement, even though both books have the word "growth" in their titles, the former continues to look backward to Keynes and to macroeconomics and general competitive analysis.

$$W(x) = \begin{cases} \frac{1}{a_C}(\theta^n x - \hat{x}) - \frac{b-a_I}{a_C b} n\hat{x} & \text{for } x \in [\frac{\hat{x}}{\theta^{n+1}}, \frac{\hat{x}}{\theta^n}) \\ 0 & \text{for } x \in [\hat{x}, \infty) \end{cases}$$

where $n = 0, 1, 2...$ and $\theta = b/a_I$ for $d = 1$. We just need to verify that the straight-down-the-turnpike policy is indeed optimal under this postulated value function. Define $W_0(x, x') \equiv u(x, x') - u(\hat{x}, \hat{x}) + W(x')$.

First, consider $x > \hat{x}/\theta$. If $x' < \zeta(\hat{x} - x) + \hat{x}$ and $x' \geq \hat{x}$, then

$$\frac{\partial W_0(x, x')}{\partial x'} = -\frac{1}{b} < 0.$$

If $x' < \zeta(\hat{x} - x) + \hat{x}$ and $x' \in [\frac{\hat{x}}{\theta^{n+1}}, \frac{\hat{x}}{\theta^n})$ for some non-negative integer n, then

$$\frac{\partial W_0(x, x')}{\partial x'} = -\frac{1}{b} + \frac{\theta^n}{a_C} > 0,$$

where the inequality follows from $\theta > 1$ ($b > a_I$) and $a_I > a_C$. If $x' \geq \zeta(\hat{x} - x) + \hat{x}$ and $x' \in [\frac{\hat{x}}{\theta^{n+1}}, \frac{\hat{x}}{\theta^n})$ for some non-negative integer n, then

$$\frac{\partial W_0(x, x')}{\partial x'} = -\frac{a_I}{a_C b} + \frac{\theta^n}{a_C} = -\frac{1}{a_C \theta} + \frac{\theta^n}{a_C} > 0,$$

where the inequality follows from $\theta > 1$. If $x' \geq \zeta(\hat{x} - x) + \hat{x}$ and $x' \geq \hat{x}$, then

$$\frac{\partial W_0(x, x')}{\partial x'} = -\frac{a_I}{a_C b} < 0.$$

Since $W_0(x, x')$ strictly increases with x' for $x' < \hat{x}$ and strictly decreases with x' for $x' \geq \hat{x}$, $W_0(x, x')$ is maximized at $x' = \hat{x}$ for $x > \hat{x}/\theta$. Similarly, we can show that $W_0(x, x')$ is maximized at $x' = \theta x$ for $x \leq \hat{x}/\theta$. Thus, the straight-down-the-turnpike policy is indeed optimal. We have obtained the desired conclusion. ■

Proof of Proposition 11: The proof for the discounted case can be found in Deng et al. [12]. We therefore focus on the undiscounted case. Let $\delta = 1$. Since $\delta < 1/\theta$, we have $\theta < 1$. Since $\theta < 1$, for any program $\{x_t, y_t\}_{t=0}^{\infty}$ starting from some $\bar{x} \in \mathbb{R}_+$, $\lim_{t\to\infty} x_t = 0$ and we can show $\sum_{t=0}^{\infty} u(x_t, x_{t+1})$ to be finite and well-defined. In this case, there does not exist any other program $\{x_t, y_t\}_{t=0}^{\infty}$ starting from $\bar{x}$ that *overtakes* a program $\{x_t^*, y_t^*\}_{t=0}^{\infty}$ starting from $\bar{x}$ if and only if

$$\sum_{t=0}^{\infty} [u(x_t, x_{t+1}) - u(x_t^*, x_{t+1}^*)] \leq 0$$

for every program $\{x_t, y_t\}_{t=0}^{\infty}$ starting from $\bar{x}$. Let $\{x_t^*, y_t^*\}_{t=0}^{\infty}$ starting from $\bar{x}$ be the program that satisfies $x_{t+1}^* = (1-d)x_t^*$ for any non-negative integer t. Fix an arbitrary program $\{x_t, y_t\}_{t=0}^{\infty}$ starting from $\bar{x}$. We know from the discounted case that

$$\sum_{t=0}^{\infty} \delta^t [u(x_t, x_{t+1}) - u(x_t^*, x_{t+1}^*)] \leq 0$$

for any $\delta \in (0, 1)$. We can show that the expression on the left-hand-side of the inequality above is continuous in δ. Thus, we must have

$$\sum_{t=0}^{\infty}[u(x_t, x_{t+1}) - u(x_t^*, x_{t+1}^*)] = \lim_{\delta \to 1} \sum_{t=0}^{\infty} \delta^t [u(x_t, x_{t+1}) - u(x_t^*, x_{t+1}^*)] \leq 0.$$

Since this holds for any arbitrary program $\{x_t, y_t\}_{t=0}^{\infty}$, $\{x_t^*, y_t^*\}_{t=0}^{\infty}$ is optimal, and thus, $(1-d)x \in h(x)$ for any x. What is left to show is that the optimal policy is unique and thus the optimal policy correspondence boils down the optimal policy function $g(x) = (1-d)x$. Since we know $(1-d)x \in h(x)$ for any x, we can construct the value function for this undiscounted case as $\lim_{\delta \to 1} V(x)$, where $V(\cdot)$ is the value function based on the dynamic programming approach for the discounted case. Then, following essentially the same argument as in the proof of Theorem 1 in Deng et al. [12], we can show that if $\delta < 1/\theta$, then $x' = (1-d)x$ is uniquely optimal for any x. Thus, we have obtained the desired conclusion. ■

References

1. Acemoglu, D., Carvalho, V.M., Ozdaglar, A., Tahbaz-Salehi, A.: The network origins of aggregate fluctuations. Econometrica **80**(5), 1977–2016 (2012)
2. Atsumi, H.: Neoclassical growth and the efficient program of capital accumulation. Rev. Econ. Stud. **32**(2), 127–136 (1965)
3. Benhabib, J., Majumdar, M., Nishimura, K.: Global equilibrium dynamics with stationary recursive preferences. J. Econ. Behav. Organ. **8**(3), 429–452 (1987)
4. Benhabib, J., Nishimura, K.: Competitive equilibrium cycles. J. Econ. Theory **35**, 284–306 (1985)
5. Boldrin, M., Deneckere, R.J.: Sources of complex dynamics in two-sector growth models. J. Econ. Dyn. Control **14**(3–4), 627–653 (1990)
6. Boldrin, M., Montrucchio, L.: On the indeterminacy of capital accumulation paths. J. Econ. Theory **40**(1), 26–39 (1986)
7. Champernowne, D.: A note on J.V. Neumann's article on "A model of economic equilibrium". Rev. Econ. Stud. **13**(1), 10–18 (1945–1946)
8. Deneckere, R., Pelikan, S.: Competitive chaos. J. Econ. Theory **40**(1), 13–25 (1986)
9. Deng, L., Fujio, M.: Optimal growth in the two-sector RSL model with capital-intensive consumption goods: a dynamic programming approach. Pure Appl. Funct. Anal. **6**(5), 917–938 (2021)
10. Deng, L., Fujio, M., Khan, M.A.: Optimal growth in the Robinson-Shinkai-Leontief model: the case of capital-intensive consumption goods. Stud. Nonlinear Dyn. Econom. **23**(4), 20190032 (2019)

11. Deng, L., Fujio, M., Khan, M.A.: Eventual periodicity in the two-sector RSL model: Equilibrium vis-à-vis optimum growth. Econ. Theor. **72**, 615–639 (2021)
12. Deng, L., Fujio, M., Khan, M.A.: On optimal extinction in the matchbox two-sector model. Econ. Theory (2022). (forthcoming)
13. Fujio, M.: The Leontief two-sector model and undiscounted optimal growth with irreversible investment: the case of labor-intensive consumption goods. J. Econ. **86**(2), 145–159 (2005)
14. Fujio, M.: Optimal Transition Dynamics in the Leontief Two-sector Growth Model. Ph.D. thesis, The Johns Hopkins University (2006)
15. Fujio, M.: Undiscounted optimal growth in a Leontief two-sector model with circulating capital: the case of a capital-intensive consumption good. J. Econ. Behav. Organ. **66**(2), 420–436 (2008)
16. Fujio, M.: Optimal transition dynamics in the Leontief two-sector growth model with durable capital: the case of capital-intensive consumption goods. Jpn. Econ. Rev. **60**(4), 490–511 (2009)
17. Fujio, M., Lei, Y., Deng, L., Khan, M.A.: The miniature two-sector model of optimal growth: the neglected case of a capital-intensive investment-good sector. J. Econ. Behav. Organ. **186**, 662–671 (2021)
18. Gale, D.: On optimal development in a multi-sector economy. Rev. Econ. Stud. **34**(1), 1–18 (1967)
19. Gale, D.: A mathematical theory of optimal economic development. Bull. Am. Math. Soc. **74**(2), 207–223 (1968)
20. Gale, D.: Nonlinear duality and qualitative properties of optimal growth. In: Integer and Nonlinear Programming, pp. 309–319. North-Holland, Amsterdam (1970)
21. Gale, D.: On equilibrium growth in dynamic economic models. In: Mathematical Topics in Economic Theory and Computation, pp. 84–98. SIAM, Philadelhia (1971)
22. Gale, D.: On the theory of interest. Am. Math. Monthly **80**(8), 853–86 (1973)
23. Gale, D.: Pure exchange equilibrium of dynamic economic models. J. Econ. Theory **6**(1), 12–36 (1973)
24. Gale, D., Sutherland, W.R.: Analysis of a one good model of economic development. In: Mathematics of the Decision Sciences, pp. 120–136. American Mathematical Society, Providence (1968)
25. Hicks, J.: Capital and Growth. Clarendon Press, Oxford (1965)
26. Khan, M.A., Mitra, T.: On choice of technique in the Robinson-Solow-Srinivasan model. Int. J. Econ. Theory **1**(2), 83–110 (2005)
27. Khan, M.A., Mitra, T.: On topological chaos in the Robinson-Solow-Srinivasan model. Econ. Lett. **88**(1), 127–133 (2005)
28. Khan, M.A., Mitra, T.: Discounted optimal growth in the two-sector RSS model: a geometric investigation. Adv. Math. Econ. **8**, 349–381 (2006)
29. Khan, M.A., Mitra, T.: Undiscounted optimal growth in the two-sector Robinson-Solow-Srinivasan model: a synthesis of the value-loss approach and dynamic programming. Econ. Theor. **29**(2), 341–362 (2006)
30. Khan, M.A., Mitra, T.: Optimal growth in a two-sector RSS model without discounting: a geometric investigation. Jpn. Econ. Rev. **58**(2), 191–225 (2007)
31. Khan, M.A., Mitra, T.: Optimal growth under discounting in the two-sector Robinson-Solow-Srinivasan model: a dynamic programming approach. J. Differ. Equations Appl. **13**(2–3), 151–168 (2007)
32. Khan, M.A., Mitra, T.: Impatience and dynamic optimal behavior: a bifurcation analysis of the Robinson-Solow-Srinivasan model. Nonlinear Anal. Theory Methods Appl. **75**(3), 1400–1418 (2012)
33. Khan, M.A., Mitra, T.: Optimal growth in a two-sector RSS model with discounting: a further geometric investigation. Adv. Math. Econ. **17**, 1–33 (2013)
34. Khan, M.A., Mitra, T.: Complicated dynamics and parametric restrictions in the Robinson-Solow-Srinivasan (RSS) model. Adv. Math. Econ. **23**, 109–146 (2020)
35. Khan, M.A., Zaslavski, A.: Preface to a memorial issue for Roy Radner. Pure Appl. Funct. Anal. 7(6), 1–23 (2022)

36. Khan, M.A., Zaslavski, A.J.: On a uniform turnpike of the third kind in the Robinson-Solow-Srinivasan model. J. Econ. **92**(2), 137–166 (2007)
37. Khan, M.A., Zaslavski, A.J.: On two classical turnpike results for the Robinson-Solow-Srinivasan model. In: Kusuoka, S., Maruyama, T. (eds.) Advances in Mathematical Economics, pp. 47–97. Springer (2010)
38. Koopmans, T.C.: Economic growth at a maximal rate. Q. J. Econ. **78**(3), 355–394 (1964)
39. Majumdar, M.K.: David Gale: Duality and Turnpikes. Mimeo, Cornell University (2022)
40. McKenzie, L.W.: Accumulation programs of maximum utility and the von Neumann facet. In: Wolfe, J.N. (ed.) Value, Capital and Growth. Edinburgh University Press, Edinburgh (1968)
41. McKenzie, L.W.: Optimal economic growth, turnpike theorems and comparative dynamics. In: Arrow, K.J., Intrilligator, M. (eds.) Handbook of Mathematical Economics, pp. 1281–1355. Elsevier, New York (1986)
42. McKenzie, L.W.: Turnpike theory. In: Eatwell, J., Milgate, M., Newman, P.K. (eds.) The New Palgrave. MacMillan, London (1987)
43. McKenzie, L.W.: Classical General Equilibrium Theory. The MIT Press, Cambridge (2002)
44. Mitra, T.: On the existence of a stationary optimal stock for a multi-sector economy with non-convex technology. In: Majumdar, M. (ed.) Equilibrium and Dynamics, pp. 214–240. Springer, New York (1992)
45. Mitra, T., Nishimura, K. (eds.): Equilibrium, Trade and Growth: Selected Papers of Lionel W. McKenzie. The MIT Press, Cambridge (2009)
46. Morishima, M.: The multi-sectoral theory of economic growth. In: de Finetti, B. (ed.) Theories of Mathematical Optimization in Economics, pp. 79–165. Centro Internazionale Matematico Estivo, Rome (1965)
47. Morishima, M.: Theory of Economic Growth. Oxford University Press, Oxford (1969)
48. Nishimura, K., Yano, M.: Nonlinear dynamics and chaos in optimal growth: an example. Econometrica **63**(4), 981–1001 (1995)
49. Nishimura, K., Yano, M.: Chaotic solutions in dynamic linear programming. Chaos, Solitons Fractals **7**(11), 1941–1953 (1996)
50. Nishimura, K., Yano, M.: Chaotic solutions in infinite-time horizon linear programming and economic dynamics. In: Kusuoka, S., Maruyama, T. (eds.) Advances in Mathematical Economics, vol. 1, pp. 115–126. Springer (1999)
51. Ramsey, F.P.: A mathematical theory of saving. Econ. J. **38**, 543–559 (1928)
52. von Neumann, J.: Über ein ökonomisches gleichungssystem und eine verallgemeinerung des brouwerschen fixpunktsatzes. In: Menger, K. (ed.) Ergebnisse eines Mathematischen Kolloquiums, vol. 8, pp. 73–83. Springer, Berlin (1938). (translated as "A model of general economic equilibrium," Review of Economic Studies, XIII, 1–9)
53. von Weizsäcker, C.C.: Existence of optimal programs of accumulation for an infinite time horizon. Rev. Econ. Stud. **32**(2), 85–104 (1965)
54. Zaslavski, A.J.: Optimal Control Problems Related to the Robinson-Solow-Srinivasan Model. Springer, New York (2021)

Production Functions with Continuum of Vintage Inputs

Natali Hritonenko and Yuri Yatsenko

Abstract We provide a thorough economic analysis of dynamic CES production functions with continuum of capital vintage inputs in a wide range of elasticity of substitution among vintages. In doing so, we identify additional constraints for the original CES function (Jovanovic and Yatsenko in J Econ Theory, 147:551–569, 2012 [29]) and prove that the constructed modified CES function converges to a Cobb–Douglas production function with continuum of vintage inputs when the elasticity of substitution approaches unity. We discuss advantages and applications of both functions.

Keywords Vintage capital models · Optimal investments in old and new vintages · Learning · Production functions with continuum of vintage inputs

1 Introduction

Early origins of problems studied in this chapter can be traced back to the famous paper of David Gale [15] who stated

> The chief problem in the theory of dynamic economic planning is that of determining what amounts of resources should be allocated to the production of current consumption goods and what amounts should be allocated to the production of capital equipment to be used for producing consumption goods in the future.

Gale [15] analyzed optimization in a very general version of the Ramsey economic model with various types of labor and no technological change. Production functions represent a cornerstone of all economic growth theories [3]. It is impossible to overestimate their importance for modern economics. This chapter explores production

N. Hritonenko (✉)
Department of Mathematics, Prairie View A&M University, Prairie View, TX, USA
e-mail: nahtironenko@outlook.com

Y. Yatsenko
Dunham College of Business, Houston Christian University, Houston, TX, USA

M. A. Khan et al. (eds.), *Matching, Dynamics and Games for the Allocation of Resources*, Monographs in Mathematical Economics 7,
https://doi.org/10.1007/978-981-96-6661-4_13

functions with many capital vintages of various ages. Such CES functions were introduced by Jovanovic and Yatsenko [29] to analyze optimal investments in old and new vintages under evolving technology and learning.

David Gale [15] identified three central issues in dynamic production models as:

(1) *optimal balanced growth programs* ("golden-rule" paths)
(2) *asymptotic properties of "good" programs* ("turnpike" theorems)
(3) *dynamic prices* associated with optimal programs.

The first issue for the dynamic CES production function with continuum of vintage inputs was resolved in [29], while the second one was studied in [19, 24, 25].

1.1 Vintage Production Functions: A Brief Literature Survey

Vintage production functions describe the optimal replacement of age-dependent technology-specific capital under embodied technical change, see Hritonenko and Yatsenko [20], Boucekkine et al. [5, 6], Cooley et al. [11], Greenwood et al. [18] and the references therein. Førsund and Vislie [14] provide a comprehensive analysis of vintage production functions as driving forces of inter-industry and intra-industry structural changes, focusing on seminal works of [26, 27]. Vintage models can be described in discrete or continuous time, the latter being presented by means of integral or partial differential equations [17, 38, 39]. Similar mathematical problems also appear in the modeling of age-dependent biological populations [22].

Vintage capital models [1, 4, 36, 39] allow different vintages to participate in the production process. In doing so, most vintage models assume perfect substitutability among different vintages and assert that all optimal investment flows into the newest vintage with highest efficiency [1, 5–7, 11, 18, 21, 26]. Older vintages with lower productivity participate in production until they are replaced by superior vintages.

Few vintage models assume a limited substitutability among vintages, which leads to coexistence of several vintages and simultaneous investment into older vintages. The elasticity of substitution in production is a measure of how easy it is to shift between different production factors, most notably, between labor and capital [10, 32].

Chari and Hopenhayn [9] analyzed a CES vintage model with a limited substitutability between two (old and new) vintages and discovered an inverted-U (bell-shaped) age-profile of optimal investment in the presence of learning effects. Jovanovic and Yatsenko [29] introduced a novel CES production function, in which inputs are different vintages of capital and a continuum of vintages coexists in production. They analyzed the optimal balanced growth in this model and obtained the bell-shaped profiles of age-dependent investment and capital along balanced growth when the elasticity of substitution is large. Hritonenko et al. [19, 25] studied transition dynamics of optimal solutions in the [29] model and showed that initially bang-bang optimal investment eventually converges to the balanced trajectory. Hritonenko et al. [24] proved that bell-shaped investment profiles in this model converge to a spiked investment into one vintage with the largest productivity when the elasticity of substitution among vintages increases.

Samaniego and Yu Sun [37] introduced prices, taxes, subsidies, and related subjects to the CES vintage model of [29], thereby starting analysis of third Gale's issue. They explored the impact of vintage-specific taxes and subsidies on vintage age-distributions. In particular, their results added a new important argument to the old embodiment controversy [12] that the usefulness of models with embodied technical progress relies on their policy relevance.

Kredler [30] introduced a modification of [29] model with human capital, in which long-lived workers accumulate human capital following learning curve and complementarity in production occurs between labor inputs inside the same vintage (rather than across different vintages). He investigated the timing decision for technology phase-out and the distribution of workers in dying technology and showed that the experience premium diminishes as a technology ages

Lanteri [33] constructed and studied a dynamic equilibrium model with imperfect substitutability between new and used capital that occurs in the investment technology rather than in the production function as in [29]. This model provides a prospective framework for analyzing movements and endogenous prices of used capital, both at macro and microeconomic scales.

Martinez [31] constructed a vintage capital model with investments in both old and new capital vintages to analyze the extent to which automation can explain the observed fall in labor's share of income in the United States in the last 30 years. Farias et al. [13] described the technology adoption process in line with [29], using a CES function with of three inputs, physical, skilled labor, and unskilled labor.

Caunedo and Keller [8] employed a vintage capital model similar to [29] that links equipment prices to the quality and composition of capital stock to analyze the role of capital-embodied technology adoption in explaining cross-country disparities. They found that one-third of the disparities in the level of productivity can be accounted to disparities in capital embodied technology.

Livdan and Nezlobin [34] extended the Q-theory of investment to capital goods with non- geometric economic depreciation and demonstrated significant vintage capital effects. When the economic depreciation is non-geometric, a firm's capital stock and asset replacement cost fundamentally depend on its vintage composition and investment history. Lu et al. [35] explored the optimal pricing problem for age-dependent perishable goods and obtained the optimal strategy to price items based on their ages.

In this chapter, we derive the Cobb–Douglas function with continuum of vintage inputs through detailed economic analysis of the CES vintage production function of [29] in a wide range of the elasticity of substitution. Specifically, we are interested in properties of the CES function when the elasticity decreases and approaches unity. Based on our analysis, we identify additional constraints on the CES vintage production function, under which it converges to the corresponding Cobb–Douglas production function when the elasticity of substitution approaches unity.

Our new contribution to the production functions theory is three-fold:

(a) the modified CES production function with continuum of vintages for all range $[0,\infty)$ of the elasticity of substitution,
(b) the Cobb–Douglas function with continuum of vintage inputs,
(c) bell-shaped profiles of balanced age-dependent investment and capital in the Cobb–Douglas case with learning.

The chapter is as follows. Section 2 briefly describes the model and summarizes key analytic results of [29]. Section 3 justifies the need for modifying the original economic model, describes the modified CES vintage function, and provides its formal analysis, including the proof that the modified function produces a Cobb–Douglas vintage function when the substitution elasticity approaches unity. Section 4 analyzes the modified vintage CES function in the case of the elasticity values below unity (when different vintages are complements rather than substitutes). Section 5 contains proof of the existence of inverted U-shaped age-profiles of optimal capital and investments in the Cobb–Douglas vintage function in presence of learning. Section 6 discusses and compares transition dynamics in vintage models with and without investment in older vintages. Section 7 concludes.

2 Model Description and Preliminary Results

Let us consider a firm that faces the CES production function [28]

$$y(t) = \left(\int_{-\infty}^{t} A(t-v) z^{\beta}(v) k^{\beta}(v,t) dv \right)^{\alpha/\beta}, \tag{1}$$

whose inputs are capital vintages introduced at different times v, $v \in (-\infty,t]$, and $y(t)$ is the aggregate product output. More exactly, the inputs are the amounts $k(v,t)$ of capital at time t embodying technology of vintage v. The given parameter $\beta < 1$ reflects the limited substitutability of different capital vintages. The *elasticity of substitution* among different vintages $\sigma = 1/(1-\beta)$ is higher at larger β and approaches ∞ when $\beta \to 1$. Thus, the range $0 < \beta < 1$ corresponds to $1 < \sigma < \infty$. The parameter $0 < \alpha \leq 1$ describes the returns to scale as in other economic growth models [3]. At the first time in economic theory, the elegant Formula (1) reflects a limited substitutability among capital vintages created at different times and is simple enough to allow for further qualitative analysis [19, 24, 25]. The given functions in (1) are the efficiency $z(v)$ of new capital vintage v and the age-dependent factor $A(t-v)$ of the efficiency of the capital vintage v at time t, $t \in [0,\infty)$, $v \in (-\infty, t]$. The age-dependent factor $A(a)$ describes such processes as physical capital deterioration, human learning–by-doing, forgetting–by-doing, spillovers, which depend on the vintage age $t-v$ [2, 17, 29].

We assume that the firm solves the following profit maximization problem

$$\max_{u,x} \int_{0}^{\infty} e^{-rt} \left(y(t) - x(t) - \int_{-\infty}^{t} u(v,t) dv \right) dt \quad x \geq 0, \quad u \geq 0. \tag{2}$$

The decision variables in (2) are the investment $x(t)$ in new capital vintage t and the investment $u(v, t)$ in the old capital vintages $v < t$ (of the age $t-v$), $t \in [0, \infty)$, $v \in (-\infty, t]$. The unknown capital amount $k(v, t)$ in (1) is determined assuming that the capital deteriorates with a given constant rate $\delta > 0$. Then, the evolution of the heterogeneous capital stock $k(v,t)$ is described by the following partial-differential equation [17, 19, 38]:

$$\frac{\partial k(v,t)}{\partial t} = -\delta k(v,t) + u(v,t), \quad v \in [0,t), \quad t[0,\infty), \tag{3}$$

with the boundary conditions

$$k(v,0) = k_0(-v), \; v \in (-\infty, 0], \tag{4}$$

$$k(t,t) = x(t), \quad t \in [0,\infty), \tag{5}$$

where $k_0\,(a)$ is a given distribution of capital over past vintages of ages $a \in [0,\infty)$ at the initial time $t = 0$. The analytic solution of the linear boundary problem (3)–(5) under given controls u and x is

$$k(v,t) = e^{-\delta(t-v)}x(v) + \int_v^t e^{-\delta(t-s)}u(v,s)ds \;\text{ for }\; 0 < v \le t, \tag{6}$$

$$k(v,t) = e^{-\delta(t-v)}k_0(-v) + \int_0^t e^{-\delta(t-s)}u(v,s)ds \;\text{ for }\; -\infty < v \le 0, \tag{7}$$

for any age-distribution $k_0(v)$ of old vintages.

The CES production function (1) generalizes traditional vintage production functions of the form [2, 5, 7, 11, 18, 20, 21].

$$y(t) = \int_{-\infty}^t A(t-v)z(v)k(v,t)dv, \tag{8}$$

which occurs at $\beta = 1$ (i.e., $\sigma = \infty$) and assumes that vintages of capital are perfect substitutes. At $\beta = 1$, (1) is formally equivalent to the Formula (8). Gamboa and Maldonado [16] analyzed feasibility and optimality of initial capital stock in a discrete version of the vintage model (8) and established upper bounds for the rate of technological progress in terms of vintage capital productivity.

Jovanovic and Yatsenko [29] studied the long-term (balanced) optimal dynamics of the optimization problem (1)–(7) at $0 < \beta < 1$. They predicted that, in the case $\beta \to 1$ ($\sigma \to \infty$) of perfectly substitutable vintages, the optimal bell-shaped investment $u(v, t)$ converges to a Dirac-type function located at the age of the largest vintage productivity. This fact was formally proven in [24].

In Sect. 3, we review outcomes of [24, 29] for the case $0 < \beta < 1$, i.e., $\sigma > 1$. In the context of production, it is a realistic case that no vintage of capital is essential and positive output can be produced with any subset of vintages. We shall consider the case $\sigma \leq 1$ in Sect. 4 that includes the Cobb–Douglas form.

2.1 Balanced Growth Path

Under natural conditions, the problem (1)–(7) possesses an interior *steady-state solution*, known as the *balanced path* in the economic growth theory [1, 3, 21]. It is a solution of the problem when given economic variables grow with the same exponential rate, but a given initial condition (4) and the related Eq. (7) are disregarded.

Lemma 1 (balanced growth path) *Under exponential technological change*

$$z(t) = \overline{z}e^{\gamma t}, \quad 0 < \gamma\alpha/(1-\alpha) < r, \tag{9}$$

the optimization problem (1)–(7) has a unique steady-state (balanced growth) trajectory:

$$k(v,t) = e^{\varsigma t}\chi(t-v), \quad \varsigma = \frac{\gamma\alpha}{1-\alpha}, \tag{10}$$

$$y(t) = e^{\varsigma t}\left(\frac{\alpha\overline{z}^{\alpha}}{r+\delta}\right)^{\frac{\alpha}{1-\alpha}}\left(\int_0^{\infty} e^{-\frac{\gamma\beta}{1-\beta}s}A^{\frac{1}{1-\beta}}(s)ds\right)^{\frac{\alpha(1-\beta)}{\beta(1-\alpha)}}, \tag{11}$$

$$u(v,t) = e^{\zeta t}\tilde{u}(t-v), \quad x(t) = \overline{k}e^{\zeta t}A^{\frac{1}{1-\beta}}(0), \quad -\infty < v < t, 0 < t < \infty, \tag{12}$$

where

$$\overline{k} = \left(\frac{\alpha\overline{z}^{\alpha}}{r+\delta}\right)^{\frac{1}{1-\alpha}}\left(\int_0^{\infty} e^{-\frac{\gamma\beta}{1-\beta}s}A^{\frac{1}{1-\beta}}(s)ds\right)^{\frac{\alpha-\beta}{\beta(1-\alpha)}}, \tag{13}$$

$$\chi(a) = \overline{k}e^{-\frac{\gamma\beta}{1-\beta}a}A^{\frac{1}{1-\beta}}(a), \quad a = t-v, \tag{14}$$

$$\tilde{u}(a) = \overline{k}e^{-\frac{\gamma\beta}{1-\beta}a}\left[\left(\delta + \frac{\gamma\alpha}{1-\alpha} - \frac{\gamma\beta}{1-\beta}\right)A^{\frac{1}{1-\beta}}(a) + \frac{d}{da}\left(A^{\frac{1}{1-\beta}}(a)\right)\right]. \tag{15}$$

The proof follows from the results of [29].

Remark. The optimal policy (10)–(12) allows using an indefinitely old capital, which is an unrealistic but clear consequence of the assumption about the exponential deterioration of capital. To avoid this, we can modify the problem (1)–(7) assuming the existence of a maximal physical age T of capital. Then the unknown investment $u(v, t)$ and capital $k(v, t)$ are positive for $t-T < v \leq t$ and are truncated to zero if $v > T - t$. Similar technique was used in an optimal control of populations [5, 20, 22] introduced and studied vintage capital models, in which the scrapping age $T(t) > 0$ of the oldest vintages in use is endogenous and depends on the amount of a given labor resource. We do not pursue such models further in the present chapter.

Lemma 1 is formally true for any value of β, not necessarily from the range (0, 1) explored in [29]. However, if we assume for a moment that β approaches 0, and $A(a)$ does not decrease, then the improper integral in (11) and (13) diverges. It means that $\overline{k} \to \infty$ and, therefore, the optimal $y(t)$, $k(v, t)$, $x(t)$, and $u(v, t)$ increase indefinitely at $\beta \to +0$. If we assume that $\beta < 0$ and $\beta \to -0$, then by (11)–(13) the optimal balanced trajectory is $y(t) = 0$, $k(v, t) = 0$, $x(t) = 0$, and $u(v, t) = 0$ (no growth at all). So, the optimization problem (1)–(7) has a little sense at small β. It indicates some deficiencies of the CES production function (1), which we are addressed below.

3 Modified Vintage Production Function: From CES to Cobb–Douglas

In the previous section, we had to rule out the possibility of small β because then the optimal balanced solution approaches zero or diverges. As Jovanovic and Yatsenko [29] demonstrate, the CES function (1) works well for a range of positive β close to 1. In this section, we reveal some deficiencies of the original production function (1) at small values of β and introduce its modified form for such β.

Indeed, following the general economic theory, a CES model should produce a Cobb–Douglas model at $\beta \to 0$. It does not happen for the vintage model (1). To demonstrate that fact, let us consider the special case $A \equiv 1$, $z \equiv 1$ of the model (1). Assuming that $k(v, t) > 0$ at $v \in (-\infty, t]$, we have $\lim_{\beta\to 0+} k^{\beta}(v, t) = 1$ for a given $k(v, t) > 0$ at fixed v and t. Then, $\lim_{\beta\to 0} \int_{-\infty}^{t} k^{\beta}(v, t)dv = \lim_{\beta\to 0} \lim_{T\to\infty} \int_{-T}^{t} k^{\beta}(v, t)dv = \lim_{T\to\infty} \int_{-T}^{t} \lim_{\beta\to 0} k^{\beta}(v, t)dv = \lim_{T\to\infty} (t + T) = \infty$.

Therefore, the one-side limit

$$\lim_{\beta\to 0^+} y(t) = \lim_{\beta\to 0^+} \left(\int_{-\infty}^{t} k^{\beta}(v, t)dv \right)^{\alpha/\beta} = \exp\left\{ \lim_{\beta\to 0^+} \ln\left(\left(\int_{-\infty}^{t} k^{\beta}(v, t)dv \right)^{\alpha/\beta} \right) \right\}$$
$$= \exp\left\{ \lim_{\beta\to 0^+} \left(\frac{\alpha}{\beta} \ln\left(\int_{-\infty}^{t} k^{\beta}(v, t)dv \right) \right) \right\} = \infty, \tag{16}$$

but

$$\lim_{\beta\to 0^-} y(t) = \lim_{\beta\to 0^-} \left(\int_{-\infty}^{t} k^{\beta}(v, t)dv \right)^{\alpha/\beta} = \exp\left\{ \lim_{\beta\to 0^-} \left(\frac{\alpha}{\beta} \ln\left(\int_{-\infty}^{t} k^{\beta}(v, t)dv \right) \right) \right\} = 0. \tag{17}$$

Hence, at $\beta > 0$ and $\beta \to 0$, the value $y(t)$ increases infinitely for any given time t, but $y(t)$ tends to zero at $\beta < 0$ and $\beta \to 0$.

In general, the production function (1) is not good for the values $\beta \to 0$, because then $k^{\beta}(v,t) \to 1$ at $k(v,t) > 0$ and the improper integral $\int_{-\infty}^{t} k^{\beta}(v,t)dv \to \infty$ at any fixed time t.

To extend the CES production function (1) to a form that will produce a Cobb–Douglas vintage production function at $\beta \to 0$, let us select

$$y(t) = b\left(\int_{-\infty}^{t} w(t-v)z^{\beta}(v)k^{\beta}(v,t)dv\right)^{\alpha/\beta}, \tag{18}$$

where the given kernel function $w(a) \geq 0, 0 \leq a < \infty$, satisfies the additional constraint

$$\int_0^{\infty} w(a)da = 1. \tag{19}$$

The key difference of the modified function (18) and (19) from the original CES function (1) is the "normalization" condition (19). The function w in (19) plays the role analogous to constant parameters of the traditional CES functions with a finite number of factors, while the constraint (19) generalizes well-known conditions for such parameters. This fact can be illustrated with the following simple example.

Illustrative Example: The standard two-factor CES function

$$y = A\left(c_1x_1^{\beta} + c_2x_2^{\beta}\right)^{\alpha/\beta}, \quad c_1 > 0, \quad c_2 > 0, c_1 + c_2 = 1, \tag{20}$$

with two production factors $x_1 > 0$ and $x_2 > 0$ and "factor share" parameters c_1 and c_2 [3]. Let us take the limit of (20) as

$$\begin{aligned}\lim_{\beta\to 0} y = A\lim_{\beta\to 0}\left(c_1x_1^{\beta} + c_2x_2^{\beta}\right)^{\alpha/\beta} &= A\exp\left\{\lim_{\beta\to 0}\ln\left(c_1x_1^{\beta} + c_2x_2^{\beta}\right)^{\alpha/\beta}\right\}\\ &= A\exp\left\{\alpha\lim_{\beta\to 0}\left(\frac{1}{\beta}\ln\left(c_1x_1^{\beta} + c_2x_2^{\beta}\right)\right)\right\}.\end{aligned} \tag{21}$$

The limit in (21) is finite and positive at the "normalization" condition $c_1 + c_2 = 1$ in (20), otherwise, it does not exist at $c_1 + c_2 > 1$ and is zero at $c_1 + c_2 < 1$. If $c_1 + c_2 = 1$, we apply L'Hopital's rule to resolve the indeterminacy 0/0 in (21) and obtain

$$\begin{aligned}\lim_{\beta\to 0}\left(\frac{1}{\beta}\ln\left(c_1x_1^{\beta} + c_2x_2^{\beta}\right)\right) &= \frac{d\ln\left(c_1x_1^{\beta} + c_2x_2^{\beta}\right)/d\beta}{d\beta/d\beta}\Big|_{\beta=0}\\ &= \left\{\frac{c_1x_1^{\beta}\ln x_1 + c_2x_2^{\beta}\ln x_2}{c_1x_1^{\beta} + c_2x_2^{\beta}}\right\}\Big|_{\beta=0} = c_1\ln x_1 + c_2\ln x_2.\end{aligned} \tag{22}$$

Substituting (22) to (21), we obtain the Cobb–Douglas function $y = A\left(x_1^{c_1} x_2^{c_2}\right)^{\alpha}$. The following theorem generalizes this outcome for the vintage production function (18) and (19).

Theorem 1 *If and only if the normalization constraint (19) holds, the CES vintage production function (18) with positive inputs $k(v, t) > 0$, $-\infty < v \le t$, $0 \le t < \infty$, converges to the Cobb–Douglas vintage production function.*

$$y(t) = be^{\alpha \int_{-\infty}^{t} w(t-v)\ln[z(v)k(v,t)]dv} \text{ at } \beta \to 0. \tag{23}$$

Proof Let us first analyze the integral $\int_{-\infty}^{t} w(t-v)z^{\beta}(v)k^{\beta}(v,t)dv$ at $\beta \to 0$. We have $\lim\limits_{\beta \to 0+} k^{\beta}(v,t) = 1$ for a given $z(v)k(v,t) > 0$ at fixed v and t. Therefore,

$$\lim_{\beta \to 0} \int_{-\infty}^{t} w(t-v)z^{\beta}(v)k^{\beta}(v,t)dv = \lim_{\beta \to 0} \lim_{T \to \infty} \int_{-T}^{t} w(t-v)z^{\beta}(v)k^{\beta}(v,t)dv$$

$$= \lim_{T \to \infty} \int_{-T}^{t} w(t-v) \lim_{\beta \to 0}\left(z^{\beta}(v)k^{\beta}(v,t)\right)dv = \lim_{T \to \infty} \int \int_{-T}^{t} w(t-v)dv$$

$$= \int_{-\infty}^{t} w(t-v)dv = 1$$

by (19). Then,

$$\lim_{\beta \to 0} \ln\left(\int_{-\infty}^{t} w(t-v)z^{\beta}(v)k^{\beta}(v,t)dv\right)$$

$$= \ln \lim_{\beta \to 0}\left(\int_{-\infty}^{t} w(t-v)z^{\beta}(v)k^{\beta}(v,t)dv\right) = \ln \quad 1 = 0.$$

To estimate the limit of (18), we transform it as

$$\lim_{\beta \to 0} \frac{1}{b} y(t) = \lim_{\beta \to 0}\left(\int_{-\infty}^{t} w(t-v)z^{\beta}(v)k^{\beta}(v,t)dv\right)^{\alpha/\beta}$$

$$= \exp\left\{\lim_{\beta \to 0} \ln\left(\left(\int_{-\infty}^{t} w(t-v)z^{\beta}(v)k^{\beta}(v,t)dv\right)^{\alpha/\beta}\right)\right\}$$

$$= \exp\left\{\lim_{\beta \to 0}\left(\frac{\alpha}{\beta} \ln\left(\int_{-\infty}^{t} w(t-v)z^{\beta}(v)k^{\beta}(v,t)dv\right)\right)\right\}.$$

The indeterminacy $\frac{0}{0}$ appears in the above limit. Next, we apply L'Hopital's rule to resolve it and obtain

$$\lim_{\beta\to 0}\frac{\alpha}{\beta}\ln\left(\int_{-\infty}^{t} w(t-v)z^{\beta}(v)k^{\beta}(v,t)dv\right)$$
$$=\alpha\frac{d}{d\beta}\ln\left(\int_{-\infty}^{t} w(t-v)[z(v)k(v,t)]^{\beta}dv\right)|_{\beta=0}$$
$$=\alpha\frac{\int_{-\infty}^{t} w(t-v)[z(v)k(v,t)]^{\beta}\ln[z(v)k(v,t)]dv}{\int_{-\infty}^{t} w(t-v)[z(v)k(v,t)]^{\beta}dv}|_{\beta=0}$$
$$=\alpha\frac{\int_{-\infty}^{t} w(t-v)\ln[z(v)k(v,t)]dv}{\int_{-\infty}^{t} w(t-v)dv}=\alpha\int_{-\infty}^{t} w(t-v)\ln[z(v)k(v,t)]dv.$$

Combing the last two formulas, we obtain the Formula (23).
Theorem is proven.

Thus, the modified CES function (18) produces the Cobb–Douglas function

$$y(t)=b\left(\int_{-\infty}^{t} w(t-v)z^{\beta}(v)k^{\beta}(v,t)dv\right)^{\alpha/\beta}\xrightarrow{\beta\to 0} y(t)=be^{\alpha\int_{-\infty}^{t} w(t-v)\ln[z(v)k(v,t)]dv}. \quad (24)$$

Cobb–Douglas production functions restrict the elasticity σ of substitution between production factors and occur when $\sigma = 1$. In many cases (see, e.g. [32]), classic two-factor Cobb–Douglas functions provide a good description of observed data, when more advanced production functions provide a poor causal explanation. In such cases, a Cobb–Douglas function still produces a remarkably good fit. It shows that the Cobb–Douglas functional form is flexible and can fit the data well even when it does not have a meaningful economic interpretation.

The given age-dependent kernel $w(a)$ in the Eq. (18) can reflect deterioration, learning, and other age-related phenomena, while it remains a formal parameter of the integral CES function (18) (similarly to c_1 and c_2 in the two-factor CES (20)). One of possible choices for the function w is, for example, $w(a) = \theta e^{-\theta a}$ at any given $\theta = \text{const} > 0$. The exponent $\theta e^{-\theta a}$ means that older vintages have a smaller "share" in production compared to newer vintages and the difference in the shares is bigger for larger parameters θ. This assumption seems to be natural, so we use this formula in the next section.

4 Modified Vintage CES Function with Complementary Inputs

Contemporary research indicates that the estimates of elasticity σ can be below unity, that is, $\beta < 0$, which means the complementarity rather that substitutability of production inputs, see, e.g., Atkeson and Kehoe [2], Chirinko [10], Jovanovic [28].

In this section, we consider the case of negative β, when the elasticity of substitution σ is smaller than 1. Then, the modified CES vintage model (18) and (19) is still valid, but at stricter assumptions on the CES parameter function $w(a)$. As the below statement explains, these assumptions are required to guarantee a positive growth.

Theorem 2 *Let $\beta < 0$ and*

$$w(a) = \theta e^{-\theta a} \quad atany \ \theta = \text{const} > 0. \tag{25}$$

Then, the optimization problem (2)–(7) with the modified CES vintage function (18) and (19) possesses a positive balanced growth trajectory:

$$\chi(a) = \bar{k}\theta^{\frac{1}{1-\beta}} e^{-\frac{\gamma\beta+\theta}{1-\beta}a}, \tag{26}$$

$$\tilde{u}(a) = \bar{k}\theta^{\frac{1}{1-\beta}}\left(\delta + \frac{\gamma\alpha}{1-\alpha} - \frac{\gamma\beta+\theta}{1-\beta}\right)e^{-\frac{\gamma\beta+\theta}{1-\beta}a}, \tag{27}$$

$$\bar{k} = \left(\frac{\alpha\bar{z}^{\alpha}}{r+\delta}\right)^{\frac{1}{1-\alpha}}\left(\frac{1-\beta}{\gamma\beta+\theta}\theta^{\frac{1}{1-\beta}}\right)^{\frac{\alpha-\beta}{\beta(1-\alpha)}}, \tag{28}$$

$$y(t) = \left(\frac{\alpha\bar{z}^{\alpha}}{r+\delta}\right)^{\frac{\alpha}{1-\alpha}}\left(\frac{1-\beta}{\gamma\beta+\theta}\theta^{\frac{1}{1-\beta}}\right)^{\frac{\alpha(1-\beta)}{\beta(1-\alpha)}} e^{\frac{\gamma\alpha}{1-\alpha}t}, \tag{29}$$

if and only if the parameter θ in (25) satisfies the following condition:

$$\theta > -\gamma\beta. \tag{30}$$

Proof We will use the above Formulas (9)–(15) for balanced growth, substituting the given function $A(a)$ for $w(a)$ and assuming $w(a) = \theta e^{-\theta a}$, $\theta > 0$. Then, the age-dependent profiles (14) and (15) of the capital and investment become (26) and (27), where

$$\bar{k} = \left(\frac{\alpha\bar{z}^{\alpha}}{r+\delta}\right)^{\frac{1}{1-\alpha}}\left(\theta^{\frac{1}{1-\beta}}\int_0^{\infty} e^{-\frac{\gamma\beta+\theta}{1-\beta}s}ds\right)^{\frac{\alpha-\beta}{\beta(1-\alpha)}}. \tag{31}$$

The optimal output growth (11) is now

$$y(t) = \left(\frac{\alpha\bar{z}^{\alpha}}{r+\delta}\right)^{\frac{\alpha}{1-\alpha}} \left(\theta^{\frac{1}{1-\beta}} \int_0^{\infty} e^{-\frac{\gamma\beta+\theta}{1-\beta}s} ds\right)^{\frac{\alpha(1-\beta)}{\beta(1-\alpha)}} e^{\frac{\gamma\alpha}{1-\alpha}t}. \qquad (32)$$

Next, one can see that the improper integral in (31) and (32) is finite if $\theta > \gamma\beta$. If (30) holds, then (28) and (29) follow from (31) and (32). Theorem is proven.

The condition (30) of Theorem 2 means that the "share factor" $w(a)$ of older vintages in the CES function (18) and (19) has to decrease in age a exponentially with a rate faster than $-\gamma\beta$.

If (30) fails, then the only optimal balanced trajectory is $y(t) = 0$, $k(v, t) = 0$, $x(t) = 0$, and $u(v, t) = 0$ (no growth at all). So, the restriction (30) on the CES parameter β is necessary for having a non-trivial balanced growth.

By (30), both the level and growth rate of the balanced age-dependent capital and investment profiles (26) and (27) depend on the choice of the parameter θ. The rate of the balanced optimal output (29) does not depend on θ, but its level does.

By Theorem 2, the optimal dynamics of the modified optimization problem (2)–(7), (18) and (19) remains qualitatively similar to the one of the original CES model (1)–(7). However, by (30), the balanced age-dependent capital and investment profiles (26) and (27) at $\beta < 0$ always decrease in age. Moreover, if we add a learning curve to the model (18), these profiles will have maximums at certain finite ages.

When a negative β indefinitely decreases, the constraint (30) requires larger and larger values of the parameter θ So, there is no way to obtain a Leontieff-type production function from the CES function (18) and (19) as its limit at $\beta \to -\infty$. In the case $\beta \to -\infty$, the elasticity of substitution σ tends toward zero (no substitution).

5 Optimal Investment with Learning in Cobb–Douglas Vintage Function

The optimal investment in the CES vintage production function (1)–(7) may flow to various (new or old) capital vintages, because different vintages are imperfect substitutes in production. As [37] states, a gradual diffusion of new capital goods is achieved via the introduction of vintage-specific learning, which accumulates gradually over time.

A key outcome of [29] is the inverted U-shaped profile of age-dependent investments in old vintages in presence of learning, which explains intra-firm adoption lags. A similar result was earlier obtained by [9] in a CES model with two (old and young) vintages. This property was analytically proven and numerically demonstrated for the original CES model (1)–(7) in the range $0 < \beta < 1$, which corresponds to the elasticity of substitution among different vintages $1 < \sigma < \infty$. The analytic proof of this result for the vintage CES model (1)–(7) follows from Lemma 1 when the learning is described by the model function $A(a) = 1 - e^{-\phi a}$ at a given $\varphi > 0$. It is natural to check if this property remains in force for the vintage Cobb–Douglas function (23), which occurs at $\sigma = 1$.

We cannot use the same learning model $w(a) = 1 - e^{-\phi a}$ in the integral Cobb–Douglas function (23) because it does not satisfy the formal condition (19) for w. So, we choose the given $w(a)$ as

$$w(a) = e^{-\theta a} - e^{-(\phi+\theta)a}, \quad \theta > 0, \quad \varphi > 0, \tag{33}$$

to describe both learning and deterioration of vintages when the vintage age a increases. The function (33) satisfies (19) at $\varphi = \theta^2/(1 - \theta)$, therefore, we must select

$$w(a) = e^{-\theta a} - e^{-\theta/(1-\theta)a}, \quad 0 < \theta < 1. \tag{34}$$

Theorem 3 *Under exponential technological change (9) and learning (34), the optimization problem (1)–(7) with the Cobb–Douglas vintage function (23) possesses the following positive balanced growth trajectory:*

$$k(v,t) = e^{\varsigma t}\chi(t-v), \quad \varsigma = \frac{\gamma\alpha}{1-\alpha}, \tag{35}$$

$$y(t) = e^{\varsigma t}\overline{k}\left(\frac{\alpha\overline{z}^{\alpha}}{r+\delta}\right)^{-1}, \tag{36}$$

$$u(v,t) = e^{\zeta t}\tilde{u}(t-v), \quad x(t) = 0, \quad -\infty < v < t, 0 < t < \infty, \tag{37}$$

where

$$\overline{k} = \left(\frac{\alpha\overline{z}^{\alpha}}{r+\delta}\right)^{\frac{1}{1-\alpha}} \exp\left(\frac{\alpha}{1-\alpha}\int_0^{\infty}(-\gamma s + \ln w(s))w(s)ds\right), \tag{38}$$

$$\chi(a) = \overline{k}w(a), \quad a = t - v, \tag{39}$$

$$\tilde{u}(a) = \overline{k}\left[\left(\delta + \frac{\gamma\alpha}{1-\alpha}\right)w(a) + w'(a)\right]. \tag{40}$$

Proof We use the Formulas (9)–(15) for balanced growth in the CES model (1)–(7), substituting the function $A(a)$ for $w(a)$ defined by (34). To obtain the Cobb–Douglas function case we must set $\beta = 0$. The only indeterminacy at $\beta \to 0$ occurs in the Formula (38) for constant $\overline{k}$. Similarly to the proofs of Theorems 1 and 2, we transform (13) as

$$\overline{k} = \left(\frac{\alpha\overline{z}^{\alpha}}{r+\delta}\right)^{\frac{1}{1-\alpha}}\left(\int_0^{\infty} e^{-\frac{\gamma\beta}{1-\beta}s}A^{\frac{1}{1-\beta}}(s)ds\right)^{\frac{\alpha-\beta}{\beta(1-\alpha)}} = \left(\frac{\alpha\overline{z}^{\alpha}}{r+\delta}\right)^{\frac{1}{1-\alpha}} e^{\frac{\alpha-\beta}{\beta(1-\alpha)}\ln\int_0^{\infty} e^{-\frac{\gamma\beta}{1-\beta}s}w^{\frac{1}{1-\beta}}(s)ds}. \tag{41}$$

Next, we apply the L'Hopital's rule to resolve the indeterminacy 0/0 in (41) at $\beta \to \infty$ and obtain.

$$\lim_{\beta \to 0} \frac{\alpha - \beta}{\beta(1-\alpha)} \ln \int_0^\infty e^{-\frac{\gamma\beta}{1-\beta}s} w^{\frac{1}{1-\beta}}(s)ds = \left(\frac{\alpha}{1-\alpha} \int_0^\infty (-\gamma s + \ln w(s))w(s)ds \right). \tag{42}$$

Combining (42) and (41), we get the Formula (38) for $\overline{k}$.

The Formulas (36), (37), (39), and (40) are obtained from their counterparts in Lemma 1 by substituting $\beta = 0$. Theorem is proven.

Theorem 3 demonstrates that major qualitative properties of the CES vintage function (1) hold for the Cobb–Douglas function (23). Namely, both balanced age-dependent capital and investment profiles are bell-shaped with a maximum at certain vintage ages. Since Cobb–Douglas functions are a special case of CES functions, those properties are described by simpler relations.

Indeed, the age-dependent profile (39) of old capital vintages in the Cobb–Douglas case coincides with the learning-deterioration profile (34). Taking the derivative of (34), we conclude that it has a maximum at

$$a*_k = -\frac{1-\theta}{\theta^2} \ln(1-\theta^2) > 0. \tag{43}$$

The investment age-dependent profile (40) in this case has a maximum at the age

$$a*_u = -\frac{1-\theta}{\theta^2} \ln\left((1-\theta^2)\left(\delta + \frac{\gamma\alpha}{1-\alpha} - \theta\right) / \left(\delta + \frac{\gamma\alpha}{1-\alpha} - \frac{\theta}{1-\theta}\right)\right) > 0, \tag{44}$$

earlier than the capital profile (39).

By Theorem 3, both bell-shaped age-distributions $\chi(a)$ and $\tilde{u}(a)$ of capital and investment are positive for any vintage age a. It means that the investment in old vintages never stops in Cobb–Douglas case (though eventually becomes quite small). It matches the outcome of [29] that the investment in old vintages in the CES model (1)–(7) continues forever if $\beta < 0.9$.

6 On Transition Dynamics of Optimal Investment and Capital

The optimal dynamics in economic growth models commonly includes two parts: balanced growth and transition dynamics. All the above results concern balanced growth trajectories in the optimization model (1)–(7). Transition dynamics is notoriously challenging to analyze in vintage capital models. The goal of the transition is to switch from an initial non-optimal distribution of pre-installed vintages to an optimal balanced trajectory in the most effective way.

First significant results in the transition dynamic analysis of vintage capital models were obtained by [4–6]. As most vintage related research, these models do not allow investments in older capital vintages. Paper [19] provides a complete analysis of transition dynamics in the optimization model (1)–(7) with linear utility. This section briefly reviews and compares nonmonotonic and bang-bang effects in the models with and without investment in older vintages.

As shown by [4–6], the optimal investment in new vintages is nonmonotonic because of replacement echoes and nonexponential asset deterioration. In the linear utility case, everlasting fluctuations of investments and output can occur. In contrast, no replacement echoes occur in the vintage model (1)–(7) because it does not assume any resource constraint as in [5, 6]. Initial bang-bang optimal investment to specific vintages in the model (1)–(7) eventually converges to an interior trajectory that represents balanced economic growth. The investment dynamics in each old vintage depends on whether it has been initially overfunded or underfunded. For underfunded vintages, the optimal investment is maximal until the vintage capital increases to the optimal level. For overfunded vintages, the optimal policy is to wait until the vintage capital decreases to the optimal level because of deterioration. After the transition ends for all old vintages, the investment policy is interior and coincides with the balanced growth trajectory.

Adding nonlinear utility to vintage models makes their transition dynamics richer and more realistic, though essentially complicates its analysis. Boucekkine et al. [6] discovered that everlasting fluctuations in investments in the latest technology are not optimal under the concave utility and the transition involves decreasing replacement echoes and fluctuations in the scrapping age. Still, some investment echoes occur to allow the lifetime of machines to converge to balanced trajectory. At the same time, the convergence of the scrapping rule is non-robust and involves oscillations in order to smooth consumption. Similar convergence properties have been proven for the Ramsey vintage capital model with nonlinear utility in [23].

Balanced growth in the model (1)–(7) with nonlinear CRRA utility and investment to old vintages is analyzed in [29], but its transition dynamics remains largely an open problem. Hritonenko et al. [25] show that the optimal investment dynamics appears to be very different for old and new vintages. Time dynamics of the optimal investment $u(v, t)$ in vintages installed during the planning horizon, at $v \in [0, \infty)$, is interior and converges to the balanced growth trajectory. The time-dependent investments $u(v, t)$ in the old vintages installed at $v < 0$ (before the planning horizon) heavily depend on shapes of initial distributions on the prehistory and their convergence to balanced growth is non-robust, lumpy and challenging to prove.

7 Discussion and Conclusion

In line with a general economic theory, production inputs are *substitutes* if the elasticity σ of substitution between inputs is >1, If the elasticity $\sigma < 1$ (and $\beta < 0$), then the inputs are *complementary*. In the original CES model (1)–(7), when all vintages become complementary, then it is optimal to invest into all of them simultaneously (see Lemma 1). As [25] demonstrates, the case $\sigma < 1$ (that is, $\beta < 0$) leads to infinite jumps of investments, capital, and production (in the absence of any external restrictions on investment in the model). This issue is the underlying reason for modifying the vintage CES function (1) in Sect. 3. As shown in Sect. 4, the model (2)–(7) with the modified CES vintage function (18) and (19) suits well for an economic analysis of the complementarity case $\sigma < 1$.

The choice of the value σ (and β) is an important empirical question. There is no direct data on the elasticity of substitution between old and new capital vintages. Chari and Hopenhayn [9] used values of β between zero and one (that imply values of σ between one and infinity). Jovanovic [28] used some data on the elasticity of substitution for human capital, which indicate that the elasticity σ may be below unity and, correspondingly, $\beta < 0$. An extensive numeric simulation in [29] at $0 < \beta < 1$ assumes $\alpha = 0.85$ for the returns to scale, $\gamma = 0.004$ for the technological change rate, and $\delta = 0.08$ for deterioration rate. At $\sigma > 26.7$ ($\beta > 0.96$), all optimal investment goes into the latest vintage and the CES model (1) produces the same pattern (spiked investment into one vintage) as the traditional vintage models (8). At $\sigma < 26.7$, the optimal investment shape (12)–(15) includes simultaneous investments into older vintages. A reasonable range of the elasticity for physical capital is identified as $10 < \sigma < 20$, which corresponds to $0.9 < \beta < 0.95$. Lanteri [33] refers to the range $1 < \sigma < 10$ of the elasticity of substitution between new and used capital in farming, construction, and aircraft industries.

We shall notice that the normalization (19) is *not necessary* in the case of large elasticities σ (i.e., at β close to 1) and the original vintage CES production function (1) is perfect for such σ. When β approaches 1, both CES models (1) and (18)–(19) converge to traditional Solow-type vintage capital models [2, 5, 7, 11, 18, 20, 21]. In this range, the presence of the constraint (19) on the given function $w(a)$ in (18) does not affect the model dynamics, so, this constraint is difficult to motivate.

In summary, the modified CES vintage model (18) describes the substitution of different capital vintages for the full range $[0, \infty)$ of the elasticity σ. The original CES model (1) works in the case of large σ, while the modified CES model (18) works for all σ including $\sigma < 1$. The Cobb–Douglas model (23) describes the case $\sigma = 1$. So, the choice of appropriate CES vintage function depends on expected values of the elasticity of substitution among capital vintages.

Acknowledgements The authors are thankful to the reviewer for careful reading of original manuscript and valuable comments that essentially improved the clarity of results.

References

1. Arrow, K.: The economic implications of learning by doing. Rev. Econ. Stud. **29**(3), 155–173 (1962)
2. Atkeson, A., Kehoe, P.: Modeling and measuring organization capital. J. Polit. Econ. **113**(5), 1026–1053 (2005)
3. Barro, R.J., Sala-i-Martin, X.: Economic Growth, 2nd edn. MIT Press, Massachusetts (2003)
4. Benhabib, J., Rustichini, A.: Vintage capital, investment, and growth. J. Econ. Theory **55**(2), 323–339 (1991)
5. Boucekkine, R., Germain, M., Licandro, O.: Replacement echoes in the vintage capital growth model. J. Econ. Theory **74**, 333–348 (1997)
6. Boucekkine, R., Germain, M., Licandro, O., et al.: Creative destruction, investment volatility, and the average age of capital. J. Econ. Growth **3**, 361–384 (1998)
7. Boucekkine, R., Hritonenko, N., Yatsenko, Y.: Optimal investment in heterogeneous capital and technology under restricted natural resource. J. Optim. Theory Appl. **163**, 310–331 (2014)
8. Caunedo, J., Keller, E.: Capital obsolescence and agricultural productivity. Q. J. Econ. **136**, 505–561 (2021)
9. Chari, V., Hopenhayn, H.: Vintage human capital. J. Polit. Econ. **99**(6), 1142–1165 (1991)
10. Chirinko, R.: σ: The long and short of it. J. Macroecon. **30**, 671–686 (2008)
11. Cooley, T., Greenwood, J., Yorukoglu, M.: The replacement problem J. . Monet. Econ. **40**(3), 457–499 (1997)
12. Denison, E.F.: The unimportance of the embodied question. Am. Econ. Rev. **54**, 90–94 (1964)
13. Farias, M.E., Scavia, J., Fuentes, R.: Bridging the gaps: credits, adoption, and inequality. J. Money,Credit, Bank (2018). https://doi.org/10.1111/jmcb.12549
14. Førsund, F.R., Vislie, J.: Leif Johansen on intra-industry structural change. J. Policy Model. **38**, 515–527 (2016)
15. Gale, D.: On optimal development in a multi-sector economy. Rev. Econ. Stud. **34**, 1–18 (1967)
16. Gamboa, F., Maldonado, W.: Feasibility and optimality of the initial capital stock in the Ramsey vintage capital model. J. Math. Econ. **52**, 40–45 (2014)
17. Goetz, R., Hritonenko, N., Yatsenko, Y.: The optimal economic lifetime of vintage capital in the presence of operating cost, technological progress, and learning. J. Econ. Dyn. Control **32**, 3032–3053 (2008)
18. Greenwood, J., Hercowitz, Z., Krusell, P.: Long-run implications of investment-specific technological change. Am. Econ. Rev. **87**, 342–362 (1997)
19. Hritonenko, N., Kato, N., Yatsenko, Y.: Optimal control of investments in old and new capital under improving technology. J. Optim. Theory Appl. **172**, 247–266 (2017)
20. Hritonenko, N., Yatsenko, Y.: Modeling and Optimization of the Lifetime of Technologies. Kluwer Academic Publishers, Dordrecht (1996)
21. Hritonenko, N., Yatsenko, Y.: Turnpike and optimal trajectories in integral dynamic models with endogenous delay. J. Optim. Theory Appl. **127**, 109–127 (2005)
22. Hritonenko, N., Yatsenko, Y.: The structure of optimal time- and age-dependent harvesting in the Lotka-McKendrik population model. Math. Biosci. **208**, 48–62 (2007)
23. Hritonenko, N., Yatsenko, Y.: Optimal control of Solow vintage capital model with nonlinear utility. Optimization **57**, 581–592 (2008)
24. Hritonenko, N., Yatsenko, Y., Boranbayev, A.: Generalized functions in the qualitative study of heterogeneous populations. Math. Popul. Stud. **26**, 146–162 (2019)
25. Hritonenko, N., Yatsenko, Y., Boranbayev, A.: Analysis of optimal investment in continuum of age-dependent vintages. Pure Appl. Funct. Anal. **6**(4), 743–759 (2021)
26. Johansen, L.: Substitution versus fixed proportions coefficients in the theory of economic growth: a synthesis. Econometrica **27**, 157–176 (1959)
27. Johansen, L.: A Multi-sectoral Study of Economic Growth. North-Holland Publishing Company, Amsterdam (1960)
28. Jovanovic, B.: When should firms invest in old capital. Int. J. Econ. Theory **5**, 107–123 (2009)

29. Jovanovic, B., Yatsenko, Y.: Investment in vintage capital. J. Econ. Theory **147**, 551–569 (2012)
30. Kredler, M.: Vintage human capital and learning curves. J. Econ. Dyn. Control **40**, 154–178 (2014)
31. Martinez, D.: Automation, growth and factor shares, 2018 Meeting Papers Society for Economic Dynamics, no. 736 (2018). https://EconPapers.repec.org/RePEc:red:sed018:736
32. Miller, E.: An Assessment of CES and Cobb-Douglas Production Functions: Working Paper 2008–05. Congressional Budget Office (2008)
33. Lanteri, A.: The market for used capital: endogenous irreversibility and reallocation over the business cycle. Am. Econ. Rev. **108**, 2383–2419 (2018)
34. Livdan, D., Nezlobin, A.: Investment, capital stock, and replacement cost of assets when economic depreciation is non-geometric. J. Financ. Econ. **142**(3), 1444–1469 (2021)
35. Lu, J., Zhang, J., Lu, F., et al.: Optimal pricing on an age-specific inventory system for perishable items. Oper. Res. Int. J. **20**, 605–625 (2020). https://doi.org/10.1007/s12351-017-0366-x
36. Solow, R.: Investment and technical progress. In: Mathematical Methods in the Social Sciences. Stanford University Press, Stanford (1960)
37. Samaniego, R.M., Yu Sun, J.: The embodiment controversy: on the policy implications of vintage capital models. J. Money, Credit, Bank **54**(5), 1175–1222 (2022)
38. Xepapadeas, A., De Zeeuw, A.: Environmental policy and competitiveness: the Porter hypothesis and the composition of capital. J. Environ. Econ. Manag. **37**, 165–182 (1999)
39. Yatsenko, Y., Hritonenko, N.: Optimization in integral model of developing systems. Optimization **31**, 179–192 (1994)

A Two Country Model of Trade with International Borrowing and Lending

Kazumichi Iwasa and Kazuo Nishimura

Abstract We investigate the properties of a two-country dynamic Heckscher–Ohlin model that allows international borrowing and lending. As is well known, international trade patterns become undecidable when international borrowing and lending is allowed. To avoid this, we assume a consumable capital good to be nontradable. A key feature of our model is the existence of a continuum of steady state levels of capital stocks, which enables us to examine how the initial amount of physical capital and assets in each country affects the amount of capital and assets in the steady state.

Keywords Two-country model · International borrowing and lending · Continuum of steady states

1 Introduction

Although many existing studies using dynamic two-country models have not considered international asset markets, globalization of financial markets has progressed rapidly over the past several decades, and international asset markets are becoming increasingly important. In this paper, we investigate the properties of a two-country dynamic Heckscher–Ohlin model with international borrowing and lending to examine the long-term impact of countries' physical capital and financial assets on their economic growth and production patterns.

As is well known, international trade patterns become undecidable when international borrowing and lending is allowed. To avoid this, Meng and Velasco [4] construct a small-open economy with nontradable capital goods in which international borrowing and lending is allowed, and demonstrate the occurrence of indeterminacy, Ono and Shibata [6] examine specialization patterns in a two-country model

K. Iwasa · K. Nishimura (✉)
Research Institute for Economics and Business Administration, Kobe University, Nada-ku, Kobe, Japan
e-mail: nishimura@rieb.kobe-u.ac.jp

K. Iwasa
e-mail: iwasa@rieb.kobe-u.ac.jp

M. A. Khan et al. (eds.), *Matching, Dynamics and Games for the Allocation of Resources*, Monographs in Mathematical Economics 7,
https://doi.org/10.1007/978-981-96-6661-4_14

with international borrowing and lending by introducing an adjustment cost to the investment.

Hu and Mino [3] consider two-country dynamic Heckscher–Ohlin models, and derive the conditions under which indeterminacy occurs. They examine two types of trade structures, one of which is a two-country version of Meng and Velasco [4].[1]

In this study, we consider two types of consumption good, one of which is a pure consumption good and the other is a consumable capital good. Then, we assume consumable capital is nontradable as in Meng and Velasco [4].

The key feature of our model is that there exists a continuum of steady state levels of capital stocks. So, we can examine how the initial amount of physical capital and assets in each country affects the amount of capital and assets in the steady state. It highly contrasts with results in the previous studies, where capital goods are nonconsumable and nontradable. In those models, the steady state level of capital stocks is unique, and hence both countries with relatively more and less capital initially will accumulate equal amounts of capital in the long run.

In Sect. 2, we present a two-sector model with international borrowing and lending. In Sect. 3, we derive the existence of a continuum of steady states by applying the diagrammatic analysis in Bond et al. [2]. In Sect. 4, we examine local stability of the steady states. Section 5 concludes the paper.

2 The Two Country Model of Trade with International Borrowing and Lending

In this section, we formulate a two-country model, where there are international borrowing and lending and a nontradable good. There are two countries, home and foreign. We assume that the home and foreign countries are symmetric except for the initial capital endowment and the initial asset in each country. They have the same population normalized to be one, with each household having an endowment of labor, L. Factors of production are assumed to be mobile between sectors within a country, but immobile internationally. However, there is a market for international borrowing and lending. There are two goods using a fixed factor (labor, L) and a reproducible factor (capital, K) under conditions of perfect competition and constant returns to scale. Good 1 is a consumable capital good, and the second good is a pure consumption good. We assume good 1 is not tradable.[2] We refer to the representative country as the home country: the corresponding behavioral relations for the other (foreign) country will be denoted by a "$*$." Let p denotes the price of good 1 in home and good 2 be numeraire. Each household has a concave felicity function u defined over consumption of goods 1 and 2, C_1 and C_2, respectively.

[1] They also consider the other trade structure in Nishimura and Shimomura [5], where all goods are tradable but international borrowing and lending is not allowed, and argue that whether trade makes economic fluctuations more or less likely depends on the structure of trade.

[2] If we assume good 2 is not tradable, the model loses its tractability.

2.1 The Production Side

On the production side, we will assume

Assumption 1 The production function in each sector is quasi-concave and linearly homogeneous. Both factors are indispensable for producing, and consumable capital good 1 is labor intensive.

The results of the static Heckscher-Ohlin model are well known, so here we provide only a brief review of properties that will be important to the dynamic model.

Letting w denote the wage rate and r the rental on capital, the technology in sector i can be characterized by the unit cost function $a_i(w, r)$, $i = 1, 2$. Under incomplete specialization, the competitive profit conditions require that

$$a_1(w, r) = p, \tag{1}$$

$$a_2(w, r) = 1, \tag{2}$$

where good 2 is chosen as numeraire. Let $w(p)$ and $r(p)$ be the solution to the system of equations, (1) and (2). Then, we have

$$w(p) = \frac{pa_{2r} - a_{1r}}{\Delta}, \quad w'(p) = \frac{a_{2r}}{\Delta}, \quad r(p) = \frac{a_{1w} - pa_{2w}}{\Delta}, \quad \text{and} \quad r'(p) = \frac{-a_{2w}}{\Delta}, \tag{3}$$

where a_{iw} and a_{ir} are the labor and capital input coefficients in sector i, respectively, and Assumption 1 implies that they satisfy

$$\Delta \equiv a_{1w}a_{2r} - a_{2w}a_{1r} > 0.$$

Factor market equilibrium requires that

$$L = a_{1w}Y_1 + a_{2w}Y_2, \tag{4}$$

$$K = a_{1r}Y_1 + a_{2r}Y_2, \tag{5}$$

where K is the stock of capital and Y_i is the output of good i. From the equations above, we see

$$Y_1(p, K) = w'(p)L + r'(p)K, \tag{6}$$

$$Y_2(p, K) = w(p)L + r(p)K - p[w'(p)L + r'(p)K]. \tag{7}$$

Notice that (6) and (7) yield

$$wL + rK = pY_1 + Y_2, \tag{8}$$

and hence we can use (4) and (8), instead of (4) and (5), to find outputs (Y_1, Y_2).

Since the assumptions to be made below will ensure that the economy is incompletely specialized in both the autarkic and trade steady states, we limit our presentation of the production side to the case of incomplete specialization. Our assumption on the factor intensity ranking of sectors is chosen for convenience due to emphasis on a diagrammatic presentation. We indicate below in cases where the factor intensity rankings play a role.

2.2 The Consumption Side

We analyze the optimization problem for a representative household that owns L units of labor. We will impose the following restrictions on this felicity function:

Assumption 2 The felicity function is strictly concave, with $u_{11} < 0$ and $D \equiv u_{11}u_{22} - u_{12}u_{21} > 0$ for any $(C_1, C_2) \in \{(C_1, C_2) \in \mathbb{R}^2_+ | u_i(C_1, C_2) > 0, \ i = 1, 2\}$, and satisfies $\lim_{C_i \to 0} u_i(C_1, C_2) = \infty \ (i = 1, 2)$ for any $C_j \ (j \neq i)$.

The home households maximize

$$\int_0^\infty u(C_1, C_2)e^{-\rho t} dt$$

$$\text{subject to } \dot{B} = RB + rK + wL - pC_1 - C_2 - pI,$$
$$\dot{K} = I - \delta K,$$
$$\lim_{t \to \infty} \exp\left(-\int_0^t R(s)ds\right) B(t) \geq 0,$$

where B, R, and I denote the stock of bonds, the interest rate, and gross investment.

The Hamiltonian function for this optimization problem is given by

$$\mathcal{H} \equiv u(C_1, C_2) + \lambda(RB + rK + wL - pC_1 - C_2 - pI) + q(I - \delta K),$$

where λ and q denote the shadow values of bonds and capital, respectively.

Then, we have

$$\frac{\partial \mathcal{H}}{\partial C_1} = u_1(C_1, C_2) - \lambda p = 0,$$
$$\frac{\partial \mathcal{H}}{\partial C_2} = u_2(C_1, C_2) - \lambda = 0,$$
$$\frac{\partial \mathcal{H}}{\partial I} = -\lambda p + q = 0,$$
$$\frac{\partial \mathcal{H}}{\partial B} = \lambda R = \rho\lambda - \dot{\lambda},$$

$$\frac{\partial \mathcal{H}}{\partial K} = \lambda r - q\delta = \rho q - \dot{q}.$$

So, we obtain

$$C_1 = C_1(p, \lambda) \text{ and } C_2 = C_2(p, \lambda), \tag{9}$$

$$q = \lambda p, \tag{10}$$

$$\dot{\lambda} = \lambda(\rho - R), \tag{11}$$

$$\dot{q} = q\left(\rho + \delta - \frac{r}{p}\right), \tag{12}$$

where consumption relations $C_i(p, \lambda)$ for $i = 1, 2$ and an expenditure relation $E(p, \lambda) \equiv pC_1(p, \lambda) + C_2(p, \lambda)$ satisfy

$$\begin{aligned} \lambda C_{1\lambda} &= pC_{1p} + C_{2p}, \\ E_\lambda &= pC_{1\lambda} + C_{2\lambda} < 0, \\ C_{1p} &< 0, \\ E_p &= C_1 + \lambda C_{1\lambda}, \end{aligned}$$

which are proved in Bond et al. [1].

From (10), (11), and (12), we have the non-arbitrage condition between bond and capital as follows.

$$R = \frac{r}{p} - \delta + \frac{\dot{p}}{p}. \tag{13}$$

2.3 Market Equilibrium

When (13) holds, households will arbitrarily divide their surplus between financial and capital investments. So, we may conclude that the level of capital investment is determined to clear the domestic market for good 1, that is,

$$I = \dot{K} + \delta K = Y_1 - C_1$$

holds. Then, from (8) we have

$$\dot{B} = RB + Y_2(p, K) - C_2(p, \lambda).$$

Let Z_i denote the excess demand for good i ($i = 1, 2$), where Z_2 is given by

$$Z_2(p, K, \lambda) = C_2(p, \lambda) - Y_2(p, K).$$

Notice that

$$\dot{B} + \dot{B}^* = R(B + B^*) - (Z_2 + Z_2^*)$$

holds, and hence we see that the international credit market will clear, if the international market for good 2 is cleared and $B_0 + B_0^* = 0$, where B_0 and B_0^* denote the initial stock of bond in home and foreign, respectively.

3 Steady States

A steady state will be characterized by the existence of a price $\tilde{p}$, capital stock $\tilde{K}$, and the stock of bonds $\tilde{B}$, such that $\dot{B} = \dot{K} = \dot{\lambda} = \dot{q} = 0$ and markets clear. From (12), a steady state with incomplete specialization will require that there is some $\tilde{p} > 0$ such that $\bar{r}(\tilde{p}) = \rho + \delta$ holds, where $\bar{r}(p) \equiv r(p)/p$. With Assumption 1, $\tilde{p}$ uniquely exists, but in the case where good 1 is capital intensive, some restrictions on production technologies must be assumed to guarantee the existence of $\tilde{p}$.[3]

3.1 Determination of an Autarkic Steady State

An autarkic steady state requires that $Z_1 = 0$ and $Z_2 = 0$ with $B = 0$. The former condition requires production of good 1 in order to sustain the steady state capital stock, and the latter condition requires production of good 2 in the steady state as a result of Assumption 2. Therefore, the autarkic steady state price must be the consistent with incomplete specialization.

The market clearing conditions in the autarkic steady state are

$$Y_1 = w'(\tilde{p})L + r'(\tilde{p})K = C_1 + \delta K \text{ and } Y_2 = C_2. \tag{14}$$

Substituting (14) into the labor market equilibrium condition, (4), we define the *steady state* Rybczynski line as follows:

$$\frac{\tilde{a}_{1w}}{\tilde{a}_{2w}} \cdot \frac{r'(\tilde{p})}{r'(\tilde{p}) - \delta} C_1 + C_2 = \left[\frac{1}{\tilde{a}_{2w}} + \frac{\tilde{a}_{1w}}{\tilde{a}_{2w}} \cdot \frac{\delta w'(\tilde{p})}{r'(\tilde{p}) - \delta} \right] L, \quad \text{for } C_1 \geq -\delta k_2(\tilde{p})L \text{ and } C_2 \geq 0, \tag{15}$$

where $\tilde{a}_{iw} \equiv a_{iw}(w(\tilde{p}), r(\tilde{p}))$ for $i = 1, 2$.

The steady state Rybczynski line is the locus of steady state consumption levels that are attainable as the stock of capital is varied, given the stock of labor and relative prices, and is illustrated by the negatively sloped line in Fig. 1. The steady state Rybczynski line coincides with the Rybczynski line from the static trade model if $\delta = 0$.

We obtain the following Lemma.

[3] With the Cobb-Douglas technologies, $\tilde{p}$ uniquely exists in that case.

Lemma 1 *Let K be the steady state capital stock. Then, the outputs of two goods at the steady state, $(\hat{Y}_1(K), \hat{Y}_2(K))$, are derived from the intersection between the steady state Rybczynski line,*

$$\frac{r'(\tilde{p}) - (\rho + \delta)}{r'(\tilde{p}) - \delta}\tilde{p}C_1 + C_2 = \left[w(\tilde{p}) + \rho\frac{\tilde{p}w'(\tilde{p})}{\delta - r'(\tilde{p})}\right]L, \tag{16}$$

and the steady state resource constraint,

$$\tilde{p}C_1 + C_2 = w(\tilde{p})L + \rho\tilde{p}K, \tag{17}$$

as $(\hat{Y}_1(K), \hat{Y}_2(K)) = (C_1 + \delta K, C_2)$.
Proof From (3), we have

$$a_{1w} = (r - pr')\Delta,\ a_{1r} = (pw' - w)\Delta,\ a_{2w} = -r'\Delta,\ a_{2r} = w'\Delta,\ \text{and}\ \Delta = \frac{1}{w'r - wr'}. \tag{18}$$

Then, it is easy to see that (15) is identical to (16). On the other hand, (17) is easily derived from (8). □

Since $r(\tilde{p}) - \delta\tilde{p} = \rho\tilde{p}$, the *steady state* resource constraint (17) also represents the budget constraint for households that have no bond ($B = 0$), but own capital stock K with investment δK. So, the intersection between the income expansion path with $\tilde{p}$ and (17) corresponds to the consumption bundle for such households.

Therefore, we may conclude that goods and bond markets will clear with $B = 0$ and $\dot{B} = \dot{K} = \dot{\lambda} = \dot{q} = 0$, when the steady state resource constraint passes through the intersection between the income expansion path and the steady state Rybczynski line.

The intersec0tion, (C_1^A, C_2^A) in Fig. 1, corresponds to the autarkic steady state. Here, the steady state values of K, B, λ and q are given by

$$K^A = \frac{C_1^A - w'(\tilde{p})L}{r'(\tilde{p}) - \delta},\ B^A = 0,\ \lambda^A = u_2(C_1^A, C_2^A)\ \text{and}\ q^A = \tilde{p}u_2(C_1^A, C_2^A).$$

Hence, we have[4]

Proposition 1 *An intersection between the steady state Rybczynski line and the income expansion path with the steady state price of good 1 corresponds to an autarkic steady state. Therefore, it uniquely exists as long as labor intensive good 1 is normal and preferences exhibit neither a satiation level nor a minimum subsistence level.*

In the rest of paper, we assume normality in consumption for simplicity.

Assumption 3 Both goods are normal in consumption.

---

[4] Notice that the intersection must be unique when good 2 is inferior at some income levels, because the slope of the steady state Rybczynski line is steeper than that of the budget constraint (17).

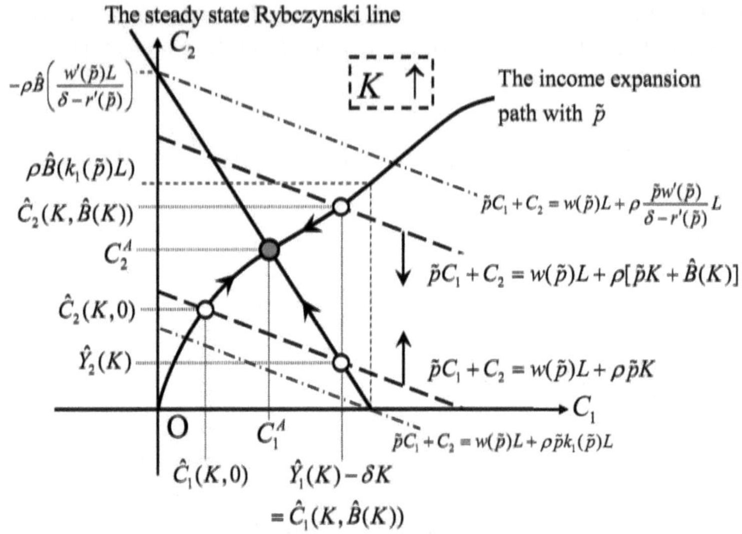

Fig. 1 The income expansion path and the steady state Rybczynski line

3.2 *Excess Demand for Good 2*

Let $(\hat{C}_1(K, B), \hat{C}_2(K, B))$ denote the intersection between the income expansion path with $\tilde{p}$ and the steady state budget constraint,

$$\tilde{p}C_1 + C_2 = w(\tilde{p})L + \rho(\tilde{p}K + B).$$

Then, it corresponds to households' consumption bundle at the steady state where the levels of capital stock and bond are K and B, respectively, since $R = \rho$ holds at steady states.

Since good 1 is not tradable, the excess demand for good 1 must be zero. When the market for good 1 is cleared, households must have some amount of bonds at non autarkic steady states.

It is apparent from Fig. 1 that for $K \in [k_1(\tilde{p})L, w'(\tilde{p})L/[\delta - r'(\tilde{p})]]$, there uniquely exists the value of B that yields market clearing for good 1 in Home. Let us denote it as $\hat{B}(K)$. Then, we have

$$\hat{Y}_1(K) - \delta K = \hat{C}_1(K, \hat{B}(K))$$

for any $K \in [k_1(\tilde{p})L, w'(\tilde{p})L/[\delta - r'(\tilde{p})]]$.

We define the steady state excess demand function as follows.

$$\hat{Z}_2(K) \equiv \hat{C}_2(K, \hat{B}(K)) - \hat{Y}_2(K),$$

which denotes the excess demand for good 2 when capital stocks in Home is K and the domestic market for good 1 is clear with $B = \hat{B}(K)$.

Under normality in consumption, excess demand will be strictly decreasing in K with $\hat{Z}_2(K^A) = 0$. Since $\hat{Y}_2(K)$ is linear in K, the shape of $\hat{Z}_2$ reflects exactly that of the income expansion path. In the case of homothetic preferences the slope of the function is constant, while the function is concave (convex) in K, when good 2 is a necessity (luxury).

3.3 *The Foreign Country and World Market Equilibrium*

First, we assume

Assumption 4 Both countries are identical except their initial capital stocks and asset holdings.

These assumptions ensure that the autarkic steady state prices are the same in each country, and will be the same as the trade steady state prices.[5]

Utilizing Fig. 1, we will show the existence of a continuum of trade steady states.

Notice that $\hat{Z}_2(K) \equiv \rho\hat{B}(K)$ holds for any $K \in [k_1(\tilde{p})L, w'(\tilde{p})L/[\delta - r'(\tilde{p})]]$. Therefore, the pair of capital stocks in Home and Foreign, (K^T, K^{T*}), can be a trade steady state, if and only if

$$\hat{Z}_2(K^T) + \hat{Z}_2(K^{T*}) = 0,$$

with $(K^T, K^{T*}) \in [k_1(\tilde{p})L, w'(\tilde{p})L/[\delta - r'(\tilde{p})]) \times [k_1(\tilde{p})L, w'(\tilde{p})L/[\delta - r'(\tilde{p})])$.[6]

If

$$\hat{B}(k_1(\tilde{p})L) + \hat{B}\left(\frac{w'(\tilde{p})}{\delta - r'(\tilde{p})}L\right) < 0 \tag{19}$$

holds as in Fig. 1, there is some value of K, say $\bar{K}$, that satisfies

$$\hat{B}(k_1(\tilde{p})L) + \hat{B}\left(\bar{K}\right) = 0 \text{ with } \bar{K} \in \left(K^A, \frac{w'(\tilde{p})}{\delta - r'(\tilde{p})}L\right).$$

On the other hand, if

---

[5] Since the factor prices are equalized across two countries at any trade steady state, capital stocks in both countries satisfy

$$K \in [k_1(\tilde{p})L, k_2(\tilde{p})L] \text{ and } K^* \in [k_1(\tilde{p})L, k_2(\tilde{p})L],$$

or both countries are completely specialized with $K = K^*$, the latter case of which can not be a steady state as in the autarkic case.

[6] If $K^T = w'(\tilde{p})L/[\delta - r'(\tilde{p})]$ holds, households in Home are unable to consume good 1, and hence such a case can not be a steady state.

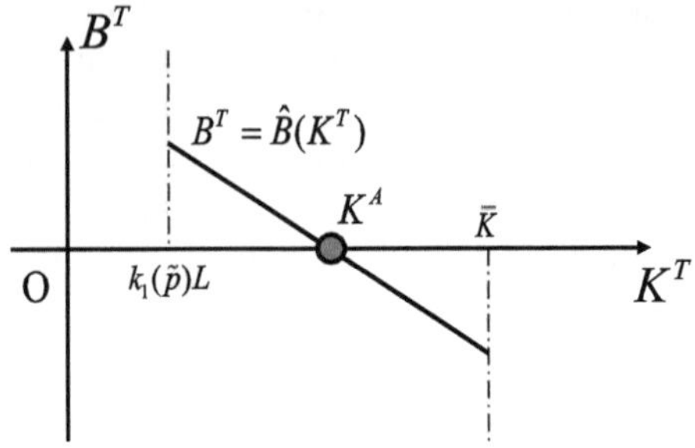

Fig. 2 The relationship between capital stocks and bond holdings

$$\hat{B}(k_1(\tilde{p})L) + \hat{B}\left(\frac{w'(\tilde{p})}{\delta - r'(\tilde{p})}L\right) > 0$$

holds, then for some value of K, say $\underline{K}$, we have

$$\hat{B}(\underline{K}) + \hat{B}\left(\frac{w'(\tilde{p})}{\delta - r'(\tilde{p})}L\right) = 0 \text{ with } \underline{K} \in \left(k_1(\tilde{p})L, K^A\right).$$

In the rest of paper, we assume (19) holds just for simplifying the description of steady states.

Then, we obtain

Proposition 2 *For each $K \in [k_1(\tilde{p})L, \bar{K}]$, there uniquely exists a pair of stock of bond in Home and capital stock in Foreign, (B, K^*), that satisfies $B = \hat{B}(K)$ and $\hat{B}(K) + \hat{B}(K^*) = 0$. It implies that there exists a continuum of steady states under trade environment.*

Since $\hat{B}(K) = \hat{Z}_2(K)/\rho$ holds, the relationship between the home country's capital stock, K^T, and its bond holdings, B^T, can be depicted as in Fig. 2, when preferences are homothetic.[7] Therefore, the capital abundant country at the steady state will be a debtor country and export capital intensive good 2 to the capital scarce country. In the case where consumable capital good 1 is capital intensive, the relationship will be reversed, and hence the capital abundant country at the steady state will be a creditor country and import labor intensive good 2 from the capital scarce country.

Also, we obtain the steady state pair of capital stocks in Home and Foreign as in Fig. 3, where preferences assumed to be homothetic. Notice that the locus becomes concave (convex) to the origin, when good 2 is a necessity (luxury), because the pair (K^T, K^{T*}) must yield the pair (B^T, B^{T*}) that satisfies $B^T + B^{T*} = 0$.

In the case of homothetic preferences, $K^T + K^{T*} = 2K^A$ holds, and the pair of capital stocks in any trade steady state will be a pair of capital stocks in some steady state under free trade environment, where both goods are tradable but there is no international credit market, that is, $B = B^* = 0$ must hold. It can be derived

[7] The graph is concave (convex), when good 2 is a necessity (luxury) as stated above.

Fig. 3 The locus of equilibrium pairs of K^T and K^{T*}

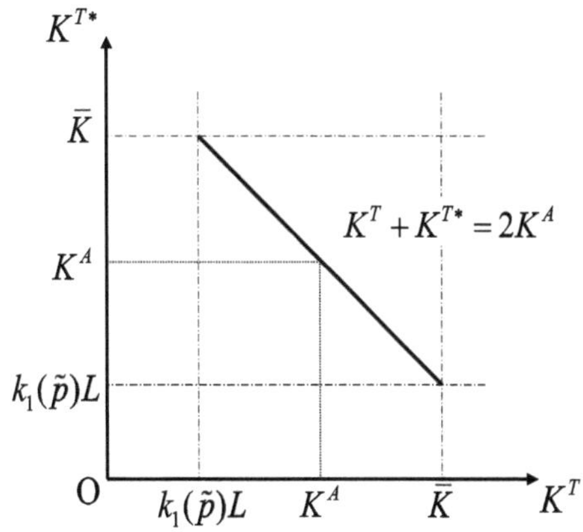

from the fact that with homothetic preferences, the income expansion path will be linear in (C_1, C_2) space, and hence $\hat{Z}_2(K^T) + \hat{Z}_2(K^{T*}) = 0$ necessarily implies that $\hat{C}_2(K^T, 0) - \hat{Y}_2(K^T) + \hat{C}_2(K^{T*}, 0) - \hat{Y}_2(K^{T*}) = 0$.

4 Stability of Steady States

Since the discount factor is the same across two countries, we see that

$$\frac{\dot{\lambda}}{\lambda} = \frac{\dot{\lambda}^*}{\lambda^*}$$

holds, and hence we have

$$\lambda^* = m\lambda$$

for some $m > 0$.

Then, the excess demand for good 2 in the world market is given by

$$Z_2^W \equiv C_2(p, \lambda) + C_2(p^*, m\lambda) - Y_2(p, K) - Y_2(p^*, K^*). \tag{20}$$

Differentiating both sides of (20) with respect to time yields

$$\begin{aligned}\dot{Z}_2^W &= C_{2p}\dot{p} + C_{2\lambda}\dot{\lambda} + C_{2p}^*\dot{p}^* + mC_{2\lambda}^*\dot{\lambda} - Y_{2p}\dot{p} - Y_{2K}\dot{K} - Y_{2p}^*\dot{p}^* - Y_{2K}^*\dot{K}^* \\ &= pZ_{2p}(R + \delta - \bar{r}) + \lambda C_{2\lambda}(\rho - R) + p^*Z_{2p}^*(R + \delta - \bar{r}^*) + m\lambda C_{2\lambda}^*(\rho - R) \\ &\quad -Y_{2K}(Y_1 - C_1 - \delta K) - Y_{2K}^*(Y_1^* - C_1^* - \delta K^*). \end{aligned} \tag{21}$$

Notice that $\dot{Z}_2^W = 0$ yields

$$R \equiv R(p, p^*, K, K^*, \lambda; m)$$
$$\equiv \frac{pZ_{2p}(\bar{r}-\delta) - \rho\lambda C_{2\lambda} + p^*Z^*_{2p}(\bar{r}^*-\delta) - \rho m\lambda C^*_{2\lambda} + Y_{2K}(Y_1 - C_1 - \delta K) + Y^*_{2K}(Y^*_1 - C^*_1 - \delta K^*)}{\Lambda},$$

where $\Lambda \equiv pZ_{2p} - \lambda C_{2\lambda} + p^*Z^*_{2p} - m\lambda C^*_{2\lambda}$ is always positive, which is shown in the proof of Lemma 2 in Appendix 1.

We have

Lemma 2 *At steady states,*

$$R_p = \frac{\tilde{p}\bar{r}'\lambda C_{1\lambda}}{\Lambda}, \quad R_{p^*} = \frac{\tilde{p}\bar{r}'m\lambda C^*_{1\lambda}}{\Lambda}, \quad R_K = R_{K^*} = \frac{-\tilde{p}^2\bar{r}'(r'-\delta)}{\Lambda},$$
$$\text{and } R_\lambda = \frac{\tilde{p}^2\bar{r}'(C_{1\lambda} + mC^*_{1\lambda})}{\Lambda},$$

where $\Lambda > 0$.

Proof See Appendix 1. □

4.1 Dynamic System

Let (K^T, K^{T*}) be a pair of steady state capital stocks in home and foreign. Then, we have

$$m = \hat{m}(K^T, K^{T*})$$
$$\equiv \frac{u_2(\hat{C}_1(K^{T*}, \hat{B}(K^{T*})), \hat{C}_2(K^{T*}, \hat{B}(K^{T*})))}{u_2(\hat{C}_1(K^T, \hat{B}(K^T)), \hat{C}_2(K^T, \hat{B}(K^T)))}.$$

Since K^T and K^{T*} move in opposite directions, if K^T increases and home households' consumption bundles move toward the origin along the income expansion, then K^{T*} decreases and foreign ones move upward. Therefore, for each value of m, the steady state pair of capital stock exists and is unique, where the steady state becomes an autarkic one with $m = 1$.[8]

The dynamic general equilibrium system can be described as

$$\dot{B} = R(p, p^*, K, K^*, \lambda; m)B - Z_2(p, K, \lambda),$$
$$\dot{K} = Y_1(p, K) - C_1(p, \lambda) - \delta K,$$
$$\dot{K}^* = Y_1(p^*, K^*) - C_1(p^*, m\lambda) - \delta K^*,$$
$$\dot{\lambda} = \lambda\left[\rho - R(p, p^*, K, K^*, \lambda; m)\right],$$
$$\dot{q} = q\left[\rho + \delta - \bar{r}(p)\right],$$
$$\dot{q}^* = q^*\left[\rho + \delta - \bar{r}(p^*)\right],$$

[8] Since K^T and K^{T*} must be in $[k_1(\tilde{p})L, \bar{K}]$, m will satisfy $\hat{m}(\bar{K}, k_1(\tilde{p})L) \le m \le \hat{m}(k_1(\tilde{p})L, \bar{K})$.

$$0 = q - \lambda p,$$
$$0 = q^* - m\lambda p^*,$$

which determines three state variables, K, K^*, B and three jump variables, λ, q, q^*. We will use this system to analyze the bond and capital accumulation on the equilibrium path, and to derive results on the dynamics in the neighborhood of the steady state with $m \in (\hat{m}(\bar{K}, k_1(\tilde{p})L), \hat{m}(k_1(\tilde{p})L, \bar{K}))$.

Differentiating the system gives the Jacobian,

$$\tilde{J} = \begin{bmatrix} \rho & BR_K - Z_{2K} & BR_{K^*} & BR_\lambda - Z_{2\lambda} & 0 & 0 & BR_p - Z_{2p} & BR_{p^*} \\ 0 & Y_{1K} - \delta & 0 & -C_{1\lambda} & 0 & 0 & Y_{1p} - C_{1p} & 0 \\ 0 & 0 & Y^*_{1K} - \delta & -mC^*_{1\lambda} & 0 & 0 & 0 & Y^*_{1p} - C^*_{1p} \\ 0 & -\lambda R_K & -\lambda R_{K^*} & -\lambda R_\lambda & 0 & 0 & -\lambda R_p & -\lambda R_{p^*} \\ 0 & 0 & 0 & 0 & 0 & 0 & -q\bar{r}' & 0 \\ 0 & 0 & 0 & 0 & 0 & 0 & 0 & -q^*\bar{r}' \\ 0 & 0 & 0 & -\tilde{p} & 1 & 0 & -\lambda & 0 \\ 0 & 0 & 0 & -m\tilde{p} & 0 & 1 & 0 & -m\lambda \end{bmatrix},$$

where each element is evaluated at the steady state.

Then, $\tilde{J}$ yields the characteristic equation as follows.

Lemma 3 *The characteristic equation is given by*

$$J(x) \equiv m\lambda^2 x(\rho - x)(r' - \delta - x)(\tilde{p}\bar{r}' + x)j(x),$$

where

$$j(x) = x^2 - \rho x + \frac{\tilde{p}\bar{r}'\lambda\left[(r' - \delta)(C_{2\lambda} + mC^*_{2\lambda}) + \tilde{p}^2\bar{r}'(C_{1\lambda} + mC^*_{1\lambda})\right]}{\Lambda}.$$

Then, $J(x) = 0$ has three positive roots, ρ, $-\tilde{p}\bar{r}'$ and x_1, two negative roots, $r' - \delta$ and x_2, and one zero root, where x_1 and x_2 denote the two roots of $j(x) = 0$.

Proof See Appendix 2. □

From the Lemma above, we see that there exists two dimensional stable manifold around each of steady states under trade environment.

Based on the above, we obtain the main result of the paper as follows.

Theorem 1 *For initially given capital stocks and bond holdings in two countries, there uniquely exists the dynamic equilibrium path that converges to one of the steady states. Thus, the saddle-point stability holds under trade environment as in the case of free trade environment, where two goods are tradable but international lending and borrowing is not allowed.*

Remark 1 The dynamic equilibrium path of this model is not Pareto efficient, despite allowing international lending and borrowing, while the path is Pareto efficient under free trade environment. This is because factor price equalization does not hold in this model due to the nontradable capital good.

5 Concluding Remarks

We have examined the dynamic property of a two-country dynamic Heckscher–Ohlin model that allows international borrowing and lending, where the consumable capital good assumed to be nontradable. We have derived the existence of a continuum of the steady state levels of capital stocks in each country, which is a key feature of our model. Then, we have proved the saddle-point stability of steady states, which implies that the initial amount of capital and bond holdings in each country determines their steady state levels and the trade patterns in the long run.

Appendix 1

First, totally differentiating $u_1(C_1, C_2) = \lambda p$ and $u_2(C_1, C_2) = \lambda$ yields

$$C_{1p} = \frac{\lambda u_{22}}{D},\ C_{1\lambda} = \frac{u_{22}p - u_{12}}{D},\ C_{2p} = \frac{-\lambda u_{12}}{D} \text{ and } C_{2\lambda} = \frac{u_{11} - u_{12}p}{D}. \tag{22}$$

Then, we have

$$\begin{aligned}\Lambda &= -\lambda\frac{u_{11} - u_{12}p}{D} + p\left(\frac{-\lambda u_{12}}{D} - Y_{2p}\right) - m\lambda\frac{u_{11}^* - u_{12}^*p^*}{D^*} + p^*\left(\frac{-m\lambda u_{12}^*}{D^*} - Y_{2p}^*\right)\\ &= -\frac{\lambda u_{11}}{D} - pY_{2p} - \frac{m\lambda u_{11}^*}{D^*} - p^*Y_{2p}^*\\ &> 0.\end{aligned}$$

Since $p = p^* = \tilde{p}, R = \rho, \bar{r} = \bar{r}^* = \rho + \delta, Y_1 = C_1 + \delta K,$ and $Y_1^* = C_1^* + \delta K^*$ hold at the steady states, we see from (21) that

$$\begin{aligned}&[\tilde{p}(Z_{2p} + Z_{2p}^*) - \lambda(C_{2\lambda} + mC_{2\lambda}^*)]dR\\ &= [\bar{r}'\tilde{p}Z_{2p} + Y_{2K}(Y_{1p} - C_{1p})]dp + [\bar{r}'\tilde{p}Z_{2p}^* + Y_{2K}^*(Y_{1p}^* - C_{1p}^*)]dp^*\\ &\quad + Y_{2K}(Y_{1K} - \delta)dK + Y_{2K}^*(Y_{1K}^* - \delta)dK^* + (Y_{2K}C_{1\lambda} + Y_{2K}^*mC_{1\lambda}^*)d\lambda.\end{aligned}$$

Then, we have

$$\begin{aligned}&\frac{dR}{dp} = \frac{\bar{r}'\tilde{p}Z_{2p} + Y_{2K}(Y_{1p} - C_{1p})}{\Lambda},\ \frac{dR}{dp^*} = \frac{\bar{r}'\tilde{p}Z_{2p}^* + Y_{2K}^*(Y_{1p}^* - C_{1p}^*)}{\Lambda},\\ &\frac{dR}{dK} = \frac{Y_{2K}(Y_{1K} - \delta)}{\Lambda},\ \frac{dR}{dK^*} = \frac{Y_{2K}^*(Y_{1K}^* - \delta)}{\Lambda},\\ \text{and } &\frac{dR}{d\lambda} = \frac{Y_{2K}C_{1\lambda} + Y_{2K}^*mC_{1\lambda}^*}{\Lambda}.\end{aligned}$$

Notice that from $p = p^*$,

$$Y_{1K} = Y_{1K}^* = r' \text{ and } Y_{2K} = Y_{2K}^* = r - \tilde{p}r' = -\tilde{p}^2\bar{r}'$$

hold, and hence we see that from (22) and $pY_{1p} + Y_{2p} = 0$,

$$\begin{aligned} &\bar{r}'\tilde{p}Z_{2p} + Y_{2K}(Y_{1p} - C_{1p}) = \bar{r}'\tilde{p}(C_{2p} - Y_{2p}) - \tilde{p}^2\bar{r}'(Y_{1p} - C_{1p}) \\ &= \bar{r}'\tilde{p}[(\tilde{p}C_{1p} + C_{2p}) - (\tilde{p}Y_{1p} + Y_{2p})] \\ &= \bar{r}'\tilde{p}\lambda C_{1\lambda} \end{aligned}$$

and

$$\bar{r}'\tilde{p}Z_{2p}^* + Y_{2K}^*(Y_{1p}^* - C_{1p}^*) = \bar{r}'\tilde{p}m\lambda C_{1\lambda}^*.$$

Thus, we have

$$R_p = \frac{\tilde{p}\bar{r}'\lambda C_{1\lambda}}{\Lambda},\ R_{p^*} = \frac{\tilde{p}\bar{r}'m\lambda C_{1\lambda}^*}{\Lambda},\ R_K = R_{K^*} = \frac{-\tilde{p}^2\bar{r}'(r' - \delta)}{\Lambda},$$
$$\text{and } R_\lambda = \frac{\tilde{p}^2\bar{r}'(C_{1\lambda} + mC_{1\lambda}^*)}{\Lambda}.$$

Appendix 2

The characteristic equation $J(x)$ is given by

$$J(x) = \begin{vmatrix} \rho - x & BR_K - Z_{2K} & BR_{K^*} & BR_\lambda - Z_{2\lambda} & 0 & 0 & BR_p - Z_{2p} & BR_{p^*} \\ 0 & Y_{1K} - \delta - x & 0 & -C_{1\lambda} & 0 & 0 & Y_{1p} - C_{1p} & 0 \\ 0 & 0 & Y_{1K}^* - \delta - x & -mC_{1\lambda}^* & 0 & 0 & 0 & Y_{1p}^* - C_{1p}^* \\ 0 & -\lambda R_K & -\lambda R_{K^*} & -\lambda R_\lambda - x & 0 & 0 & -\lambda R_p & -\lambda R_{p^*} \\ 0 & 0 & 0 & 0 & -x & 0 & -q\bar{r}' & 0 \\ 0 & 0 & 0 & 0 & 0 & -x & 0 & -q^*\bar{r}' \\ 0 & 0 & 0 & -\tilde{p} & 1 & 0 & -\lambda & 0 \\ 0 & 0 & 0 & -m\tilde{p} & 0 & 1 & 0 & -m\lambda \end{vmatrix}.$$

Then,

$$J(x) = (\rho - x)\begin{vmatrix} r' - \delta - x & 0 & -C_{1\lambda} & 0 & 0 & Y_{1p} - C_{1p} & 0 \\ 0 & r' - \delta - x & -mC_{1\lambda}^* & 0 & 0 & 0 & Y_{1p}^* - C_{1p}^* \\ -\lambda R_K & -\lambda R_K & -\lambda R_\lambda - x & 0 & 0 & -\lambda R_p & -\lambda R_{p^*} \\ 0 & 0 & 0 & -x & 0 & -\lambda\tilde{p}\bar{r}' & 0 \\ 0 & 0 & 0 & 0 & -x & 0 & -m\lambda\tilde{p}\bar{r}' \\ 0 & 0 & -\tilde{p} & 1 & 0 & -\lambda & 0 \\ 0 & 0 & -m\tilde{p} & 0 & 1 & 0 & -m\lambda \end{vmatrix}$$

$$
= (\rho - x)\begin{vmatrix}
r'-\delta-x & 0 & -C_{1\lambda} & 0 & 0 & Y_{1p}-C_{1p} & 0 \\
-(r'-\delta-x) & r'-\delta-x & -mC^*_{1\lambda} & 0 & 0 & 0 & Y^*_{1p}-C^*_{1p} \\
0 & -\lambda R_K & -\lambda R_\lambda - x & 0 & 0 & -\lambda R_p & -\lambda R_{p^*} \\
0 & 0 & 0 & -x & 0 & -\lambda\tilde{p}\tilde{r}' & 0 \\
0 & 0 & 0 & 0 & -x & 0 & -m\lambda\tilde{p}\tilde{r}' \\
0 & 0 & -\tilde{p} & 1 & 0 & -\lambda & 0 \\
0 & 0 & -m\tilde{p} & 0 & 1 & 0 & -m\lambda
\end{vmatrix}
$$

$$
= (\rho - x)(r'-\delta-x)\begin{vmatrix}
r'-\delta-x & -(C_{1\lambda}+mC^*_{1\lambda}) & 0 & 0 & Y_{1p}-C_{1p} & Y^*_{1p}-C^*_{1p} \\
-\lambda R_K & -\lambda R_\lambda - x & 0 & 0 & -\lambda R_p & -\lambda R_{p^*} \\
0 & 0 & -x & 0 & -\lambda\tilde{p}\tilde{r}' & 0 \\
0 & 0 & 0 & -x & 0 & -m\lambda\tilde{p}\tilde{r}' \\
0 & -\tilde{p} & 1 & 0 & -\lambda & 0 \\
0 & -m\tilde{p} & 0 & 1 & 0 & -m\lambda
\end{vmatrix}
$$

$$
= (\rho - x)(r'-\delta-x)\begin{vmatrix}
r'-\delta-x & -(C_{1\lambda}+mC^*_{1\lambda}) & 0 & 0 & Y_{1p}-C_{1p} & Y^*_{1p}-C^*_{1p} \\
-\lambda R_K & -\lambda R_\lambda - x & 0 & 0 & -\lambda R_p & -\lambda R_{p^*} \\
0 & -\tilde{p}x & -x & 0 & -\lambda(\tilde{p}\tilde{r}'+x) & 0 \\
0 & -m\tilde{p}x & 0 & -x & 0 & -m\lambda(\tilde{p}\tilde{r}'+x) \\
0 & 0 & 1 & 0 & 0 & 0 \\
0 & 0 & 0 & 1 & 0 & 0
\end{vmatrix}
$$

$$
= (\rho - x)(r'-\delta-x)\begin{vmatrix}
r'-\delta-x & -(C_{1\lambda}+mC^*_{1\lambda}) & Y_{1p}-C_{1p} & Y^*_{1p}-C^*_{1p} \\
-\lambda R_K & -\lambda R_\lambda - x & -\lambda R_p & -\lambda R_{p^*} \\
0 & -\tilde{p}x & -\lambda(\tilde{p}\tilde{r}'+x) & 0 \\
0 & 0 & m\lambda(\tilde{p}\tilde{r}'+x) & -m\lambda(\tilde{p}\tilde{r}'+x)
\end{vmatrix}
$$

$$
= -m\lambda(\rho - x)(r'-\delta-x)(\tilde{p}\tilde{r}'+x)\begin{vmatrix}
r'-\delta-x & -(C_{1\lambda}+mC^*_{1\lambda}) & Y_{1p}-C_{1p}+Y^*_{1p}-C^*_{1p} \\
-\lambda R_K & -\lambda R_\lambda - x & -\lambda(R_p+R_{p^*}) \\
0 & -\tilde{p}x & -\lambda(\tilde{p}\tilde{r}'+x)
\end{vmatrix}
$$

$$
= -m\lambda(\rho - x)(r'-\delta-x)(\tilde{p}\tilde{r}'+x)\begin{vmatrix}
r'-\delta-x+\frac{R_K}{R_\lambda}(C_{1\lambda}+mC^*_{1\lambda}) & -(C_{1\lambda}+mC^*_{1\lambda}) & Y_{1p}-C_{1p}+Y^*_{1p}-C^*_{1p} \\
-\lambda R_K+\frac{R_K}{R_\lambda}(\lambda R_\lambda+x) & -\lambda R_\lambda - x & -\lambda(R_p+R_{p^*}) \\
\frac{R_K}{R_\lambda}\tilde{p}x & -\tilde{p}x & -\lambda(\tilde{p}\tilde{r}'+x)
\end{vmatrix}
$$

$$
= -m\lambda(\rho - x)(r'-\delta-x)(\tilde{p}\tilde{r}'+x)\begin{vmatrix}
-x & -(C_{1\lambda}+mC^*_{1\lambda}) & Y_{1p}-C_{1p}+Y^*_{1p}-C^*_{1p} \\
\frac{R_K}{R_\lambda}x & -\lambda R_\lambda - x & -\lambda(R_p+R_{p^*}) \\
\frac{R_K}{R_\lambda}\tilde{p}x & -\tilde{p}x & -\lambda(\tilde{p}\tilde{r}'+x)
\end{vmatrix}
$$

$$
= -m\lambda x(\rho - x)(r'-\delta-x)(\tilde{p}\tilde{r}'+x)\begin{vmatrix}
-1 & -(C_{1\lambda}+mC^*_{1\lambda}) & Y_{1p}-C_{1p}+Y^*_{1p}-C^*_{1p} \\
0 & -\lambda R_\lambda - x - \frac{R_K}{R_\lambda}(C_{1\lambda}+mC^*_{1\lambda}) & -\lambda(R_p+R_{p^*}) + \frac{R_K}{R_\lambda}(Y_{1p}-C_{1p}+Y^*_{1p}-C^*_{1p}) \\
0 & \tilde{p}\lambda R_\lambda & -\lambda[(\tilde{p}\tilde{r}'+x)-\tilde{p}(R_p+R_{p^*})]
\end{vmatrix}
$$

$$
= m\lambda^2 x(\rho - x)(r'-\delta-x)(\tilde{p}\tilde{r}'+x)\begin{vmatrix}
-\lambda R_\lambda - x + r' - \delta & -\lambda(R_p+R_{p^*}) + \frac{R_K}{R_\lambda}(Y_{1p}-C_{1p}+Y^*_{1p}-C^*_{1p}) \\
\tilde{p}R_\lambda & -[(\tilde{p}\tilde{r}'+x)-\tilde{p}(R_p+R_{p^*})]
\end{vmatrix},
$$

where

$$
\begin{vmatrix}
-\lambda R_\lambda - x + r' - \delta & -\lambda(R_p+R_{p^*}) + \frac{R_K}{R_\lambda}(Y_{1p}-C_{1p}+Y^*_{1p}-C^*_{1p}) \\
\tilde{p}R_\lambda & -[(\tilde{p}\tilde{r}'+x)-\tilde{p}(R_p+R_{p^*})]
\end{vmatrix}
$$

$$
= \left[\lambda R_\lambda + x - r' + \delta\right][(\tilde{p}\tilde{r}'+x)-\tilde{p}(R_p+R_{p^*})] + \tilde{p}R_\lambda\left[\lambda(R_p+R_{p^*}) + \frac{R_K}{R_\lambda}(C_{1p}-Y_{1p}+C^*_{1p}-Y^*_{1p})\right]
$$

$$
= x^2 + \left[\lambda R_\lambda - r' + \delta + \tilde{p}\tilde{r}' - \tilde{p}(R_p+R_{p^*})\right]x
$$

$$
+\tilde{p}\tilde{r}'\lambda R_\lambda - (r'-\delta)[\tilde{p}\tilde{r}' - \tilde{p}(R_p+R_{p^*})] + \tilde{p}R_K(C_{1p}-Y_{1p}+C^*_{1p}-Y^*_{1p}).
$$

From Lemma 2, we see that

$$\lambda R_\lambda - r' + \delta + \tilde{p}\bar{r}' - \tilde{p}(R_p + R_{p^*}) = -r' + \delta + \tilde{p}\bar{r}' = -\rho$$

and

$$\begin{aligned}
&\tilde{p}\bar{r}'\lambda R_\lambda - (r' - \delta)[\tilde{p}\bar{r}' - \tilde{p}(R_p + R_{p^*})] + \tilde{p}R_K(C_{1p} - Y_{1p} + C^*_{1p} - Y^*_{1p}) \\
&= \frac{1}{\Lambda}[(\tilde{p}\bar{r}' + r' - \delta)\tilde{p}^2\bar{r}'\lambda(C_{1\lambda} + mC^*_{1\lambda}) - (r' - \delta)\tilde{p}\bar{r}'(\tilde{p}Z_{2p} - \lambda C_{2\lambda} + \tilde{p}Z^*_{2p} - m\lambda C^*_{2\lambda}) \\
&\quad -\tilde{p}^3\bar{r}'(r' - \delta)(C_{1p} - Y_{1p} + C^*_{1p} - Y^*_{1p})] \\
&= \frac{1}{\Lambda}\{(\tilde{p}\bar{r}' + r' - \delta)\tilde{p}\bar{r}'\lambda(\tilde{p}C_{1\lambda} + m\tilde{p}C^*_{1\lambda}) + (r' - \delta)\tilde{p}\bar{r}'\lambda(C_{2\lambda} + mC^*_{2\lambda}) \\
&\quad -\tilde{p}^2\bar{r}'(r' - \delta)[\tilde{p}C_{1p} + C_{2p} - (\tilde{p}Y_{1p} + Y_{2p}) + \tilde{p}C^*_{1p} + C^*_{2p} - (\tilde{p}Y^*_{1p} + Y^*_{2p})]\} \\
&= \frac{\tilde{p}\bar{r}'\lambda\left[(r' - \delta)(C_{2\lambda} + mC^*_{2\lambda}) + \tilde{p}^2\bar{r}'(C_{1\lambda} + mC^*_{1\lambda})\right]}{\Lambda}.
\end{aligned}$$

Thus, we have

$$J(x) = m\lambda^2 x(\rho - x)(r' - \delta - x)(\tilde{p}\bar{r}' + x)j(x),$$

where

$$j(x) = x^2 - \rho x + \frac{\tilde{p}\bar{r}'\lambda\left[(r' - \delta)(C_{2\lambda} + mC^*_{2\lambda}) + \tilde{p}^2\bar{r}'(C_{1\lambda} + mC^*_{1\lambda})\right]}{\Lambda}.$$

Under Assumptions 1 and 3, we obtain

$$\frac{\tilde{p}\bar{r}'\lambda\left[(r' - \delta)(C_{2\lambda} + mC^*_{2\lambda}) + \tilde{p}^2\bar{r}'(C_{1\lambda} + mC^*_{1\lambda})\right]}{\Lambda} < 0.$$

Acknowledgements This work was supported by JSPS KAKENHI Grant Numbers 22K01477 and 20H05633.

References

1. Bond, E., Iwasa, K., Nishimura, K.: A dynamic two country Heckscher-Ohlin model with non-homothetic preferences. Econ. Theory **48**, 171–204 (2011)
2. Bond, E., Iwasa, K., Nishimura, K.: The dynamic Heckscher-Ohlin model: a diagrammatic analysis. Int. J. Econ. Theory **8**, 197–211 (2012)
3. Hu, Y., Mino, K.: Trade structure and belief-driven fluctuations in a global economy. J. Int. Econ. **90**, 414–424 (2013)
4. Meng, Q., Velasco, A.: Market imperfections and the instability of open economies. J. Int. Econ. **64**, 503–519 (2004)

5. Nishimura, K., Shimomura, K.: Trade and indeterminacy in a dynamic general equilibrium model. J. Econ. Theory **105**, 244–259 (2002)
6. Ono, Y., Shibata, A.: Time patience and specialization patterns in the presence of asset trade. J. Money Credit Bank. **42**, 93–112 (2010)

David Gale: Duality and Turnpikes

Mukul Majumdar

Thanks to the work of that genius, John von Neumann, game theory was already knocking on the door when World War II ended. The fruitful theories of linear and non-linear programming, of George Dantzig and Gale-Kuhn-Tucker, were waiting in the wings. ⋯ But still it is true that, whether in Washington, Rotterdam or Moscow, we today live in a golden age for analytical economics: duality, turnpike theorems, stability of dynamic equilibrium–was there ever a time of more fruitful progress in conquering such new frontiers of knowledge?
Paul Samuelson (1965)

Abstract This paper presents a review of some of the developments in mathematical economics that were influenced by contributions of David Gale. It begins with the fundamental duality theorems of linear programming and its application to two person zero sum matrix games. Turning to dynamic economics, the multi-sector analysis of a closed model of production by Gale is considered in detail. The applications of von Neumann equilibrium prices to establish the Samuelson conjecture on turnpike properties of a class of optimal programs, to derive Malinvaud prices supporting efficient programs and to throw light on the need for substitution possibilities to sustain a positive level of consumption are elaborated. Finally, Gale's multi-sector model with a primary factor of production is recalled. The golden rule equilibrium and the use of golden rule prices to establish a turnpike property of programs that are optimal according to the overtaking criterion are presented with an emphasis on the value loss technique.

Keywords Duality · Linear programming · Matrix games · Optimal growth · Turnpike theorems

M. Majumdar (✉)
Department of Economics, Cornell University, Ithaca, NY, USA
e-mail: mkm5@cornell.edu

M. A. Khan et al. (eds.), *Matching, Dynamics and Games for the Allocation of Resources*, Monographs in Mathematical Economics 7,
https://doi.org/10.1007/978-981-96-6661-4_15

1 Introduction

In his tribute to Kuhn and Gale [37] reminded us that Gale's "achievements were many and varied, but the citation for the von Neumann Theory Prize that David won got it just right when it said: 'Gale's research is known for its elegance.' A mathematician can receive no higher praise than having his work called elegant.".

In remembering David Gale I touch upon *only a few* of the themes in mathematical economics that he explored in a significant number of papers.

I begin by enumerating some of David Gale's enduring contributions to static and dynamic equilibrium theory. But my primary focus is on linear programming, matrix game theory and turnpike theorems. In choosing specific results from these areas, I have been partially guided by his *Linear Economic Models*, Gale [19], to be abbreviated as, *LEM:* my first exposure to David Gale in 1964 as an undergraduate at Presidency College. It was also used as the primary text in the course on *Linear Models* in the Fall (1965) semester at Berkeley by Karl Vind. The first topic is the fundamental duality in linear programming problems. Next, I turn to matrix games, saddle points and minimax theorems. In his enjoyable informal account of five great mathematical theories of the twentieth century (the simplex algorithm being one) Casti [6] tells us : "It is no wonder that von Neumann could later remark that—As far as I can see, there would be no theory of games...without that theorem... I thought that there was nothing worth publishing until the "Minimax Theorem" was proved." The rest of the paper deals with the multi-sector models of production starting with the duality of the von Neumann equilibrium. The analytical framework for dealing with many sectors presented in Gale [18] was a building block in contexts far removed from the original. I provide three such applications: the Samuelson conjecture on a turnpike property proved by Radner [52] (and strengthened by Nikaido [48]) by using his value loss lemma, a characterization of inter-temporal efficiency in terms of a price system represented by a linear functional on an infinite dimensional commodity space, and a connection between technological substitution and sustainability. Finally, I turn to the problem of optimal development in a multi-sector economy with a primary factor of production introduced in Gale [22]. The main result is a turnpike property of the program that is optimal according to the overtaking criterion. The approach taken here is a direct value loss argument (in contrast with the route taken by Gale), which can be easily adapted in the one good model to derive a catenary turnpike theorem (studied by a number of authors: Gale [23] is a crisp exposition).

Needless to say that "duality" has been explored thoroughly in many settings: in the theories of consumption and production, dynamic programming..... In each, the word has its connotation, its twist. To take just one example, a group of researchers in Berkeley led by McFadden explored "duality" in production theory in depth (see Fuss and McFadden [16]). The Shephard-Uzawa duality theorem, noted (McFadden [44], p. 29), provides a basis for relating the structural properties of production possibilities to structural properties of cost function. In application, it is useful to have a large family of duality relationships of the form "the production possibility set has property 'P' if and only if the cost function has property 'Q'." The reader is referred to the extensive survey of duality in demand theory and other applications Diewert [13], an outstanding contributor to the development of the literature.

Here is a short outline of the organization of topics. In Sect. 2, I collect some mathematical results that are used in subsequent exposition. Just to remind the reader of the expanse of Gale's research, I take a casual look in Sect. 3 at some of his influential papers on static and dynamic equilibrium in the Walrasian tradition. In Sects. 4–5, our focus is on the duality theorems of linear programming, and characterization of 'dual' variables as prices. In Sect. 6, we turn to saddle points and two person zero sum matrix games. The duality theorem of linear programming is used to prove that every two person zero sum matrix game has a solution in mixed strategies. This approach is particularly insightful from a computational point of view. The existence of a saddle point is proved by invoking the minmax theorem of Kakutani. Sects. 7–8 deal with issues in inter-temporal economics. The fundamental building block of Sect. 6 is Gale [18] and that of Sect. 7 is Gale [21]. The use of dual prices and the value loss technique to prove turnpike theorems are elaborated in these sections.

2 Technicalities

Here the basic notation and definitions are summarized. A non-negative (respectively, positive, negative) real number a is denoted by $a \geqq 0$ (respectively, $a > 0$, $a < 0$). A vector $x = (x_i) \in \mathbb{R}^n$ is called *non-negative* (written, $x \geqq 0$), if $x_i \geqq 0$ for all $i = 1, 2, \cdots, n$. It is called *semi-positive* (written, $x \geq 0$), if $x \geqq 0$ and $x_i > 0$ for some i, and *positive* (written, $x > 0$) if $x_i > 0$ for all $i = 1, 2, \cdots, n$. We write:

$$| x |= \sum_{i=1}^{n} | x_i |, \ \| x \|= \left(\sum_{i=1}^{n} x_i^2 \right)^{\frac{1}{2}}, \ | x |_{\max}= \max_i | x_i | .$$

We denote by $\mathcal{P}_n = \{x \in \mathbb{R}^n : x \geq 0, \ | x |= 1\}$. The set of all non-negative n - vectors is $\mathbb{R}^n_+$. For an $m \times n$ real matrix $A = (\alpha_{ij})$, $a_i = (\alpha_{i1}, \ldots, \alpha_{in})$ is the i-th row vector, and $a^j = (\alpha_{1j}, \cdots, \alpha_{mj})$ is the j-th column vector.

Theorem 2.1 *Let $\mathfrak{D}$ be a real $k \times l$ matrix and $\mathfrak{e}$ an l–vector. Exactly one of the following alternatives holds: Either the inequalities*

$$x\mathfrak{D} \leqq \mathfrak{e} \tag{1}$$

have a non-negative solution, or the inequalities

$$\mathfrak{D}y \geqq 0, \mathfrak{e}y < 0 \tag{2}$$

have a non-negative solution.

Proof See (Gale [19], Theorem 2.8, p. 47). □

A (non-empty) subset K of $\mathbb{R}^n$ is a *convex cone* if it satisfies:

(i) $x + y \in K$ for any $x, y \in K$;
(ii) $\lambda x \in K$ for any $x \in K$ and $\lambda \geqq 0$.

The *dual* convex cone K^* of a convex cone K is defined as:

$$K^* = \{y \in \mathbb{R}^n : x \cdot y \geqq 0 \text{ for any } x \in K\}.$$

Observe that even though K may not be closed, K^* is always closed. Now, K^* has its dual $(K^*)^*$, usually denoted simply by K^{**}. A convex cone K is *polyhedral* if it is generated by a finite set of vectors $\{a^1, a^2, \cdots, a^s\}$, i.e., $K = \{y = \sum_{i=1}^{s} \lambda_i a^i : \lambda_i \geqq 0\}$. A polyhedral convex cone K is closed. A convex cone K is polyhedral if and only if K is the set of all solutions to a finite system of linear inequalities (see Nikaido [49], Moore [45]).

Theorem 2.2 *For a convex cone K, we have (i) $K^{**} \supset K$; (ii) $K^{**} = K$ if and only if K is closed.*

Proof See (Nikaido [49], Theorem 3.4, p. 34). Also see (Gale [19], Theorem 2.12, p. 56): note that his finite cone is necessarily closed. □

We appeal to the following "separation theorem" several times (an important point is that the p is semi-positive, we cannot claim that p is positive).

Theorem 2.3 *Let X be a convex set in $\mathbb{R}^n$ that contains no positive vector. Then there is a semi-positive vector $p \geq 0$ such that $p \cdot x \leqq 0$ for all $x \in X$.*

Proof See (Nikaido [49], Theorem 3.5, p. 35). □

In his papers on optimal growth, versions of Kuhn-Tucker theorem are invoked by Gale (see, for example Gale [21, 22]) to assert the existence of "competitive" prices. To understand this "duality" consider a constrained maximization problem:

$$\max \quad f(x) \text{ subject to } g_i(x) \geq 0, \ \text{ for } i = 1, 2, ...m \tag{3}$$

where f and g_i are real valued functions on a non-empty subset X in $\mathbb{R}^n$. Introducing the Lagrangian multipliers λ_i $(i = 1, 2, ..., m)$, we set up:

$$f(x) + \sum_{i=1}^{m} \lambda_i g_i(x) \tag{4}$$

the Lagrangian associated with the maximization problem.

Theorem 2.4 *Assume that X is convex, and the objective function f and the constraint functions g_i are concave (for $i = 1, 2, \cdots, m$). In addition assume (the Slater condition) that there is some $c \in X$ such that $g_i(c) > 0$ (for $i = 1, 2, \cdots, m$). Then if $\hat{x}$ is a solution to the constrained maximization problem, there are non-negative λ_i $(i = 1, 2, \cdots, m)$ such that:*

(a) $\lambda_i \cdot g_i(\hat{x}) = 0$.
(b) $\hat{x}$ *maximizes the Lagrangian over* X *for these* λ_i.

Proof The elegant proof of Uzawa [62] first introduces a function $\varphi : X \to \mathbb{R}^{m+1}$ as follows:

$$\varphi_0(x) = f(x) - f(\hat{x})$$
$$\varphi_i(x) = g_i(x)(\text{ for } i = 1, 2, \cdots, m).$$

Then define

$$Y = \{y : y \leqq \varphi(x) \text{ for some } x \in X\}.$$

One shows that the set Y does not contain a positive vector. Theorem 2.3 is invoked and the conclusion follows after some calculations. □

Theorem 2.5 *Let* K *be a closed convex cone. If* K *contains no semi-positive vector* $u \geq 0$, *then* $-K^*$ *contains a positive* $p > 0$, *and vice versa.*

Proof See (Nikaido [49], Theorem 3.6, p. 35). □

Now, we come to the Kakutani fixed point theorem:

Theorem 2.6 *Let* S *be a non-empty compact convex subset of* R^n *and* ϕ *be an upper semi-continuous correspondence from* S *into* S *such that for all* $x \in S$, *the set* $\varphi(x)$ *is convex (non-empty), then there is some* $\hat{x} \in S$ *such that* $\hat{x} \in \varphi(\hat{x})$.

Proof See Kakutani [31] or Nikaido [49]. □

3 Gale: A Researcher for All Seasons

The resurgence of interest in the Walrasian framework some years after Walrasian [66] led to a deep exploration (introducing new tools) of a number of issues: (a) the existence of an "equilibrium"; (b) the uniqueness of an equilibrium; (c) the stability of the Walras-Samuelson tatonnement process; (d) the comparative static problems on the movement of equilibrium prices and quantities; (e) the link between Walrasian equilibrium allocations and Pareto optimal allocations; (f) the role of prices in coordinating independent decisions involving utility maximizing consumers and profit maximizing firms ("decentralization and informational efficiency"). Gale's primary contribution to (a) were his early paper (Gale [17]), and a subsequent influential paper with Mas-Colell (Gale and Mas-Colell [27], Corrections Gale and Mas-Colell [28] in response to counter examples by James Foster, then a graduate student at Cornell, and others). In connection with (b), the paper with Nikaido [29] is a landmark, a definitive contribution to the global univalence problem (for a summary from the perspective of analysis, see Parthasarathy [50]), as well as the literature on Samuelson's factor price equalization theorem (see (Nikaido [49], Chap. 7)). Regarding (c),Gale [20] suggested possible complicated behavior of a tatonnement by means of simple examples: Scarf had earlier given an example of periodic behavior in his celebrated paper (Scarf [59]). In fact the behavior could be chaotic as noted later by Bala and Majumdar [2]. Regarding (d), Gale [25] is a contribution to the transfer paradox, added a new twist to the comparative static analysis of general equilibrium (see Aumann and Peleg [1], a comprehensive review is given in Majumdar [39].

Gale [17] asserted that the existence of a Walrasian equilibrium was a *simple corollary* to the following "Principal Lemma"–often referred to as the Gale-Nikaido-Debreu Lemma, which is stated here in a modified form (Debreu [9]);

Lemma 3.1 (Principal Lemma) *Let Z be a non-empty compact subset of* $\mathbb{R}^n$. *If* ς *is an upper semi-continuous correspondence from* $P_{(n)}$ *to Z such that for every* $p \in P_{(n)}$, *the set* $\varsigma(p)$ *is non- empty, convex and satisfies*

$$p.\varsigma(p) \leqq 0 \tag{5}$$

then there is some $\hat{p} \in P_{(n)}$ *such that*

$$\varsigma(p) \cap (-\mathbb{R}^n_+) \neq \emptyset. \tag{6}$$

As a second application we note that Kakutani's fixed point theorem (Theorem 2.6) can itself be proved by using the Principal Lemma: It is due to Uzawa [63] (see the sketch in Debreu [11] or the elaboration in Mukherji [47]. elaboration). I am struck by the scant appreciation of this deep result in the literature on general equilibrium in the decades following its publication.

Theorem 3.1 *Assume that the Principal Lemma holds. Let* φ *be an upper semi-continuous correspondence from* $P_{(n)}$ *into itself such that* $\varphi(x)$ *is (non-empty) convex for each* $x \in P_{(n)}$. *Then there is some* $\hat{x} \in P_{(n)}$ *such that* $\hat{x} \in \varphi(\hat{x})$.

Yet another application is due to Kuhn [36] who noted that the Principal Lemma can be used to prove that a two person zero sum matrix game has a solution in mixed strategies. This problem is explored in depth in Sect. 6.

The difficulties of deriving unambiguous "qualitative" comparative static results in Walrasian models with many commodities became more and more apparent, due to a notable line of research initiated by Sonnenschein [61] (see Debreu's important contribution (Debreu [10]) and the references cited there). Thus, to figure out the impact of a change in a parameter on equilibrium prices, one turns to the possibility of *direct computation of equilibrium prices* and carry out the sensitivity analysis. The two theorems stated above enable us to appreciate the remarkable contributions of Scarf on computing fixed points. With the stunning advance of computing power, it became possible to handle a "fairly large" number of commodities. Gale was not involved in the breakthrough. But in a paper (Gale [26]) written in an engaging style, addressed to mathematicians, he said: "the paper has the dual purpose of, first, showing the equivalence of the Hex and Brouwer's Theorems, and, secondly, introducing the reader to the subject of fixed point computations". The exposition had the usual imprint: elegant proofs through "elementary" arguments.

Going now to dynamic equilibrium theory, Gale [24] provided a masterly exposition of a pure exchange model with overlapping generations. Here we have a stylist who matched Hicks at his best: formal statements of *assumptions* relevant for *theorems* conspicuously absent. Yet, he convincingly explored a number of fundamental themes: (a) non-optimality of competitive equilibria; (b) a complete account of stationary competitive programs; (c) a study of dynamical systems generated by non-stationary competitive programs, in particular of the possibility of cyclical behavior.

After this admittedly casual look at some of Gale's research in the Walrasian tradition (my book (Majumdar [39]) and earlier tribute (Majumdar [38]) deal with some of the topics in depth), I turn to a formal review of the first topic enriched by one of his fundamental contributions.

4 Linear Programming

Before proceeding to the formal presentation of the results, we can take a quick look at the history of development of linear programming. Let us first hear from Gale (*LEM*):

- "The term 'linear programming' came into being about 1947–1948. The earliest published work containing these words in the title is due to Dantzig (1949). ⋯. The duality theorem was known to von Neumann at least as early as 1947, and is contained in a set of privately circulated notes, but the proof as given there is incomplete. The first published proof based on von Neumann's notes is due to Gale, Kuhn and Tucker".

Dantzig himself has reflected on the early days of linear programming (Dantzig [8]). He specifically mentioned a meeting with von Neumann on October 3, 1947 in which von Neumann "proceeded to give me a lecture on the mathematical theory of linear programming". During a subsequent visit to Princeton. (June 1948), he met Tucker. "Soon Tucker and his students H. Kuhn and D. Gale began their historic work on game theory, nonlinear programming and duality theory. The Princeton group became the focal point among mathematicians working in the field. ⋯. Today everyone cites von Neumann as the originator of the duality theorem and credits Tucker, Kuhn and Gale as the publishers of the first rigorous proof."

Gale (*LEM*), Dorfman et al. [14] or the masterly Dantzig [7] had a collection of celebrated examples of linear programming problems (the diet problem, the transportation and assignment problems, network-flows, ...) and a detailed exposition of the simplex algorithm. After a formal review of the duality theorems, I turn to one such example involving the activity analysis model of production.

5 Duality Theorems of Linear Programming

We now turn to the basic duality in linear programming problems. Given an $m \times n$ matrix $A = (\alpha_{ij})$, an n-vector $b = (\beta_j)$, and an m-vector $c = [\gamma_i]$, the "primal" problem (M) is: find an m-vector $x = (\xi_i)$ such that

$$xc \text{ is a maximum} \tag{7}$$

$$\text{subject to } xA \leqq b,\ x \geqq 0. \tag{8}$$

Associated with this maximization problem, the "dual" problem (m) is : find an n-vector $y = (\eta_j)$ such that

$$by \text{ is a minimum} \tag{9}$$

$$\text{subject to } Ay \geqq c,\ y \geqq 0. \tag{10}$$

An m-vector x [resp. an n-vector y] satisfying (8) [resp. (10)] is called a *feasible solution* to the primal [resp. dual] problem. An *optimal solution* to the primal problem (M) is a feasible m-vector $\hat{x}$ such that $\hat{x}c \geq xc$ for all feasible solutions to the primal. Similarly, an optimal solution to the dual problem (m) is a feasible n-vector $\hat{y}$ such that $by \geq b\hat{y}$ for all feasible solutions to the dual. If $\hat{x}$ is an *optimal solution* to the primal problem, the number $\hat{x}c$ is the *value* of the primal problem. If $\hat{y}$ is a solution to the dual problem, the number $b\hat{y}$ is the value of the dual problem.

Lemma 5.1 *Let $x = (\xi_i)$ and $y = (\eta_j)$ be feasible solutions to the primal maximization problem and the dual minimization problem respectively. Then,*

$$xc \leqq by. \tag{11}$$

Proof

$$xc \leqq x(Ay) = (xA)y \leqq by.$$

□

An implication of the above Lemma 5.1 is:

Theorem 5.1 (Optimality criterion) *If $\hat{x} = (\xi_i)$ and $\hat{y} = (\eta_j)$ are feasible solutions to the primal (maximization) problem and the dual (minimization) problem such that*

$$\hat{x}c = b\hat{y} \tag{12}$$

then these feasible solutions are optimal solutions to the respective problems.

Proof Let x be any feasible solution to the primal problem. Since $\hat{y}$ is a feasible solution to the dual,

$$xc \leqq b\hat{y} = \hat{x}c \tag{13}$$

showing that $\hat{x}$ is an optimal solution to the primal problem. A similar argument proves the optimality of $\hat{y}$. □

Theorem 5.2 (Fundamental Duality Theorem) *If both the primal and dual problems have feasible solutions, then both have optimal solutions, and the values of the two problems are the same. If either problem does not have a feasible solution, then neither has an optimal solution.*

Proof Suppose first that both the maximization problem and its dual have feasible solutions. Then we have non-negative solutions $\hat{x}$ and $\hat{y}$ to the following inequalities;

$$xA \leqq b \tag{14}$$
$$Ay \geqq c. \tag{15}$$

It is enough to show that there are solutions to (14) and (15) that also satisfy

$$xc - by \geqq 0 \tag{16}$$

and then use Lemma 5.1 and Theorem 5.1. Suppose then that the system (14) – (16) has no non-negative solution. Writing out the inequalities explicitly, we get:

$$\sum_i \xi_i \alpha_{ij} \leqq \beta_j \ \text{ for } j = 1, ...n \tag{17}$$

$$\sum_j \eta_j(-\alpha_{ij}) \leqq -\gamma_i \ \text{ for } j = 1, \ldots, m \tag{18}$$

$$\sum_j \eta_j \beta_j - \sum_i \xi_i \gamma_i \leqq 0. \tag{19}$$

If these inequalities have no non-negative solution, then Theorem 2.1 (with the appropriate $\mathfrak{D}$) states that there are (i) a non-negative n-vector $z = (\zeta_i)$, (ii) a non-negative m-vector $w = (\varpi_i)$, and (iii) a number $\theta \geqq 0$ such that :

$$Az \geqq \theta c \tag{20}$$
$$wA \leqq \theta b \tag{21}$$
$$zb - wc < 0. \tag{22}$$

First, consider the case $\theta = 0$. Here, considering the feasible solutions $\hat{x}$ and $\hat{y}$ we have from (8), (13), (20) and (21);

$$0 \leqq \hat{x}Az \leqq zb, \tag{23}$$

and

$$0 \geqq wA\hat{y} \geqq wc \tag{24}$$

contradicting (22). Hence, we must have $\theta > 0$. From (20) and (21) we see that w/θ and z/θ are a pair of feasible solutions, so from Lemma 5.1,

$$(w/\theta)x \leqq (z/\theta)b \tag{25}$$

and (25) contradicts (22). It follows that the set of inequalities (14)–(16) has a solution, and this solution provides the pair of optimal solutions.

Finally, suppose, that the primal problem does not have a feasible solution, i.e., there is no non-negative x satisfying (8). Then, certainly, there is no optimal solution to the primal. Again, by appealing to Theorem 2.1, there is a non-negative vector $\tilde{z}$ such that

$$A\tilde{z} \geqq 0 \text{ and } \tilde{z}b < 0. \tag{26}$$

But if the dual problem has a feasible solution, then from (10), there is some non-negative $\tilde{y}$ such that

$$A\tilde{y} \geqq c. \tag{27}$$

Using (26) and (27), $\tilde{y} + \lambda\tilde{z}$ is a feasible solution to the dual for all positive λ as seen from:

$$A(\tilde{y} + \lambda\tilde{z}) = A\tilde{y} + \lambda(A\tilde{z}) \geqq c$$

But since $\tilde{z}b < 0$ (from (26)), the expression $b(\tilde{y} + \lambda\tilde{z})$ can be made arbitrarily small. Hence, the dual problem of minimization has no solution. A similar argument can be constructed if the dual problem has no feasible solution. □

Our next result is of particular interest in interpreting dual variables.

Theorem 5.3 (Equilibrium Theorem) *Feasible solutions $x = (\xi_i)$ and $y = (\eta_j)$ of the primal and the dual (satisfying (8) and (10) respectively) are optimal solutions if and only if*

$$\eta_j = 0 \textit{ whenever } \sum_{i=1}^{m} \xi_i \alpha_{ij} < \beta_j \tag{28}$$

$$\textit{and } \xi_i = 0 \textit{ whenever } \sum_{j=1}^{n} \eta_j \alpha_{ij} > \gamma_i. \tag{29}$$

Proof Suppose that (28) and (29) hold for a pair of feasible solutions x and y. Then we have,

$$\sum_{i=1}^{m} \xi_i \alpha_{ij} \leqq \beta_j \text{ for } j = 1, 2, \cdots, n. \tag{30}$$

Multiplying the j-th inequality of (30) by η_j, and summing over j and making use of (30), we get

$$\sum_{j=1}^{n}\eta_j\beta_j=\sum_{j=1}^{n}\eta_j\sum_{i=1}^{m}\xi_i\alpha_{ij}=\sum_{i,j}\xi_i\eta_j\alpha_{ij}.$$

Similarly, from

$$\sum_{j=1}^{n}\eta_j\alpha_{ij}\geqq\gamma_i \text{ for } i=1,2...m \tag{31}$$

and (29) we get

$$\sum_{i=1}^{m}\xi_i\gamma_i=\sum_{i=1}^{m}\xi_i\sum_{j=1}^{n}\eta_j\alpha_{ij}=\sum_{i,j}\xi_i\eta_j\alpha_{ij}. \tag{32}$$

Hence, from (31) and (32) we get:

$$\sum_{i=1}^{m}\xi_i\gamma_i=\sum_{j=1}^{n}\eta_j\beta_j. \tag{33}$$

Using Theorem 5.1, we conclude that x, y are optimal solutions.
Conversely, if (ξ_i) and (η_j) are optimal solutions, then from Theorem 5.2 we know that:

$$\sum_{i=1}^{m}\xi_i\gamma_i=\sum_{i,j}\xi_i\eta_j\alpha_{ij}=\sum_{j=1}^{n}\eta_j\beta_j. \tag{34}$$

From the first equation we have

$$\sum_{i=1}^{m}\xi_i(\gamma_i-\sum_{j=1}^{n}\eta_j\alpha_{ij})=0 \tag{35}$$

but since $y=(\eta_j)$ is feasible, it follows that the terms $(\gamma_i-\sum_{j=1}^{n}\eta_j\alpha_{ij})$ are non-positive, and, hence for each i

$$\xi_i(\gamma_i-\sum_{j=1}^{n}\eta_j\alpha_{ij})=0 \tag{36}$$

from which (29) follows. A similar argument proves (28). □

5.1 *The Simple Linear Model: Dual Variables as Prices*

Consider a model of production with m perfectly divisible goods, each produced by one and only one activity (a technique of production). With a chosen unit of measurement, the jth activity, when operated at unit intensity, produces one unit of the jth good by using a semi-positive m-vector $a^j = (a_{1j}, a_{2j}, \cdots, a_{mj})$ of inputs. By requiring a^j to be semi-positive we rule out the possibility of "free" production. Inputs are required in fixed proportions and we assume constant returns to scale. This means that if the jth activity is operated at a non-negative intensity ρ, the inputs required are given by the m-vector $\rho a^j = (\rho a_{1j}, \rho a_{2j}, \cdots, \rho a_{mj})$. The activities are operated independently. So, if $x = [\xi_1, \cdots, \xi_m]$, is a planned vector of intensities (a production schedule), the *total* quantity of good i required as an input is given by $\sum_{j=1}^{m} \xi_j a_{ij}$, and the output vector, given our convention, is also x. Let $A = [a_{ij}]$, be the *input requirement matrix* where $a_{ij} (\geqq 0)$ the quantity of good i required to operate the jth activity at unit intensity (at which a unit of good j is produced as output).

Assume that when operated at unit intensity the jth activity has a net market return or profit $\gamma_j > 0$. The total supply of available inputs is given by a positive vector $b = [\sigma_1, \cdots, \sigma_m]$. The objective of the "firm" (the "producer" who owns b) is to maximize its total profit subject to the input requirements described by A, and the constraints imposed by the given supply of inputs b. More precisely, let $c = (\gamma_j)$ be the vector of net (market) returns or profits, and the firm solves the following linear programming problem (M)

$$\max \quad cx \tag{37}$$

$$\text{subject to } Ax \leqq b, \ x \geqq 0. \tag{38}$$

Observe that since the matrix A is non-negative, and b is positive, the set of all feasible solutions is non-empty, compact and the maximization problem has a solution as the objective function is continuous in x. The dual problem (m) consists in assigning "accounting" ("shadow or "efficiency") prices $y = (\pi_j)$ to the goods in such a way as to:

$$\min \quad yb \tag{39}$$

$$\text{subject to } yA \geqq c, \ y \geqq 0. \tag{40}$$

A convenient characterization of optimality comes from Theorem 2.4. A production schedule $x = (\xi_i)$ and a system of accounting prices $y = (\pi_j)$ are optimal solutions to (M) and (m) if:

$$\sum_{j=1}^{m} \alpha_{ij} \xi_j < \sigma_i \text{ implies that } \pi_i = 0, \tag{41}$$

$$\sum_{i=1}^{m} \alpha_{ij} \pi_i > \gamma_j \text{ implies that } \xi_j = 0. \tag{42}$$

The interpretation of (41) is straightforward: if according to a production schedule x, the total quantity of good i required as an input is less than its available supply σ_i

(in other words there is a surplus or excess supply of good i) its efficiency price π_i must be zero; and (42) asserts that if the cost of operating activity j calculated at the accounting prices y exceeds γ_j, the market return from activity j, its intensity level must be equal to 0.

5.2 Dual Variables and Comparative Statics

Let us consider the problems (M) and (m). Let us keep the matrix A as a constant and denote their sets of feasible solutions by $X(b)$ and $Y(c)$ respectively. As long as the sets $X(b)$ and $Y(c)$ are non-empty, we can define:

$$\begin{aligned} f(b,c) &= \max\ xc \quad \text{over} \quad x \in X(b), \\ &= \min\ by \quad \text{over} \quad y \in Y(c). \end{aligned}$$

We observe that if we define $B = \{b : X(b) \neq \emptyset\}$ and $C = \{c : Y(c) \neq \emptyset\}$, these are closed convex cones with non-void interiors in R^n and R^m respectively (see (Nikaido [49], pp. 140–141). Then, $f(b,c)$ is a function on $B \times C$ with the following properties:

Theorem 5.4 *(i) For a fixed $c \in C$, $f(b,c)$ is concave on B, and for a fixed $b \in B$, $f(b,c)$ is convex on C.*
(ii) Hence, $f(b,c)$ has left and right hand partial derivatives

$$\frac{\partial f_+}{\partial b_j}, \frac{\partial f_+}{\partial c_i}; \frac{\partial f_-}{\partial b_j}, \frac{\partial f_-}{\partial c_i}$$

over $B^\circ \times C^\circ$ where B° and C° are the interiors of B and C respectively.
(iii) If $\hat{x} = (\hat{\xi}_i)$ and $\hat{y} = (\hat{\eta}_j)$ are optimal solutions to (M) and (m) respectively corresponding to $b \in B^\circ$ and $c \in C^\circ$ respectively, then:

$$\begin{aligned} \frac{\partial f_-}{\partial b_j} &\geqq \hat{\eta}_j \geqq \frac{\partial f_+}{\partial b_j} \quad (j = 1, 2, \cdots, n) \\ \frac{\partial f_+}{\partial c_i} &\geqq \hat{\xi}_i \geqq \frac{\partial f_-}{\partial c_i} \quad (i = 1, 2, \cdots, m) \end{aligned}$$

where the derivatives are evaluated at (b,c).

Going back to our example of activity analysis model, we see (a bit informally) that the dual variable $\hat{\eta}_j$ is precisely the rate of change of maximum profit with respect to a change in the constraining value σ_j of the resource. Comparative statics results of this type (including changes in A) were discussed in Samuelson [55] and Beckman [3].

6 Saddle Points, Matrix Games and Linear Programming

In his paper on balanced growth von Neumann [64] (translation of the paper published in 1935-36), pointed out that the mathematical proof of his primary result is possible only by means of a generalization of Brouwer's Fix-Point Theorem, i.e., by the use of very fundamental *topological* facts. This generalized fix-point theorem (the "lemma" of paragraph 7) is also interesting in itself. The connection with topology may be very surprising at first, but the author thinks that it is natural in problems of this kind. The immediate reason for this is the occurrence of a certain "minimum - maximum" problem, familiar from calculus of variations. In our present question the minimum - maximum problem has been formulated in paragraph 5. It is clearly related to another problem occurring in the theory of games (see footnote 1, paragraph "6").

In that footnote, sketching the connection, he stressed that "in contrast with a simple maximum or minimum problem, the possibility of a solution of which would be evident", the question of a possible solution to the saddle point or minimum - maximum problem is "far more profound".

6.1 Linear Programming and Saddle Points

Corresponding to the dual problems (M) and (m), we define a function $\phi : R^m_+ \times R^n_+ \to R$ by

$$\phi(x, y) = xc + by - xAy. \tag{43}$$

We can readily rearrange (43) as:

$$\begin{aligned}\phi(x, y) &= xc + (b - xA)y \\ &= by + x(c - Ay).\end{aligned}$$

A point $(\hat{x}, \hat{y}) \in R^m_+ \times R^n_+$ is called a *saddle point* of ϕ if

$$\hat{x} \text{ maximizes } \phi(x, \hat{y}) \text{ over } R^m_+, \tag{44}$$

$$\hat{y} \text{ minimizes } \phi(\hat{x}, y) \text{ over } R^n_+, \tag{45}$$

so that

$$\phi(x, \hat{y}) \leqq \phi(\hat{x}, \hat{y}) \leqq \phi(\hat{x}, y), \; x \in R^m_+, \; y \in R^n_+. \tag{46}$$

The property of simultaneous "maximization and minimization" (maximin/minimax) played a central role–as we shall see–in interpreting the notion of an equilibrium in game theory. Let us note a connection between saddle points and optimal solutions to (M) and (m):

Theorem 6.1 *A pair $(\hat{x}, \hat{y}) \in R^m_+ \times R^n_+$ is a saddle point of ϕ if and only if $(\hat{x}, \hat{y})$ solves (M) and (m) respectively.*

Proof Suppose that $(\hat{x}, \hat{y}) \in R^m_+ \times R^n_+$ is a pair of optimal solutions to (M) and (m), we have:

$$\hat{x}c = b\hat{y},$$
$$b - \hat{x}A \geqq 0, \ c - A\hat{y} \leqq 0$$
$$(b - \hat{x}A)\hat{y} = \hat{x}(c - A\hat{y}) = 0.$$

Hence,

$$\phi(\hat{x}, \hat{y}) = b\hat{y} = b\hat{y} + \hat{x}(c - A\hat{y}) \geqq b\hat{y} + x(c - A\hat{y}) = \phi(x, \hat{y}), \ (x \in R^m_+),$$

$$\phi(\hat{x}, \hat{y}) = \hat{x}c = \hat{x}c + (b - \hat{x}A)\hat{y} \leqq \hat{x}c + (b - \hat{x}A)y = \phi(\hat{x}, y), \ (y \in R^n_+).$$

Conversely, suppose that $(\hat{x}, \hat{y})$ is a saddle point of ϕ. Then from (46):

$$\phi(\hat{x}, \hat{y}) \leqq \phi(\hat{x}, y), \ \text{for} \ y \in R^n_+$$
$$\hat{x}c + (b - \hat{x}A)\hat{y} \leqq \hat{x}c + (b - \hat{x}A)y \ \text{for} \ y \in R^n_+$$
$$\text{leading to} \ (b - \hat{x}A)y \geqq 0 \ \text{for} \ y \in R^n_+. \tag{47}$$

Similarly,

$$x(c - A\hat{y}) \leqq 0 \ \text{for} \ x \in R^m_+. \tag{48}$$

From (47) and (48), we get $\hat{x}A \leqq b$, $\hat{x} \geqq 0$; $A\hat{y} \geqq c$, $\hat{y} \geqq 0$. Hence, $(\hat{x}, \hat{y})$ is a pair of feasible solutions to (M) and (m). On the other hand, from (46) we get, at $x = 0$, $y = 0$,

$$b\hat{y} = \phi(0, \hat{y}) \leqq \phi(\hat{x}, \hat{y}) \leqq \phi(\hat{x}, 0) = \hat{x}c.$$

Hence, using Lemma 5.1 and Theorem 2.2, $(\hat{x}, \hat{y})$ solves (M) and (m). □

6.2 *Linear Programming and Matrix Games*

A two-person zero sum matrix game is described as follows. Player 1 (resp. Player 2) has a finite set S (resp. a finite set T) of *pure* strategies. The number a_{ij} is the *payoff* to Player 1 from Player 2 when Player 1 plays the strategy i (=1, 2, ..., m) and Player 2 plays the strategy j(= 1, 2, ...n) (the payoff to Player 2 is $-a_{ij}$: hence "zero sum"). A *mixed* strategy $x = (x_i)$ for Player 1 (resp. $y = (y_j)$ for Player 2) is a non-negative *m-vector* satisfying $\sum_{i=1}^m x_i = 1$ (resp. a non-negative *n-vector* satisfying $\sum_{j=1}^n y_j = 1$). One interprets x_i (resp. y_j) as the probability that P_1 (resp. P_2) plays the pure strategy i (resp. pure strategy j). The sets of all mixed strategies of players 1 and 2 are denoted by $< S >$ and $< T >$ respectively.

When the two players play x and y, the formula for the expected pay off $f(x, y)$ is given by

$$f(x, y) = xAy. \tag{49}$$

A game $\{< S >, < T >, f\}$ has a solution $(\bar{x}, \bar{y}, \omega)$ if

$$\begin{aligned} &f(\bar{x}, y) \geqq \omega \text{ for all } y \in < T >;\\ &f(x, \bar{y}) \leqq \omega \text{ for all } x \in < S >;\\ &\text{or, } \bar{x}Ay \geqq \bar{x}A\bar{y} \geqq xA\bar{y} \text{ for all } x \in < S >, y \in < T >, \quad \text{where } \bar{x}A\bar{y} = \omega. \end{aligned}$$

In view of the extensive use of matrix games in introducing equilibrium and efficiency, it is not surprising that there are many proofs of the following fundamental theorem (Kuhn [36] has already been mentioned earlier). The proof using the duality theorem of linear programming is of particular interest from the computational point of view.

Theorem 6.2 *Every matrix game has a solution in mixed strategies.*

Proof This is done for the case where the pay off matrix $A = (\alpha_{ij})$ is strictly positive (i.e., $\alpha_{ij} > 0$ for all i, j). The proof for the general case follows easily. Let $A = (\alpha_{ij})$ be any $m \times n$ matrix and let $A' = (\alpha_{ij} + \Theta\ )$ be a strictly positive matrix obtained by adding a sufficiently large positive number Θ. Now assume that the fundamental theorem is true when the pay-off matrix is A'. Then there are mixed strategies $\bar{x}$ and $\bar{y}$ such that

$$\bar{x}A'y \geqq \omega' \geqq xA'\bar{y} \ \text{ for all mixed strategies } \ x \ \text{ and } \ y. \tag{50}$$

Here, $\omega' = \bar{x}A'\bar{y}$ is the value of the game. But a routine calculation leads to $\bar{x}A'y = \bar{x}Ay + \Theta$ and $xA'\bar{y} = xA\bar{y} + \Theta$ so that

$$\bar{x}Ay \ \geqq \omega \ \equiv \omega' - \Theta \ \geqq xA\bar{y}.$$

This means that $(\bar{x}, \bar{y}; \omega)$ is a solution to the original game.

Let A be the strictly positive payoff matrix of a game. Let $u = [1, 1, \cdots, 1]$ and $v = (1, 1, \cdots, 1)$ be the sum vectors in R^m and R^n respectively. Consider the linear programming problem:

$$\begin{aligned} &\min \quad xu\\ &\text{subject to } \ xA \geqq v, x \geqq 0. \end{aligned}$$

Since A is strictly positive, the problem has a feasible solution. The dual problem

$$\begin{aligned} &\max \quad vy\\ &\text{subject to } \ Ay \leqq u, y \geqq 0. \end{aligned}$$

also has a feasible solution (consider $y = 0$). From the duality theorem, therefore, there exist optimal solutions x_0 and y_0 such that:

$$x_0 u = v y_0 = \mu \tag{51}$$

where μ is clearly positive. We now show that a solution $(\bar{x}, \bar{y}, \omega)$ of the game is given by:

$$\bar{x} = \frac{x_0}{\mu}, \bar{y} = \frac{y_0}{\mu}, \omega = \frac{1}{\mu}. \tag{52}$$

The fact that $\bar{x}$ and $\bar{y}$ are mixed strategies follows from (51). □

6.3 A Minimax Theorem

Here is Gale (LEM, p. 211) talking about the fundamental existence theorem of matrix games and the minimax theorem:

"The history of the minimax theorem is interesting. Von Neumann's original proof was highly involved and depended on a rather difficult fixed point theorem of topology. In 1941, Kakutani gave a short and elegant proof of the result, which still depended on the fixed point theorem. Meanwhile, Ville (1938) presented the first "elementary" proof of the result, requiring nothing more advanced than a simple fact about convex sets. The proof given by von Neumann and Morgenstern is essentially a modification of Ville's proof, which in turn is much like the geometric proof presented in the appendix of this chapter. The proof via symmetrization is due to Gale, Kuhn and Tucker."

I prefer brevity, and Gale himself appreciated the elegance of Kakutani's proof:

Theorem 6.3 *Let $f(x, y)$ be a real valued continuous function defined for $x \in K$, and $y \in L$ where K and L are two non-empty compact convex sets in R^m and R^n respectively. If for every $x_0 \in K$, and for every real number α, $\{y \in L : f(x_0, y) \leqq \alpha\}$ is convex, and if for every $y_0 \in L$, and for every real number β, $\{x \in K : f(x, y_0) \geqq \beta\}$ is convex , then we have*

$$\max_{x \in K} \min_{y \in L} f(x, y) = \min_{y \in L} \max_{x \in K} f(x, y).$$

Proof (Sketch) For $x \in K$, consider $f(x, \cdot)$ as a function of y. Define for $x \in K$,

$$\phi(x) = \{y \in L, f(x, y) = \min_{y \in L} f(x, y)\}.$$

It is easily verified that $\phi(x)$ is non-empty, compact and convex. By the maximum theorem, $\phi : K \longrightarrow L$ is an upper semi-continuous correspondence. Similarly, for each $y \in L$, consider $f(\cdot, y)$ as a function of x. Define for each $y \in L$,

$$\psi(y) = \{x \in K, f(x, y) = \max_{x \in K} f(x, y)\}.$$

Again, one can verify that $\psi(y)$ is non-empty, compact and convex, and, by the maximum theorem, $\psi : L \longrightarrow K$ is an upper semi-continuous correspondence. Let $M = K \times L$ be a non-empty, compact, convex set in R^{m+n}. Define the correspondence $\Psi : M \longrightarrow M$ by

$$\Psi(x, y) = \psi(y) \times \phi(x).$$

The Kakutani fixed point theorem applied to Ψ gives us the existence of $(x^0, y^0) \in \psi(y^0) \times \phi(x^0)$. Hence,

$$f(x^0, y^0) \geqq f(x, y^0) \text{ for all } x \in K, \text{ i.e., } f(x^0, y^0) = \max_{x \in K} f(x, y^0)$$
$$f(x^0, y^0) \leqq f(x^0, y) \text{ for all } y \in L, \text{ i.e., } f(x^0, y^0) = \min_{y \in L} f(x^0, y).$$

Consequently we have:

$$\min_{y \in L} \max_{x \in K} f(x, y) \leqq \max_{x \in K} f(x, y^0) = f(x^0, y^0) = \min_{y \in L} f(x^0, y) = \max_{x \in K} \min_{y \in L} f(x, y).$$

The inequality in the other direction can be easily verified. □

It is clear that Theorem 5.2 is a simple consequence of Theorem 6.3. For a detailed investigation of the related literature see Simons [60] and the other articles in Du and Pardalos [15].

7 Closed Linear Model of Production: The Von Neumann Duality

A general framework for dealing with the von Neumann problem of balanced growth at the maximal rate was presented by Gale [18]. In what follows I rely heavily on the succinct exposition of Karlin [32]. Consider an economy with m producible goods (no primary factor of production, no possibility of obtaining non-producible goods obtained through trade with the outside world). The *production possibility* (or, technological transformation) is described by a set $\mathcal{T}$ in $R^m \times R^m_+$:

$$\mathcal{T} = \{(x, y) \in R^m_+ \times R^m_+ : y \text{ is producible from } x\}.$$

One refers to $x = (x_i)$ as an *input* vector that gets transformed into an *output* vector $y = (y_i)$ in the next period. A typical input-output vector (x, y) is referred to as an *activity* or *technical process*. The set $\mathcal{T}$ is assumed to have the following properties:

(*T.1*) $\mathcal{T}$ is a closed convex cone in $R^m_+ \times R^m_+$; $(0, 0) \in \mathcal{T}$ (continuity, convexity, constant returns to scale and the possibility of inaction).

($T.2$) If $(x, y) \in \mathcal{T}$, and $x' \geqq x, 0 \leqq y' \leqq y$, then $(x', y') \in \mathcal{T}$ (free disposal).
($T.3$) $(0, y) \in \mathcal{T}$ implies $y = 0$ (impossibility of "free" production).
($T.4$) There is some (x^0, y^0) such that $y^0 > 0$ (all goods are producible).

We call a process $(x, y) \in \mathcal{T}$ non-trivial if $x \neq 0$. For any non-trivial process (x, y) define the growth factor $\lambda(x, y)$ as:

$$\lambda(x, y) = \max\{\lambda : y \geqq \lambda x\}.$$

It is clear that:

$$0 \leqq \lambda(x, y) < \infty, \text{ and, } \lambda(\alpha x, \alpha y) = \lambda(x, y) \text{ for any } \alpha > 0.$$

Define

$$\lambda^* = \sup\{\lambda(x, y) : (x, y) \in \mathcal{T}, \; x \neq 0\} > 0 \text{ (from}(T.4)).$$

Then there is a sequence of processes $(x^q, y^q) \in \mathcal{T}$, (with $|x^q| = \sum_{i=1}^m x_i^q = 1$), such that $\lambda^q \equiv \lambda(x^q, y^q) \to \lambda^*$ as $q \to \infty$. By using (T.1) and (T.3) one can prove that (y^q) is a bounded sequence. Now, $y^q \geqq \lambda^q x^q$. Note that $\sum_{i=1}^m y_i^q \geqq \lambda^q$. Taking a subsequence (retain notation) one concludes that (x^q, y^q, λ^q) converges to some (x, y, λ^*). Then, by (T.1), $(x, y) \in \mathcal{T}$ and $y \geqq \lambda^* x$. By using (T.3) we get the following:

Theorem 7.1 *There is a non-trivial von Neumann process $(x^*, y^*) \in \mathcal{T}$ such that $y^* = \lambda^* x^*$, $\lambda^* = \lambda(x^*, y^*)$, and $\lambda^* \geqq \lambda(x, y)$ for all $(x, y) \in \mathcal{T}$.*

One refers to λ^* as the *maximal balanced growth (or, expansion)* factor of $\mathcal{T}$. It should be stressed that the words "growth" or "expansion" should never be interpreted as a claim that $\lambda^* > 1$. The next result is a characterization of a von Neumann process in terms of a price system. Gale's elegant proof is, according to (Debreu [12], p. 1262) one of the early applications of a separation theorem in mathematical economics.

Theorem 7.2 *There is $p^* \geq 0$ such that*

$$p^* y \leq \lambda^*(p^* x) \textit{ for all } (x, y) \in \mathcal{T}.$$

Proof Consider the set M in R^m defined by

$$M = \{y - \lambda^* x : (x, y) \in \mathcal{T}\}.$$

Observe that M is convex and does not contain a positive vector. Use Theorem 2.3 to get the p^* with the desired property. □

One refers to $(x^*, y^*, \lambda^*, p^*)$ as a *von Neumann equilibrium* and p^* as a *von Neumann price system.*

T

he von Neumann model (A, B) may be considered a special case. Here, there are m goods and n activities (also interpreted as *sectors* of an economy). Activity j is described by two non-negative vectors a^j and b^j $(j = 1, 2, \cdots)$: the jth activity, operated at the unit level, transforms an input vector $a^j = [a_{1j}, a_{2j}, \ldots, a_{mj}]$ into an output vector $b^j = [b_{1j}, b_{2j}, \ldots, b_{mj}]$. The inputs are required in fixed proportions and outputs are produced in fixed proportions. Let $A = (a^j) \equiv (a_{ij})$ be the (non-negative input requirement matrix and $B = [b^j] \equiv (b_{ij})$ be the (non-negative) output matrix. We assume that the activities can be operated independently and constant returns to scale prevail. Hence, if the activities are operated at intensities $v = (v_j)$, the total quantity of commodity i used as inputs is given by $\sum_{j=1}^n a_{ij} v_j$ and the total quantity of commodity i produced is $\sum_{j=1}^n b_{ij} v_j$. The transformation set $\mathcal{T}$ is now defined as:

$$\mathcal{T} = \{(x, y) : x \geqq Av, 0 \leqq y \leqq Bv, \text{ for some } v \geqq 0\}. \tag{53}$$

We assume

(i) For every activity j, $a^j \geq 0$ (for each j, there is some i such that $a_{ij} > 0$).
(ii) For every commodity i, $b_i \geq 0$ (for each i, there is some j such that $b_{ij} > 0$).

The relaxation of the restrictive assumption of von Neumann that $a_{ij} + b_{ij} > 0$ for *all* i, j was one of the primary motivations behind Gale [18] and Kemeny et al. [33]. A succinct treatment was offered in Howe [30]. By using the results above we get:

Theorem 7.3 *If the non-negative matrices A and B satisfy (i) and (ii), then there exist* (i) *an activity level* $v^* \geq 0$, (ii) *a price system* $p^* \geq 0$ *and* (iii) *an expansion factor* $\lambda^* > 0$ *such that* $Bv^* \geq \lambda^* Av^*$; $p^* B \leq \lambda^* p^* A$.

Now we return to the simple activity analysis (Leontief) model as a special case of the von Neumann (A, B) model. Here, we have $m = n$, and each good i is produced by one and only one activity (sector). The unique activity producing good j can thus be labeled as "the jth activity or sector". So, A is an $n \times n$ non-negative matrix, a_{ij} being the quantity of good i required as input to produce one unit of good j, and $a^j \geq 0$. The output matrix B is the $n \times n$ identity matrix.

An assessment of the scope, originality and some limitations of the original von Neumann model (A, B) as well as the significance of the concept of a von Neumann equilibrium was offered by Koopmans [34]. The paper contains, he said, the first explicit statement of the activity analysis of production. It also has "an explicit statement of the relationship between commodity prices and production coefficients" and distinguishes processes by means of profit maximization. "It is the first rigorous, formal, and fully explicit model in non-aggregative capital theory." Finally, along with Wald [66], von Neumann's paper "became the starting point for a systematic development of abstract models of competitive equilibrium". But there were some obvious limits to the approach to studying the evolution of an economy over time. First, was the absence of a *treatment of consumption as an end in itself*. Next, "It seems arbitrary and contrary to any experience about economics growth that all production activities grow in time at the same proportional rate". While von Neumann explicitly recognized that "we are interested in those states where the whole economy expands without change of structure" (meaning that the inter-sectoral proportions remain the same), studies of economic development suggests that long run growth is a process which is marked by a restructuring of an economy, with dramatic changes in the *relative size and rate of growth* of various sectors (see, for example, Bell [4]).

The Gale model has been subsequently reinterpreted/applied in many contexts in dynamic economics, far removed from the original. Some the deficiencies were taken care of. I shall provide three such examples where the framework turned out to be particularly useful.

7.1 The Samuelson Conjecture and the Radner Turnpike

Few development planners can afford to ignore the initial stocks $\tilde{x}$ of goods in the economy (in many cases, inherited as a leftover after colonial loots). Going back to the closed linear framework of Gale, consider a planning problem where the economy starts from a given initial vector of stocks $\tilde{x}$, and the objective is to maximize a welfare function defined on the "target" stocks in the terminal period. A conjecture by Samuelson (quoted from his statement (1983, p.486) enhanced the significance of the von Neumann equilibrium as follows:

"Then our knowledge that the von Neumann ray is the ray of fastest balanced growth leads us to suspect that anyone planning for the *far* future should move *initially near to* it, catch a fast ride there for *most of the time*, planning to move *penultimately away* from it towards one's ultimate goal ray. I conjectured this in Samuelson [57] as did Dorfman et al. [14], but not until Radner [52] was the conjecture elevated into a known theorem." Radner's proof rested on a value-loss lemma. The conjecture on the "initial" and "penultimate stages" of an optimal path was captured by Nikaido [48] in an important refinement of the Radner proof. Here the objective is to show how the main steps of the proof provide an insight into the corresponding consumption turnpike theorem of Gale that will be elaborated later. This exposition is somewhat simpler than Radner's, proofs are only sketched. A thorough examination of the assumptions, and possible extensions can be found in (Nikaido [49]).

Consider a planning problem over N (a positive, finite integer) periods. Given an initial stock vector $\tilde{x} > 0$, a sequence $(x_t)_{t=0}^{N}$ of m- vectors is a growth *program* from $\tilde{x}$ if $x_0 = \tilde{x}$, and $(x_t, x_{t+1}) \in \mathcal{T}$ for $t = 0, 1, \cdots, N-1$. For any two non-zero vectors (x, z) in R^m define the "angular distance" $d(x, z)$ as;

$$d(x, z) = \left\| \frac{x}{\|x\|} - \frac{z}{\|z\|} \right\|$$

where $\|x\| = \sqrt{x \cdot x}$.

The technology $\mathcal{T}$ satisfies (T.1) through (T.4) and let $(x^*, y^*, \lambda^*, p^*)$ be a *von Neumann equilibrium*. Assume that

(*T.5*) $x^* > 0$, $p^* > 0$. We normalize and set $\|x^*\| = 1$.
(*T.6*) $p^*(y - \lambda^* x) < 0$, *for all* $(x, y) \in \mathcal{T}$ *that are not proportional to* (x^*, y^*).

(T.6) is called the *unique profitability condition*. We note that (T.6) is typically not satisfied when $\mathcal{T}$ is polyhedral (for example, the von Neumann (A, B) model. The contribution of Morishima Morishima [46] is particularly important in this context). (T.6) implies that the normalized balanced growth vector x^* is unique. Given its importance in the proof to be sketched below, we identify a property of $\mathcal{T}$ that guarantees (T.6).

(*TSC*) If $(x, y) \in \mathcal{T}$, and $(u, v) \in \mathcal{T}$, and if x and u are linearly independent, then for any positive numbers α, β with $\alpha + \beta = 1$, there is an output vector z such that

$$(\alpha x + \beta u, z) \in \mathcal{T}, \text{ and } z > \alpha y + \beta v. \text{ (Strict convexity)}$$

One can show that if (TSC) holds, then the unique profitability condition (T.6) holds. The planner's *objective* (welfare, felicity) function $w : R^m \to R$ satisfies :

(W.1) w is *continuous, homogeneous of degree one.*
(W.2) $w(x^*) > 0$.
(W.3) $w(x) \leqq \mathcal{K} p^* x$ *for some* $\mathcal{K} > 0$ *and for all* $x \in R^m_+$.

The assumption (W.3) holds if, for example, $w(x) = q \cdot x$ where q is a positive vector. A growth program $(\bar{x}_t)_{t=0}^N$ from an initial $\tilde{x} > 0$ is *optimal* if $w(\bar{x}_N) \geqq w(x_N)$ for all growth programs $(x_t)_{t=0}^N$ from $\tilde{x}$. First, we state the value loss lemma:

Lemma 7.1 *For any* $\varepsilon > 0$, *there is some* ρ *such that*
(i) $\lambda^* > \rho > 0$.
Writing $\delta \equiv \lambda^* - \rho$,
(ii) For any $(x, y) \in \mathcal{T}$ *for which* $d(x, x^*) \geqq \varepsilon$, *one has*

$$p^* y \leq (\lambda^* - \delta) p^* x.$$

Proof This is a sketch from Nikaido [49]. Since $\mathcal{T}$ is a cone, it is enough to prove the assertion for processes $(x, y) \in \mathcal{T}$, with $\|x\| = 1$. Let

$$V = \{(x, y) : (x, y) \in \mathcal{T}, \|x\| = 1, \|x - x^*\| \geqq \varepsilon\}.$$

V may or may not be empty. If V is empty, any ρ fulfilling (i) will do. If V is non-empty, one can show that it is closed and bounded. For any $(x, y) \in V$, we have, in particular, $x \neq 0$, and x is not proportional to x^*, so that $p^* y < \lambda^* p^* x$. Hence, $p^* x > 0$ for $(x, y) \in \mathcal{T}$, and

$$f(x, y) = \frac{p^* y}{p^* x}$$

is a well-defined, continuous function on the non-empty, compact set V. This function attains a maximum value, say, τ, which must be smaller than λ^* by (T.6). Then any $\rho \in (\tau, \lambda^*)$ is a desired positive number. □

Theorem 7.4 *Let $(x^*, y^*, \lambda^*, p^*)$ be a von Neumann equilibrium satisfying (T.5) and (T.6). Let $\tilde{x} > 0$ be given and let the objective function w satisfy (W.1)-(W.3). Then for any $\varepsilon > 0$, there is a positive integer S such that for any N and any optimal program $(x_t)_{t=0}^N$ from $\tilde{x}$, the number of periods in which $d(x_t, x^*) \geq \varepsilon$ cannot exceed S.*

Proof First, since $(x^*, \lambda^* x^*) \in \mathcal{T}$, for some (small enough!) $\chi > 0$, $(\chi x^*, \chi\lambda^* x^*) \in \mathcal{T}$ and $\chi x^* \leq \tilde{x}$. Writing $L \equiv \chi\lambda^*(> 0)$, and using free disposal, $(\tilde{x}, Lx^*) \in \mathcal{T}$. We define a growth program $(\breve{x}_t)_{t=0}^N$ from $\tilde{x}$, by:

$$\breve{x}_0 = \tilde{x},\ \breve{x}_1 \equiv Lx^*, \ldots,\ \breve{x}_n = L(\lambda^*)^{n-1}\tilde{x},\ n = 1, 2, \cdots, N.$$

Consider any $\varepsilon \gg 0$, and any growth program $(x_t^\wedge)_{t=0}^N$ *from* $\tilde{x}$. For any period t such that $d(x_t^\wedge, x^*) \geqq \varepsilon$, one has

$$p^* x_{t+1}^\wedge \leq (\lambda^* - \delta) p^* x_t^\wedge.$$

On the other hand, by definition of a von Neumann equilibrium, for all t,

$$p^* x_{t+1}^\wedge \leq \lambda^* p^* x_t^\wedge.$$

Now assume that $d(x_t^\wedge, x^*) \geq \varepsilon$ holds for P periods. Then one has:

$$p^* x_N^\wedge \leq (\lambda^* - \delta)^P (\lambda^*)^{N-P} p^* \tilde{x}.$$

Hence,

$$w(x_N^\wedge) \leq \mathcal{K} p^* x_N^\wedge \leq \mathcal{K}(\lambda^* - \delta)^P (\lambda^*)^{N-P} p^* \tilde{x}.$$

By homogeneity of w ,

$$w(\breve{x}_N) = L(\lambda^*)^{N-1} w(\tilde{x}) > 0.$$

Hence,

$$\frac{w(x_N^\wedge)}{w(\breve{x}_N)} \leq C \left[\frac{\lambda^* - \delta}{\lambda^*}\right]^P,$$

$$\text{where } C = \frac{\mathcal{K}\lambda^* p^* \tilde{x}}{L w(\tilde{x})}.$$

□

Hence, for $(\breve{x}_t)_{t=o}^{N}$ from $\tilde{x}$ to be optimal,

$$C\left[\frac{(\lambda^* - \delta)}{\lambda^*}\right]^P \geq 1, \text{ or } P \leq \frac{\log\ C}{\log \lambda^* - \log(\lambda^* - \delta)}.$$

Hence, we can take S to be the smallest positive integer greater than or equal to

$$\max\left[1, \quad \frac{\log C}{\log \lambda^* - \log(\lambda^* - \delta)}\right].$$

7.2 *Inter-temporal Efficiency*

We briefly turn to the closed model with an explicit treatment of consumption and initial stocks relying on the detailed exploration of Majumdar [42] (to be referred as M-M-M). Assume (T.1) through (T.4). Furthermore, assume, in this subsection), without loss of generality that the von Neumann growth factor $\lambda^* = 1$ to *simplify notation* (see M-M-M, p. 29 and (Winter [67], p. 68 - 69) for more details).
Now keeping in mind that $\lambda^* = 1$, we make the following assumption:

(*TE1*) There is some $x^* > 0$, such that $(x^*, x^*) \in \mathcal{T}$.
(*TE2*) The von Neumann price vector $p^* > 0$.

A *feasible production program* from an initial stock vector $X^{\sim} > 0$ is a sequence $(\mathbf{x}, \mathbf{y}) = (x_t, y_{t+1})$ such that

$$x_0 = X^{\sim}, x_t \leqq y_t \text{ for all } t \geqq 1;$$
$$(x_t, y_{t+1}) \in \mathcal{T}, \text{ for all } t \geqq 0.$$

The *feasible consumption program* $\mathbf{c} = (c_t)$ generated by $(\mathbf{x}, \mathbf{y})$ is defined as:

$$c_t = y_t - x_t \text{ for all } t \geqq 1.$$

We refer to $(\mathbf{x}, \mathbf{y}, \mathbf{c})$ as a feasible program from $X^{\sim}$. A feasible program $(\mathbf{x}^*, \mathbf{y}^*, \mathbf{c}^*)$ from $X^{\sim}$ is *efficient* if there is no other feasible program $(\mathbf{x}, \mathbf{y}, \mathbf{c})$ from $X^{\sim}$ such that

$$c_t \geqq c_t^* \text{ for all } t \text{ and } c_t > c_t^* \text{for some} t. \tag{54}$$

We write C to denote the set of all consumption programs. By using the (l_1, l_∞) duality from the theory of Banach spaces, we get:

Theorem 7.5 *Let* $(\mathbf{x}^*, \mathbf{y}^*, \mathbf{c}^*)$ *from* $X^{\sim}$ *be efficient. Then there exists a non-null sequence* (p_t) *of non-negative price vectors such that*

$$\sum_{t=1}^{\infty} p_t c_t^* \geqq \sum_{t=1}^{\infty} p_t c_t \textit{ for all consumption programs } \mathbf{c} = (c_t). \tag{55}$$

Proof (sketch) First, define

$$\mathcal{G} = \{\mathbf{c} = (c_t) : c_t = y_t - x_t \text{ for all } t \geqq 1, x_0 = X^{\sim}; (x_t, y_{t+1}) \in \mathcal{T}\}.$$

Clearly, $\mathcal{G}$ contains all feasible consumption programs $\mathbf{c} = (c_t)$ which satisfy the properties defining the set and, in addition, requires $c_t \geqq 0$ for all $t \geqq 1$. Also, for any feasible consumption program $\tau \geqq 1$.

$$\sum_{t-1}^{\tau} p^* c_t \leqq p^* X^{\sim}$$

Since $p^* > 0$, one has, for any feasible consumption program $\mathbf{c} = (c_t)$,

$$\mid \mathbf{c} \mid_{l_1} = \sum_{t=1}^{\infty} |c_t| \leqq \alpha p^* X^{\sim} \quad \text{where } \alpha \text{ is determined by } p^*.$$

Let Λ be the linear space of all sequences of m-vectors $\mathbf{c} = (c_t)$ such that $\mid \mathbf{c} \mid_{l_1}$ is finite. An element p of Λ^*, the set of all continuous linear functionals on Λ can be represented by a sequence $p = (p_t)$ such that $\mid p_t \mid_{\max}$is finite. Let $\mathcal{F} = (\mathcal{G} \cap \Lambda) - \Lambda^+$ where Λ^+ is the set of all non-negative sequences in Λ^+.

The next few steps are (i) to show that $\mathcal{F}$ is a convex, closed (under point-wise convergence) subset of Λ, (ii) $\mathcal{F}$ has an interior point, and finally (iii) if $((\mathbf{x}^*, \mathbf{y}^*, \mathbf{c}^*)$ from $X^{\sim}$ is *efficient*, c^* is a boundary point of $\mathcal{F}$.

Now, invoking a separation argument completes the proof. □

It should be stressed that the condition of *present value maximization is a result of independent interest, and does not* follow from the well-known approaches leading to the existence of Malinvaud prices–in a closed model such prices can be shown to exist under (T1)–(T4) (see Malinvaud [43] or Radner [53], footnote on p. 60)). Gale's "competitive" prices were in the Malinvaud tradition. Present value maximization was not a prominent issue in Gale's work.

7.3 Sustainability and Substitution Possibilities in the Technology

We now turn to a variation (reinterpretation?) of the closed linear model of production. Our aim is to get an *impossibility theorem* on the possibility of sustaining a stationary consumption program when the technology does not permit adequate input substitution.

A good k is an *exhaustible resource* if for all $(x, y) \in \mathcal{T}$, $y_k \leqq x_k$. In other words, *no* known process can augment the stock of an exhaustible resource. Of the m goods, there are m_1 producible goods that can be either used as an input or consumed. There are m_2 exhaustible resources, these are important inputs that are not consumed directly $(m_1 + m_2 = m)$. For an input or output vector x or y, we have the following ordering convention: the first m_1 coordinates of x (or y) denote the quantities of producible goods (to be written $\tilde{x} = (\tilde{x}_i)$ or $\tilde{y} = (\tilde{y}_i)$, where $i = 1, 2, ...m_1$). The following m_2 coordinates will denote quantities of exhaustible resources, to be written $s = (s_k)$ as input stock, and $s' = (s'_k)$ as an output stock where $k = m_1 + 1, m_1 + 2, ..., m_1 + m_2$. We write $x = (\tilde{x}, s)$ and $y = (\tilde{y}, s')$ to explicitly distinguish between producible goods and resources. Formally,

T.1. *For any process* $(x, y) \in \mathcal{T}$,*wherex* $= (\tilde{x}, s)$, *and* $y = (\tilde{y}, s')$,*the inequality* "$s \geqq s'$" *holds.*

We assume that all producible capital goods, as well as the resource can be stored costlessly (no depreciation when stored):

T.2. *For any* $x = (\tilde{x}, s) \geqq 0$, $(x, x) \in \mathcal{T}$. *Also,* "$(0, y) \in \mathcal{T}$" *implies that* "$y = 0$" (impossibility of free production).

T.3. "$\big((x = (\tilde{x}, s), y = (\tilde{y}, s')) \in \mathcal{T}\big)$" *implies* "$\big((\tilde{x}_\sim, s_\sim), (\tilde{y}_\sim, s'_\sim) \in \mathcal{T}\big)$ *where* $(\tilde{x}_\sim, s_\sim) \geqq (\tilde{x}, s)$, $0 \leqq \tilde{y}_\sim \leqq \tilde{y}$, *and* $0 \leqq s'_\sim \leqq s'$" *(free disposal).*

To think of the future consumption possibilities and resource exhaustion, it is important to introduce the set Z of *net outputs* generated by $\mathcal{T}$ defined by:

$$
\begin{aligned}
Z &= \{z : z = (y - x), \text{ where } (x, y) \in \mathcal{T}\}. \\
&= \{z : z = (\tilde{y} - \tilde{x}, s' - s) \text{ where } ((\tilde{x}, s), (\tilde{y}, s')) \in \mathcal{T}\}.
\end{aligned} \tag{56}
$$

Note that the storage process (x, x) is assumed to be technologically feasible. So $0 \in Z$. The structure of Z is determined by the properties of $\mathcal{T}$.

T.4. $\mathcal{T}$ *is a closed convex cone.*

The importance of the role of resources in converting x to y is captured by the following assumption:

T.5. *Let* "$\big(x = (\tilde{x}, s), y = (\tilde{y}, s')\big) \in \mathcal{T}$". *If for any* i, $\tilde{y}_i > \tilde{x}_i$, *there is some (one or more)* k *such that* $s_k > s'_k$.

In other words, for a positive net production of any producible good i, a positive quantity of some (one or more) resource is needed.

Lemma 7.2 *Under T.4 and T.5,* Z *is a convex cone that does not contain any* $u \geq 0$.

Proof Z is clearly a convex cone since $\mathcal{T}$ *is*. Consider $z = (\tilde{y} - \tilde{x}, s' - s) \in Z$. Observe that $s' - s \leqq 0$. On the other hand, $\tilde{y} - \tilde{x} \geq 0$ implies, by T.5, that $s' - s \leq 0$. □

We assume a stronger property of Z than that ensured by this Lemma 7.2.

Let $cl(Z)$ be the closure of Z. $cl(Z)$ is a convex cone ((Nikaido [49], Theorem 2.6). We *assume*:

T.6 *$cl(Z)$ is a convex cone that does not contain any $u \geq 0$.*

We now turn to the basic "duality"result: At a price system $p \geqq 0$, define the profit of a process $(x, y) \in \mathcal{T}$ as $\pi[(x, y), p] = py - px = pz$. We prove the existence of a *positive* $p > 0$ such that no process in $\mathcal{T}$ generates a positive profit.

Lemma 7.3 *There is $p > 0$ such that $p \cdot z \leqq 0$ for all $z \in cl(Z)$.*

Proof Since $cl(Z)$ is a closed convex cone that does not contain any $u \geq 0$, it follows from Theorem 2.2 that $-(cl(Z))^*$ contains a positive $p > 0$. Hence, $-pz \geqq 0$ for all $z \in cl(Z)$, or $pz \leqq 0$ for all $z \in cl(Z)$. □

We write $p = (p_1, p_2) > 0$ to stress that the m_1-vector $p_1 > 0$ denotes the vector of prices of producible goods and $p_2 > 0$ is the m_2-vector denoting the prices of exhaustible resources. We can write more explicitly

$$pz = p_1(\tilde{y} - \tilde{x}) + p_2(s' - s) \leqq 0. \tag{57}$$

The initial stocks of the producible goods $\tilde{X} > 0$ and the resources $S > 0$ are given. To avoid misunderstandings, at the risk of repetition of the obvious, let me write out some formal definitions fully. A *program from* $(\tilde{X}, S)$ is a sequence $(\mathbf{x}, \mathbf{y}, \tilde{\mathbf{c}}) = (x_t, y_{t+1}, \tilde{c}_{t+1})$ satisfying:

$$\begin{aligned}
\tilde{x}_0 &= \tilde{X}, s_0 = S, \\
(x_t, y_{t+1}) &= (x_t = (\tilde{x}_t, s_t), y_{t+1} = (\tilde{y}_{t+1}, s'_{t+1})) \in \mathcal{T} \text{ for } t \geqq 0, \\
s_{t+1} &= s'_{t+1} \text{ for } t \geqq 0, \\
\tilde{c}_{t+1} &= (\tilde{y}_{t+1} - \tilde{x}_{t+1}) \geqq 0 \text{ for } t \geqq 0.
\end{aligned} \tag{58}$$

A program specifies a sequence of decisions described as follows. The initial stocks $\tilde{X}$ and S are used in period 0 as inputs in a chosen process to generate the first-period output vector $(\tilde{y}_1, s'_1)$, where $s'_1 \leqq S$. A part $\tilde{c}_1$ of $\tilde{y}_1$, the vector of producible goods available, is chosen for consumption. The rest, the vector $\tilde{x}_1 (= \tilde{y}_1 - \tilde{c}_1)$ and the available stocks of the resources $s_1 = s'_1$ are used as inputs in a chosen process to generate the output vector $(\tilde{y}_2, s'_2)$ where s'_2 is the stock of resource, $s'_2 \leqq s'_1$, and the story is repeated.

Observe that as resources are not consumed, by setting $s_{t+1} = s'_{t+1}$ (use of the entire stocks of resources as inputs in period $t + 1$), the formulation introduces a "conservation" or (in Hurwicz's words) "non-wastefulness" clause in the definition. A program defines a corresponding sequence (s'_t) of resource vectors that is non-increasing.

We now come to the main result on the long-run behavior of consumptions in our model. Recall from Sect. 2 that for any $c \in \mathbb{R}^{m_1}$, we have

$$| c | = \sum_{k=1}^{m_1} |c_k| . \tag{59}$$

Theorem 7.6 *Let* $(\mathbf{x}, \mathbf{y}, \mathbf{c}) = (x_t, y_{t+1}, \tilde{c}_{t+1})$ *be a program from* $(\tilde{X}, S) > 0$. *Then*

$$\lim_{t\to\infty} \mid c_t \mid = 0. \tag{60}$$

Proof We use (57) repeatedly. First, by the definition of a program in (58) we have

$$\tilde{c}_1 = \tilde{y}_1 - \tilde{x}_1 \text{ and } \Big((\tilde{X}, S), (\tilde{y}_1, s_1')\Big) \in \mathcal{T}. \tag{61}$$

Hence, using (57) and the price vector $p = (p_1, p_2) > 0$

$$\begin{aligned} p_1(\tilde{y}_1 - \tilde{X}) + p_2(s_1' - S) \leqq 0 \text{ or,} \\ p_1\tilde{y}_1 \leqq p_1\tilde{X} + p_2(S - s_1'). \end{aligned} \tag{62}$$

It follows from (61) and (62) that

$$p_1\tilde{c}_1 = p_1(\tilde{y}_1 - \tilde{x}_1) \leqq p_1\tilde{X} + p_2(S - s_1') - p_1\tilde{x}_1. \tag{63}$$

Similarly,

$$\tilde{c}_2 = \tilde{y}_2 - \tilde{x}_2, \text{ and } ((\tilde{x}_1, s_1), (\tilde{y}_2, s_2')) \in \mathcal{T} \text{ with } s_1 = s_1'. \tag{64}$$

Again, using (57),

$$\begin{aligned} p_1(\tilde{y}_2 - \tilde{x}_1) + p_2(s_2' - s_1') \leqq 0 \text{ or,} \\ p_1\tilde{y}_2 \leqq p_1\tilde{x}_1 + p_2(s_1' - s_2'). \end{aligned} \tag{65}$$

From (64) and (65),

$$p_1\tilde{c}_2 = p_1(\tilde{y}_2 - \tilde{x}_2) \leqq p_1\tilde{x}_1 + p_2(s_1' - s_2') - p_2\tilde{x}_2. \tag{66}$$

From (63) and (66) we get, after simplifying,

$$p_1\tilde{c}_1 + p_1\tilde{c}_2 \leqq p_1\tilde{X} + p_2(S - s_2') - p_1\tilde{x}_2.$$

Repeating the steps we get, for any finite T,

$$\sum_{t=1}^{T}(p_1\tilde{c}_t) \leqq p_1\tilde{X} + p_2(S - s_T') - p_2\tilde{x}_T \leqq p_1X + p_2S.$$

It follows that:

$$\sum_{t=1}^{\infty}(p_1\tilde{c}_t) = \lim_{T\to\infty}\left[\sum_{t=1}^{T}(p_1\tilde{c}_t)\right] \leqq p_1X + p_2S.$$

In particular,

$$\lim_{t\to\infty}(p_1\tilde{c}_t) = 0. \tag{67}$$

Let $\alpha = \min_k(p_{1k}) > 0$. Since $(p_1\tilde{c}_t) \geqq \alpha \mid \tilde{c}_t \mid \geqq 0$, (60) follows from (67). □

Remark 7.1 *We now turn to the "polyhedral" case. Replace T.4 by*

T.4′ $\mathcal{T}$ is a polyhedral convex cone.

It follows that (under T.4′ and T.5):

Z is a polyhedral (hence, closed) convex cone that does not contain any $u \geq 0$.

Now, from Theorem 2.2 we get that $-Z^*$ contains a positive $p > 0$. Hence, $-pz \geqq 0$ for all $z \in Z$, or $pz \leqq 0$ for all $z \in Z$. Using the same arguments as in the above proof, we see that for any program $(\mathbf{x}, \mathbf{y}, \mathbf{c}) = (x_t, y_{t+1}, \tilde{c}_{t+1})$ from $(\tilde{X}, S) > 0$, $\lim_{t\to\infty} \mid \tilde{c}_t \mid= 0$.

In a more restricted situation (a single resource that is essential in *all* processes), we analyzed the polyhedral case earlier (Majumdar and Bar [40]). The point that we should stress is this: under T.5, *any property* of $\mathcal{T}$ that ensures that Z *is a closed* convex cone enables us to invoke Theorem 2.5 and get the positive p satisfying $pz \leqq 0$ for all $z \in Z$. The *polyhedral* case T.4' merely identifies a class of prominent technologies.

8 Golden Rule, Overtaking and the Gale Turnpike

In his influential paper, Gale [21] began:

> There have been, as far as I am aware, two main approaches to the infinite time horizon problem. The first is to discount future consumptions by some arbitrary factor, and then try to maximize the resulting infinite series or integral. The second is the approach pioneered by Ramsey in Ramsey [54], reworked quite recently by Samuelson and Solow [58], Koopmans [35] and Weizsacker [65] among others. The present paper continues along the latter lines and extends some of the Ramsey results to general technologies of the type that are usually considered in modern multi-sector analysis.

The primary contribution was, of course, the existence of an optimal program according to variations of the overtaking criterion of Weizsacker [65]. The road was somewhat tortuous; however, the subsequent decisive contribution of Brock [5] provides an enjoyable ride. But Gale noted:

> To obtain the existence proof we have found it necessary to investigate first a number of subjects which have been central to much of the recent literature on dynamic production models including (i) optimal balanced growth programs ("golden rule" paths) (2) asymptotic properties of "good" programs ("turnpike theorems") (3) dynamic prices associated with optimal programs (inter-temporal optimality).

My plan is to present a multi-sector model with a single primary factor of production (labor), review the properties of the golden rule programs, and to establish a turnpike theorem for any program that is optimal according to the overtaking criterion.

In the concluding section, I shall take up Gale's "one good model of development" and briefly provide an exposition of the existence issue and results on "competitive" (dual) prices and optimal programs, and a catenary turnpike theorem.

In the model of this section, there are m producible goods that can be used as inputs or can be directly consumed. There is a primary factor of production which is not directly consumed, cannot be produced as outputs, but is an essential input in production.

A triplet $(L, x, y) \in R_+ \times R_+^m \times R_+^m$ represents an *activity or production process* describes the transformation of the vector (L, x) of inputs into the output vector y. The set of all activities, the technology, T is assumed to satisfy the following properties:

(*B.1*) $\mathcal{T}$ is a closed convex cone in $R_+ \times R_+^m \times R_+^m$ with $(0, 0, 0) \in \mathcal{T}$ (continuity, convexity and constant returns to scale; possibility of inaction).
(*B.2*) $(0, x, y) \in \mathcal{T}$ implies $y = 0$ (indispensability of labor as an input).
(*B.3*) If $(L, x, y) \in \mathcal{T}$ and if (L^1, x^1, y^1) satisfies $L^1 \geqq L$, $x^1 \geqq x$, $0 \leqq y^1 \leqq y$, then $(L^1, x^1, y^1) \in \mathcal{T}$ (free disposal).

To establish uniqueness, we need the following strict convexity properties:

(*B4.i*) If $(L, x^1, y^1) \in \mathcal{T}$ and $(L, x^2, y^2) \in \mathcal{T}$ satisfying $y^1 \neq y^2$, $L > 0$, then for any positive numbers $\beta_1 > 0$, $\beta_2 > 0$ with $\beta_1 + \beta_2 = 1$, there is an input vector x such that:

$$(L, x, \beta_1 y^1 + \beta_2 y^2) \in \mathcal{T}, x < \beta_1 x^1 + \beta_2 x^2.$$

(*B4.o*) If $(L, x^1, y^1) \in \mathcal{T}$ and $(L, x^2, y^2) \in \mathcal{T}$ satisfying $x^1 \neq x^2$, $L > 0$, then for any positive numbers $\alpha_1 > 0$, $\alpha_2 > 0$ with $\alpha_1 + \alpha_2 = 1$, there is an output vector y such that:

$$(L, \alpha_1 x^1 + \alpha_2 x^2, y) \in \mathcal{T}, \alpha_1 y^1 + \alpha_2 y^2 < y.$$

The growth rate of the labor force is an exogenously given constant. Formally:

(*L*) For all $t = 0, 1, 2..., L_t = L_0 \mu^t$ where $\mu > 0$. We set $L_0 = 1$.

Given $\hat{y} > 0$, the initial stocks of producible goods, a *(feasible) program* is an infinite sequence $(\mathbf{x}, \mathbf{y}, \mathbf{c}) = (L_t, x_t, y_t, c_t)$ satisfying:

$$y_0 = \hat{y},$$
$$y_t - x_t = c_t \geqq 0, t = 0, 1...$$
$$(L_t, x_t, y_{t+1}) \in \mathcal{T}, \ t = 0, 1...$$

We refer to **x** as the *input program* (sequence), **y** as the *output* program and **c** as the *consumption* program. It has been convenient to recast the model in per capita terms, using the cone property (see B1), feasibility conditions are stated as follows:

$$y_0 = y_0/L_0 = \hat{y},$$
$$\left(\frac{y_t}{\mu^t}\right) - \left(\frac{x_t}{\mu^t}\right) = \left(\frac{c_t}{\mu^t}\right) \geqq 0, \; t = 0, 1, \cdots$$
$$\left(1, \left(\frac{x_t}{\mu^t}\right), \left(\frac{\mu y_{t+1}}{\mu^{t+1}}\right)\right) \in \mathcal{T}, \; t = 0, 1, \cdots.$$

The utility function $u(\cdot) : R^m_+ \to R$ involved in defining a criterion to evaluate alternative programs is assumed to satisfy:

(*U1*) $u(\cdot)$ is continuous and strictly concave;
(*U2*) $v \geqq w$ implies $u(v) \geqq u(w)$, with strict inequality holding if $v \geq w$.

A feasible program $(\mathbf{x}, \mathbf{y}, \mathbf{c})$ from a vector $\hat{y} > 0$ of initial stocks is said to *overtake* another feasible program $(\mathbf{x}', \mathbf{y}', c')$ from (the same) $\hat{y}$ if there exists a positive integer τ such that

$$\sum_{t=0}^{T} \left[\frac{u(c_t)}{L_t}\right] \geqq \sum_{t=0}^{T} \left[\frac{u(c_t')}{L_t}\right] \quad \text{for all } T \geqq \tau.$$

A feasible program $(\mathbf{x}^*, \mathbf{y}^*, \mathbf{c}^*)$ from an initial $\hat{y} > 0$ is (Weizsacker) *optimal* if it overtakes all feasible programs from the same initial $\hat{y}$.

8.1 Golden Rule Equilibrium

We now turn to balanced growth programs in which x_t, y_t, c_t all have the same growth factor μ. Such a program is generated by a triplet of vectors x, y, c satisfying:

$$y - x = c \geqq 0, \tag{68}$$
$$(1, x, \mu y) \in \mathcal{T}. \tag{69}$$

A triplet (x, y, c) is trivial if $x = y = c = 0$. If μ is too large, there may be only a trivial triplet $x = y = c = 0$. Our interest is in non-trivial triplets . To this effect we make the following assumption on productivity of $\mathcal{T}$

(*P*) There is an activity $(1, x^{\smile}, y^{\smile}) \in \mathcal{T}$ such that $y^{\smile} - -x^{\smile} = c^{\smile} > 0$.

Based on (B1) and (B2), we can show that

$$\Gamma_+ \equiv \{y - x : y - x \geqq 0, (1, x, \mu y) \in \mathcal{T}\}$$

is a compact, convex set containing $c^{\smile} > 0$. Hence, $u(y - x)$ is maximized at some $y^g - x^g = c^g$ over Γ_+.

Lemma 8.1 *(i) A triplet (x, y, c) is either trivial or satisfies $x > 0,\ y > 0$.*
(ii) The triplet (x^g, y^g, c^g) maximizing $u(y–x)$ over Γ_+ is nontrivial and unique.

Proof (Sketch: see (Nikaido [49], Lemma 14.1) for details) (i) Since $y = x + c$, $x \geqq 0, c \geqq 0$, the supposition that $y = 0$ implies the triviality of (x, y, c). On the other hand, suppose that $y \geq 0$. Observe now that $(1, x, \mu y)$, $(1, 0, 0) \in \mathcal{T}$ and $\mu y \neq 0$. Therefore, using (B4i) with $\beta_1 = \beta_2 = 1/2$, there is an input vector z such that

$$\left(1, z, \frac{\mu y}{2}\right) \in \mathcal{T}, \frac{x}{2} > z. \tag{70}$$

Since $z \geqq 0$, we must have $x > 2z \geqq 0$. We also have $y = x + c \geqq x > 0$.

(ii) If $c^g = 0$, $u(y^g - x^g) < u(y^{\smile} - x^{\smile})$, a contradiction to the property that $(y^g - x^g)$ maximizes $u(y - x)$ over Γ_+. Hence, (x^g, y^g, c^g) is nontrivial. For uniqueness use (B4.i) and strict concavity of u. □

The triplet (x^g, y^g, c^g) generates the *golden rule program* $(\mu^t x^g, \mu^t y^g, \mu^t c^g)$ consistent with the exogenous growth factor of labor. There is a price system $p > 0$ "supporting" the golden rule triplet.

Theorem 8.1 *Let (x^g, y^g, c^g) be the golden rule triplet. There is a price vector $p^g > 0$ fulfilling the following three conditions;*

(α) $u(c) - u(c^g) \leqq p^g(c - c^g)$ for all $c \geqq 0$.

(β) $p^g(y - x) \leqq p^g(y^g - x^g)$ for all processes $(1, x, \mu y) \in \mathcal{T}$, with the strict ineqality.

(γ) $p^g(y - x) < p^g(y^g - x^g)$ holding if $(x, y) \neq (x^g, y^g)$.

Proof (sketch from (Nikaido [49], Lemma 14.2). The proof of (Peleg [51], Lemma 3.6) can be adapted). Let

$$\Gamma = \{y - x : (1, x, \mu y) \in \mathcal{T}\ ).$$

Clearly $\Gamma \supset \Gamma_+$. Define

$$Q = \{c : c \geqq 0, u(c) \geqq u(c^g)).$$

Then $M = \Gamma$ - Q is a convex set containing no positive vector. By Theorem 2.3 there is $p \geq 0$ such that $pz \leqq 0$ for all $z \in M$. Hence,

$$p(y - x) \leqq pc \text{ for } (y - x) \in \Gamma_+, c \in Q. \tag{71}$$

From (71) since $c^g (= y^g – x^g) \in Q$, we get (β). The inequality condition (γ) follows from (B4.i) and (B4.o).

It remains to prove (α). To this end , it is shown that for the above p we can find a positive scalar $\theta > 0$ such that

(α') $u(c)–u(c^g) \leqq \theta p(c - c^g)$ for all $c \geqq 0$.

With such a scalar, we can satisfy (α), (β) and (γ) with p replaced by $\theta p = p^g$.

Now, consider the constrained optimization problem:

$$\begin{aligned} \max \quad & u(c) \text{ on } R^m_+ \\ \text{subject to} \quad & g(c) = pc^g - pc \geqq 0. \end{aligned}$$

It is shown that c^g is a solution to this problem. Hence, by Theorem 2.5, there is a Lagrangian multiplier θ such that:

$$\theta g(c^g) = 0$$
$$u(c^g) + \theta g(c^g) \geqq u(c) + \theta g(c), c \in R^m_=$$

It is easy to see that $\theta > 0$ and $p > 0$. □

We refer to $((x^g, y^g, c^g), p)$ satisfying (α), (β), (γ) as a *golden rule equilibrium.* Next comes a value loss lemma.

Lemma 8.2 *Let* $((x^g, y^g, c^g), p)$ *be a golden rule equilibrium. For every* $\varepsilon > 0$, *there is a* $\delta > 0$ *such that :*

$$\max (\| x - x^g \|, \| y - y^g \|) \geqq \varepsilon, \max (1, x, \mu y) \in T$$

implies

$$p(y - x) \leqq p(y^g - x^g) - \delta.$$

The proof is analogous to the Radner lemma, so we move on. Henceforth, we assume that :

(G) The golden rule equilibrium satisfies $c^g > 0$.

The next result can be called an "accessibility" lemma: it says that the golden rule can be reached from any initial $\hat{y} > 0$ in a finite number of periods. More formally:

Lemma 8.3 *For any* $\hat{y} > 0$ *there is a (finite) positive integer* N_0 *such that there is a feasible program* $(\mathbf{x}^\sim, \mathbf{y}^\sim, \mathbf{c}^\sim) = (x_t^\sim, y_t^\sim, c_t^\sim)$ *with an initial* $y_0^\sim = \hat{y}$ *which coincides with the golden rule program* $(\mu^t x^g, \mu^t y^g, \mu^t c^g)_0^\infty$ *for all* $t \geqq N_0$.

The proof (Nikaido [49], Lemma 14.4) is long and involves some delicate calculations. Note that this program $(x_t^\sim, y_t^\sim, c_t^\sim)$ is *good* according to Gale's classification and his proof of the existence of a good program is also based on a subtle calculation.

8.2 Gale's Turnpike

We come to a turnpike property of an optimal program in our model.

Theorem 8.2 *Let $\hat{y} > 0$ be given, and $(\mathbf{x}^*, \mathbf{y}^*, \mathbf{c}^*)$ be the optimal program from $\hat{y}$. Then for any $\varepsilon > 0$, there is a positive integer S such that:*

$$\max\left(\left\|\left(\frac{x_t^*}{\mu^t}\right) - x^g\right\|, \left\|\left(\frac{y_{t+1}^*}{\mu^{t+1}}\right) - y^g\right\|\right) \geqq \varepsilon$$

holds except possibly for S periods.

Proof The proof rests on a comparison of the optimal $(\mathbf{x}^*, \mathbf{y}^*, \mathbf{c}^*)$ with with program $(\mathbf{x}^\sim, \mathbf{y}^\sim, \mathbf{c}^\sim)$ in Lemma 8.3. From the definition of overtaking, there is some τ (which can be chosen to be greater than N_0) such that for *all* $N \geqq \tau$

$$\sum_{t=0}^{N}(u(c_t^*)/\mu^t) \geqq \sum_{t=0}^{N}(u(c_t^\sim)/\mu^t)$$

or

$$\sum_{t=0}^{N}[(u(c_t^*)/\mu^t) - u(c^g)] \geqq \sum_{t=0}^{N}[(u(c_t^\sim)/\mu^t) - u(c^g)]. \tag{72}$$

To simplify typing, we write $u_t^* = (u(c_t^*)/\mu^t)$ and $u_t^\sim = (u(c_t^\sim)/\mu^t)$. We rewrite (72) as:

$$\sum_{t=0}^{N}(u_t^* - u(c^g)) \geqq \sum_{t=0}^{N_0-1}[(u_t^\sim - u(c^g)] \tag{73}$$

(recall that $u_t^\sim = u(c^g)$ for $N_0 \leqq t \leqq N$.). Denote $\sum_{t=0}^{N_0-1}[(u_t^\sim - u(c^g)]$ by Θ. By feasibility of the optimal $(\mathbf{x}^*, \mathbf{y}^*, \mathbf{c}^*)$, we get:

$$(1, x_t/\mu^t, \mu y_{t+1}/\mu^t) \in \mathcal{T}, \quad t = 0, 1, ...N-1.$$

Given $\varepsilon > 0$, choose the $\delta > 0$ the existence of which is the claim of the Lemma 8.2,

and define the set of integers:

$$\mathcal{L} = \{t : \max(\ \|(x_t^*/\mu^t) - x^g\|, \|(y_{t+1}^*/\mu^{t+1}) - y^g\|) \geqq \varepsilon, N \geqq t \geqq 0.\}.$$

The number of elements in $\mathcal{L}$ is denoted by $\#(\mathcal{L})$.

$$p^g\left[(-x_t/\mu^t) + (y_{t+1}/\mu^{t+1}) - (y^g - x^g)\right] \leqq 0 \text{ for all } t = 0, 1, ...N-1,$$
$$\leqq -\delta \text{ for } t \in \mathcal{L}. \tag{74}$$

From Theorem 8.1 and (74) we get, for all $N \geqq \tau$,

$$
\begin{aligned}
\Theta \equiv \sum_{t=0}^{N_0-1} [(u_t^{\sim} - u(e^g)] &\leqq \sum_{t=0}^{N} [(u(c_t^*)/\mu^t) - u(c^g)] \\
&\leqq \sum_{t=0}^{N} p^g [(c_t^*/\mu^t) = e^g] \\
&= \sum_{t=0}^{N} p^g [(y_t^*/\mu^t) = (x_t^*/\mu^t) - (y^g - x^g)] \\
&= \sum_{t=0}^{N-1} p^g [-(x_t^*/\mu^t) + (y_{t+1}^*/\mu^{t+1}) - (y^g - x^g)] + p^g\, \hat{y} - p^g (x_N^*/\mu^N) - p^g (y^g - x^g) \\
&\leqq -\delta[\#(\mathcal{L})] + p^g\, \hat{y}
\end{aligned}
$$

and the conclusion follows easily. □

Remark 8.1 In a typical one good model the results outlined above can be derived in a relatively simple manner (see Majumdar and Mitra [41] for a comprehensive account of the discrete time infinite horizon optimization models (both the discounted and undiscounted cases)). In a typical finite horizon (over a finite N periods) aggregative model, a planner 's optimization problem is described as follows.

Given initial stock $a > 0$ and an attainable target stock $z > 0$,

$$
\begin{aligned}
\max \quad & \sum_{t=0}^{N} u(c_t) \\
subjectto\ c_0 + x_0 &= a, \\
c_t + x_t &= f(x_{t-1}), t = 1, ...N - 1 \\
c_N + z &= f(x_{N-1}).
\end{aligned}
$$

Here f is the gross output function satisfying the appropriate differentiability, strict concavity and endpoint (Uzawa-Inada) condition on its derivative for Theorem 8.2 can be established in a less complicated manner to establish that for any $\varepsilon > 0$, there is a a positive integer S such that the number of periods for which $| x_t^* - x^g |> \varepsilon$ cannot exceed S (here x_t^* is the optimal input in period t and x^g is the golden rule input satisfying $f'(x^g) = 1$). Of course, the result has a "bite" if N is large. This is at the heart of the "catenary" turnpike theorem studied in Gale [23].

9 Linear Economic Models

LEM captures the readers with a conscious attempt to choose a "clean-looking page of symbols" that "will have a good psychological effect on the reader: ..a tangled symbolism suggests a tangled argument and is likely to frighten rather than entice.".

(*LEM*, Preface, p.xvii), but it is "a text not in economics, but in applied mathematics. Nevertheless, the theorems we prove are about economics, and are used by economists, and in many cases were first discovered by economists.". In terms of mathematical prerequisites, he stressed that "calculus" is never used. Our principal tool is matrix algebra, but no previous knowledge is required here as all necessary facts are developed in the text. What is required is the ability to follow a moderately involved mathematical argument, an ability which generally comes with a fair amount of experience and is often characterized by the illusive phrase "mathematical maturity". Every time I look at such sentences, I cannot but think–with a smile-about (Debreu [9], Chap. 1): "This chapter presents *all* the mathematical concepts and results which will be used later. .. Its reading requires, in principle, *no* knowledge of mathematics.". Chapter 2 (Real Linear Algebra) contained all the results frequently used. Interestingly, there are no explorations of determinants or characteristic roots and the book has scant references to basic topological concepts He managed to keep the presentation "self-contained" by invoking appropriate results from Chapter 2 to complete long proofs based on ingenious manipulations. However, on page 312, he had to acknowledge that "a standard "compactness" argument is needed" to complete a rigorous proof. And on page 317 he tells the reader; "Thus we are proving the classical result that a nonnegative matrix always has a non-negative eigen value and eigen vector".

Much that has been said about "first love" and "true love" holds for the books we read. In my case, Samuelson's *Foundations* (Samuelson [56]), Gale's *Linear Economic Models* and Debreu's *Theory of Value* were all loves at first sight: some imperfections I did note in my pensive mood as I grew older. But true love and first love never die.

Acknowledgements I wish to thank Professor Ali Khan for his guidance in completing this paper. For suggestions on style, organization and substance, I am very much indebted to Professors Arnab Basu,Talia Bar, Nancy Chau, Erwin Diewert, Ram Sewak Dubey and Yaw Nyarko.

References

1. Aumann, R., Peleg, B.: A note on Gale's example. J. Math. Econ. **1**(1), 209–211 (1974)
2. Bala, V., Majumdar, M.: Chaotic tatonnement. Econ. Theor. **2**, 437–446 (1992)
3. Beckman, M.J.: Comparative statics in linear programming. Rev. Econ. Stud. **23**, 232–235 (1955)
4. Bell, C.: Development economics. In: Majumdar, M., Willis, I., Sgro, M.P., Gowdy, J.M. (eds.) Fundamental Economics, vol. 2, pp. 1–35. EOLSS Publishing, Oxford (2011)
5. Brock, W.A.: On the existence of weakly maximal programmes in a multi-sector economy. Rev. Econ. Stud. **37**(110), 275–280 (1970)
6. Casti, J.L.: Five Golden Rules: Great Thories of 20th Century and Why They Matter. Wiley, New York (1996)
7. Dantzig, G.B.: Linear Programming and Extensions. Rand and Princeton University Press, Princeton (1963)

8. Dantzig, G.B.: Reminiscences about the origins of linear programming. Oper. Res. Lett. **1**, 43–48 (1982)
9. Debreu, G.: Theory of Value. Wiley, New York (1959)
10. Debreu, G.: Excess demand functions. J. Math. Econ. **1**(1), 15–21 (1974)
11. Debreu, G.: Existence of competitive equilibrium. In: Arrow, K.J., Intriligator, M. (eds.) Handbook of Mathematical Economics, vol. 2. Noth Holland, Amsterdam (1982)
12. Debreu, G.: Theoretic models: mathematical form and economic content. Econometrica **54**(6), 1259–1270 (1986)
13. Diewert, W.E.: Applications of duality theory. In: Intriligator, M.D., Kendrick, D. (eds.) Frontiers of Qualitative Economics, pp. 106–176. Noth Holland, Amsterdam (1974)
14. Dorfman, R., Samuelson, P.A., Solow, R.M.: Linear Programming and Economic Analysis. McGraw-Hill, New York (1958)
15. Du, D., Pardalos, P. (eds.): Minimax and Applications. Kluwer Academic Publishers, Dordrech (1995)
16. Fuss, M., McFadden, D. (eds.): Production EconomicsâĿ¯: A Dual Approach to Theory and Applications, vol. 1. Noth Holland, Amsterdam (1978)
17. Gale, D.: The law of supply and demand. Math. Scand. **3**, 155–169 (1955)
18. Gale, D.: The closed linear model of production. In: Kuhn, H.W., Tucker, A.W. (eds.) Linear Inequalities and Related Systems, pp. 265–304. Princeton University Press, Princeton (1956)
19. Gale, D.: The Theory of Linear Economic Models. McGraw Hill, New York (1960)
20. Gale, D.: A note on global instability of competitive equilibrium. Naval Research Logistics Quarterly **10**(1), 81–87 (1963)
21. Gale, D.: On optimal development in a multi-sector economy. Rev. Econ. Stud. **34**(1), 1–18 (1967)
22. Gale, D.: A geometric duality theorem with economic applications. Rev. Econ. Stud. **34**, 19–24 (1967)
23. Gale, D.: Nonlinear duality and qualitative properties of optimal growth. In: Abadie, J. (ed.) Integer and Nonlinear Programming, pp. 309–319. Noth Holland, Amsterdam (1970)
24. Gale, D.: Pure exchange equilibrium of dynamic economic models. J. Econ. Theory **6**(1), 12–36 (1973)
25. Gale, D.: Exchange equilibrium and coalitions: an example. J. Math. Econ. **1**(1), 63–66 (1974)
26. Gale, D.: The game of hex and the Brouwer fixed-point theorem. Am. Math. Mon. **86**(10), 818–827 (1979)
27. Gale, D., Mas-Colell, A.: An equilibrium existence theorem for a general model without ordered preferences. J. Math. Econ. **2**(1), 9–15 (1975)
28. Gale, D., Mas-Colell, A.: Corrections to an equilibrium existence theorem for a general model without ordered preferences. J. Math. Econ. **6**, 297–298 (1979)
29. Gale, D., Nikaido, H.: The Jacobian matrix and global univalence of mappings. Math. Ann. **159**, 81–93 (1965)
30. Howe, C.W.: An alternative proof of the existence of genaral equilibriumin a von Neumann model. Econometrica **28**, 635–659 (1960)
31. Kakutani, S.: A generalization of Brouwer's fixed point theorem. Duke Math. J. **8**, 457–459 (1941)
32. Karlin, S.: Mathematical Methods and Theory in Ganmes, Programming and Economics, vol. 1. Addison-Wesley, Reading, MA (1959)
33. Kemeny, J.G., Moregenstern, O., Thompson, G.L.: A generalization of the von Neumann model of an expanding economy. Econometrica **24**, 115–135 (1958)
34. Koopmans, T.C.: Economic growth at a maximal rate. Q. J. Econ. **78**, 355–394 (1964)
35. Koopmans, T.C.: On the Concept of Optimal Growth, pp. 228–288. Pontificia Academia Scientiarum, Vatican City (1965)
36. Kuhn, H.W.: A Note on the law of supply and demand. Math. Scand. **4**, 143–146 (1956)
37. Kuhn, H.W., Gale, D.: The Berkeley memorial. In: Games and Economic Behavior, vol. 66 (2009)

38. Majumdar, M.: Equilibrium and optimality: some imprints of David Gale. Games Econom. Behav. **66**, 607–626 (2009)
39. Majumdar, M.: Equilibrium, Welfare and Uncertainty: Beyond Arrow-Debreu. Routledge, London (2009)
40. Majumdar, M., Bar, T.: Some theoretical issues in sustainable development: an exposition. J. Econ. Theory Soc. Dev. **1**, 37–56 (2013)
41. Majumdar, M., Mitra, T.: Intertemporal decentralization. Finn. Econ. Pap. **6**, 79–101 (1991)
42. Majumdar, M., Mitra, T., McFadden, D.: On efficiency and Pareto optimality of competitive programs in closed multi-sector models. J. Econ. Theory **13**, 26–46 (1976)
43. Malinvaud, E.: Efficient capital accumulation: a corrigendum. Econometrica **30**(3), 570–573 (1962)
44. McFadden, D.: Cost, revenue and profit functions. In: Fuss, M., McFadden, D. (eds.) Production Economics: A Dual Approach to Theory and Applications, vol. 1, pp. 3–110. Noth Holland, Amsterdam (1978)
45. Moore, J.: Genral Equilibrium and Welfare Economics. Springer, Berlin (2007)
46. Morishima, M.: Equilibrium, Stability and Growth. Oxford University Press, London (1964)
47. Mukherji, A.: An implication of the existence of competitive equilibrium. Contemp. Issues Ideas Soc. Sci. **1**, 1–10 (2005)
48. Nikaido, H.: Persistence of continual growth near the von Neumann ray: a strong version of the Radner turnpike theorem. Econometrica **32**, 151–162 (1964)
49. Nikaido, H.: Convex Structures and Economic Theory. Academic Press, New York (1968)
50. Parthasarathy, T.: On Global Univalence Theorems. Springer Verlag, Berlin (1983)
51. Peleg, B.: On competitive prices for optimal consumption plans. J. SIAM Appl. Math. **26**, 239–254 (1974)
52. Radner, R.: Paths of economic growth that are optimal with regard only to the final states: a turnpike theorem. Rev. Econ. Stud. **28**, 98–104 (1961)
53. Radner, R.: Efficiency prices for infinte horizon production programmes. Rev. Econ. Stud. **34**, 51–67 (1967)
54. Ramsey, F.P.: A mathematical theory of saving. Econ. J. **38**, 543–59 (1928)
55. Samuelson, P.A.: Comparative statics and the logic of economic maximizing. Rev. Econ. Stud. **14**, 41–43 (1946)
56. Samuelson, P.A.: Foundations of Economic Analysis, Harvard University Press, Cambridge (1965). (with a New Introduction, Atheneum, New York; (1983), Enlarged Edition, Harvard University Press, Cambridge, 1947)
57. Samuelson, P.A.: Market mechanism and maximization, part iii: dynamics and linear programming, publisher The Rand Corporation, reprinted. In: Stiglitz, J. (ed.) Collected Scientific Papers of Paul A. Samuelson, vol. 1, pp. 607–626. Mit Press, Cambridge (1949)
58. Samuelson, P.A., Solow, R.: A complete capital model involving heterogeneous capital goods. Q. J. Econ. **38**, 537–562 (1956)
59. Scarf, H.: Some examples of global instability of the competitive equilibrium. Int. Econ. Rev. **1**, 157–172 (1960)
60. Simons, S.: Minimax theorems and their proofs. In: Du, D., Pardalos, P. (eds.) Minimax and Applications, pp. 1–24. Kluwer Academic Publishers, Dordrech (1995)
61. Sonnenschein, H.F.: Market excess demand functions. Econometrica **40**, 549–563 (1972)
62. Uzawa, H.: The Kuhn-Tucker theorem in concave programming. In: Arrow, K.J., Hurwicz, L., Uzawa, H. (eds.) Studies in Linear and Non-linear Programming, vol. 2. Stanford University Press, Stanford (1958)
63. Uzawa, H.: Walras' existence theorem and Brouwer's fixed point theory. Econ. Stud. Q. **8**, 59–62 (1962)
64. von Neumann, J.: A model of general economic equilibrium. Rev. Econ. Stud. **13**, 1–9 (1945)
65. von Weizsacker, C.C.: Existence of optimal programs of accumulation for an infinte time horizon. Rev. Econ. Stud. **32**, 85–104 (1965)

66. Wald, A.: On some systems of equations of mathematical economics. Econometrica **19**(4), 368–403 (1936–1951). (English translation (published in 1951) of Über einige Gleichungssysteme der mathematischen Ökonomie *published in* Zeitschrift für Nationalökonomie, vol. 7, pp. 637–670, 1936)
67. Winter, S.G.: The norm of a closed technology and the straight-down-the-Turnpike theorem. Rev. Econ. Stud. **34**, 67–89 (1967)

Stochastic Dynamical Framework for Risk Managements of SMEs' Growth: Model, Approach and Quantitative Analysis

Huiqi Wang, Lin Meng, and George Xianzhi Yuan

Abstract This study aims to establish an interpretable prototype system by describing the development and growth mechanisms of small and micro enterprises (SMEs) under a tech-economic paradigm by applying the stochastic dynamical system approach with the associated stochastic resonance theory. It will also develop an associated quantitative method of risk management and growth optimization under the measurement of systematic response risk, bilateral cooperation risk, and value at response risks (VaRR) performance. From the perspective of bistable stochastic resonance (SR), as the inherent regularity of capital-product switches is applied together with both capital-related driving and input risk stimulating, the system is driven by the combination of the forces that compete and cooperate to produce an optimal control on response risk compression or VaRR improvement. Furthermore, based on the real-world supporting profiles from the manufacturing sector in China, we investigate the validity by developing different strategies from micro profile adjustment, mesoscopic structure control, and macro climate regulation. It is found that any strategies are not always effective, and some non-monotonic phenomena are even against the intuitions, which are results of stochastic resonance features caused by the stochastic dynamical system. These findings suggest that, with the integrative consideration of specialization level, management orientation, scale preference, and strategic partner, they can play a more critical role, which provides the crucial quantitative support, and

H. Wang (✉) · L. Meng
College of Mathematics and Statistics, Chongqing University, Chongqing, China
e-mail: wanghuiqi@cqu.edu.cn

L. Meng
e-mail: menglin0528@163.com

G. X. Yuan (✉)
School of Economics and Finance, Chongqing University of Technology, Chongqing, China
e-mail: george_yuan99@yahoo.com

Business School, Chengdu University, Chengdu, China

Business School, Sun Yat-Sen University, Guangzhou, China

Business School, East China University of Science and Technology, Shanghai, China

Shanghai Hammer Digital Technology Co., Ltd., Shanghai, China

M. A. Khan et al. (eds.), *Matching, Dynamics and Games for the Allocation of Resources*, Monographs in Mathematical Economics 7,
https://doi.org/10.1007/978-981-96-6661-4_16

further leads us to create useful decision rules in risk management practices. Finally, we hope the framework established in this study may provide some mathematical fundamentals to explain SME's growth with innovation for Schumpeter's economic development theory introduced around 1910's.

Keywords SME's growth · Risk management · Stochastic dynamical framework · Bilateral cooperation risk · Systematic response risk · Value at response risks (VaRR)

1 Introduction

Small and micro enterprises (SMEs) have been essential in fostering economic growth, promoting employment growth, and stimulating innovation vitality [1, 2]. Generally, they are privately owned corporations, partnerships, or sole proprietorships with fewer employees or less annual revenue than regular-sized businesses or corporations. According to the information provided by the People's Bank of China, by the end of 2017, there were about 28 million legal persons and more than 65 million individual industrial and commercial households in China. They have become the vital carriers of mass entrepreneurship and innovation. Various programs for stimulating and extending loans to SMEs have been developed in the past several years to construct inclusive financial services [3]. These programs strengthen the support of multi-level capital markets, broaden orderly financing channels for SMEs, improve the environment of SME finance [4], and maintain financial stability [5].

However, numerous difficulties and challenges still exist on the road to development. With changes in policy and financial settings, more commercial banks are becoming actively interested in exploring new ways to meet increasing demands for credit; banks have introduced dozens of new loan products designed for SMEs. For example, the innovations in loan products and the changes in commercial banking in China have alleviated the problem of access to loans for SMEs [6]. Consequently, the availability of external finance for SMEs has become a topic of significant research interest and an issue of great importance to policymakers. Meanwhile, previous research has suggested various statistical approaches to score the SME's credit. These primarily include relationship-based and transactions-based lending technologies [3, 6–8], which involve soft quantitative and hard qualitative information. Much of the existing literature on this topic notes that these technologies have been rising in popularity among Chinese banks since the early 21st century and have proven to be relatively effective, to a certain extent, in reducing information asymmetries and improving the accuracy of risk-based pricing of the loans [6]. But the latest data showed that the average life expectancy of SMEs is about three years, and only about one-third are in regular operation after 3 years.

Moreover, they have obtained loans for the first time after four years and four months on average since their establishment. Therefore, they must get through the average three-year death period to successfully get support from financial institutes.

The main reason is that SMEs are typically much more information-opaque than large enterprises [7, 9]. They often need certified audited financial statements to yield credible financial information regularly. These result in many difficulties for the institutes to employ standard financial statement criteria in the loan decision process, especially for SMEs at start-up and growth stages [10, 11]. Thus, a thorough examination of the state and development potential requires ample time and the awareness of factors such as expected future loan performance from various possible information dimensions, such as enterprise characteristics, financial data, business behavior, credit/loan record, external evaluation, industry, regional policy, and relationships with outsiders. Large banks have performed some competitive advantages in the channels of information acquisition.

Nevertheless, recent data shows that the non-performing loans (NPLs) ratio of sector-wide SMEs still reaches 2.75%, which is 1.7% higher than that of regular-sized enterprises. Therefore, we are inspired to think of some details about inherent action mechanisms (which must be addressed or reflected fully in many assessment approaches) among different information dimensions. The purpose is to establish a dynamical prototype system, incorporating these dimensions, to measure capital-related risk profiles, which can be regarded as the approach for SME state assessment to support "quick loan" practices.

On the other hand, since Schumpeter [12] introduced his economic development theory based on the growth due to innovation around 1910s, a lot of scholars have tried many ways to construct the mathematical framework in supporting the development of economic theory due to innovation. For example, from the micro perspective of intra-enterprise, SMEs can be characterized by the partnership systems [13–15], typically including technology-side and capital-side. They join efforts, including their time, technologies, resources, etc., to achieve goals that each side, acting alone, could not attain easily [16, 17]. The system has two essential aspects, competition, and cooperation; it evolves randomly within the utility equilibria of each side. Specifically, in the process of capital to product, the enterprise state easily traps into the technology equilibrium (maximum utility of technology-side), which generally deviates from the marketing-side utility. On the other hand, in product-to-capital processes, enterprises must adjust and transfer new resources to inspire the marketing side to make a further effort, which is achievable through the input of more external capital from financial institutes or strategic partners. Accordingly, it involves cooperation from capital equilibrium (maximum utility of marketing side). Actually, the process contains many intrinsic periodic structures, leading to varied research and development (R&D) and market transformation cycles. Simultaneously, motivated by external capital, the rhythm synchronism between the internal and external environments can affect or magnify the efficiency of capital-product switches [18, 19]. Further, we actually could interpret the process from the perspective of tech-economic paradigms (see [20–26], too) by following the key idea in two aspects: first, technology innovation produces technology equilibrium and the related operation laws and associated ecosystem, which mismatch with the old paradigm; and second, how to describe the stochastic effect by the so-called stochastic (random) forces? Indeed, the contradictions accumulate to a certain degree, and the enterprise, as an evolution system, collapses in the form of recession or stress until it overcomes the barrier with the help of capital driving or stimulus, and pressure is gradually released by the establishment of the new paradigm. Thereby, a capital equilibrium forms; meanwhile, the enterprise begins to gather its resources for a new round of capital-product switches. If this process stalls in any equilibrium with higher probability, the enterprise has insufficient ability and power for further development, usually resulting from the unbalanced or inadequate distribution of internal resources. Inevitably, it is difficult for the other side to reach its maximum utility during an extended period, which leads to a high rupture risk of bilateral cooperation. For example, excessive exposure to bilateral cooperation risk can be effectively controlled by introducing more external capital from strategic partners or decreasing the bilateral utility barrier.

Conversely, suppose the enterprise performs sufficient power or over-capitalization, although the bilateral risk remains low. In that case, any random fluctuations from the uncertainty of the internal or external environment will again decrease the switching regularity of state evolution. The enterprise is then subject to a high systematic response risk. A healthy state of an enterprise's response to risk involves the following factors: internal elements allocation, external survival environment, and strategic capital from the mesoscopic structure coordinate, which enable its sustainable and regular capital-product switches. Examining these factors facilitates the evaluation of SMEs' development and growth (see [23, 26] and references wherein) to identify the possibility for SMEs as stochastic dynamical systems to appear the stochastic resonance phenomena.

Putting all discussion above together, it seems that using the Langevin (dynamic) equation to describe the stochastic dynamical behaviors for the tech-economic paradigms is a proper way to build the possible mathematical structures for Schumpeter's economic development theory based on the growth due to the innovation around 1910s under the framework of stochastic dynamical system approach, as supported by a series of studies given in [13, 15, 21, 23, 25, 26]) for which the stochastic dynamical system is associated with stochastic resonance theory developed around 1980's as a powerful tool (see [27, 28] and reference wherein), because it allows us to describe the stochastic effect of the so-called stochastic (random) forces by the phenomena of stochastic resonance caused by the stochastic dynamical system associated with uncertainty environment for the process of the growth during SMEs' evolution and development.

That is, regarding the focused entity as a Brownian particle, the fluctuation-dissipation effects stimulate the state as follows: (i) potential effect resulting from state mismatch with an old paradigm or original operation laws, which produces a dragging force to local equilibria; (ii) inherent capital-product switching effect, achieved by the internal and external capital; (iii) stochastic effect from the uncertainty of the internal or external environment, synthetically regarded as an input risk; (iv) damping effect that acts on the interface between the entity and environment, performing obstinate rejection to any change of a development state.

Furthermore, a stochastic differential equation can describe the state evolution over time as a generic system model that combines all the evolving effects. Therefore, we develop a quantitative method for risk management and performance optimization under the measurements of system response risks and the value at response risks (VaRR) performance. We hope the framework established in this paper may provide some initial fundamental mathematical structures to explain Schumpeter's economic development theory based on SMEs' growth with innovation, for which the mathematical framework is not there since 1910s [12].

2 The Evolution Mechanism of SME's Growth

From the perspective of tech-economic paradigms, we start considering an SME characterized by the partnership system, including the technology-side (denoted as S_T, who performs the ability of transforming capital into product) and capital-side (denoted as S_C, who performs the ability of transforming product into capital). The enterprise raises the internal initial capital and recruits external strategic partners to pursue the development of a new product, which consists of a business opportunity during the capital-to-product period, when S_T plays a pivotal role in doing the most R&D work. More capital is adjusted or involved to break the routine trajectories and make radical changes, as it faces the constriction of operation laws under the original internal allocation. Despite the high risk, this profitability can be translated into the final return if the project thrives during the product-to-capital period, when S_C plays a leading role. With the increase in capital return, technological advantage gradually disappears due to the competition. In the established paradigm, the marginal return with the technology advantage-related low-risk begins to diminish. Further, the returned capital is considered profitable uses and new technology ventures. Thus, the exhaustion of a paradigm brings with it both the need for radical entrepreneurship and the idle capital to take the high risks of trial and error. Under these conditions, several strands of acting effects associated with the many underutilized or marginalized innovations drive the state to continually switch between developing innovative technology for new products and pursuing capital return to conceive the next capital-product evolution.

The following factors are primary theoretical elements for us to establish the framework of SME's growth: their establishment with registered capital C_R, their real operating capital, originating from the internal paid-in capital C_P and external capital C_E, and the scale m comprising development elements (i.e., staffing, facilities, material resources). It reflects all the scale-related capital inputs C_S, allowing for careful allocation and adjustment between both sides to stimulate their efforts at different periods. In the following description, C_R is used to be the normalization factor, that is, the enterprise raises internal capital $c_0 = C_P/C_R$ and recruits external capital $\Delta_c = C_E/C_R$ to achieve the capital-product switches. Inevitably, the process requires the flows of efforts (a, b) from S_T and S_C, respectively. The effort each side achieves is due to the bilateral inputs under capital-constraint condition; thus, the state variable $x = a - b$ reflects the bilateral competition and cooperation level and also forms the driving force of the state evolution.

Based on the initial inputs between S_T and S_C, it could stimulate both sides to be willing to make efforts (a_0, b_0), which corresponds to the initial state $x_0 = a_0 - b_0$. In Fig. 1, the dynamical evolution of state x is subject to the utility potential with lower convex form $U_T(x)$ (blue line) or $U_C(x)$ (red line) that play the role of inherent regulation on bilateral competition and cooperation relationship within the SME. Any deviation from the maximum utility of one side could result in the internal force from the side driving the state back to its local equilibrium atx_T or x_C. Specifically, in the process of capital-to-product process (i.e., the start-up and product improvement stages), in order to stimulate S_T to make more effort to play a dominative role, the essential elements should be adjusted to firstly ensure the R&D requirements, which leads to increasing a and decreasing b. Thus the state x increases until it traps into the maximum utility U_T^{opt} of $U_T(x)$, and evolves around $x = x_T$, at which U_T^{opt} might have seriously deviated from the maximum utility U_C^{opt} of $U_C(x)$ at $x = x_C$. Fortunately, in the process of product to capital (i.e., the market expansion and cash-out stages), the enterprise must cut down on technology investments to facilitate marketing expenditures; thus, the effort of S_C overweighs that of S_T, and capital allocation fluctuates due to market expanding demands. This interaction makes the state x subject to $U_C(x)$, and x decreases until it resides into the maximum utility U_C^{opt} of $U_C(x)$, evolving around $x = x_C$.

Regarding the focused SME as a Brownian particle, in Fig. 1a, we reversely plot the bilateral utility curves to facilitate the description from Langevin dynamics [29, 30]. Based on the above description, accompanying the bilateral cooperative process within competitive capital constraint, the state variable x can be interpreted as a dynamical evolution subject to an aligned utility potential $U(x)$, as shown in Fig. 1d, which has a potential barrier at $x = x_b$ and two wells at $x = x_T$ and x_C, respectively. Here $U(x)$ is considered as an symmetric bistable structure, which has been widely observed in physical, biological, and financial systems [31], and here it is typically expressed as

$$U(x) = -\nu_0 \left(\frac{1}{2}x^2 - \frac{1}{4}x^4 \right), \tag{1}$$

which shows the barrier with height $\Delta U = \frac{1}{4}\nu_0$ located at $x_b = 0$ and two symmetric wells respectively located at $x_T = -1$ and $x_C = 1$ (corresponding to two local equilibria of the maximum utilities of S_T's and S_C's).

The bistable utility potential represents the powerful and inherent guiding rule and the prevailing paradigms used by long-standing SMEs. The symmetric assumption is based on the consideration of capital-product switches, i.e., capital-to-product and product-to-capital, are equally important and inevitable for SME's survival and development, and it is a poor status for SMEs only or more easily trapping in any one equilibrium. The non-equivalent capital-product switches can be further described by the extended asymmetric bistable structures but are not included in this study [13, 32].

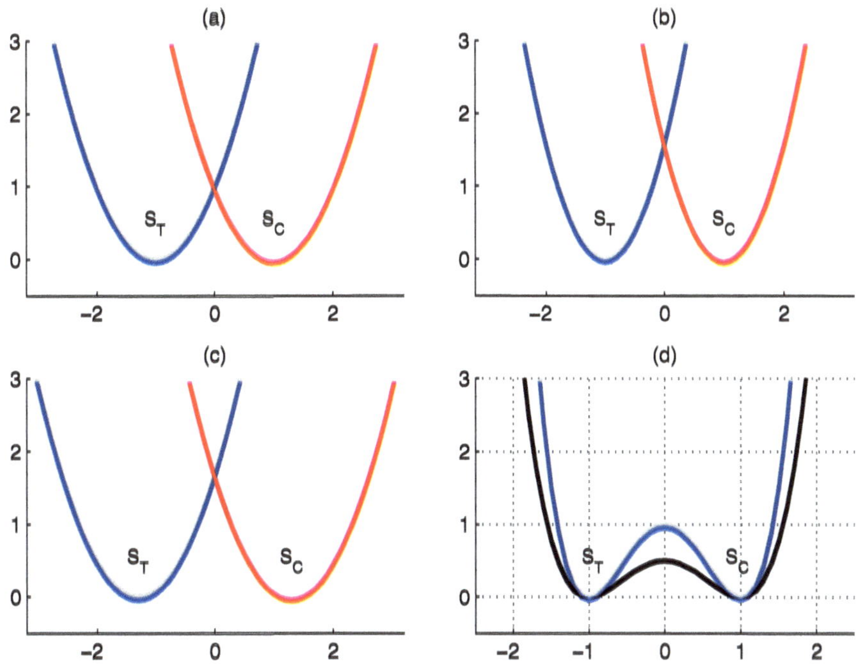

Fig. 1 The formation of two local equilibria in SMEs characterized by bilateral competition and cooperation

Here it is noted that parameter ν_0 in Eq. (1) determines the height of the utility potential barrier, which reflects the industry access difficulty or specialization level of S_T and S_C. It is because if the focused SME belongs to the industry with a relatively high degree of entry, one or both sides, S_T or S_C, often pride themselves on their professionalism (specialization level in technology or marketing) in the competition and cooperation process. One side is probably reluctant to share the professional information with the other side, or the other side is hard to understand the shared information, which leads to a yawning information gap emerging between both sides. It also means that each side has a higher expectation of the state x to locate at the equilibrium position of maximum utility. Once deviating from the position, the more significant loss of utility performs an invisible hand, leading to a more potent dragging force to drive the state back to equilibrium. As a result, the utility curve shows a higher change rate to the deviation, as shown in Fig. 1b. Naturally, the combined action from both sides forms a higher utility barrier, which corresponds to a more significant value of ν_0 in Eq. (1). Thus it is involved to describe the industry access difficulty or specialization level; also, we can understand the barrier with larger ν_0 from the position deviate of two equilibria to reflect the increase of information asymmetry, as shown in Fig. 1c.

Furthermore, in technological terms, it is worth emphasizing that some powerful enterprises, at the time of the exhaustion of a particular paradigm, are likely to become the most conservative forces. Their investment in mature technologies makes them prefer to avoid genuinely revolutionary change, and this pattern efficiently traps them in a static equilibrium. However, since their productivity, market, and profit growth rates are likely to be stagnating, the main hope for their revitalization lies in radical change. Thus, these enterprises are likely to be the victims of paradigm closure. Breaking out of a dormant state is bound to demand cooperation with outsiders, and a better choice is to invest idle capital in those prospective SMEs. Under the present situation, large enterprises might play the role of strategic partners, and the introduced capital becomes the external driving force to magnify the capital-product switching efficiency within SMEs.

With the introduction of external capital, capital-product switches, as inherent periodic regulation, build momentum and cooperatively stimulate the state x to evolve within the bistable utility potential $U(x)$ described by Eq. (1). Sources for external capital include loans that financial institutes provide to SMEs and idle capital from strategic partners. Here we involve a commonly used periodic function $f(t) = A\sin(\omega t)$ to describe the capital-related driving force with the amplitude A and driving period $T = 2\pi/\omega$, which shows high consistency with the capital-product switching periodically [19, 33]. Generally, different industries may have the different cycle times of product development and market exploitation. However, most enterprises implement the responsible person of the annual performance index of evaluation, and the strategies of external investment and internal product planning always match the annual cycle [18, 34, 35]. Furthermore, another inevitable effect of capital-product switches is the random fluctuations from uncertain factors, including configuration instability, internal stimulation mechanism, macro policy fluctuation, and even the bilateral fluctuations resulting from subjective uncertainty to technology or marketing prospects. All these effects, synthetically regarded as input risk, can be described by a random noise process [36–38].

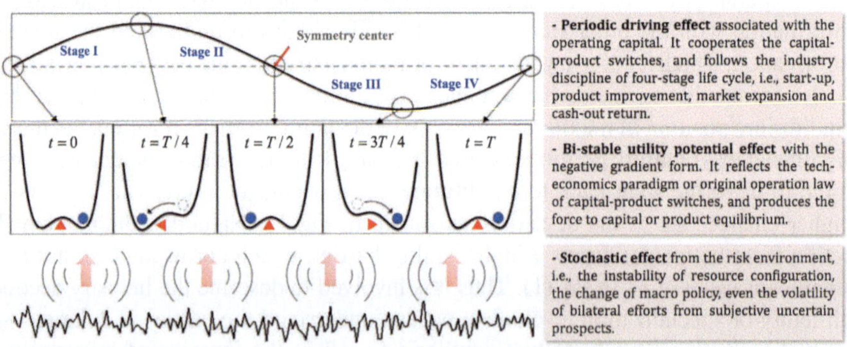

Fig. 2 The dynamics of capital-product transitions in SMEs with bistable utility potential stimulated by periodic force and random effect

Specifically, in the process of capital-to-product, i.e., at $t \in (0, T/2]$, S_T takes more resources and makes a further effort at the cost of S_C's utility, which, under the condition of capital constraint, leads to the increase of S_T's utility and the decrease of S_C's. Thus, we observe in Fig. 2 that the state at $t = T/4$ easily traps into the local equilibrium S_T under the effect of random fluctuations. Specifically, the probability W_T escaping from the left equilibrium is lower than that of W_C escaping from the right. For process product-to-capital, the enterprise must adjust and transfer more resources to inspire S_C to increase its efforts, achieved by continuous capital input; accordingly, S_C's utility increases, and S_T's utility decreases. At $t = 3T/4$ with capital-related driving force $f(t) = -A$, it reaches the maximum effect obviously with $W_T < W_C$. It becomes more difficult for the enterprise, as a particle, to climb over the utility barrier from the state x_C to x_T, and it easily traps into the right equilibrium to make state evolution. The above two processes occur alternately with the timing of events, and the transitional states with $W_C = W_T$ are also shown in Fig. 2.

Based on the state theory of nonlinear stochastic resonance (SR) [28], we noted that random fluctuation from input risk is not always unfavorable to increase the system disorder within SMEs. It has the potential to play a cooperative role in capital-product switches, especially for those SMEs with a state of under-capitalization (lack of capital-related driving force) or SMEs with a state of over-specialization/radical innovations (shown as a higher utility barrier) [39, 40]. A practical illustration will be given and further emphasized in the following sections, both in theory and practice.

3 SME's Mesoscopic Structure

In this section, we further consider that the focused SMEs interact, directly or indirectly, with others, or at least that it has the opportunity to do so in order to achieve a desired outcome. Drawing from economic sociology, social network researchers have long predicted that certain network positions are more advantageous than others [41–43]. An effective network structure allows the enterprise to control business processes, balance risks, and even absorb resources to cooperate with the achievement of capital-product switches. Thus there is an increasing interest in understanding how and why individuals harvest value from the network [44–46].

To achieve this goal, we reconstruct the multi-layered network, which consists of the relationships of inter-enterprise's business behaviors [41]. That is, we don't care about the generic complex networks among entities, only layered structures with one core and multiple layers, as shown in Fig. 3. Thus, we define the focused enterprise as the core P_0, which directly connects with the 1st degree strategic partners $P_{1,1}, P_{1,2}, \cdots, P_{1,N_1}$ on layer $\mathcal{L}_1$, where the enterprises, as strategic partners, are always bond together by a formal agreement for contributing initial funding to P_0. $\mathcal{L}_2 = \{P_{2,1}, P_{2,2}, \cdots, P_{2,N_2}\}$ represents the 2nd degree (indirect/non-strategic) partners to P_0 on layer 2. We apply the same approach to more layers, $\mathcal{L}_j = \{P_{j,1}, P_{j,2}, \cdots, P_{j,N_j}\}$.

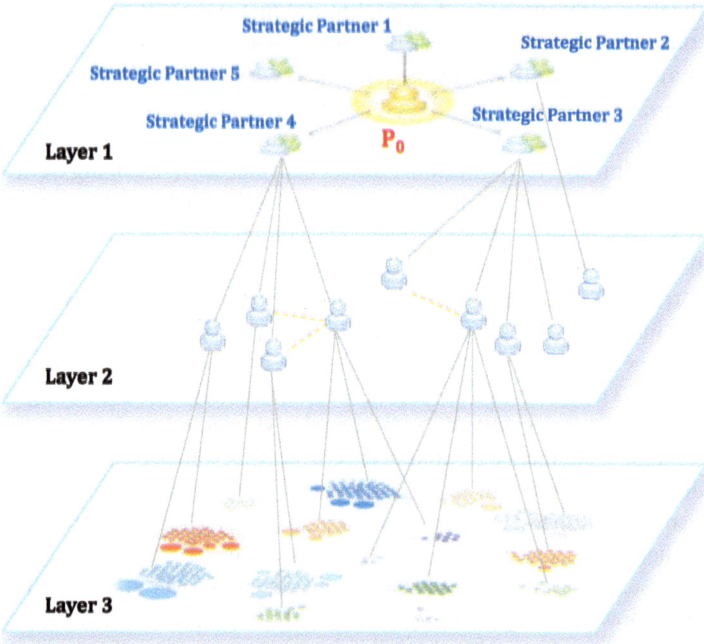

Fig. 3 The layered structure of an enterprise's network

With the help of the enterprise big data platform (EBP), we collected 300,000 samples and associated them with layered networks from the manufacturing sector in China. Here, we outline the following levels for enterprises: an SME is an enterprise with registered capital C_R no more than 1 million in RMB, a medium-sized enterprise has 1 million to 10 million in RMB, and a large-sized enterprise has more than 10 million RMB. To limit the layers to four (beyond the core), we first explore the average number of direct or indirect partners correlated with the size. Table 1 shows that partners significantly increase with the increase of degree. Meanwhile, we also observe the increase of direct and indirect partners for the enterprise's growth from the small-sized to the large-sized. From the perspective of partners, on average, the networks with which SMEs are associated are much more straightforward.

Further, we investigate some characteristics correlated with the enterprise scale and capital-path degrees, and mainly focus on the correlation coefficient φ_j on each layer, and the path complexity ψ_j from each layer to the core. The notes of Table 1 provide detailed definitions. We find that, in the SME-centered networks, the investment relationships within the enterprises on each layer are much less than those in medium-sized and large-sized enterprise-centered networks. Based on surveys on SMEs, the sample ratio of existing investment relationships between direct partners is no more than 0.1%. Case analysis shows these SMEs do not always perform better in operating situations.

Table 1 The enterprise's network profiles based on the large samples from the manufacturing sector in China

	Average partners				Correlation coefficient [a]			Path complexity [b]		
Scale [c]	N_1	N_2	N_3	N_4	φ_1	φ_2	φ_3	ψ_1	ψ_2	ψ_3
S&M	1.51	2.30	7.64	14.08	0.0030	0.0066	0.0034	0	0.0857	0.1380
Medium	2.59	4.91	22.11	72.34	0.0077	0.0127	0.0060	0	0.1595	0.1943
Large	3.65	23.09	139.21	756.44	0.0387	0.0179	0.0063	0	0.1563	0.2634

[a]Capital correlation coefficient, between the jth degree partners on layer j, is defined by $\varphi_j = \frac{n_j}{N_j(N_j-1)/2}$, where N_j is the number of partners on layer j (i.e., the capacity of $\mathcal{L}_j$), and n_j is the total number of connections (i.e., existed investment relationships within $\mathcal{L}_j$). All the results φ_j are the statistical average values based on the samples with $N_j \neq 0$

[b]Capital-path complexity is defined by $\psi_j = \frac{1}{N_j}\sum_{i=1}^{N_j} \kappa_{j,i} - 1$, which is the average investment path number of partners $\mathrm{P}_{j,1}, \mathrm{P}_{j,2}, \cdots, \mathrm{P}_{j,N_j} \in \mathcal{L}_j$ to P_0 not including one necessary capital path between $\mathrm{P}_{j,i}$ to P_0 (otherwise, $\mathrm{P}_{j,i}$ will be not included in the network), $\kappa_{j,i} \geq 1$ is the number of j-degree investment relationship between $\mathrm{P}_{j,i} \in \mathcal{L}_j$ and P_0. All the results ψ_j are the statistical average values based on the samples with $N_j \neq 0$

[c]The enterprises are divided into three types (i.e., S&M, medium and large) based on the registered capital C_{R} (in unit RMB yuan), that is, SME (no more than 1 million), medium enterprise (1∼10 millions), and large enterprise (more than 10 millions)

Moreover, Table 1 also shows the simple capital path from indirect partners to the core in the SME-centered networks. Except for one necessary capital path between $\mathrm{P}_{j,i}$ to P_0 (otherwise, $\mathrm{P}_{j,i}$ will not be included in the network), few other capital paths exist, which is particularly noticeable compared with the grown-sized enterprises. The results are beyond our original expectations of SME-centered networks. Thus we are motivated to use a simplified star-shaped structure to describe the SME network, in which the core is the focused SME, surrounded by the direct strategic partners on layer 1.

Here, the simplified network, as a whole, is regarded as an SME's mesoscopic structure. All the external capital is introduced to the focused P_0 from the direct strategic partners $\mathrm{P}_{1,i}$, $i = 1, 2, \cdots, N_1$ on $\mathcal{L}_1$, and the investment relationships, between $\mathrm{P}_{1,i}$ to P_0 are supposed to be mutually independent, approximately. Additionally, as a strategic selection of direct partners, for potential cooperators, who are extremely useful but unable to satisfy the requisite conditions currently, it is also an excellent choice to establish relations with direct partners or higher degree partners.

4 Stochastic Dynamical Model

In this section, we model the dynamical behaviors of SME state evolution under a mesoscopic structure. Regarding the focused entity as a Brownian particle, the following fluctuation-dissipation effects should be considered to demonstrate the evolution mechanism from Langevin dynamics (e.g., see Eq. (2)):

- Utility potential effect with the negative gradient form of symmetric bistable utility potential field $U(x)$, which reflects the tech-economics paradigm or original operation laws, and produces the driving force $-\frac{\partial}{\partial x}U(x)$ to equilibria, x_T or x_C.
- Periodic driving effect $f(t)$ associated with the inherent capital-product regulation of enterprise operation, and the driving frequency ω is relative to the industry discipline of the four-stage life cycle, i.e., start-up, product improvement, market expansion, and cash-out return.
- Stochastic effect from the uncertainty of internal/external environment, including configuration instability, internal stimulation mechanism, macro policy fluctuation, and even the volatility of bilateral efforts from subjective uncertainty to technology/marketing prospects. They are synthetically regarded as input risk and can be described by a random process $\varepsilon(t)$.
- Damping effect associated with the contradiction, disorder, and competition from the mismatch with the old paradigm or original operation laws. It acts on the interface between the entity and environment and performs obstinate rejection to any change of state x, and it is generally supposed to be proportional to the change rate based on Stokes' law [47].

For a generic system model that combines all evolving effects, the evolution of state x with time t can be described by an over-damped nonlinear Langevin equation [29, 30, 48], which is a simplified version based on implicit Newton's law, and corresponds to the limit where no average acceleration takes place. That is, we regarded the focused SME with scale m_0 as a Brownian particle (with mass m_0) moving in the utility potential field $U(x)$, deriving its periodic force $f(t)$ from the internal operation and external cooperation perturbed by stochastic fluctuation $\varepsilon(t)$. We describe the following system model for an SME without strategic partners:

$$\gamma_S \cdot \frac{\mathrm{d}}{\mathrm{d}t}x(t) = -\frac{\partial}{\partial x}U(x) + f(t) + \sqrt{2D_0} \cdot \xi(t), \tag{2}$$

where $x(t)$ is the state variable and it reflects the cooperation and competition process at time t. $\gamma_S > 0$ is the damping constant of friction, and it can be normally set to unity without loss of generality. Capital-related driving effect $f(t)$ follows the process of capital-product switches; thus, we employ $f(t) = d_0 \sin(\omega t)$ to describe the capital-product periodic force, which is characterized by amplitude d_0 and angular frequency $\omega = 2\pi/T$ (T is the capital-product cycle, typically including four stages, i.e., start-up, product improvement, market expansion and cash-out return, described in Sect. 2). Specifically, we consider an SME with initial capital c_0, divided into two parts: $c_0 = d_0 + p_0$ with lever coefficient $l_0 \in (0, 1)$.

Accordingly, $d_0 = l_0 c_0 > 0$ is the operating capital; $p_0 = (1 - l_0)c_0 > 0$ is the scale-related capital, contributing to the enterprise scale $m_0 = \theta_0(1 - l_0)c_0$, measured by capital input p_0 and scale transforming factor θ_0, which reflects the scale expansion of unit input capital. $\xi(t)$ is the standard white Gaussian noise process, and $\sqrt{2D_0} \cdot \xi(t)$ mimics the stationary random excitation caused by an intra-SME environment in a standard operation regime. This process follows the fluctuation-dissipation relationship [49] to measure the input risk level during the SME growth:

$$D_0 = \frac{K_B I_C \gamma_S}{m_0}, \tag{3}$$

in which I_C is the macro climate index of the relevant industry the SME. K_B bridges the internal and external system parameters by the fluctuation-dissipation assumption. It is important to note that, different industries have varying cycle times of product R&D and market exploitation; however, most enterprises employ a responsible person for the annual performance index of evaluation. As a result, the external investment and internal product planning strategies always match the annual cycle. Alternatively, the cycle structure in some unique industries, i.e., pharmaceutical engineering, may be divided into different stages and span more extended periods of capital-product switches. In other words, the switching frequency is low enough to allow the enterprise to reach the local statistic equilibrium during one period. Thus, the SME system is supposed to satisfy the adiabatic approximation condition: the driving frequency is much slower than the inverse value of the relaxation time or the time for probability within one well to equalize. Therefore, the adiabatic limit can describe the asymptotic long-time distribution function.

Further, considering the focused SME under mesoscopic structure, all the first-degree strategic partners are directly re-denoted by $\mathcal{L}_1 = \{P_1, P_2, \cdots, P_N\}$, and it is easy to achieve the external introduced capital information $\mathcal{C}_N = \{c_1, c_2, \cdots, c_N\}$ from each partners. Similarly, as stipulated in the contract about purposes, external capital c_j from P_j is generally divided into two parts $c_j = d_j + p_j$ with lever coefficient $l_j \in (0, 1)$, that is, $d_j = l_j c_j \geq 0$ is the remained operating capital, $p_j = (1 - l_j) c_j \geq 0$ is the scale expanding or input capital, which contributes to the enterprise scale $\Delta m_j = \theta_j (1 - l_j) c_j$, also measured by capital input p_j and scale transforming factor θ_j. Thus, under the framework of the mesoscopic structure, the enterprise scale and operating capital-related driving amplitude are given by $m = m_0 + \sum_{j=1}^{N} \Delta m_j$ and $A = d_0 + \sum_{j=1}^{N} d_j$, respectively. The dynamical model is further expressed as:

$$\gamma_S \cdot \frac{\mathrm{d}}{\mathrm{d}t} x(t) = -\frac{\partial}{\partial x} U(x) + f(t) + \sqrt{\frac{2 K_B I_C \gamma_S}{m}} \cdot \xi(t), \tag{4}$$

where, $f(t) = A \sin(\omega t)$ is the periodic driving force, cooperatively formed by all partners from $\mathcal{L}_1$ without regard to legal persons, enterprises or financial institutions. According to the fluctuation-dissipation relationship, the input risk level of a mesoscopic SME system with expanded scale m satisfies the relationship: $D = \frac{K_B I_C \gamma_S}{\sum_{j=0}^{N} \theta_j (1 - l_j) c_j} = \frac{m_0}{m} \cdot D_0$. Obviously, the level D is synthetically dependent on the micro, macro, and mesoscopic profiles, and it can be controlled by adjusting the related parameters, i.e., the level can be decreased by decreasing the capital lever l_j or added-investing from strategic partners. However, it only sometimes benefits from decreasing the system disorder with lower D. To a certain extent, it also reflects the enterprise's vitality, which is essential for innovations. Thus, sometimes it might cooperate with capital-product switches for those SMEs with under-capitalization or over-specialization.

5 Performance Measurement

To further explore the meaning of performance improvement associated with SME's development and growth, we first need to look closer into the performance definitions. Therefore, in this section, we seek to incorporate the different forms of response risks for measuring the corresponding value at risks.

5.1 System Stationary Response

We confine the analysis to a one-dimensional Markovian process of state $x(t)$ subjected to the perturbation of periodic effect and input-risk-related random fluctuation.

In the absence of capital-related driving effect, the fluctuation causes state x transitions between capital equilibrium $x_{\mathrm{T}}=-1$ and technology equilibrium $x_{\mathrm{C}}=1$, respectively with probabilities $n_{\mathrm{T}}(t)$ and $n_{\mathrm{C}}(t)$, which satisfy the normalization condition $n_{\mathrm{T}}(t)+n_{\mathrm{C}}(t)=1$, and the transition rate, with the Kramers escaping form [50], is given by

$$r_{\mathrm{K}} = \frac{\nu_0}{\sqrt{2}\pi m \gamma_{\mathrm{S}}} \exp\left(-\frac{\nu_0}{4D}\right). \tag{5}$$

Further, with the driving effect of $f(t)=A\sin(\omega t)$, two local equilibria are biased alternately, as shown in Fig. 2, and the transition rates $W_{\mathrm{T/C}}(t)$ escaping from two equilibria x_{T} and x_{C} at time t, are given by

$$W_{\mathrm{T/C}}(t) = r_{\mathrm{K}} \exp\left[\mp\frac{A}{D}\sin(\omega t)\right], \tag{6}$$

where $W_{\mathrm{T/C}}(t)$ is the uniform and simplified expression of $W_{\mathrm{T}}(t)$ or $W_{\mathrm{C}}(t)$, and the relevant master equation for $n_{\mathrm{T/C}}(t)$ reads: $\frac{\mathrm{d}}{\mathrm{d}t}n_{\mathrm{T/C}}(t)=-W_{\mathrm{T/C}}(t)n_{\mathrm{T/C}}(t)+W_{\mathrm{C/T}}(t)n_{\mathrm{C/T}}(t)$. Based on linear response theory [51, 52], the system response can be analyzed in the long-time limit. Up to first-order form, the expectation of system response process $x(t)$ can be expressed as

$$\langle x(t)\rangle = A|\chi(\Omega)|\sin(\omega t-\phi), \tag{7}$$

where $\chi(\Omega)$ is the system susceptibility, which is the Fourier transform of system response $h(t)$, that is, $\chi(\Omega) = \int_{-\infty}^{\infty} h(\tau)e^{-j\Omega\tau}d\tau$ with the modulus $|\chi(\Omega)| = \sqrt{(\chi_R(\Omega))^2 + (\chi_I(\Omega))^2}$, and the retarded positive phase shift reads $\phi = \arctan\{\chi_I(\Omega)/\chi_R(\Omega)\}$, that is, the time delay of system response to inputs is given by $\tau_0 = \frac{T}{2\pi}\phi$, which measures the strategy implementation time delay in theory. It might be a useful reference for scheduling some control strategies making the SME's response as expected time frame.

Furthermore, the power spectral density (PSD) of system response has the following form:

$$S_X(\Omega) = \frac{\pi}{2}A^2|\chi(\Omega)|^2 \cdot [\delta(\Omega - \omega) + \delta(\Omega + \omega)] + S_{X,\xi}(\Omega), \tag{8}$$

where $S_{X,\xi}(\Omega)$ is the PSD of unperturbed system, and the contributions of random fluctuation to $S_{X,\xi}(\Omega)$ can be approximately decomposed into two parts, one is the (inter-equilibrium) disorder of capital-product switches with relaxation rate $2r_K$, and the other is (intra-equilibrium) local perturbations around capital or technology equilibrium. Their contributions produce

$$S_{X,\xi}(\Omega) = \frac{4r_K}{4r_K^2 + \Omega^2} + \frac{2m^2\gamma_S^2 D}{4\nu_0^2 + m^2\gamma_S^2\Omega^2}. \tag{9}$$

Submitting Eq. (9) into Eq. (8), the PSD can be written as

$$S_X(\Omega) = \frac{2\pi A^2 r_K^2}{D^2(4r_K^2 + \omega^2)} \cdot [\delta(\Omega - \omega) + \delta(\Omega + \omega)] + \frac{4r_K}{4r_K^2 + \Omega^2} + \frac{2D}{4\nu_0^2 + \Omega^2} \tag{10}$$

which indicates that the state process $x(t)$, as system output, includes three parts: (i) the periodic stationary response, which shows a delta function at the capital-product switching frequency ω; (ii) the zero-frequency output, which shows nonlinear dependence on transition rate $2r_K$, and can be effectively compressed by adjusting $2r_K = \omega$; and (iii) the broadband noise outputs, which show energy bumps centered at frequency $\Omega = 0$, and produce the perturbations at $\Omega = \omega$. The intensity can be linearly controlled by adjusting input risk level D and nonlinearly controlled by specialization level ν_0.

5.2 *Risk Measurement*

Under the framework of tech-economic paradigms, which perceive the exposure to bilateral cooperation risk as the enterprise state performing insufficient ability and power for further development and growth, an indicator relative to the transition rate can function as a measurement. Specifically, the state $x(t)$ efficiently traps into one local equilibrium at $x_T(t)$ (or $x_C(t)$). Thus transition rate, i.e., $r_K \in (0, \omega/2)$ keeps at a lower level, and it is difficult for S_C (or S_T) to reach its maximum utility for an extended period, which inevitably leads to a high rupture risk of bilateral cooperation.

Systematic response risk is the state volatility within one local equilibrium and the state uncertainty between two local equilibria. With the increase of input risk level D from 0, the intra-equilibrium volatility increases firstly, and the inter-equilibrium uncertainty also increases but remains at a negligible level. As a result, the enterprise usually performs insufficient ability and power due to under-capitalization, over-specialization, or under-stimulus. Suppose the input risk with level D increases within an interval, and it begins to play the important role and cooperate with a capital-related driving force to produce capital-product switches, the intra-equilibrium volatility will decrease instead, and the inter-equilibrium uncertainty will increase significantly. Here the enterprise quickly shows the optimal switching regularity with the joint consideration of bilateral cooperation risk and systematic response risk. If the input risk level D further increases, the enterprise with over-stimulus can be regarded as the over-capitalization or under-specialization from the other point of view. Although it easily achieves capital-product switches, both increase of the intra-equilibrium volatility and inter-equilibrium uncertainty results in a decrease in switching efficiency.

Regardless of the evolution state, two types of response risks are inevitable and coexist in the SME system driven by input risk with level $D > 0$. From the power spectral density $S_X(\Omega)$ of state process $x(t)$ at the capital-product switching frequency $\omega = 2\pi/T$, we can precisely extract the levels as follows:

(1) **Bilateral cooperation risk** level is measured by R_B:

$$R_B^2 = \frac{r_K T^2}{r_K^2 T^2 + \pi^2}, \tag{11}$$

which directly shows nonlinear relevance to transition rate r_K, and indirect dependence on input risk level D. It can be effectively compressed by adjusting $r_K = \pi/T$, a comprehensive result of multiple parameters; it exactly corresponds to the system control to make one capital-product switch occur during the driving period in a statistical sense.

(2) **Systematic response risk** level is measured by R_S:

$$R_S^2 = \frac{K_B I_C \gamma_S}{2\sum_{j=0}^{N} \theta_j (1 - l_j) c_j} \cdot \frac{T^2}{\nu_0^2 T^2 + \pi^2}, \tag{12}$$

which is interpreted as the intra-equilibrium volatility and inter-equilibrium uncertainty, respectively demonstrated as nonlinear behaviors with the increase of input risk level D. Taken together, the systematic response risk level can be linearly controlled by D, which is also the combined effect from different dimensions.

Obviously, micro profiles (i.e., capital, lever, scale, management orientation, and specialization level), mesoscopic structure (i.e., capital, lever, number), and macro environment (industry, climate) should be mutually coordinated to enable the enterprise to achieve the sustainable and regular switches between two local equilibria. As a result, it is also regarded as the health indicator to evaluate the state of SMEs' development and growth.

5.3 VaRR Performance

To introduce the definition of VaRR performance, three typical cases should first be considered:
(i) The enterprise state stabilizes at one local equilibrium, x_T or x_C, to satisfy its maximum utility. It seems to remain relatively stable at a reasonable level, as shown in Fig. 4a. However, this stability renders the other party unwilling to apply the effort they should. Thus, the rupture risk of bilateral cooperation remains relatively

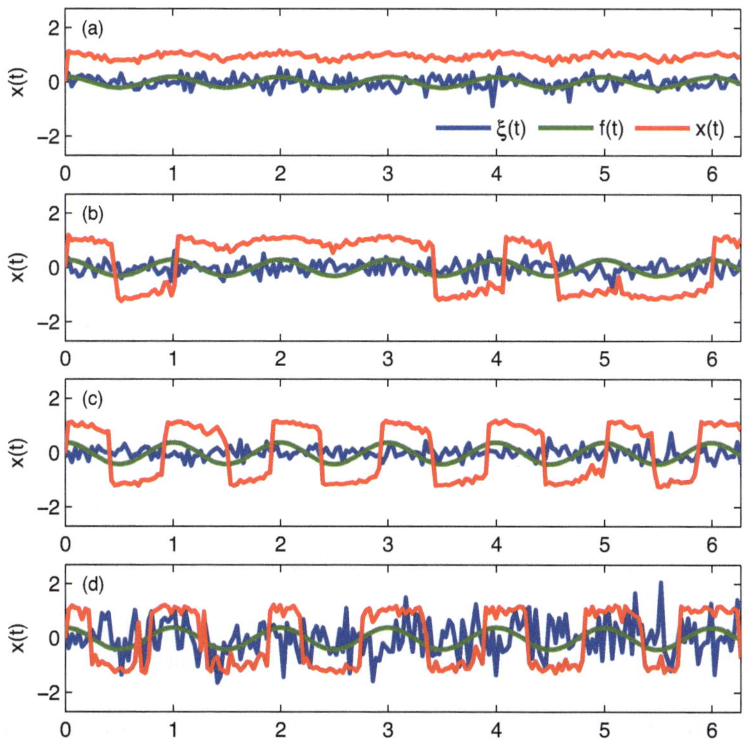

Fig. 4 The stochastic evolution of SME's growth state with different operating capital and macro climate index: **a** $A=0.1$, $I_C=1.0$; **b** $A=0.3$, $I_C=1.0$; **c** $A=0.4$, $I_C=1.0$; **d** $A=0.4$, $I_C=5.0$. The other parameters are set to be $m=1.0$, $\gamma_S=1.0$, $\nu_0=1.0$, $K_B=0.025$

high if without appropriate driving force or random fluctuation in the system. The perspective of local stabilization is too "narrow" to evaluate the state adequately.
(ii) The enterprise state switches between two equilibria, x_T and x_C that seem unstable and shows a reduced level of bilateral cooperation risk. Actually, if the capital-related periodic effect, cooperated by random fluctuations, guarantees the continuation of the project on to the next developing stage, the state is inspired to randomly switch between x_T and x_C, as shown in Fig. 4b. The enterprise will face the higher uncertainty of capital-product switches during one driving period. Nevertheless, with the excessive stimulus, as shown in Fig. 4d, the enterprise further exposed to a much higher systematic response risk instead. Thus, the perspective of capital-product switches is also too "narrow".
(iii) The periodic force's effect tilts the utility potential back and forth, thereby alternately raising and lowering the right and left barriers asymmetrically. Based on SR theory, a suitable dose of input risk can make the state achieve one capital-product switch during one driving period. One can clearly observe the optimal regularity of capital-product switches as shown in Fig. 4c. Strictly speaking, this holds true only in the statistical average and is regarded as the globally stable state.

Under the framework of tech-economic paradigms, as the inherent effect of capital-product switches applied together with strategic capital from mesoscopic structure, input risk might magnify the switching efficiency, activate the system for maximum competition and cooperation, and achieve global optimum. However, the performance (development quality of enterprise state) is challenging to evaluate comprehensively and objectively using a single indicator, which cannot adequately describe the risk dimensions. Thus, it is essential to clarify what performance index should be proposed to evaluate the state from the dynamic perspective of system evolution.

From the separate components of PSD at the driving frequency ω, the system regularity indicator is reflected in the power of stationary periodic response, and it is measured by

$$R_{\text{out}}^2 = \frac{\pi\left(\sum_{j=0}^{N} l_j c_j\right)^2 \left(\sum_{j=0}^{N} \theta_j (1 - l_j) c_j\right)^2}{2K_{\text{B}}^2 I_{\text{C}}^2 \gamma_{\text{S}}^2} \cdot \frac{r_{\text{K}}^2 (DT^2}{r_{\text{K}}^2 T^2 + \pi^2}, \tag{13}$$

where R_{out} essentially measures the sustainability of an SME's growth with capital-product switches and corresponds to the first component in the PSD of growth process. The capital lever l_j can be employed within fixed capital input to balance the inputs between operating capital and scale-related capital as an optimization dimension.

Suppose we interpret system performance as the willingness to take on risky evolution in pursuit of the regular switches between capital and technology equilibria [53, 54]. Then we may further define the value at response risks (VaRR) as follows,

$$\mathrm{VaRR} = \frac{R_{\mathrm{out}}}{R_{\mathrm{S}} + R_{\mathrm{B}}}. \tag{14}$$

which is a thorough consideration, and the potential influence of extreme consequences in Fig. 4 is reflected. This assessment refines a stationary unit-risk return by measuring how much risk is involved in producing the return over a given period. It is widely used in individual securities, investment funds, and portfolios [55, 56].

6 The Quantitative Analysis on Risk Managements of Strategies

In this section, we discuss the effect of risk management from various possible dimensions. Based on the results of system simulations, we observe the stationary dynamical behaviors, including the systematic response risk, bilateral cooperation risk, and VaRR performance, plotted as a function of the following dimensions, respectively:

(S1) capital and lever (C, L), balancing the operating capital and lever;
(S2) damping friction γ_{S}, reflecting the management orientation;
(S3) specialization level ν_0, controlling the utility barrier height;
(S4) macro climate index I_{C}, impacting the input risk level.

Based on Monte Carlo method and Runge-Kutta algorithm, the results are obtained by using 100,000 trajectories of $x(t)$ realizations (within 5 driving periods) for ensemble averaging, and the basic parameters involved in different strategies S1~S4 are listed in Table 2, in which the total input capital C and total capital lever L are defined in the notes. If not specified, we consider the most common annual capital-product switches and set the driving frequency to be $\omega = 2\pi/365$. Further, some available indicators of implementing methods and tactics about how to adjust the variation trend of these dimensional parameters, have been considered in Ref. [25]. It is worth emphasizing that the implementation of any one strategy might have the systemic influence on multiple parameters, thus the methods and tactics for S1~S4 are proposed only based on the consideration of the most sensitive one.

6.1 *Capital and Lever*

In this idealized setting, we ignore the internal and external approaches of capital access and directly investigate the effect of input capital C and lever L, which hence determine the operating capital $A = LC$ and enterprise scale $m = \theta(1 - L)C$. S1-row of Table 2 lists the related parameters. By system simulations, two types of response risk levels are plotted as a function of two variables, C and L, in Fig. 5. For any capital-level situation $\{C, L\}$ in $C - L$ plane with $C \in [0.1, 1.0]$ and $L \in [0.1, 0.9]$, systematic response risk $R_S(C, L)$ level and bilateral cooperative risk $R_B(C, L)$ are measured by the color bar located at the right of each figure.

The bottom convex region (less capital and lower lever, or more capital and lower lever) illustrates two types of risks with opposite levels (i.e., the lower systematic risk level R_S and higher bilateral risk level R_B). Here, the state x traps in the local equilibrium, S_C or S_T. Specifically, within this region, the enterprise performs a situation of under-capitalization, which means that it cannot afford operational expenses within current scale. Sometimes, it is financially sound but needs more funds to expand to meet market demand or further R&D expenses. The operating capital $A = LC$ is relatively inadequate to drive the entity with scale $m = \theta(1 - L)C$ to surmount the utility barrier $\nu_0/4 = 0.25$. Thus the state performs intra-equilibrium evolution with

Table 2 The parameter setting for strategies of different dimensions

Strategy	C^a	L^b	θ^c	γ_S	ν_0	K_B	I_C
S1: capital and lever	0.1 ∼ 1.0	0.1 ∼ 0.9	1.0	1.0	1.0	0.025	1.0
S2: damping friction	0.4, 0.6, 0.8	0.5	1.0	0.4 ∼ 1.2	1.0	0.025	1.0
S3: utility barrier	0.4, 0.6, 0.8	0.5	1.0	1.0	0.8 ∼ 1.6	0.025	1.0
S4: macro climate index	0.4, 0.6, 0.8	0.5	1.0	1.0	1.0	0.025	0.5 ∼ 2.0

[a] Total input capital $C = c_0 + \sum_{j=1}^{N} c_j$ includes the internal initial investment c_0 and introduced external capital c_j from all strategic partners, here c_j has been comparably set to relative values by applying the registered capital C_R to be normalized factor

[b] Total capital lever is defined by $L = \frac{A}{C} = \frac{1}{C}\sum_{j=0}^{N} l_j c_j$

[c] All the transforming factors θ_j, from internal or external expanding capital d_j contributing to enterprise scale Δm_j, are set to be $\theta_j = \theta$, $j = 0, 1, \cdots, N$

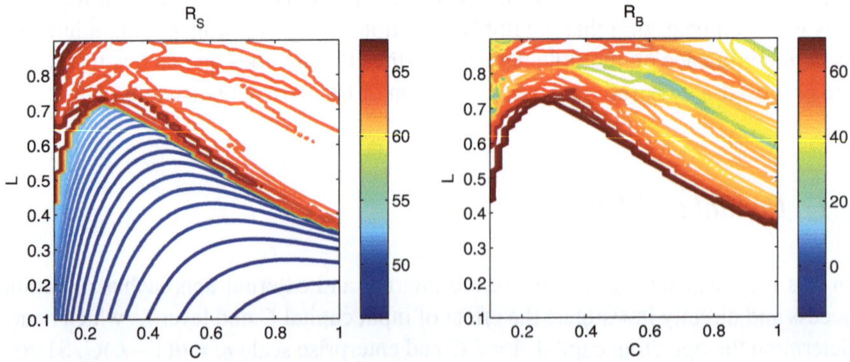

Fig. 5 The influences of regulating total input capital and lever on systematic response risk and bilateral cooperation risk

lower capital-product switching efficiency, even if it is stimulated by the input risk associated with random fluctuation with intensity $D = \frac{1}{m} K_B I_C \gamma_S = 0.025/m$. From the perspective of bilateral cooperation, the state traps in a local equilibrium (i.e., the maximum utility of S_T), which easily makes S_C unwilling to apply the effort they should. Following this, the enterprise is over-exposed to bilateral risk if without additional periodic force by optimizing capital structure or balancing input between operating capital A and scale m (decreasing scale-related input to increase operating capital). As the capital lever profile $\{C, L\}$ approaches the convex boundary by decreasing both capital and lever ($C < 0.27$) factors, or increasing capital and decreasing lever ($C > 0.27$), it gradually overcomes the under-capitalization problem by balancing A and m, which leads to the increase of rate r_K, escaping from the local equilibrium. It is a sensitive process from a trapped state to an optimal capital-product switching state. Thus, we observed the narrow transition belt, especially for $C > 0.27$. As $\{C, L\}$ upward crosses the boundary, the enterprise gradually gets into the situation of relative over-capitalization, and the role of input risk changes from cooperator to obstructor. To clarify, operating capital is relatively sufficient under the random stimulus of input risk with increased levels. The entity with a smaller scale becomes more irregular than one with a larger scale when it reaches the optimal switching points. Thus, we observe that systematic risk decreases and then increases again with the increase of L for $C > 0.27$.

On the other hand, for fixed capital C, we first observe the bilateral risk R_B remains at a high level with a much lower L, which means that almost no capital-product switches occur due to the highly inadequate operating capital. Then, R_B gradually decreases with increasing L. The enterprise might go through a capital-product switch during several annual cycles. As L increases until the capital-product switching rate equals the driving frequency (once a year), we observe the minimal R_B at the green belt. However, the much higher L (the situation of extreme over-capitalization) may lead to the utility barrier's disappearance in effective potentials $U(x) - Ax$ and $U(x) + Ax$. It means that even without the cooperation of random stimulus, the entity (with much smaller scale m or adequately operating capital A) could achieve at least one capital-product switch each year. Thus, the intra-equilibrium volatility and inter-equilibrium uncertainty gradually become significant with the increase of L, that is, the switching irregularity, measured by bilateral risk, increases again.

Further, as L reduces to a certain degree, the significant noise is crucial in stimulating the state to approximately perform random hops absolutely. Instead, we observe R_B deceases and R_S increases. Finally, we emphasize that, in practical implementations, lever control is an irreversible process, if not increased capital input to compress the relative ratio of expenses on scale expansion.

From the profile of capital and lever's control effect on two types of response risks, we may interpret the VaRR performance shown in Fig. 6. Based on SR theory, the inherent effect of capital-product switches applied together with random stimulus might magnify the switching efficiency, and competes and cooperates to make the system achieve global optimum in the statistical sense. Here, for $C < 0.27$, we observe that VaRR shows non-significant SR behavior with the increase of L, and for higher capital input $C > 0.27$, the phenomenon disappears. Instead, it monotonously

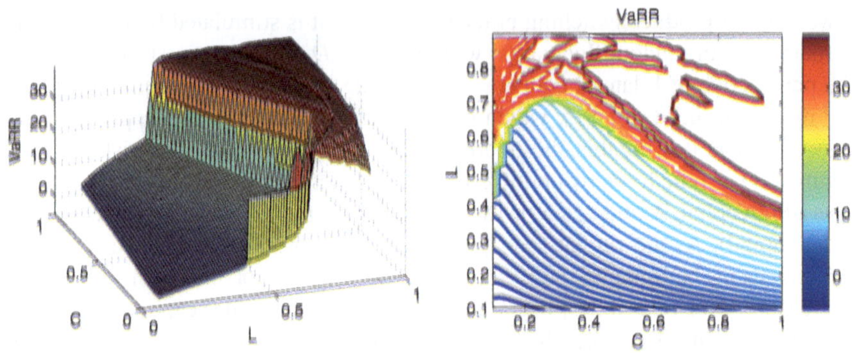

Fig. 6 The influences of regulating total input capital and lever on VaRR performance

increases and asymptotically tends to a maximum value of VaRR performance. On the other hand, if L is fixed within [0.58, 0.69], we observe the significant area of VaRR suppression, which is non-negligible for risk management by optimizing capital structure.

6.2 The Management Orientation

Next, we are interested in the control effect of management orientation from loose to delicacy, which is reflected on the adjustment of decreasing damping friction γ_S. In Fig. 7, we plot the curves of response risks and VaRR performance as a function of increasing γ_S, respectively, with total input capital $C=0.40$, 0.60, 0.80 and the same lever $L=0.5$. One can clearly observe that two types of response risks perform non-monotonic behaviors. It is because of the fact that the adjustment of decreasing friction causes the decrease of damping force in the evolution process, which makes the state overcome the utility barrier more easily, and magnifies the capital-product switches.

Nevertheless, two aspects can aid in the interpretation of management orientation. One is that of over-damping; here, the control of delicacy management could effectively cause the enterprise to reduce vicious resource competition from internal contradictions and conflicts, and thus the capital-product switching efficiency might be magnified. Delicacy management, for reference, is an approach to developing a harmonious culture to strengthen the centripetal force, increasing management control level to form the driving force of solidarity and unity. The other aspect is under-damping; here, the delicacy management control further magnifies the effect of disordering the normal periodic operation of inherent capital-product switches. Thus, many switches randomly occur during one period, causing the system to lose its remarkable regularity. Therefore, it is seen from Fig. 7 that, as to insufficient capital $C=0.4$, systematic risk shows SR behavior of one-peak at $\gamma_S=0.65$. However, this risk is significantly suppressed as γ_S reaches the region [1.0, 1.2]. As expected, the phenomenon of SR behavior disappears for sufficient capital $C=0.8$ when the non-significant increase replaces it with increasing γ_S. Moreover, the input capital $C=0.6$ is a sensitive case, in which both systematic risk and bilateral risk display the SR behaviors of double-peak, i.e., two peaks of systematic risk respectively at $\gamma_S=0.70$ and 0.85, and two peaks of bilateral risk respectively at $\gamma_S=0.60$ and 0.80, which are formed by the competition of opposite interactions between under-damping and over-damping mentioned above. Here γ_S can be controlled to avoid the risk resonance effectively.

Based on the above analysis of two types of response risks, in general, the VaRR performance is gradually improved with the control of management orientation from loose to delicacy (with the decrease of γ_S). However, the strategy does not always effective. For example, we may observe the apparent performance loss within $\gamma_S \in [0.80, 0.85]$ for $C=0.6$. In this case, increasing and decreasing γ_S are adequate controls to get rid of VaRR deterioration. Of course, in practice, the control to increase γ_S does not inherently mean a decrease in management level, which can be adjusted by implementing some methods and tactics, i.e., stimulating the internal vitality and advocating the spirit of opening and equalization, respecting and encourage employees to innovate. Therefore, the optimal management level is a systematic consideration under the measurement of response risks or VaRR performance.

6.3 Specialization Level

In Fig. 8, we investigate the control effect of specialization level, which is reflected in the height of utility barrier $\frac{1}{4}\nu_0$, in which ν_0 is the control parameter. The increase of ν_0 can be influenced by introducing skilled technicians to promote the core competence of products, or improving professional market and sales channels. Conversely, some implementations, i.e., establishing a training system, encouraging common employees to upgrade themselves, avoiding information asymmetry, strengthening

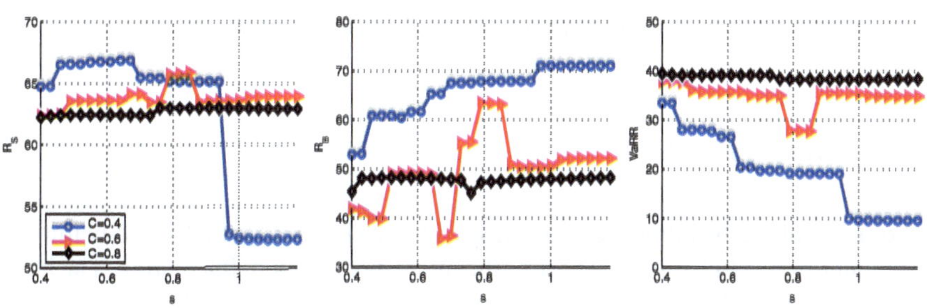

Fig. 7 The influences of regulating management orientation (from loose to delicacy) on systematic response risk, bilateral cooperation risk and VaRR performance

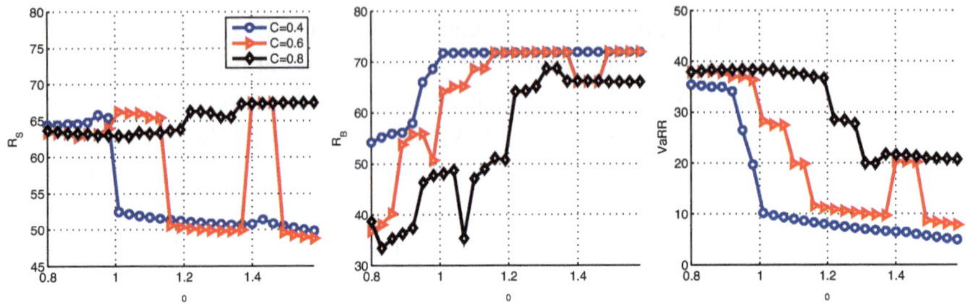

Fig. 8 The influences of regulating specialization level on systematic response risk, bilateral cooperation risk and VaRR performance

the control of bilateral moral hazards, and improving or engaging in the corporate social responsibility, have the effect of decreasing barrier height to a certain extent.

Here, how the barrier height influences the state evolution should be also interpreted from two aspects: one is the under-specialization, corresponding to a lower barrier. As a result, the capital-product switches achieve more easily, even with less capital input or random stimulus, and the more capital input, the more chaotic it performs. Consequently, increasing ν_0 could control the evolution from inter-equilibrium disordered to inter-equilibrium ordered. The other is the over-specialization (known as radical innovation), generally occurring in high-tech enterprises, where the evolution from capital to product should take a high risk to overcome the utility barrier. Thus, increasing ν_0 will lead to the system degeneration from inter-equilibrium ordered to intra-equilibrium disordered or from intra-equilibrium disordered to intra-equilibrium ordered. Therefore, it is necessary to effectively estimate the specialization situation before the control strategy is designed and implemented. For example, it is observed from Fig. 8 that VaRR performance is improved significantly within $\nu_0 \in [1.40, 1.45]$ for $C = 0.6$. Thus, ν_0 is an effective method to improve the performance by controlling it around the position of the SR peak. However, for $C = 0.8$, the SR peak at the position $\nu_0 = 1.4$ is weak, and the improvement is invalid or non-significant for strategy S3 only if ν_0 can be control to reach the region [0.80, 1.25]. Another intuitive result is the case for inadequate capital $C = 0.4$, and obviously, it is not a wise decision to blindly pursue technology dominance.

6.4 *Macro Climate*

Next, we analyze the influence of the macro climate index I_C, which directly impacts the input risk level $D = \frac{1}{m} K_B I_C \gamma_S$, i.e., the intensity of random fluctuation. The index control is usually based on the related policies that the nation or local government formulates and implements. For example, I_C might increase by formulating relevant policies to activate the vigor of technology innovations. On the other hand,

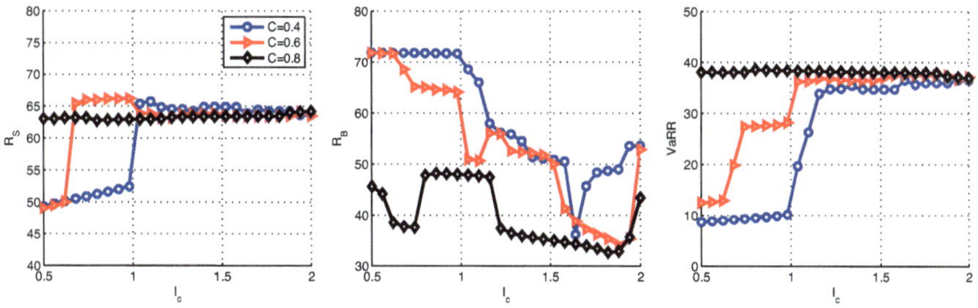

Fig. 9 The influences of regulating macro climate index on systematic response risk, bilateral cooperation risk and VaRR performance

the decrease in I_C might result from strengthened supervision or some restraint mechanisms on SMEs.

In Fig. 9, we observe the weak one-peak SR phenomenon of systematic risk for $C=0.4$ and 0.6 with increased I_C. The peak position shifts obviously to the left with more input capital C, but for $C=0.8$, SR phenomenon disappears, and the systematic risk level shows a slight increase, which is because of the occurrence of an over-capitalization situation. That is, the adjustment of increasing index within $I_C \in [0.5, 2.0]$ (corresponding to the input risk level $D \in [0.03125, 0.125]$) does not significantly impact on systematic risk. Nevertheless, for bilateral risk, we observe the apparent inhibition at $I_C=0.7$ and 1.8, and bell-shape SR within $I_C \in [0.8, 1.1]$. However, overall the bilateral risk level remains depressed compared to $C=0.4$ and 0.6.

Further, based on the VaRR curves, we conclude that macro climate index can be controlled to effectively improve VaRR performance for those enterprises trapping in under-capitalization situation, and there exists a threshold value (i.e., 1.2 for $C=0.4$, 1.0 for $C=0.6$), over which the performance rapidly tends to be stable. Still, for the enterprises with over-capitalization (i.e., $C=0.8$), their performance shows a non-significant decrease when implementing strategy S41 to increase I_C. Therefore, it is not wise to implement such a strategy here, and other replaceable strategies mentioned above should be seriously reconsidered.

This section has provided some quantitative results for SME system control under the measurements of system response risks and VaRR performance. It is found that any strategies are not always effective, particularly some non-monotonic phenomena against the intuition are instructive in practice.

7 The Real Case Studies from the Practice

The mathematical framework comprising the model and method in Sects. 4 and 5 have provided crucial quantitative support for risk management and performance optimization, and in this idealized setting, Sect. 6 have investigated the associated control effects under various possible adjustable dimensions. Further, the case studies, in this section, are carried out in the SMEs with the typical real-world profiles, which extracted from the large samples of manufacturing sector in China. In recent years, SMEs in this sector have been playing an important role in fostering economic growth, promoting employment growth, and stimulating the vitality of innovation. Here by generalizing to more complex and realistic settings, such as the number of strategic partners, the distribution of strategic capitals, and linearly increased capital lever, will further lead us to create useful strategies or decision rules, and also hep us comprehend some abnormal phenomena against the intuition and traditional judgement in risk management practices.

7.1 Empirical Data Description

Here, we define an SME as an enterprise with registered capital C_R less than 1 million in RMB yuan, and C_R is used to normalize the paid-in capital C_P, and forms the normalized internal capital $c_0 \triangleq \frac{C_P}{C_R}$. Similarly, the real external introduced capital C_j (from the strategic partner P_j, $j=1,2,\cdots,N$) is normalized by C_R, and forms the normalized external capital $c_j \triangleq \frac{C_j}{C_R}$.

Since some data and details cannot be gathered from financial institutes for confidentiality and proprietary reasons, we explored the support from enterprise big data platform. We completed the collection of 100,000 samples of SMEs and their associated layered networks from the manufacturing sector in China. Table 1 lists the results on mesoscopic profiles. The following cases focus on SMEs with typical profiles from the actual samples. By employing the quantitative control strategies mentioned in Sect. 6, we investigate the validity and reveal in practice how to stimulate the state to produce capital-product switches associated with SMEs' success from the perspectives of the micro profile, the mesoscopic structure, and the macro climate.

7.2 Mesoscopic Structure Control

It is supposed that SMEs are characterized by the profiles listed in Tables 1 and 3, where three types of internal capital levels, i.e., $l_0=0.5$, 0.4 and 0.3, are considered. Actually, under the condition of fixed internal capital $c_0=0.3904$, initial scale $m_0=\theta_0(1-l_0)c_0$ and operating capital $d_0=l_0c_0$ reflect the enterprise scale preferences: prudent ($l_0=0.5$), balanced ($l_0=0.4$) and radical ($l_0=0.3$), respectively. Moreover, once the strategic partners are recruited, the strategic capital c_j from partner P_j is also input into scale expansion $\Delta m_j=\theta_j(1-l_j)c_j$ and enterprise operation $\Delta c_j=l_jc_j$. Consequently, we suppose that the external capital lever l_j is linearly increased (LI) in order with an increasing number of strategic partners. Based on the results of

Refs. [13, 15], c_j is regarded as a random variable satisfying the power-law (PL) distribution with the estimated parameters $\widehat{\phi}=3.2368$, $\widehat{k}=4.6727\times10^{-4}$ and the average value $\overline{c}_j=0.1546$.

In cases 1~3, we consider the mesoscopic structure N, i.e., the number of strategic partners, as an adjustable dimension varying from 0 to 10, and the strategic capitals c_j, $j=1,2,\cdots,N$, are characterized by the independent and identical PL distribution. The results, as shown in Fig. 10, are described as follows. (i) Prudent scale: An enterprise with prudent scale preference involves systematic risk, as shown in the non-monotonic behavior of first increase then decrease with the increase of N; the risk peak occurs at $N=1$. The curve of bilateral risk performs the upside-down "N-shape", i.e., the valley and peak occur at $N=5$ and 6, respectively. However, overall, the bilateral risk is suppressed, compared with the case without any partners $N=0$. (ii) Balanced scale: An enterprise with a balanced scale reference behaves such that as N increases, systematic risk shows the non-monotonic behavior with one weak valley and peak each, non-significantly occurring at $N=1$ and 3 respectively. Furthermore, we observe that the valley and peak of bilateral risk shift to the right, and occur at $N=7$ and 8, respectively. (iii) Radical scale: In this scale, systematic

Table 3 The parameter setting for different strategies to control mesoscopic structure

	Micro profile							Mesoscopic structure				Macro climate	
Strategy	c_0	l_0	θ_0	m_0	d_0	γ_S	ν_0	N	c_j^a	l_j^b	θ_j	I_C	K_B
Case 1	0.3904	0.5	1.0	0.1952	0.1952	1.0	1.0	0~10	PL	LI: 0.5~ 0.8	1.0	1.0	0.025
Case 2	0.3904	0.4	1.0	0.2342	0.1562	1.0	1.0	0~10	PL	LI: 0.4~ 0.7	1.0	1.0	0.025
Case 3	0.3904	0.3	1.0	0.2733	0.1171	1.0	1.0	0~10	PL	LI: 0.3~ 0.6	1.0	1.0	0.025

[a]The normalized investment c_j from partner P_j satisfies the power-law (PL) distribution $P(c_j)=k\times c_j^{-\phi}$ with estimated parameters $\widehat{\phi}=3.2368$, $\widehat{k}=4.6727\times10^{-4}$, based on the SME large samples
[b]The capital lever l_j is considered as the linear increase (LI) with the increase of strategic partners

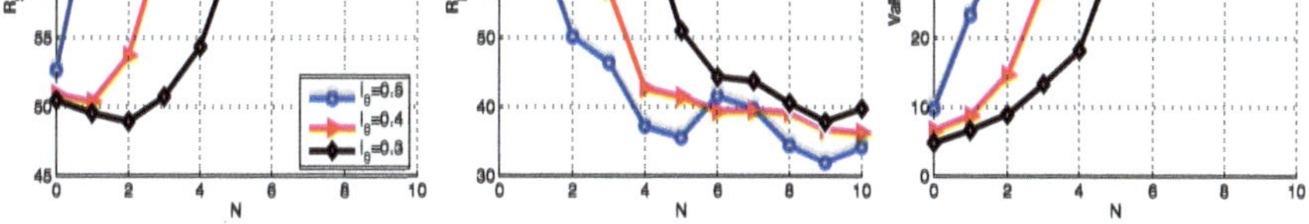

Fig. 10 The influences of regulating strategic partners on systematic response risk, bilateral cooperation risk and VaRR performance in typical SMEs with different scale preferences: **a** prudent scale $l_0=0.5$, **b** balanced scale $l_0=0.4$, and **c** radical scale $l_0=0.3$

risk first decreases and then monotonically increases; it is optimally suppressed at $N=2$. Bilateral risk shows a monotonic decrease with the increase of N.

Overall, if the enterprise is typical of the profiles listed in Table 3, the more significant capital lever leads to higher systematic risk and lower bilateral risk, and as more partners are introduced, the lever-related difference is negligible. Furthermore, based on the control effect on two types of response risks, it is an effective method to attract more strategic partners, which can improve the VaRR performance significantly. Thus, we observe that a larger capital lever causes better performance, especially for less N, and the marginal return gradually diminishes over a threshold number N_{th}, i.e., $N_{\text{th}}=3$ for prudent scale, $N_{\text{th}}=4$ for balanced scale, and $N_{\text{th}}=6$ for radical scale.

7.3 Micro Profile Regulation

In cases 4∼6, the focused SMEs are firstly described by the typical profiles illustrated in Table 4, where three types of scale preference respectively, with $l_0=0.5$, 0.4 and 0.3, are considered. We also suppose that the strategic capitals c_j, $j=1,2,\cdots,N$ are the independent random variables with identical PL distribution. With the increase of N, the capital ratio l_j of scale-related expanding input from c_j is supposed to be linearly increased (LI). Here, it is worth emphasizing that the number of strategic partners is further considered as the random variable with discrete values within 0∼10. For distinguishing the deterministic N, the number with random control is denoted as N_R. In the note of Table 4, the practical distribution has been illustrated based on the sample profiles of SMEs' mesoscopic structure, and $N_R>10$ are ignored, due to the sample ratio $P\{N_R>10\}<0.1\%$.

Following the description mentioned in Sect. 6, the strategies from the micro profile, including specialization level ν_0 and management orientation γ_S, are jointly considered to examine the suppressing or improving effect on response risks and

Table 4 The parameter setting for different strategies to adjust micro profile

	Micro profile							Mesoscopic structure				Macro climate	
Strategy	c_0	l_0	θ_0	m_0	d_0	γ_S	ν_0	N	c_j^a	l_j	θ_j	I_C	K_B
Case 4	0.3904	0.5	1.0	0.1952	0.1952	0.4∼ 1.2	0.6∼ 1.4	N_R	PL	LI: 0.5∼ 0.8	1.0	1.0	0.025
Case 5	0.3904	0.4	1.0	0.2342	0.1562	0.4∼ 1.2	0.6∼ 1.4	N_R	PL	LI: 0.4∼ 0.7	1.0	1.0	0.025
Case 6	0.3904	0.3	1.0	0.2733	0.1171	0.4∼ 1.2	0.6∼ 1.4	N_R	PL	LI: 0.3∼ 0.6	1.0	1.0	0.025

[a]The number of strategic partners is considered as the random variable with discrete number 0∼10, and it is re-denoted as N_R. It is supposed to satisfy the sample distribution from 0 to 10: {0.1849, 0.5743, 0.1289, 0.0421, 0.0369, 0.0146, 0.0072, 0.0054, 0.0025, 0.0021, 0.0010}

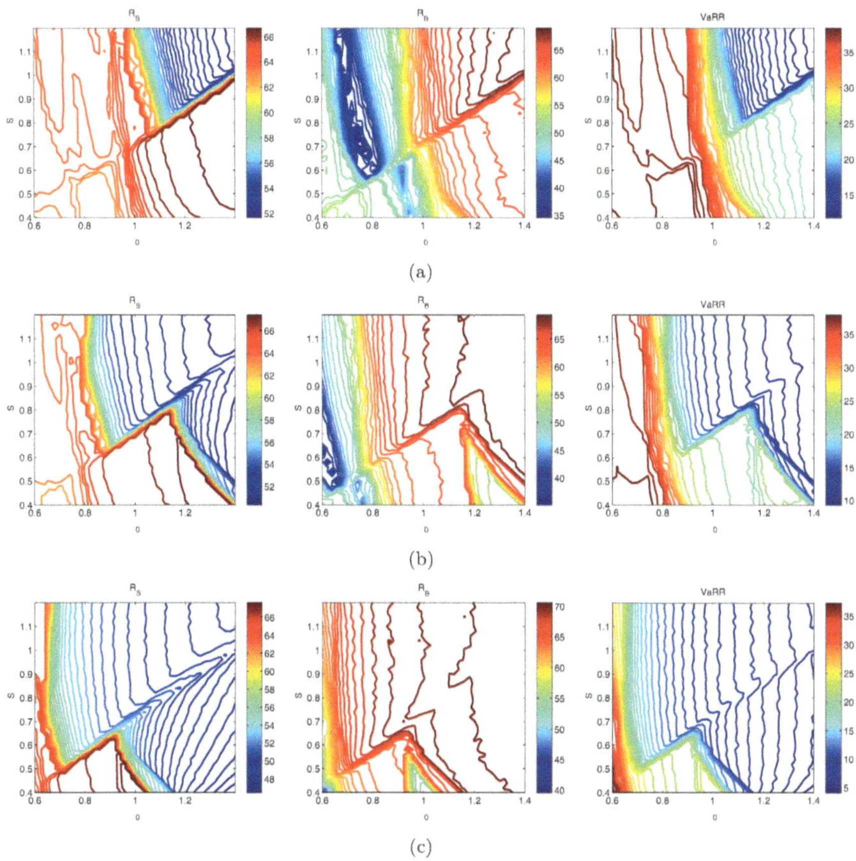

Fig. 11 The influences of jointly regulating management orientation and specialization level on systematic response risk, bilateral cooperation risk and VaRR performance in typical SMEs with different scale preferences: **a** prudent scale $l_0 = 0.5$, **b** balanced scale $l_0 = 0.4$, and **c** radical scale $l_0 = 0.3$

VaRR performance. Figure 11a~c shows these results, and we further explain them as follows.

Based on the levels of ν_0 and γ_S, the ν_0-γ_S two-dimension control plane is divided into four irregular regions, i.e., upper-left (U-L), upper right (U-R), lower left (L-L), and lower right (L-R) corresponding to the enterprise states characterized by specialization level (under/over-specialization, or interpreted as prudent/radical innovation) and management orientation (delicacy/loose). The response risk levels are measured by the color bar located at the right of each figure. Overall, the U-L and L-L regions perform higher systematic risk and lower bilateral risk, and the area shrinks to the left, varying with the scale preference from prudent to radical. Conversely, the U-R and L-R regions show lower systematic risk and higher bilateral risk, and the area expands to the left.

Further, one observes that an obsessive increase of specialization level with either technology or marketing (shown as the increase of ν_0) is an unwise decision due to the relatively loose management orientation (larger γ_S). We should thus focus on the higher region in Fig. 11. Here, we observe the monotonous decrease of VaRR performance, especial for prudent scale preference. Unless the enterprises adopt specialized delicacy management since their foundation (which is uncommon in practice), in the lower region for balanced and radical scale preference, i.e., $\gamma_S = 0.4$, VaRR performance shows weak SR behaviors. One weak peak occurs around at $\nu_0 = 1.26$ for the balanced scale and $\nu_0 = 0.99$ for the radical scale. Meanwhile, it is also foolish to obsessively emphasize on improving the management level from loose to delicacy, i.e., delicacy management of top-down command execution probably leads to the lack of enthusiasm and creativeness. Thus, in the left regions with under-specialization, we observe that the decrease of γ_S does not necessarily perform a significant improvement in VaRR performance, i.e., $\nu_0 \in [0.60, 0.92]$ for the prudent scale preference, $\nu_0 \in [0.60, 0.70]$ for the balanced scale preference. However, in some regions, i.e., specialization level $\nu_0 \in [1.04, 1.4]$ for the prudent scale preference, $\nu_0 \in [0.85, 1.4]$ for the balanced scale preference, $\nu_0 \in [0.70, 1.08]$ for the radical scale preference, if the management level can break through a threshold value (which is piecewise linear increase/decrease with the increase of ν_0, and forms a clear demarcation), the VaRR performance significantly improves, then stabilizes, and marginal return gradually fades to 0.

Based on the cases 4~6, we may conclude that the micro profiles, from management orientation and specialization level, should be jointly considered in risk management to effectively and accurately pursue the performance as expected; likewise, scale preference is an essential aspect. However, the strategies, i.e., blindly increasing specialization level, obsessively emphasizing on delicacy management, are not always effective and could be irrational to implement, regardless of an enterprise's status in practice.

7.4 Macro Climate Regulation

In the following cases 7~10, the focused SMEs are typical of the profiles included in Table 5 , where the industry climate index $I_C \in [0.8, 2.0]$ is regarded as a macro-control strategy. Specifically, **case 7** is to examine the control effect of I_C on system performance in the SMEs, with different specialization levels $\nu_0 = 0.8$, 1.0 and 1.2; **Case 8** considers the SMEs with more increased partners $N_R + \Delta_N$ (where $\Delta_N = 0$, 2 and 4), i.e., some governmental platforms are established to strengthen the mutual cooperation and help SMEs to recruit more strategic partners. In **case 9**, we consider more capital $0.3904 + \Delta_c$ (where $\Delta_c = 0$, 0.1 and 0.2) is involved and adjusted to increase both the scale-related input $0.2733 + \Delta_m$ and operating capital $0.1171 + \Delta_d$, according to the internal capital lever $l_0 = 0.3$, i.e., various SME lending programs are developed by financial institutions. Further, the increased capital Δ_c can be allocated between the scale-related input and operating capital, but in the extreme **Case 10**,

Table 5 The parameter setting for different strategies to regulate macro climate.

	Micro profile							Mesoscopic structure				Macro climate	
Strategy	c_0	l_0	θ_0	m_0	d_0	γ_S	ν_0	N	c_j	l_j	θ_j	I_C	K_B
Case 7	0.3904	0.3	1.0	0.2733	0.1171	1.0	0.8,1.0,1.2	N_R	PL	LI	1.0	0.8~2.0	0.025
Case 8	0.3904	0.3	1.0	0.2733	0.1171	1.0	1.0	$N_R+\Delta_N$	PL	LI	1.0	0.8~2.0	0.025
Case 9	0.3904+ Δ_c	0.3	1.0	0.2733+ Δ_m	0.1171+ Δ_d	1.0	1.0	N_R	PL	LI	1.0	0.8~2.0	0.025
Case 10	0.3904	0.3	1.0	0.2733	0.1171+ Δ_d	1.0	1.0	N_R	PL	LI	1.0	0.8~2.0	0.025

it is supposed that all the increased capital is input into the enterprise operation, only with the change $0.1171+\Delta_d$ (where $\Delta_d=0$, 0.1 and 0.2). For comparison and reference, the SME with the same parameters is included in all cases, i.e., the same results, VaRR performance with green curves, are plotted as a function of I_C in all four subfigures.

Figure 12 shows that the VaRR performance performs a significant or insignificant increase with the improvement of the macro climate in all four cases. Specifically, under the SME state of relative under-capitalization with $c_0=0.3904$ and under-cooperation with $\overline{N_R}=1.3089$ (refer to the note of Table 4), Fig. 12a demonstrates that the performance can be improved obviously, as I_C is controlled to increase within [1.36, 1.84] for under-specialization $\nu_0=0.8$. With the increase of specialization level, i.e., $\nu_0=1.0$, the control effect degenerates significantly, even if stimulated to reach $I_C=2$ by regulating macro climate. The higher the specialization level, the more inconspicuous the effect is. Based on the capital-product switching mechanism, there exists a cooperation relationship between the macro climate index and operating capital. However, it is a weak link due to the higher specialization-related technology barrier.

Naturally, another issue concerns with what profiles SMEs, the macro climate can be controlled to effectively improve the performance, in other words, as macro climate strategies, I_C is sensitive and effective to what kind of SME profiles. Therefore, in **case 8**, we consider the SMEs with $N_R+\Delta_N$, that is, Δ_N more strategic partners are recruited, and Δ_N is controlled to be 0, 2 and 4, corresponding to the blue, red and black curves respectively. One clearly observes that VaRR performance shows a stable increase with increasing I_C, and specially with the lower level of I_C, more partners conduct higher VaRR, as shown in Fig. 12b. Although more introduced partners lead to the increase of total input capital, actually the allocation between scale input and operation input can be adjusted and optimized by the external capital levers, to a certain extent. Here if we directly suppose that more capital Δ_c as internal capital, is involved to support the scale expansion and enterprise operation completely according to the internal lever $l_0=0.3$, as shown in Fig. 12c, the increased input capital cannot effectively stimulate capital-product switches to improve VaRR performance, that is, the enterprise traps into the dilemma of under-capitalization, mainly because of the radical scale preference ($l_0=0.3$) leading to the over-scaled situation, and with the increase of internal capital, the dilemma is still not improved. Therefore, in the extreme **case 10**, we suppose that all the increased capital is only input into the enterprise operation. It is seen from Fig. 12d that the macro climate strategy is sensitive and effective. More capital input to the operation conducts the higher VaRR improvement. This is because of the fact that, due to the cooperation of $\Delta_d=0.1$ or 0.2, the internal capital lever is adjusted from $l_0=0.3$ to 0.4427 and 0.5371, the enterprise gradually improves the dilemma of under-capitalization/over-scaled, thus the cooperative effect of macro climate regulation becomes significant by the control of I_C from 0.8 to 2.0. Of course, for the over-regulation reason, we observe that the marginal return fades to 0, and the performance increases and trends to a constant value. It is not hard to imagine that if more Δ_d is further introduced, the internal lever l_0 will be over-adjusted, thus the enterprise has the profile of extremely

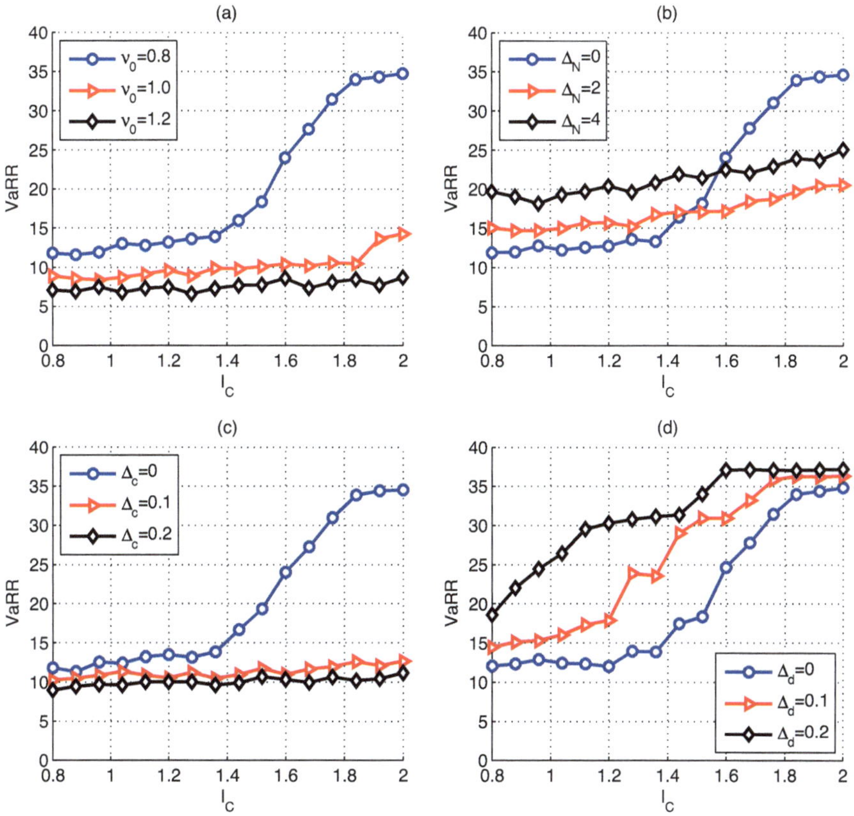

Fig. 12 The influences of regulating macro climate on VaRR performance in typical SMEs with different strategies: **a** adjusting specialization level ν_0, **b** introducing more strategic partners Δ_N, **c** increasing internal capital Δ_c, and **d** increasing operating capital Δ_d

prudent scale preference, and instead, and VaRR performance probably deteriorates with an increase of I_C.

Based on the above results and analysis, it is at present an effective strategy (see Figs. 10 and 12b), to a certain extent, for local/national government to improve the macro climate of SMEs, under the background of strengthening the cooperation or forming the industrial cluster. Nevertheless, it is more important to note that, for the SMEs themselves, some appropriate control methods and tactics from micro, mesoscopic, and macro dimensions are vital tools to ensure steady and coordinated development. Therefore, their situations should be fully recognized, including internal and external capital structures, specialization levels, and management orientations.

8 The Conclusion with Remarking

Before the end of this chapter, as mentioned in the beginning, our aim of this study is to establish an interpretable prototype system for modeling the development and growth mechanism of SMEs'growth under the tech-economic paradigm by incorporating Langevin dynamics by applying stochastic dynamical system approach with stochastic resonance theory developed in the 1980s ([27, 28], and reference wherein). In particular, by introducing the quantitative concept of the Value at Response Risks (VaRR), it allows us to describe the stochastic effect of the stochastic (random) forces by the phenomena of stochastic resonance caused by the stochastic dynamical system associated with an uncertainty environment for the process of the growth during SMEs' evolution and development.

At the same time, we also develop an associated quantitative method of risk management and growth optimization under the measurement of systematic response risk, bilateral cooperation risk, and VaRR performance. In addition, from the perspective of bistable stochastic resonance, as the inherent regularity of capital-product switches is applied together with both capital-related driving and risk environment stimulating, the discussion shows that the system is driven by the combination of evolution forces that compete and cooperate to produce an optimal control on response risk compression or VaRR improvement. Thus, by putting all work together given in this study (see Ref. [13, 15, 21, 25, 26, 45] for more discussion in detail), it seems that we establish an initial mathematical framework in explaining the dynamical mechanism of SMEs' growth with innovation for Schumpeter's economic development theory introduced around 1910's.

This research is sponsored by the National Natural Science Foundation of China (Nos. 72071019, 12226102, U1811462, and 71971031), Natural Science Foundation of Chongqing (No. CSTB2024NSCQ-MSX0551), and Fundamental Research Funds for the Central Universities (No. 2020CDJ-LHZZ-026).

Finally, we also like to thank two anonymous referees for their comments and suggestions which lead to the present version of this paper.

References

1. Gonzalez-Loureiro, M., Castelo, J.P.: A model for assessing the contribution of innovative smes to economic growth: the intangible approach. Econ. Lett. **116**(3), 312–315 (2012)
2. Tu, C., Hwang, S.N., Wong, J.Y.: How does cooperation affect innovation in micro-enterprises. Manag. Decis. **52**(8), 1390–1409 (2014)
3. Song, S., Zhang, X.: Lending technology and credit risk under different types of loans to smes: evidence from china. Int. Rev. Econ. Financ. **57**, 43–69 (2018)
4. Beck, T., Demirguc-Kunt, A., Peria, M.: Bank financing for smes: evidence across countries and bank ownership types. J. Financ. Serv. Res. **39**, 35–54 (2011)
5. Morgan, P., Pontines, V.: Financial stability and financial inclusion: case of sme lending. Singap. Econ. Rev. **63**(1), 111–124 (2018)
6. Zhang, X., Song, Z., Zhong, Z.: Does small bank advantage really exist? evidence from china. Int. Rev. Econ. Financ. **42**, 368–384 (2016)

7. Berger, A., Cowan, A., Frame, W.: The surprising use of credit scoring in small business lending by community banks and the attendant effects on credit availability, risk, and profitability. J. Financ. Serv. Res. **39**(1–2), 1–17 (2011)
8. Chen, Y., Huang, R.J., Tsai, J., et al.: Soft information and small business lending. J. Financ. Serv. Res. **47**(1), 115–133 (2015)
9. Altman, E., Sabato, G.: Modeling credit risk for smes: evidence from the us market. Abacus **43**(3), 332–357 (2014)
10. Bailey, W., Huang, W., Yang, Z.: Bank loans with chinese characteristics: some evidence on inside debt in a state-controlled banking system. J. Financ. Quant. Anal. **46**(6), 1795–1830 (2009)
11. Duan, H., Han, X., Yang, H.: An analysis of causes for smes financing difficulty. Int. J. Bus. Manag. **4**(6), 73–75 (2009)
12. Schumpeter, J.: The Theory of Economic Development: An Inquiry into Profits, Capital, Credit, Interest, and the Business Cycle (Translated by Opie R). Harvard University Press, Cambridge (1934)
13. Lin, L., Yuan, G., Wang, H., et al.: The stochastic incentive effect of venture capital in partnership systems with the asymmetric bistable cobb-douglas utility. Commun. Nonlinear Sci. **66**, 109–128 (2019)
14. Mohr, J., Spekman, R.: Characteristics of partnership success: partnership attributes, communication behavior, and conflict resolution techniques. Strateg. Manag. J. **15**(2), 135–152 (1994)
15. Yu, L., Wang, H., Lin, L., et al.: The incentive effect of venture capital in bilateral partnership systems with the bias mono-stable cobb-douglas utility. Nonlinear Dyn. **95**, 3127–3147 (2019)
16. Kale, P., Singh, H.: Building firm capabilities through learning: the role of the alliance learning process in alliance capability and firm-level alliance success. Strateg. Dir. **28**(2), 981–1000 (2008)
17. Schreiner, M., Kale, P., Corsten, D.: What really is management capability and how does it impact alliance outcomes and success. Strateg. Manag. J. **30**(13), 1395–1419 (2009)
18. Barlevy, G.: On the cyclicality of research and development. Am. Econ. Rev. **97**(4), 1131–1164 (2007)
19. Matha, T., Pierrard, O.: Search in the product market and the real business cycle. J. Econ. Dyn. Control **35**(8), 1172–1191 (2011)
20. Elbaz, J., Laguir, I., den Besten, M., et al.: Sowing the seeds: the impact of initial ties on growth and innovation among micro and small firms. Econ. B **34**(1), 970–983 (2017)
21. Li, X., Chen, K., Yuan, G., et al.: The decision-making of optimal equity and capital structure based on dynamical risk profiles: a langevin system framework for sme growth. Int. J. Intell. Syst. **36**(7), 3500–3523 (2021)
22. Perez, C.: Technological revolutions and techno-economic paradigms. Camb. J. Econ. **34**(1), 185–202 (2010)
23. Ren, R., Yuan, G.: The mechanics of dynamic behaviors for smes' growth by using fractional-order dynamic system approach. Int. J. Financ. Eng. **9**(1), 2250004 (2022)
24. Rosenberg, N.: Invention and economic growth. Econ. J. **84**(133), 90–108 (1974)
25. Wang, H., Yuan, W., Yuan, G.: The mechanism for smes growth by applying stochastic dynamical approach. Financ. Res. Lett. **48**, 102850 (2022)
26. Yuan, G., Wang, H.: The general dynamic risk assessment for the enterprise by the hologram approach in financial technology. Int. J. Financ. Eng. **6**(1), 1950001 (2019)
27. Benzi, R., Parisi, G., Sutera, A., Vulpiani, A.: A theory of stochastic resonance in climatic change. SIAM J. Appl. Math. **43**, 565–578 (1983)
28. Casado-Pascual, J., Gomez-Ordonez, J., Morillo, M., et al.: Two-state theory of nonlinear stochastic resonance. Phys. Rev. Lett. **91**, 210601 (2003)
29. Miyaguchi, T., Akimoto, T., Yamamoto, E.: Langevin equation with fluctuating diffusivity: a two-state model. Phys. Rev. E **94**(1), 012109 (2016)
30. Nava-Sedeno, J., Hatzikirou, H., Peruani, F., et al.: Extracting cellular automaton rules from physical langevin equation models for single and collective cell migration. J. Math. Bio. **75**(5), 1075–1100 (2017)

31. Reenbohn, W., Pohlong, S., Mahato, M.: Periodically driven underdamped periodic and washboard potential systems: dynamical states and stochastic resonance. Phys. Rev. E **85**(3), 031144 (2012)
32. Guo, Y., Xi, B., Shen, Y., et al.: Mean first-passage time of second-order and under-damped asymmetric bistable model. Appl. Math. Model. **40**(21–22), 9445–9454 (2016)
33. Bos, J., Economidou, C., Sanders, M.: Innovation over the industry life-cycle: evidence from eu manufacturing. J. Econ. Behav. Organ. **86**, 78–91 (2013)
34. Allanson, P., Montagna, C.: Multiproduct firms and market structure: an explorative application to the product life cycle. Int. J. Ind. Organ. **23**(7), 587–597 (2005)
35. Makridis, C.: Time to research? The cyclicality of time use in R&D among private and public sector workers. SSRN (2017). https://doi.org/10.2139/ssrn.3087060
36. Chung, T., Ji, U.: Gaussian white noise and applications to finance. Q. Inf. Complex (2004). https://doi.org/10.1142/9789812702449_0008
37. Elliott, R., Hoek, J.: A general fractional white noise theory and applications to finance. Math. Financ. **13**(2), 301–330 (2003)
38. Shah, S.A.: How risky is a random process? J. Math. Econ. **72**, 70–81 (2017)
39. Morgan, T., Anokhin, S., Wincent, J.: When the fog dissipates: the choice between value creation and value appropriation in a partner as a function of information asymmetry. J. Bus. Res. **88**, 498–504 (2018)
40. Rese, A., Baier, D.: Networks of promotors as success factor for radical innovations: theoretical considerations and empirical findings. In: Innovation Communities, Springer, Berlin, Heidelberg (2011). https://doi.org/10.1007/978-3-642-22128-6_2
41. Gao, S., Wang, H., Yuan, X., et al.: Cooperative mechanism of sme growth in the mesoscopic structure with strategic and nonstrategic partners. IEEE Intell. Syst. **35**(3), 7–18 (2020)
42. Rojas, M., Solis, E., Zhu, J.: Innovation and network multiplexity: r&d and the concurrent effects of two collaboration networks in an emerging economy. Res. Policy **47**(6), 1111–1124 (2018)
43. Wincent, J., Anokhin, S., Ortqvist, D.: Does network board capital matter? a study of innovative performance in strategic sme networks. J. Bus. Res. **63**(3), 265–275 (2010)
44. Basole, R., Rouse, W., Mcginnis, L., et al.: Models of complex enterprise networks. J. Enterp. Transform. **3**, 208–230 (2011)
45. Gao, S., Gao, N., Kan, B., et al.: Stochastic resonance in coupled star-networks with power-law heterogeneity. Phys. A **580**, 126155 (2021)
46. Peng, X., Song, W., Duan, Y.: Framework of open innovation in smes in an emerging economy: firm characteristics, network openness, and network information. Int. J. Technol. Manag. **62**, 223–250 (2013)
47. Cukier, R., Kapral, R., Lebenhaft, J., et al.: On the microscopic origin of stokes' law. J. Chem. Phys. **73**(10), 5244–5253 (1980)
48. Shaw, W., Schofield, M.: A model of returns for the post-credit-crunch reality: hybrid brownian motion with price feedback. Quant. Financ. **15**(6), 975–998 (2015)
49. Reichl, L.: A modern course in statistical physics. Am. J. Phys. **67**(12), 1285 (1999)
50. Gammaitoni, L., Hanggi, P., Jung, P., et al.: Stochastic resonance. Rev. Mod. Phys. **70**(1), 223–287 (1998)
51. Dykman, M., Haken, H., Hu, G., et al.: Linear response theory in stochastic resonance. Phys. Lett. A **180**, 332–336 (1993)
52. Pavliotis, G.: Linear response theory for diffusion processes. In: Stochastic Processes and Applications. Springer, New York (2014). https://doi.org/10.1007/978-1-4939-1323-7_9.
53. Aven, T.: On the meaning and use of the risk appetite concept. Risk Anal. **33**(3), 462–468 (2013)
54. Holm, H.J., Opper, S.: Entrepreneurs under uncertainty: an economic experiment in china. Manag. Sci. **59**(7), 1671–1687 (2013)
55. Choi, Y., Ye, X., Zhao, L., et al.: Optimizing enterprise risk management: a literature review and critical analysis of the work of wu and olson. Ann. Oper. Res. **237**(1–2), 281–300 (2016)
56. Fletcher, K.C., Abbas, A.E.: A value measure for public-sector enterprise risk management: a tsa case study. Risk Anal. **38**(5), 991–1008 (2018)

Two Turnpike Results for a Generalized von Neumann-Gale Model

Alexander J. Zaslavski

Abstract In this paper we study the turnpike phenomenon for a generalized von Neumann-Gale model and obtain two turnpike results which describe the structure of approximate optimal trajectories.

Keywords Convex process · Optimal trajectory · Turnpike · von Neumann-Gale model

1 Introduction

The study of the existence and the structure of solutions of variational problems, optimal control problems, and dynamic games defined on infinite intervals and on sufficiently large intervals has been a rapidly growing area of research [1, 5, 8, 17, 19, 45, 47, 49, 50] which has various applications in engineering [5, 45], in models of economic growth [5, 7, 10, 11, 16, 18, 21, 22, 25, 30, 38, 42, 44, 45], in infinite discrete models of solid-state physics related to dislocations in one-dimensional crystals [2, 43], in model predictive control [6] and in the theory of thermodynamical equilibrium for materials [23, 27–29]. Discrete-time problems optimal control problems were considered in [1, 3, 4, 9], finite-dimensional continuous-time problems were analyzed in [21, 24, 26, 33, 48], infinite-dimensional optimal control was studied in [5, 12, 13, 31, 32, 34, 37, 40, 41, 51] while solutions of dynamic games were discussed in [14, 15, 20, 35, 46].

In this paper we study the turnpike phenomenon for a generalized von Neumann-Gale model and obtain two turnpike results which describe the structure of approximate optimal trajectories. To have the turnpike property means, roughly speaking, that the approximate solutions of the problems are determined mainly by the objective function and are essentially independent of the choice of interval and endpoint conditions.

A. J. Zaslavski (✉)
Department of Mathematics, The Technion – Israel Institute of Technology, Haifa, Israel
e-mail: ajzasl@technion.ac.il

M. A. Khan et al. (eds.), *Matching, Dynamics and Games for the Allocation of Resources*, Monographs in Mathematical Economics 7,
https://doi.org/10.1007/978-981-96-6661-4_17

The turnpike property was discovered by P. Samuelson in 1948 when he showed that an efficient expanding economy would spend most of the time in the vicinity of a balanced equilibrium path (also called a von Neumann path). It is well known in the economic literature, where it was studied for various models of economic growth and in particular for the von Neumann-Gale model [25, 44, 45].

For each metric space Q denote by $\Pi(Q)$ the collection of nonempty compact subsets of Q endowed with the Hausdorff metric $d(\cdot, \cdot)$.

Let $K \subset R^n$ be a closed convex cone satisfying

$$K + K \subset K. \lambda K \subset K, \ \lambda > 0$$

with a nonempty interior in R^n which does not contain any nontrivial subspace of R^n. A set-valued mapping $G : K \to \Pi(R^n)$ is a convex process it satisfies

$$\lambda G(x) + \mu G(y) \subset G(\lambda x + \mu y)$$

for each $x, y \in K$ and each $\lambda, \mu \geq 0$, or equivalently, if its graph is a convex cone [25, 36].

Convex processes are the set-valued analogs of linear mappings and certain properties of the latter can be also established for the former. Denote by $\mathcal{M}$ the set of all convex processes $G : K \to \Pi(R^n)$ such that the set

$$\text{graph}(G) = \{(x, y) \in R^n \times R^n : \ x \in K, \ v \in G(x)\}$$

is closed and $G(x) \cap K \neq \emptyset$ for each $x \in K$. Denote by $\langle \cdot, \cdot \rangle$ the inner product in R^n and by $\| \cdot \|$ the Euclidean norm in R^n. We say that $x \leq y$ $(x, y \in R^n)$ if and only if $y - x \in K$. Set

$$K^* = \{\eta \in R^n : \langle \eta, x \rangle \geq 0 \text{ for every } x \in K\}.$$

It is a standard result (see [22]) that K^* has a nonempty interior. The interior of a set S is denoted by int$S$ and its boundary is denoted by ∂S.

Let $G \in \mathcal{M}$. A real number λ_0 is called an eigenvalue of G is there exists $x_0 \in K \setminus \{0\}$ such that $\lambda_0 x_0 \in G(x_0)$. We then call x_0 an eigenvector of G. A sequence $\{x_t\}_{t=0}^{T} \subset K$ where $T \in \{1, 2, \dots\} \cup \{\infty\}$ is called a trajectory of G if $x_{t+1} \in G(x_t)$ for all nonnegative integers $t < T$.

In this paper we denote by Card(A) the cardinality of a set A and assume that the sum over an empty set is zero.

2 Asymptotic Turnpike Property

Let K be a convex closed cone with a nonempty interior in R^n which does not contain any nontrivial subspace of R^n. Consider a convex process $G : K \to \Pi(R^n)$ which is a continuous mapping. It may happen that there are no eigenvalues of G in K at all.

However, it is easy to see that the set of eigenvalues of G in K is compact. Denote by α_G the maximal eigenvalues of G. Set

$$\alpha_0 \equiv \inf_{\eta \in K^* \setminus \{0\}} \sup_{x \in K \setminus \{0\}} \{\langle \eta, v \rangle \langle \eta, x \rangle^{-1} : v \in G(x)\}. \tag{1}$$

Leizarowitz [22] studied the existence of eigenvectors of convex processes and established the following result.

Proposition 1 *Suppose that if $x \in \partial K \setminus \{0\}$ and $\eta \in \partial K^* \setminus \{0\}$ are such that $\langle \eta, x \rangle = 0$, then $\langle \eta, v \rangle > 0$ for some $v \in G(x)$. Then the infimum in (1) is attained inside of $\mathrm{int} K^*$ and there exists $x_0 \in K \setminus \{0\}$ for which $\alpha_0 x_0 \in G(x_0)$ and $\alpha_0 = \alpha_G$.*

Note that the existence of Perron-Frobenius eigenvalues under different assumptions was obtained in [39].

Assume that $(X, \alpha_G X) \in \mathrm{graph}(G)$. We say that G has the asymptotic turnpike property if for any trajectory $\{x_t\}_{t=1}^{\infty}$ of G there exists a number $c_0 \geq 0$ such that

$$\lim_{t \to \infty} \alpha_G^{-t} x_t = c_0 X.$$

The asymptotic turnpike property for convex processes was studied in [22, 25, 44, 45].

The following result which is a discrete-time analog of Theorem 6.3 in [22] gives the necessary and sufficient condition for the asymptotic turnpike property.

Theorem 1 *Suppose that there is $y \in \partial K \setminus \{0\}$ for which $G(y) \cap (K \setminus \{0\}) \neq \emptyset$ and if $x \in \partial K \setminus \{0\}$ and $\eta \in \partial K^* \setminus \{0\}$ are such that $\langle \eta, x \rangle = 0$, then $\langle \eta, v \rangle > 0$ for some $v \in G(x)$. Assume that λ_0, the maximal eigenvalue of G, corresponds to a unique (up to multiplication by positive numbers) eigenvector denoted X. Then for any trajectory $\{x_t\}_{t=0}^{\infty}$ of G there exists a number $c_0 \geq 0$ such that*

$$\lim_{t \to \infty} \alpha_G^{-t} x_t = c_0 X$$

if and only if for any trajectory $\{z_t\}_{t=0}^{\infty}$ of G in the form

$$z_t = \lambda_0^t (\cos(\omega t) a + \sin(\omega t) b + X), \ t = 0, 1, \ldots$$

for some $a, b \in R^n$, $\omega \in R^1$ the following relation holds:

$$z_t = c_0 \lambda_0 X, \ t = 0, 1,$$

with some constant $c_0 \geqq 0$.

Roughly speaking, according to Theorem 1, the necessary and the sufficient condition for the asymptotic turnpike property is the absence of periodicity.

The proof of Theorem 1 is analogous to the proof of Theorem 6.3 in [22]. One of the key ingredients in this proof is provided by the following result which is a discrete-time analog of Theorem 3.2 in [21].

Proposition 2 *Let $F \subset R^n \times R^n$ be a convex compact set,*

$$domF = \{x \in R^n : \text{ there exists } y \in R^n \text{ such that } (x, y) \in F\},$$

$$F(x) = \{y \in R^n : (x, y) \in F\}, \ x \in domF.$$

Assume that the following properties hold:
(i) there is a unique $Y \in domF$ such that $Y \in F(Y)$;
(ii) if $a, b \in R^n$, $\alpha \neq 0$ and a sequence

$$z_j = Y + \cos(\alpha j)a + \sin(\alpha j)b, \ j = 0, 1, \dots$$

satisfies

$$(z_j, z_{j+1}) \in F, \ j = 0, 1, \dots,$$

then $z_j = Y$, $j = 0, 1, \dots$.
Then for each $\epsilon > 0$ there exists an integer $N(\epsilon) \geq 1$ such that for each sequence $\{z_j\}_{j=0}^{\infty} \subset R^n$ which satisfies $(z_j, z_{j+1}) \in F$, $j = 0, 1, \dots$ the inequality $\|z_j - Y\| \leq \epsilon$ holds for all $j \geq N(\epsilon)$.

3 Turnpike Results

Let $G \in \mathcal{M}$. For each $x \in K$ and each integer $T \geq 1$ denote

$$a^T(x) = \{y \in R^n : \text{ there exists a trajectory } \{x_t\}_{t=0}^T \text{ of } G$$

$$\text{for which } x_0 = x, \ x_T = y\}.$$

Evidently, $a^T(x)$ is a convex compact set for each $x \in K$ and each integer $T \geq 1$.

Let T be a natural number. A trajectory $\{x_t\}_{t=0}^T$ of G is optimal if there exists $f \in K^* \setminus \{0\}$ such that

$$\langle f, x_T \rangle = \sup\{\langle f, y \rangle : y \in a^T(x)\}.$$

Suppose that

$$G \in \mathcal{M}, \ \alpha_G > 0, \ X \in \text{int}K, \ \|X\| = 1, \ \alpha_G X \in G(X), \ p \in \text{int}K^*, \ \langle p, X \rangle = 1, \tag{2}$$

$$\alpha_G \text{ is a maximal eigenvalue of } G,$$

$$\langle \alpha_G p, x\rangle \geq \langle p, y\rangle \text{ for each } x \in K \text{ and each } y \in G(x).$$

We have the following result (see Proposition 11.3.1 of [45]).

Proposition 3 *Assume that $x_0 \in intK$, $T \geq 1$ is an integer and that $\{x_t\}_{t=0}^T$ is a trajectory of G. Then this trajectory is optimal if and only if for each $\lambda > 1$, the relation $\lambda x_T \notin a^T(x_0) - K$ holds.*

Suppose that for each trajectory $\{x_t\}_{t=0}^\infty$ of G there exists a number $c_0 \geq 0$ such that

$$\alpha_G^{-t} x_t \to c_0 X \text{ as } t \to \infty. \tag{3}$$

For each $x \in K$ we define

$$\lambda(x) = \sup\{\beta \in [0, \infty) : \text{ there exists a trajectory } \{x_t\}_{t=0}^T \text{ of } G$$

$$\text{for which } x_0 = x, \lim_{t\to\infty} \alpha_G^{-t} x_t = \beta X\}. \tag{4}$$

It is easy to verify that for each $x \in K$ the number $\lambda(x)$ is finite and there exists a trajectory $\{x_t\}_{t=0}^T$ of G for which $x_0 = x$ and $\alpha_G^{-t} x_t \to \lambda(x)X$ as $t \to \infty$. It is not difficult to see that $\lambda(cx) = c\lambda(x)$ for each $x \in K$ and each $c > 0$, each $x \in \text{int}K$ is a continuity point of the function $\lambda : K \to [0, \infty)$ and $\lambda(x) \leq \lambda(y)$ for each $x, y \in K$ satisfying $x \leq y$.

In [44] we proved the following two results (see also Theorems 11.3.1 and 11.3.2 of [45]) which describes the structure of optimal trajectories of G.

Theorem 2 *Assume that a compact nonempty set $Q \subset intK$ and $\epsilon > 0$. Then there exists an integer $L \geq 1$ such that for each integer $T \geq 2L$ and each optimal trajectory $\{x_t\}_{t=0}^T$ of G satisfying $x_0 \in Q$ the following inequality holds:*

$$\|\lambda(x_0)X - \alpha_G^{-t} x_t\| \leq \epsilon, \; t = L, \ldots, T - L.$$

Theorem 3 *Let $\epsilon > 0$. Then there exist an integer $L \geq 1$ and a number $\delta > 0$ such that for each integer $T \geq L$ and each optimal trajectory $\{x_t\}_{t=0}^T$ of G satisfying $\|x_0 - X\| \leq \delta$ the following inequality holds:*

$$\|X - \alpha_G^{-t} x_t\| \leq \epsilon, \; t = 0, \ldots, T - L.$$

In this paper we prove the following two results which describe the structure of approximate optimal trajectories of G.

Theorem 4 *Assume that a compact nonempty set $Q \subset K$,*

$$\Delta_0 = \inf\{\langle p, z\rangle : z \in Q\} > 0, \; \Delta_1 = \sup\{\langle p, z\rangle : z \in Q\} \tag{5}$$

$\epsilon > 0$ *and* $c \in (0, 1)$. *Then there exist integers* $L \geq 1, \tilde{q} \geq 1$ *such that for each integer* $T > L$ *and each trajectory* $\{x_t\}_{t=0}^T$ *of* G *satisfying* $x_0 \in Q$ *and*

$$c\langle \alpha_G^T p, x_0 \rangle \leq \langle p, x_T \rangle$$

there exist an integer $q \in \{1, \dots, \tilde{q}\}$, *integers* $a_i < b_i$, $i = 1, \dots, q$ *belonging to the interval* $[0, T]$ *and numbers*

$$\lambda_i \in [\Delta_0/4, \Delta_1 + 1], \; i = 1, \dots, q$$

such that for every $i \in \{1, \dots, q\}$,

$$\|\alpha_G^{-t} x_t - \lambda_i X\| \leq \epsilon, \; t = a_i, \dots, b_i$$

and

$$Card(\{0, \dots, T\} \setminus \cup_{i=1}^q \{a_i, \dots, b_i\}) \leq L.$$

Theorem 5 *Assume that a compact nonempty set* $Q \subset K$ *satisfies*

$$\inf\{\langle p, z \rangle : z \in Q\} > 0$$

and $\gamma, c \in (0, 1)$. *Then there exist integers* $L \geq 1$ *such that for each integer* $T > L$ *and each trajectory* $\{x_t\}_{t=0}^T$ *of* G *satisfying* $x_0 \in Q$ *and*

$$c\langle \alpha_G^T p, x_0 \rangle \leq \langle p, x_T \rangle$$

the inequality

$$Card(\{0, \dots, T\} : \| \|x_t\|^{-1} x_t - X\| > \gamma\}) \leq L$$

holds.

It is not difficult to see that in the proofs of these two results, we may assume without loss of generality that

$$\alpha_G = 1.$$

It should be mentioned that Theorems 2 and 3 describe the structure of optimal trajectories while Theorems 4 and 5 establish the turnpike property for approximate optimal trajectories. The parameters γ, ϵ, c in Theorems 4 and 5 are fixed and the constant $L, \tilde{q}$ depends on them.

4 An Auxiliary Result

Lemma 1 *Let* $0 < c_0 \leq c_1$, $\epsilon \in (0, c_0/4)$. *Then there exist*

$$\delta \in (0, c_0/2)$$

and an integer $L_0 \geq 1$ *such that for each integer* $T > 2L_0$ *and each trajectory* $\{x_t\}_{t=0}^{T}$ *of* G *satisfying*

$$\langle p, x_0 \rangle \in [c_0, c_1], \leq \langle p, x_T \rangle \geq \langle p, x_0 \rangle - \delta$$

there exists $\lambda \geq c_0/4$ *such that*

$$\|x_t - \lambda X\| \leq \epsilon, \ t \in \{L_0, \ldots, T - L_0\}.$$

Proof We may assume without loss of generality that

$$\epsilon(\|p\| + 1) < c_0/4.$$

Choose $\epsilon_0 \in (0, \epsilon)$ such that

$$2\epsilon_0(1 + 2\|p\|) < \epsilon. \tag{6}$$

First we show that there exist a natural number L_0 and $\delta \in (0, c_0/2)$ such that the following property holds:

(a) for each integer $T > 2L_0$ and each trajectory $\{x_t\}_{t=0}^{T}$ of G satisfying

$$\langle p, x_0 \rangle \in [c_0, c_1], \leq \langle p, x_T \rangle \geq \langle p, x_0 \rangle - \delta$$

for every $t \in \{L_0, \ldots, T - L_0\}$ there exists $\lambda_t \geq 0$ such that

$$\|x_t - \lambda_t X\| \leq \epsilon.$$

Assume the contrary. Then for each integer $k \geq 1$ there exists an integer

$$T_k > 2k \tag{7}$$

and a trajectory $\{x_t^{(k)}\}_{t=0}^{T_k}$ of G satisfying

$$\langle p, x_0^{(k)} \rangle \in [c_0, c_1], \tag{8}$$

$$\langle p, x_{T_k}^{(k)} \rangle \geq \langle p, x_0^{(k)} \rangle - (2k)^{-1} c_0 \tag{9}$$

and there exists

$$\tau_k \in \{k, \ldots, T_k - k\} \tag{10}$$

such that

$$\inf\{\|x^{(k)}_{\tau_k} - \lambda X\| : \lambda \in [0, \infty)\} > \epsilon_0. \tag{11}$$

Let $k \geq 1$ be an integer. Define

$$y^{(k)}_t = x^{(k)}_{t+\tau_k},\ t \in \{-\tau_k, \dots, T_k - \tau_k\}. \tag{12}$$

By (7)–(10) and (12),

$$\langle p, y^{(k)}_{-\tau_k} \rangle \in [c_0, c_1], \tag{13}$$

$$\langle p, y^{(k)}_{T_k-\tau_k} \rangle \geq \langle p, y^{(k)}_{-\tau_k} \rangle - (2k)^{-1} c_0. \tag{14}$$

By (11) and (12),

$$\inf\{\|y^{(k)}_0 - \lambda X\| : \lambda \in [0, \infty)\} > \epsilon_0. \tag{15}$$

In view of (8) and (9), the set

$$\{x^{(k)}_t : t \in \{0, \dots, T_k\},\ k = 1, 2, \dots\}$$

is bounded in the norm topology. Using (8), (10), (12), the diagonalization process and extracting subsequences we may assume without loss of generality that for each integer t there exists

$$y_t = \lim_{k \to \infty} y^{(k)}_t. \tag{16}$$

It follows from (8), (9), (12), (15) and (16) that for each integer t,

$$y_{t+1} \in G(y_t), \tag{17}$$

$$\langle p, y_t \rangle \in [c_0/2, c_1], \tag{18}$$

$$\langle p, y_{t+1} \rangle = \langle p, y_t \rangle, \tag{19}$$

$$\inf\{\|y_0 - \lambda X\| : \lambda \in [0, \infty)\} \geq \epsilon_0. \tag{20}$$

Fix a positive number ϵ_1 satisfying

$$\epsilon_1 < \min\{2^{-1}, \epsilon_0\},\ (\|p\| + 1)\epsilon_1 \|y_0 - X\| \leq \epsilon_0/8. \tag{21}$$

For each integer t set

$$\tilde{y}_t = \epsilon_1 X + (1 - \epsilon_1) y_t. \tag{22}$$

By (2), (13), (14), (17), (19), (21), (22) the equation $\alpha_G = 1$, for each integer t,

$$\langle p, \tilde{y}_t \rangle \in [c_0/4, c_1 + 1], \tag{23}$$

$$\tilde{y}_{t+1} \in G(\tilde{y}_t),\ \langle p, \tilde{y}_{t+1}\rangle = \langle p, \tilde{y}_t\rangle. \tag{24}$$

By (22),

$$\inf\{\|\tilde{y}_0 - \lambda X\| :\ \lambda \in [0, \infty)\} \ge \inf\{\|(1-\epsilon_1)y_0 + \epsilon_1 X - \lambda X\| :\ \lambda \in [0, \infty)\}$$

$$\ge \inf\{\|y_0 - \lambda X\| :\ \lambda \in [0, \infty)\} - \epsilon_1\|y_0 - X\| \ge \epsilon_0 - \epsilon_1\|y_0 - X\| \ge \epsilon_0 - \epsilon_0/8. \tag{25}$$

Denote by Q the closure of the set $\{\tilde{y}_t :\ t = 0, \pm 1, \pm 2, ...\}$. In view of (22), the set Q is bounded and

$$Q \subset \mathrm{int}(K). \tag{26}$$

Theorem 2 implies that there exists an integer $L_1 \ge 1$ such that the following property holds:

(b) for each integer $T \ge 2L_1$ and each optimal sequence $\{x_t\}_{t=0}^T$ of G satisfying $x_0 \in Q$,

$$\|\lambda(x_0)X - x_t\| \le \epsilon_0/4,\ t = L_1, \dots, T - L_1.$$

In view of (24), property (b) and the definition of Q,

$$\tilde{y}_{-L_1} \in Q,$$

$$\|\lambda(\tilde{y}_{-L_1})X - \tilde{y}_0\| \le \epsilon_0/4.$$

This contradicts (25). The contradiction we have reached proves that there exist $\delta \in (0, c_0/2)$ and a natural number L_0 such that property (a) holds. We may assume that $\delta < \epsilon_0$.

Assume that an integer $T > 2L_0$ and a trajectory $\{x_t\}_{t=0}^T$ of G satisfies

$$\langle p, x_0\rangle \in [c_0, c_1],\ \le \langle p, x_T\rangle \ge \langle p, x_0\rangle - \delta. \tag{27}$$

Property (a) and (27) implies that for every $t \in \{L_0, \dots, T - L_0\}$ there exists $\lambda_t \ge 0$ such that

$$\|x_t - \lambda_t X\| \le \epsilon_0. \tag{28}$$

Let $t \in \{L_0, \dots, T - L_0\}$. By (2) and (27),

$$\langle p, x_T\rangle \le \langle p, x_t\rangle \le \langle p, x_{L_0}\rangle \le \langle p, x_T\rangle + \delta \le \langle p, x_t\rangle + \delta. \tag{29}$$

It follows from (2), (28) and (29) that

$$\mid \lambda_t - \lambda_{L_0} \mid = \mid \langle p, \lambda_t X\rangle - \langle p, \lambda_{L_0} X\rangle \mid$$

$$\le \mid \langle p, x_t\rangle - \langle p, x_{L_0}\rangle \mid$$

$$+ \mid \langle p, x_t - \lambda_t X \rangle \mid + \mid \langle p, x_{L_0} - \lambda_{L_0} X \rangle \mid$$

$$\leq 2 \mid p \mid \epsilon_0 + \delta. \tag{30}$$

By (2), (6), (30) and the inequality $\delta < \epsilon_0$,

$$\|x_t - \lambda_{L_0} X\| \leq \|x_t - \lambda_t X\| + \mid \lambda_t - \lambda_{L_0} \mid$$

$$\leq \epsilon_0 + 2\|p\|\epsilon_0 + \delta < \epsilon. \tag{31}$$

Equations (2), (27) and (31) imply that

$$\langle p, x_{L_0} \rangle \geq \langle p, x_T \rangle \geq \langle p, x_0 \rangle - \delta \geq c_0 - \delta \geq c_0/2,$$

$$\lambda_{L_0} = \langle p, \lambda_{L_0} X \rangle \geq \langle p, x_{L_0} \rangle - \langle p, x_{L_0} - \lambda X \rangle \geq c_0/2 - \epsilon \mid p \mid \geq c_0/4.$$

Lemma 1 is proved. □

5 Proof of Theorem 4

Recall that

$$\Delta_0 = \inf\{\langle p, z \rangle : z \in Q\} > 0, \tag{31}$$

$$\Delta_1 = \sup\{\langle p, z \rangle : z \in Q\} < \infty. \tag{32}$$

We may assume without loss of generality that

$$\epsilon(\|p\| + 1) < 1.$$

Lemma 1 implies that there exist

$$\delta \in (0, \epsilon)$$

and an integer $L_0 \geq 1$ such that the following property holds:

(a) for each integer $T > 2L_0$ and each trajectory $\{x_t\}_{t=0}^T$ of G satisfying

$$\langle p, x_0 \rangle \in [c\Delta_0, \Delta_1], \ \langle p, x_T \rangle \geq \langle p, x_0 \rangle - \delta$$

there exists

$$\lambda > c\Delta_0/4 \tag{33}$$

such that

$$\|x_t - \lambda X\| \leq \epsilon, \ t \in \{L_0, \ldots, T - L_0\}.$$

Choose integers

$$\tilde{q} > 1 + \delta^{-1}\Delta_1, \tag{34}$$

$$L \geq 2\tilde{q}(2L_0 + 6). \tag{35}$$

Assume that $T > L$ is an integer and a trajectory $\{x_t\}_{t=0}^T$ of G satisfies

$$x_0 \in Q,\ c\langle p, x_0\rangle \leq \langle p, x_T\rangle. \tag{36}$$

By (31), (32) and (36),

$$\langle p, x_0\rangle \leq \Delta_1,\ c\Delta_0 \leq c\langle p, x_0\rangle \leq \langle p, x_T\rangle. \tag{37}$$

In view (2), (37) and the equation $\alpha_G = 1$, for each $t \in \{0, \dots, T\}$,

$$c\Delta_0 \leq \langle p, x_t\rangle \leq \Delta_1. \tag{38}$$

Set

$$t_0 = 0.$$

If

$$\langle p, x_T\rangle \geq \langle p, x_0\rangle - \delta,$$

then we set

$$t_1 = T.$$

Assume that

$$\langle p, x_T\rangle < \langle p, x_0\rangle - \delta.$$

Then there exists an integer $t_1 \in (0, T]$ such that

$$\langle p, x_{t_1}\rangle < \langle p, x_0\rangle - \delta$$

and if $t_1 > 1$, then

$$\langle p, x_{t_1 - 1}\rangle \geq \langle p, x_0\rangle - \delta.$$

Assume that $k \geq 1$ is an integer, integers $t_0 < t_1 < \cdots < t_k$ belong to the interval $[0, T]$, $t_0 = 0$, for each integer $k \geq 1$,

$$\langle p, x_{t_k - 1}\rangle > \langle p, x_{t_k}\rangle + \delta$$

and if $t_k > t_{k-1} + 1$, then

$$\langle p, x_{t_k - 1}\rangle > +\delta \geq \langle p, x_{t_{k=1}}\rangle.$$

(Clearly, our assumption holds for $k = 1$.) If $t_k = T$, then the construction is completed. Assume that $t_k < T$. If

$$\langle p, x_T \rangle \geq \langle p, x_{t_k} \rangle - \delta,$$

then we set

$$t_{k+1} = T$$

the construction is completed. Assume that

$$\langle p, x_T \rangle < \langle p, x_{t_k} \rangle - \delta.$$

Then there exists an integer $t_{k+1} \in (t_k, T]$ such that

$$\langle p, x_{t_{k+1}} \rangle < \langle p, x_{t_k} \rangle - \delta$$

and if $t_{k+1} > t_k + 1$, then

$$\langle p, x_{t_{k+1}-1} \rangle \geq \langle p, x_{t_k} \rangle - \delta.$$

Clearly, the assumption made for k also holds for $k + 1$. Therefore by induction we showed the existence of an integer $q \geq 1$ and integers $t_0 < t_1 \cdots < t_q$ belonging to $[0, T]$ such that

$$t = 0, \ t_q = T,$$

for each $i \in \{1, \ldots, q\} \setminus \{q\}$,

$$\langle p, x_{t_i} \rangle < \langle p, x_{t_i-1} \rangle - \delta \tag{39}$$

and for each $i \in \{1, \ldots, q\}$, if $t_i > t_{i-1} + 1$, then

$$\langle p, x_{t_i-1} \rangle \geq \langle p, x_{t_{i-1}} \rangle - \delta. \tag{40}$$

It follows from (2), (37) and (39) that

$$\Delta_1 \geq \langle p, x_0 \rangle - \langle p, x_T \rangle \geq \sum_{i=0}^{q-1} (\langle p, x_{t_i} \rangle - \langle p, x_{t_{i+1}} \rangle)$$

$$\geq \sum \{(\langle p, x_{t_i} \rangle - \langle p, x_{t_{i+1}} \rangle) : \ i \in \{0, \ldots, q-1\} \setminus \{q-1\}\}$$

$$\geq \delta(q-1)$$

and

$$q \leq 1 + \delta^{-1}\Delta_1 < \tilde{q}. \tag{41}$$

Set

$$E = \{i \in \{0, \ldots, q-1\} : \ t_{i+1} - t_i > 2L_0 + 4\}. \tag{42}$$

Let

$$i \in E. \tag{43}$$

Consider the trajectory $\{x_t\}_{t=t_i}^{t_{i+1}-1}$. By (38), (40), (42) and (43),

$$\langle p, x_{t_{i+1}-1}\rangle \geq \langle p, x_{t_i}\rangle - \delta, \tag{44}$$

$$\langle p, x_{t_i}\rangle \in [c\Delta_0, \Delta_1]. \tag{45}$$

Property (a) and Eqs. (33), (42)–(45) applied to the trajectory $\{x_t\}_{t=t_i}^{t_{i+1}-1}$ imply that there exists $\lambda_i \geq c\Delta_0/4$ such that

$$\|x_t - \lambda_i X\| \leq \epsilon, \ t \in \{t_i + L_0, \ldots, t_{i+1} - 1 - L_0\}. \tag{46}$$

By (2), (38), (46) and the inequality $\epsilon \mid p \mid < 1$,

$$\lambda_i = \lambda_i\langle p, X\rangle \leq \langle p, x_{t_i-L_0}\rangle + \langle p, \lambda_i X - x_{t_i-L_0}\rangle$$

$$\leq \Delta_1 + \epsilon\|p\| \leq \Delta_1 + 1. \tag{47}$$

Set for every $i \in E$,

$$a_i = t_i - L_0, \ b_i = t_{i+1} - 1 - L_0. \tag{48}$$

By (48),

$$\{0, \ldots, T\} \setminus \cup_{i\in E}\{a_i, \ldots, b_i\}$$

$$\subset \cup\{\{t_i, t_{i+1}\} : \ i \in \{1, \ldots, q\} \setminus E\} \cup \{\{t_i, \ldots, t_i + L_0 - 1\} \cup \{t_{i+1} - L_0, \ldots, t_{i+1}\} : \ i \in E\}$$

and by (35), (41) and (42),

$$\mathrm{Card}(\{0, \ldots, T\} \setminus \cup_{i\in E}\{a_i, \ldots, b_i\})$$

$$\leq (2L_0 + 6)q + q(2L_0 + 2) \leq 2(2L_0 + 6)q \leq 2\tilde{q}(2L_0 + 6) \leq L.$$

Theorem 4 is proved.

6 Proof of Theorem 5

Choose a positive number

$$\epsilon < 32^{-1}\Delta_0\gamma. \tag{49}$$

Let $\Delta_1 > \Delta_0 > 0$ be defined by (31), (32) and integers $\tilde{q}, L$ be as guaranteed by Theorem 4.

Assume that an $T > L$ and that a trajectory $\{x_t\}_{t=0}^T$ of G satisfies

$$x_0 \in Q$$

and

$$c\langle p, x_0\rangle \le \langle p, x_T\rangle. \tag{50}$$

Theorem 4 and Eqs. (31), (32) and (50) imply that there exist $q \in \{1, \ldots, \tilde{q}\}$ such that

$$a_i < b_i,\ i \in \{1, \ldots, q\},\ a_i, b_i \in \{0, \ldots, T\},\ i = 1, \ldots, q,$$

$$\lambda_i \in [\Delta_0/4, \Delta_1 + 1],\ i = 1, \ldots, q \tag{51}$$

such that for every $i \in \{1, \ldots, q\}$,

$$\|x_t - \lambda_i X\| \le \epsilon,\ t = a_i, \ldots, b_i \tag{52}$$

and

$$\mathrm{Card}(\{0, \ldots, T\} \setminus \cup_{i=1}^q \{a_i, \ldots, b_i\}) \le L.$$

Let $i \in \{1, \ldots, q\}$. In view of (2), (51) and (52), for every $t \in \{a_i, \ldots, b_i\}$,

$$\|\lambda_i^{-1} x_t - X\| \le \epsilon\lambda_i^{-1} \le 4\epsilon\Delta_0^{-1}, \tag{53}$$

$$\|\lambda_i^{-1} x_t\| \in [1 - 4\epsilon\Delta_0^{-1}, 1 + 4\epsilon\Delta_0^{-1}]. \tag{54}$$

By (49) and (54),

$$\| \|x_t\|^{-1} x_t - \lambda_i^{-1} x_t\| = \|\lambda_i^{-1} x_t - \|\lambda_i^{-1} x_t\|^{-1} \lambda_i^{-1} x_t\|$$

$$= \|\lambda_i^{-1} x_t\| \mid 1 - \|\lambda_i^{-1} x_t\|^{-1} \mid \le 2\|\lambda_i^{-1} x_t\|^{-1} \mid 1 - \|\lambda_i^{-1} x_t\| \mid$$

$$\le \|4 \mid 1 - \|\lambda_i^{-1} x_t\| \mid \le 16\epsilon\Delta_0^{-1}. \tag{55}$$

By (49) and (53),

$$\| \|x_t\|^{-1} x_t - X\| \le \|\lambda_i\|^{-1} x_t - X\| + \| \|x_t\|^{-1} x_t - \lambda_i^{-1} x_t\| \le 20\epsilon\Delta_0 < \gamma.$$

Theorem 5 is proved.

Acknowledgements The author is grateful to the anonymous referee for useful comments.

References

1. Aseev, S.M., Krastanov, M.I., Veliov, V.M.: Optimality conditions for discrete-time optimal control on infinite horizon. Pure Appl. Function. Anal. **2**, 395–409 (2017)
2. Aubry, S., Le Daeron, P.Y.: The discrete Frenkel-Kontorova model and its extensions I. Phys. D. **8**, 381–422 (1983)
3. Bachir, M., Blot, J.: Infinite dimensional multipliers and Pontryagin principles for discrete-time problems. Pure Appl. Function. Anal. **2**, 411–426 (2017)
4. Blot, J., Hayek, N.: Infinite-horizon optimal control in the discrete-time framework. SpringerBriefs in Optimization, New York (2014)
5. Carlson, D.A., Haurie, A., Leizarowitz, A.: Infinite horizon optimal control. Springer-Verlag, Berlin (1991)
6. Damm, T., Grune, L., Stieler, M., Worthmann, K.: An exponential turnpike theorem for dissipative discrete time optimal control problems. SIAM J. Control Optim. **52**, 1935–1957 (2014)
7. Evstigneev, I.V., Flam, S.D.: Rapid growth paths in multivalued dynamical systems generated by homogeneous convex stochastic operators. Set-Valued Anal. **6**, 61–81 (1998)
8. Gaitsgory, V., Grune, L., Thatcher, N.: Stabilization with discounted optimal control. Syst. Control Lett. **82**, 91–98 (2015)
9. Gaitsgory, V., Parkinson, A., Shvartsman, I.: Linear programming formulations of deterministic infinite horizon optimal control problems in discrete time. Discrete Contin. Dyn. Syst. Ser. B **22**, 3821–3838 (2017)
10. Gale, D.: The theory of linear economic models. McGraw-Hill Book Co., Inc, New York-Toronto-London (1960)
11. Gale, D.: On optimal development in a multi-sector economy. Rev. Econ. Stud. **34**, 1–18 (1967)
12. Gugat, M.: A turnpike result for convex hyperbolic optimal boundary control problems. Pure Appl. Function. Anal. **4**, 849–866 (2019)
13. Gugat, M., Trelat, E., Zuazua, E.: Optimal Neumann control for the 1D wave equation: finite horizon, infinite horizon, boundary tracking terms and the turnpike property. Syst. Control Lett. **90**, 61–70 (2016)
14. Guo, X., Hernandez-Lerma, O.: Zero-sum continuous-time Markov games with unbounded transition and discounted payoff rates. Bernoulli **11**, 1009–1029 (2016)
15. Hernandez-Lerma, O., Lasserre, J.B.: Zero-sum stochastic games in Borel spaces: average payoff criteria. SIAM J. Control Optim. **39**, 1520–1539 (2001)
16. Hritonenko, N., Yatsenko, Y.: Turnpike and optimal trajectories in integral dynamic models with endogenous delay. J. Optim. Theory Appl. **127**, 109–127 (2005)
17. Jasso-Fuentes, H., Hernandez-Lerma, O.: Characterizations of overtaking optimality for controlled diffusion processes. Appl. Math. Optim. **57**, 349–369 (2008)
18. Khan, M.A., Zaslavski, A.J.: On two classical turnpike results for the Robinson-Solow-Srinivisan (RSS) model. Adv. Math. Econom. **13**, 47–97 (2010)
19. Khlopin, D.V.: On Lipschitz continuity of value functions for infinite horizon problem. Pure Appl. Function. Anal. **2**, 535–552 (2017)
20. Kolokoltsov, V., Yang, W.: The turnpike theorems for Markov games. Dyn. Games Appl. **2**, 294–312 (2012)
21. Leizarowitz, A.: Convergence of viable solutions of differential inclusions with convex compact graphs. SIAM J. Control Optim. **22**, 514–522 (1985)
22. Leizarowitz, A.: Eigenvalues of convex processes and convergence properties of differential inclusions. Set-Valued Anal. **2**, 505–527 (1994)
23. Leizarowitz, A., Mizel, V.J.: One dimensional infinite horizon variational problems arising in continuum mechanics. Arch. Rational Mech. Anal. **106**, 161–194 (1989)
24. Lykina, V., Pickenhain, S., Wagner, M.: Different interpretations of the improper integral objective in an infinite horizon control problem. J. Math. Anal. Appl. **340**, 498–510 (2008)
25. Makarov, V.L., Rubinov, A.M.: Mathematical theory of economic dynamics and equilibria. Springer-Verlag, New York (1977)

26. Mammadov, M.: Turnpike theorem for an infinite horizon optimal control problem with time delay. SIAM J. Control Optim. **52**, 420–438 (2014)
27. Marcus, M., Zaslavski, A.J.: On a class of second order variational problems with constraints. Israel J. Math. **111**, 1–28 (1999)
28. Marcus, M., Zaslavski, A.J.: The structure of extremals of a class of second order variational problems, Ann. Inst. H. Poincaré, Anal. non linéaire **16**, 593–629 (1999b)
29. Marcus, M., Zaslavski, A.J.: The structure and limiting behavior of locally optimal minimizers, Ann. Inst. H. Poincaré, Anal. non linéaire **19**, 343–370 (2002)
30. McKenzie, L.W.: Turnpike theory. Econometrica **44**, 841–866 (1976)
31. Mordukhovich, B.S.: Optimal control and feedback design of state-constrained parabolic systems in uncertainly conditions. Appl. Anal. **90**, 1075–1109 (2011)
32. Mordukhovich, B.S., Shvartsman, I.: Optimization and feedback control of constrained parabolic systems under uncertain perturbations, Optimal control, stabilization and nonsmooth analysis, pp. 121–132. Springer, Lecture Notes Control Inform. Sci. (2004)
33. Pickenhain, S., Lykina, V., Wagner, M.: On the lower semicontinuity of functionals involving Lebesgue or improper Riemann integrals in infinite horizon optimal control problems. Control Cybernet. **37**, 451–468 (2008)
34. Porretta, A., Zuazua, E.: Long time versus steady state optimal control. SIAM J. Control Optim. **51**, 4242–4273 (2013)
35. Prieto-Rumeau, T., Hernandez-Lerma, O.: Bias and overtaking equilibria for zero-sum continuous-time Markov games. Math. Methods Oper. Res. **61**, 437–454 (2005)
36. Rockafellar, R.T.: Convex analysis. Princeton University Press, Princeton, NJ (1969)
37. Sagara, N.: Recursive variational problems in nonreflexive Banach spaces with an infinite horizon: an existence result. Discrete Contin. Dyn. Syst. Ser. S **11**, 1219–1232 (2018)
38. Samuelson, P.A.: A catenary turnpike theorem involving consumption and the golden rule. Amer. Econom. Rev. **55**, 486–496 (1965)
39. Tateishi, T.: Perron-Frobenius theorem for multi-valued mappings. Kodai Math. J. **15**, 155–164 (1992)
40. Trelat, E., Zhang, C., Zuazua, E.: Optimal shape design for 2D heat equations in large time. Pure Appl. Function. Anal. **3**, 255–269 (2018)
41. Trelat, E., Zhang, C., Zuazua, E.: Steady-state and periodic exponential turnpike property for optimal control problems in Hilbert spaces. SIAM J. Control Optim. **56**, 1222–1252 (2018)
42. von Weizsacker, C.C.: Existence of optimal programs of accumulation for an infinite horizon. Rev. Econ. Stud. **32**, 85–104 (1965)
43. Zaslavski, A.J.: Ground states in Frenkel-Kontorova model. Math. USSR Izvestiya **29**, 323–354 (1987)
44. Zaslavski, A.J.: Structure of optimal trajectories of convex processes. Numer. Function. Anal. Optim. **20**, 175–200 (1999)
45. Zaslavski, A.J.: Turnpike properties in the calculus of variations and optimal control. Springer, New York (2006)
46. Zaslavski, A.J.: The existence and structure of approximate solutions of dynamic discrete time zero-sum games. J. Nonlinear Convex Anal. **12**, 49–68 (2011)
47. Zaslavski, A.J.: Nonconvex optimal control and variational problems. Springer Optimization and Its Applications, New York (2013)
48. Zaslavski, A.J.: Structure of approximate solutions of optimal control problems. SpringerBriefs in Optimization, New York (2013)
49. Zaslavski, A.J.: Turnpike phenomenon and infinite horizon optimal control. Springer Optimization and Its Applications, New York (2014)
50. Zaslavski, A.J.: Discrete-time optimal control and games on large intervals. Springer, Springer Optimization and Its Applications (2017)
51. Zaslavski, A.J.: Turnpike conditions in infinite dimensional optimal control. Springer, Springer Optimization and Its Applications (2019)

Matching, Contracts and Markets

Stability and Weak Setwise Stability in Many-to-Many Matching with Contracts

Keisuke Bando and Toshiyuki Hirai

Abstract We consider a relationship between stable outcomes and weakly setwise stable outcomes in two-sided many-to-many matching with contracts. We show that the set of stable outcomes coincides with the set of weakly setwise stable outcomes if the choice function of every agent on both sides is unitary, and the choice function of every agent on one side is substitutable, though the latter is generally larger than the former. We also show that both conditions are essential by demonstrating counterexamples each of which violates one of the conditions. Further, we discuss our results in the context of existing literature.

Keywords Many-to-many matching with contracts · Stability · Weak setwise stability · Unitarity · One-side substitutability

1 Introduction

Since the seminal work by Gale and Shapley [6], the theory of two-sided matching has been developed and used to solve many practical problems.[1] They considered a college admission problem where each student is assigned to a college while each college can admit multiple students. They also considered a marriage problem as a special case of the former problem where each woman (man, resp.) can get married to at most one man (woman, resp.). The former problem is formulated as a many-to-one matching problem (without contracts), and the latter is formulated as a one-to-one matching (without contracts). During the development of the matching theory, more general models have been proposed. A matching problem in which each

[1] See Roth and Sotomayor [18] for a survey.

K. Bando (✉)
Keio University, Kanagawa, Japan
e-mail: bando@ae.keio.ac.jp

T. Hirai
Hosei University, Tokyo, Japan
e-mail: hirai@hosei.ac.jp

M. A. Khan et al. (eds.), *Matching, Dynamics and Games for the Allocation of Resources*, Monographs in Mathematical Economics 7,
https://doi.org/10.1007/978-981-96-6661-4_18

agent is allowed to match multiple partners on the other side is called a many-to-many matching problem. For example, this model is investigated by Sotomayor [19], Echenique and Oviedo [5], and Konishi and Ünver [16]. A (one-to-one, many-to-one, many-to-many) matching problem with salaries or contracts allows a pair of agents to agree on a condition such as salaries, job content, and so on by signing a contract. For example, this model is investigated by Kelso and Crawford [13], Roth [17], Blair [4], and Hatfield and Milgrom [9].

The most important concept in the matching theory is stability, which was proposed by Gale and Shapley [6]. Their stability concept is called pairwise stability, which requires being robust against a deviation by a pair of agents, one from each side. A stable outcome can be guaranteed to exist in plausible conditions, and it can be found by an algorithm called the Gale=Shapley algorithm or deferred-acceptance algorithm in an efficient time. Therefore, the concept of stability has been playing an important role in both theoretical development and practical applications.

Several extensions of the stability concept have been proposed so that they are more appropriate to analyze more general settings of two-sided matching. A stable outcome defined by Hatfield and Milgrom [9] has become the most standard concept. However, the definition of stability might be unreasonable in many-to-many matching with contracts as it sometimes requires being robust against infeasible blocks. For example, consider a simple situation consisting of one doctor d and one hospital h. There are two contracts x and y between d and h where x is the contract of working in the morning and y is the contract of working in the afternoon. The hospital h prefers to have both contracts, whereas the doctor wants to work only half a day, and prefers working in the morning (x) to the afternoon (y). Suppose that the hospital and doctor are agreeing on contract y. They are willing to sign a new contract x, that is, the doctor-hospital pair can block the contract y via x in the sense of Hatfield and Milgrom [9]. However, they cannot agree on whether they retain the current contract y because the hospital prefers to retain it whereas the doctor prefers to abandon it.

The concept of weak setwise stability, first formulated in a many-to-many matching with contracts model by Klaus and Walzl [14], overcomes such a deficiency. A weak setwise stable outcome is defined as a set of contracts that is robust against a weak setwise block, and a weak setwise block requires that agents involved in the block agree on the resulting outcome, which rules out the disagreement among the agents after a block. In general, weak setwise stability is a weaker notion than stability.

Klaus and Walzl [14] showed that if the choice functions of both hospitals and doctors satisfy the conditions called unitarity and substitutability, the set of weakly setwise stable outcomes coincides with the set of (pairwise) stable outcomes. Unitarity is a condition that requires an agent to prefer choosing at most one contract with the same agent from the other side, first formulated by Kominers [15].[2] Substi-

[2] To be more precise, unitarity condition has essentially been assumed in earlier papers on many-to-many matching with contracts, such as Roth [17] and Klaus and Walzl [14]. They stated unitarity as a condition on the feasibility of outcomes. Kominers [15] explicitly formulated as a condition on a choice function.

tutability is a well-known condition for the existence of a stable outcome in two-sided matching.

This study considers a weaker condition where the set of stable outcomes coincides with the set of weakly setwise stable outcomes. In other words, we consider when blocks do not involve infeasible blocks in many-to-many matching with contracts when the substitutability of agents on both sides is unnecessarily satisfied. To be more precise, we show that the set of stable outcomes coincides with the set of weakly setwise stable outcomes if every agent's choice function satisfies unitarity and the choice functions of agents on one side satisfy substitutability.

Remarkably, our result holds in a model where substitutability is only assumed for agents on one side. Substitutability has been playing an important role to guarantee the existence of a stable outcome. However, Hatfield and Kojima [7] showed that it is not a necessary condition for many-to-one matching with contracts. Based on this fact, several authors such as Hatfield and Kojima [8], Hatfield and Kominers [10] and Hatfield et al. [12] have proved the existence of a stable outcome by assuming weaker notions than substitutability for many-to-one matching with contracts. Yenmez [20] and Bando et al. [3] extended these results into many-to-many matching with contracts. Our result applies to many-to-many matching with contracts where substitutability is unnecessarily satisfied. Particularly, our result implies that the set of stable outcomes coincides with the set of weakly setwise stable outcomes in the setting considered in Bando et al. [3].[3]

This paper proceeds as follows. Section 2 introduces the model of two-sided many-to-many matching with contracts and the conditions on choice functions. Section 3 introduces the three stability concepts: pairwise stability, stability, and weak setwise stability. Section 4 states and proves the main result. We also provide examples to demonstrate that each of our conditions is essential. Section 5 concludes the study with a remark.

2 Many-to-Many Matching with Contracts

We consider a many-to-many matching problem with contracts between a finite set of doctors $D \neq \emptyset$ and a finite set of hospitals $H \neq \emptyset$ where $D \cap H = \emptyset$. The doctors and hospitals are sometimes called agents and denoted by $F = D \cup H$. A doctor and a hospital form a relationship by signing a bilateral contract. Each of the doctors and hospitals can sign contracts with multiple agents on the other side. A finite set of contracts is denoted by X. Each contract $x \in X$ involves a doctor $x_D \in D$ and a hospital $x_H \in H$. For each contract $x \in X$, we sometimes denote by $x_F = \{x_D, x_H\}$ a doctor-hospital pair involved in x. There are multiple possible contracts between

[3] Bando et al. [3] showed two existence results of stable outcomes for many-to-many matching with contracts. Both results assume unitarity of agents on both sides and substitutability of agents on one side while weaker substitutability conditions and a weaker IRC condition are assumed for agents on the other side. Therefore, the coincidence result holds under their existence conditions.

the same pair. In other words, there are possibly distinct contracts x, x' such that $x_F = x'_F$. A subset of contracts $Y \subseteq X$ is called an outcome. For each agent $i \in F$ and each outcome $Y \subseteq X$, we denote by $Y_i = \{x \in Y | i \in x_F\}$ the set of contracts in Y that involve i. Moreover, for each outcome $Y \subseteq X$, we denote by $Y_D = \bigcup_{y \in Y} \{y_D\}$ the set of doctors involved in at least one contract in Y, and by $Y_H = \bigcup_{y \in Y} \{y_H\}$ the set of hospitals involved in at least one contract in Y.

Each agent i has preferences over the subsets of X_i that are represented by a choice function $C^i : 2^{X_i} \to 2^{X_i}$ such that $C^i(Y) \subseteq Y$ for all $Y \subseteq X_i$ where $C^i(Y)$ represents the most preferred contracts in Y. Because a choice of agent i does not depend on contracts that i is not involved in, we can extend the domain of C^i to 2^X by defining $C^i(Y) = C^i(Y_i)$ for all $Y \subseteq X$.

We introduce some conditions on choice functions. The first is substitutability, which is standard in matching theory.

Definition 1 For each $i \in F$, choice function C^i is substitutable if for any $Y \subseteq X$ and any $x, z \in X$ with $x, z \notin Y$, $z \notin C^i(Y \cup \{z\})$ implies $z \notin C^i(Y \cup \{z, x\})$.

Hatfield and Kominers [11] showed that a stable outcome exists when every agent's choice function satisfies substitutability and the condition of irrelevance of rejected contracts (IRC) introduced by Aygün and Sönmez [1].[4]

We further introduce unitarity, which prohibits agents from choosing multiple contracts with the same agent.

Definition 2

- For each $d \in D$, choice function C^d is unitary if $x, x' \in C^d(Y)$ and $x \neq x'$ imply $x_H \neq x'_H$.
- For each $h \in H$, choice function C^h is unitary if $x, x' \in C^h(Y)$ and $x \neq x'$ imply $x_D \neq x'_D$.

Unitarity has been imposed on many-to-many matching problems with contracts by several authors such as Klaus and Walzl [14], Kominers [15], and Bando et al. [3], whereas others, such as Hatfield and Kominers [11] and Yenmez [20], did not impose the same condition. Note that unitarity is naturally assumed in a many-to-one matching problem (with and without contracts) and a many-to-many matching problem without contracts, and not in many-to-many matching with contracts. Note also that unitarity can be regarded as a variation of the substitutable condition in the sense that a hospital has a unit demand among contracts between the same pair.

---

[4] A choice function C^i satisfies IRC if for any outcome $Y \subseteq X$ and any $x \in Y$, $x \notin C^i(Y)$ implies $C^i(Y \setminus \{x\}) = C^i(Y)$. This is a standard condition in the literature. But our result holds without IRC.

3 Weak Setwise Stability

This section introduces several solution concepts. We begin with the definition of individual rationality, which is required in defining each solution concept.

Definition 3 An outcome $A \subseteq X$ is individually rational if $C^i(A) = A_i$ for all $i \in F$.

We turn to the definitions of several types of blocking concepts and the corresponding stability concepts. First, we introduce a stable outcome, which is a natural extension of a stable outcome in many-to-one matching with contracts. We follow the formulation given by Hatfield and Kominers [11].

Definition 4

- An outcome $A \subseteq X$ is blocked via $Z \subseteq X$ if $\emptyset \neq Z \subseteq X \setminus A$ and $Z_i \subseteq C^i(A \cup Z)$ for all $i \in Z_F$.
- An outcome is stable if it is individually rational and not blocked.

We sometimes refer to Z in Definition 4 as a blocking set to A.

We further define two weaker notions of stability. The first one is pairwise stability which requires robustness only to blocks with a single contract. The pairwise stability is a natural extension of that by Gale and Shapley [6] that was originally defined in a many-to-one matching problem.

Definition 5

- An outcome $A \subseteq X$ is pairwise blocked via $Z \subseteq X$ if $Z = \{x\}$ for some $x \in X \setminus A$ and $x \in C^i(A \cup \{x\})$ for each $i = x_D, x_H$.
- An outcome is pairwise stable if it is individually rational and not pairwise blocked.

The second notion is weak setwise stability proposed by Klaus and Walzl [14]. A weakly setwise stable outcome is defined as a robust outcome to a weak setwise block in which blocking agents are required to agree on some outcome after block, whereas this property is not required in the definition of block.

Definition 6

- An outcome $A \subseteq X$ is weakly setwise blocked via $Z \subseteq X$ if $\emptyset \neq Z \subseteq X \setminus A$ and there exists Y^* satisfying (i) $Z \subseteq Y^*$, (ii) $Y^* \subseteq A \cup Z$, (iii) $Y^*_i = C^i(A \cup Z)$ for all $i \in Z_F$.
- An outcome is weakly setwise stable if it is individually rational and not weakly setwise blocked.

We sometimes refer to Z in Definition 6 as a weakly setwise blocking set to A.

It is well-known that the set of pairwise stable outcomes coincides with the set of stable outcomes if every agent's choice function is substitutable. Additionally, it is clear that any stable outcome is a weakly setwise stable outcome because a weakly setwise block requires that the blocking agents agree on a resulting outcome, in addition to the definition of a block. However, there may be a weakly setwise stable outcome that is not stable.

The following is the formal description of the simple example between a doctor and a hospital introduced in the introduction. Let $D = \{d\}$, $H = \{h\}$, and $X = \{x, y\}$. Preferences of agents are given by the following list.

$$\succ_d \ : \{x\}\ \{y\}\ \emptyset,$$
$$\succ_h \ : \{x, y\}\ \{y\}\ \{x\}\ \emptyset.$$

Such a list as $\succ_h$ means $\{x\} \succ_h \{y\} \succ_h \emptyset$, and $\emptyset \succ_h Y$ for any nonempty $Y \subseteq X$ with $Y \neq \{x\}, \{y\}$. Then, $\{y\}$ is blocked via $\{x\}$ since $x \in C^d(\{x, y\})$ and $x \in C^h(\{x, y\})$. However, the doctor and hospital cannot agree on whether to maintain y since $y \notin C^d(\{x, y\}) = \{x\}$ but $y \in C^h(\{x, y\}) = \{x, y\}$. This means that $\{y\}$ is not weakly setwise blocked via $\{x\}$. Thus, $\{y\}$ is not stable but weakly setwise stable. In this example, C^d is not unitary. One might think that the unitarity of every agent's choice function is sufficient to obtain the equivalence between stable outcomes and weakly setwise stable outcomes. However, in the next section, we demonstrate that an additional condition is necessary to obtain the equivalence.

4 Equivalence

Theorem 1 *Suppose that (i) every agent's choice function is unitary and (ii) every doctor's choice function is substitutable. Then, the set of weakly setwise stable outcomes coincides with the set of stable outcomes.*

We first show the following two lemmas. Doctor-side substitutability is utilized in the first lemma, and unitarity is utilized in the second lemma.

Lemma 1 *Suppose that every doctor's choice function is substitutable. Let Z be a blocking set to an outcome A. Then, Z_h is a blocking set to A for any $h \in Z_H$.*

Proof Let Z be a blocking set to A. Fix an arbitrary $h \in Z_H$. Then, $Z_h \subseteq C_h(Z_h \cup A) = C_h(Z \cup A)$. For each doctor d and $z \in (Z_h)_d$, $z \in C_d(Z \cup A)$ implies $z \in C_d(Z_h \cup A)$ by substitutability of C_d. Thus, Z_h is a blocking set to A.

Lemma 2 *Suppose that every agent's choice function is unitary. Let Z be a blocking set to an outcome A. If $|Z_H| = 1$, then Z is a weakly setwise blocking set to A.*

Proof Define $Y^* = \bigcup_{i \in Z_F} C^i(A \cup Z)$. Note that $Z \subseteq Y^* \subseteq A \cup Z$ holds since Z is a blocking set to A. Therefore, it remains to show that $Y^*_i = C^i(A \cup Z)$ for all $i \in Z_F$.

Let $i \in Z_F$. Suppose that $Y_i^* \neq C^i(A \cup Z)$. Then, $C^i(A \cup Z) \subsetneq Y_i^*$ because $C^i(A \cup Z) \subseteq Y_i^*$ holds from the definition of Y^*. Consider any $y \in Y_i^*$ with $y \notin C^i(A \cup Z)$. Note that $y \in A$ holds by $y \notin C^i(A \cup Z)$. By $y \in Y^*$ and $y \notin C^i(A \cup Z)$, there exists $j \in Z_F \setminus \{i\}$ such that $y \in C^j(A \cup Z)$. By $\{i, j\} \subseteq Z_F$ and $Z_H = \{h\}$, there exists a contract $z \in Z$ such that $z_F = \{i, j\}$. By $z \in Z$, we have $z, y \in Y^j = C^j(A \cup Z)$. By $z \in Z$ and $y \in A$, we have $y \neq z$ contradicting unitarity.

We now show Theorem 1

Proof of Theorem 1. We show that any weakly setwise stable outcome is stable because the converse direction follows from the definition. To obtain this, it is sufficient to show that for any outcome A, if A is blocked, then it is weakly setwise blocked. Let Z be a blocking set to A. By Lemma 1, Z_h is also a blocking set to A. Note that $|(Z_h)_H| = |\{h\}| = 1$. Thus, Z_h is a weakly blocking set to A from Lemma 2. □

Note that Theorem 1 can derive Theorem 2(1) of Klaus and Walzl [14], which proves the equivalence between weak setwise stability and pairwise stability under unitarity and substitutability of every agent's choice function, together with the fact that the set of stable outcomes coincides with the set of pairwise stable outcomes if every agent's choice function is substitutable.

Compared with the results of Klaus and Walzl [14], we impose only *one-side* substitutability for obtaining the equivalence between *stable* and weakly setwise stable outcomes, whereas they require *both-side* substitutability for obtaining the equivalence between *pairwise* and weakly setwise stable outcome. We show that the equivalence between pairwise and weakly setwise stable outcomes cannot be obtained by one-side substitutability via an example. Although Klaus and Walzl [14] contains an example showing this fact, we construct another example, in which an agent's choice function violates substitutability but satisfies unilateral substitutability.

Definition 7 For each $h \in H$, choice function C^h is unilaterally substitutable if for any $Y \subseteq X$ and any $x, z \in X$ with $z_D \notin Y_D$, $z \notin C^h(Y \cup \{z\})$ implies $z \notin C^h(Y \cup \{z, x\})$.

Unilateral substitutability of a doctor's choice function can be defined similarly.

Unilateral substitutability was introduced by Hatfield and Kojima [8] for many-to-one matching with contracts. They showed the existence of a doctor-optimal stable outcome if every hospital's choice function satisfies unilateral substitutability and IRC. Moreover, strategy-proofness of doctors in the cumulative offer process is preserved under unilateral substitutability with some additional assumptions. In this sense, unilateral substitutability is a close concept to substitutability. Indeed, it is the strongest condition among the weaker conditions of substitutability proposed in the literature. The following example shows that even the presence of one agent with a unilateral substitutable choice function, but not substitutable, can prevent the equivalence between pairwise stable and weakly setwise stable outcomes.

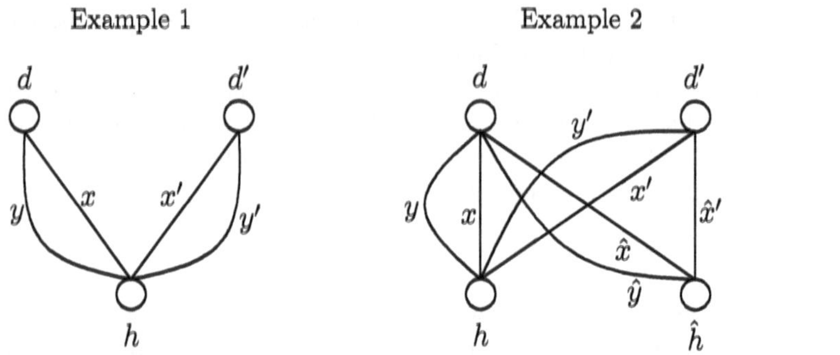

Fig. 1 Descriptions of Examples 1 and 2

Example 1 Let $D = \{d, d'\}$ and $H = \{h\}$. The set of contracts is given as $X = \{x, y, x', y'\}$, where $x_D = y_D = d$, $x'_D = y'_D = d'$, and $x_H = y_H = x'_H = y'_H = h$ (see Fig. 1).

The preferences of the agents are given as follows:

$\succ_d$: $\{x\}\ \{y\}\ \emptyset$.
$\succ_{d'}$: $\{x'\}\ \{y'\}\ \emptyset$.
$\succ_h$: $\{x, x'\}\ \{y, y'\}\ \{y, x'\}\ \{x, y'\}\ \{y\}\ \{y'\}\ \{x\}\ \{x'\}\ \emptyset$.

We proceed with the choice functions C^d, $C^{d'}$, and C^h that are derived from these preferences where all of them satisfy unitarity. Note that C^d and $C^{d'}$ satisfy substitutability, while C^h does not since $x \notin C^h(\{x, y\})$ but $x \in C^h(\{x, x', y\})$. Nonetheless, C^h satisfies unilateral substitutability. To see this, we show that there is no pair of $z \in X$ and $Y \subseteq X$ such that $z_D \notin Y_D$ and $z \notin C^h(Y \cup \{z\})$. Assume that (z, Y) satisfies $z_D \notin Y_D$.

- Assume that $z = x$. Then, $y \notin Y$ by $z_D \notin Y_D$. If $x' \in Y$, then $\{x, x'\} = C^h(Y \cup \{x\})$. If $x' \notin Y$, then $Y = \{y'\}$ or $\emptyset$. For either case, $x \in C^h(Y \cup \{x\})$.
- Assume that $z = y$. Then, $x \notin Y$ by $z_D \notin Y_D$. If $y' \in Y$, then $\{y, y'\} = C^h(Y \cup \{y\})$ by $x \notin Y$. If $y' \notin Y$, then $Y = \{x'\}$ or $\emptyset$. For either case, $y \in C^h(Y \cup \{y\})$.
- Assume that $z = x'$. Then, $y' \notin Y$ by $z_D \notin Y_D$. If $x \in Y$, then $\{x, x'\} = C^h(Y \cup \{x'\})$. If $x \notin Y$, then $Y = \{y\}$ or $\emptyset$. For either case, $x' \in C^h(Y \cup \{x'\})$.
- Assume that $z = y'$. Then, $x' \notin Y$ by $z_D \notin Y_D$. If $y \in Y$, then $\{y, y'\} = C^h(Y \cup \{y'\})$ by $x' \notin Y$. If $y \notin Y$, then $Y = \{x\}$ or $\emptyset$. For either case, $y' \in C^h(Y \cup \{y'\})$.

Hence, for either case, $z \in C^h(Y \cup \{z\})$.

Consider an outcome $\{y, y'\}$. Obviously, $\{y, y'\}$ is individually rational. By $x \notin C^h(\{x, y, y'\})$ and $x' \notin C^h(\{x', y, y'\})$, $\{y, y'\}$ is a pairwise stable outcome. However, $\{y, y'\}$ is weakly setwise blocked via $\{x, x'\}$ since

$$C^d(\{x, x', y, y'\}) = \{x\} = \{x, x'\}_d$$
$$C^{d'}(\{x, x', y, y'\}) = \{x'\} = \{x, x'\}_{d'}$$
$$C^h(\{x, x', y, y'\}) = \{x, x'\} = \{x, x'\}_h.$$

Hence, $\{y, y'\}$ is not weakly setwise stable. □

We further demonstrate that both conditions assumed in Theorem 1 are essential via counterexamples. The example in Sect. 3 shows that unitarity is crucial to obtain Theorem 1. One might think that unitarity for only one side is sufficient for obtaining the coincidence result because it is sufficient to exclude individually rational outcomes containing a pair choosing multiple contracts between them. However, the previous example also shows that both-side unitarity is necessary even though every agent's choice function is substitutable.

Substitutability assumption for agents on one side is also crucial for Theorem 1. The following example shows that even if every agent's preferences satisfy unilateral substitutability and unitarity, the set of stable outcomes does not coincide with the set of weaky setwise stable outcomes.

Example 2 Let $D = \{d, d'\}$ and $H = \{h, \hat{h}\}$. The set of contracts is given as $X = \{x, y, x', y', \hat{x}, \hat{y}, \hat{x}'\}$, where $x_D = y_D = \hat{x}_D = \hat{y}_D = d, x'_D = y'_D = \hat{x}'_D = d', x_H = y_H = x'_H = y'_H = h$, and $\hat{x}_H = \hat{y}_H = \hat{x}'_H = \hat{h}$ (see Fig. 1).

The preferences of the agents are given as follows.

$\succ_d$: $\{x, \hat{x}\}$ $\{y, \hat{y}\}$ $\{y, \hat{x}\}$ $\{x, \hat{y}\}$ $\{y\}$ $\{\hat{y}\}$ $\{x\}$ $\{\hat{x}\}$ $\emptyset$.
$\succ_{d'}$: $\{x', \hat{x}'\}$ $\{y', \hat{x}'\}$ $\{x'\}$ $\{\hat{x}'\}$ $\{y'\}$ $\emptyset$.
$\succ_h$: $\{x, x'\}$ $\{y, y'\}$ $\{y, x'\}$ $\{x, y'\}$ $\{y\}$ $\{y'\}$ $\{x\}$ $\{x'\}$ $\emptyset$.
$\succ_{\hat{h}}$: $\{\hat{x}\}$ $\{\hat{y}, \hat{x}'\}$ $\{\hat{y}\}$ $\{\hat{x}'\}$ $\emptyset$.

We proceed with the choice functions C^d, $C^{d'}$, C^h, and $C^{\hat{h}}$ that are derived from these preferences where all of them satisfy unitarity. Note that $C^{d'}$ and $C^{\hat{h}}$ satisfy substitutability, while C^d and C^h do not. Nonetheless, C^d and C^h satisfy unilateral substitutability, which can be confirmed in the same way as unilateral substitutability of C^h in Example 1.

Consider an outcome $Y = \{y, y', \hat{y}, \hat{x}'\}$. Obviously, Y is individually rational. By $x \notin C^d(Y \cup \{x\})$, $x' \notin C^h(Y \cup \{x'\})$, and $\hat{x} \notin C^d(Y \cup \{\hat{x}\})$, neither singleton consisting of one of $x, x', \hat{x}$ is a (weakly setwise) block to Y. Further, by $x \notin C^h(Y \cup \{x, \hat{x}\})$, $x \notin C^d(Y \cup \{x, x'\})$, and $x' \notin C^h(Y \cup \{x', \hat{x}\})$, no set of two contracts among $x, x', \hat{x}$ is a (weakly setwise) block to Y. Thus, $Z = \{x, x', \hat{x}\}$ is the only possible block to Y. Note that $Y \cup Z = X$. Indeed, Z is a block to Y because

$$Z_d = \{x, \hat{x}\} = C^d(Y \cup Z),$$
$$Z_{d'} = \{x'\} \subsetneq \{x', \hat{x}'\} = C^{d'}(Y \cup Z),$$
$$Z_h = \{x, x'\} = C^h(Y \cup Z),$$
$$Z_{\hat{h}} = \{\hat{x}\} = C^{\hat{h}}(Y \cup Z),$$

However, Z is not a weakly setwise block to Y since $\hat{x}' \in C^{d'}(Y \cup Z)$ while $\hat{x}' \notin C^{\hat{h}}(Y \cup Z)$. Hence Y is weakly setwise stable, but not stable. □

5 Concluding Remark

This study shows that the set of stable outcomes coincides with the set of weakly setwise stable outcomes if every agent's choice function satisfies unitarity and every doctor's choice function satisfies substitutability. This result extends the coincidence between the set of pairwise stable outcomes and the set of weakly setwise stable outcomes obtained by Klaus and Walzl [14]. We also show that our conditions are tight in the sense that the coincidence cannot be obtained if either condition is violated. Particularly, it is shown that even unilateral substitutability, which is a weak notion of substitutability, is not sufficient to obtain the coincidence result.

We conclude this paper with a remark. Some related studies show that a stable outcome exists in many-to-many matching without unitarity where the set of stable outcomes is generally strictly smaller than that of weakly setwise outcomes as shown in this paper. Hatfield and Kominers [11] showed the existence of a stable outcome under both-side substitutability. Yenmez [20] proposed a weaker notion of substitutability called substitute modifiability, and showed that a stable outcome exists when the choice functions of the agents on one side satisfy substitute modifiability and the choice functions of the agents on the other side satisfy substitutability. We may obtain a weaker condition to guarantee the existence of a weakly setwise outcome than that of Yenmez [20] because our results show that the set of weakly setwise outcomes is larger than that of stable outcomes. To the best of our knowledge, such an existence result for a weakly setwise stable outcome has not yet been found in two-sided many-to-many matching. Meanwhile, Bando and Hirai [2] showed that a weakly setwise stable outcome exists under a weaker condition than that guarantees the existence of a stable outcome in multilateral matching, in which we proved the existence under a restricted structure of possible contracts and general preferences. It may extend the domain of matching theory if we show the existence of a weakly setwise stable outcome in a matching problem admitting no stable outcome and make it possible to employ a weakly setwise stable outcome as a solution instead of a stable outcome. We leave this problem for future research.

Acknowledgements This research is financially supported by JSPS KAKENHI Grant Numbers 22K01396 (Hirai) and JP19K13647 (Bando). We would like to thank Editage (www.editage.com) for English language editing.

References

1. Aygün, O., Sönmez, T.: Matching with contracts: comment. Am. Econ. Rev. **103**, 2050–2051 (2013)
2. Bando, K., Hirai, T.: Stability and venture structures in multilateral matching. J. Econ. Theory **196**, 105292 (2021)
3. Bando, K., Hirai, T., Zhang, J.: Substitutes and stability for many-to-many matching with contracts. Game Econ. Behav. **129**, 503–512 (2021)
4. Blair, C.: The lattice structure of the set of stable matchings with multiple partners. Math. Oper. Res. **13**, 619–628 (1988)
5. Echenique, F., Oviedo, J.: A theory of stability in many-to-many matching markets. Theor. Econ. **1**, 233–273 (2006)
6. Gale, D., Shapley, L.: College admissions and the stability of marriage. Am. Math. Mon. **69**, 9–15 (1962)
7. Hatfield, J.W., Kojima, F.: Matching with contracts: comment. Am. Econ. Rev. **98**, 1189–1194 (2008)
8. Hatfield, J.W., Kojima, F.: Substitutes and stability for matching with contracts. J. Econ. Theory **145**, 1704–1723 (2010)
9. Hatfield, J.W., Milgrom, P.R.: Matching with contracts. Am. Ecom. Rev. **95**, 913–935 (2005)
10. Hatfield, J.W., Kominers, S.D.: Hidden substitute. Working Paper (2016)
11. Hatfield, J.W., Kominers, S.D.: Contract design and stability in many-to-many matching. Game Econ. Behav. **101**, 78–97 (2017)
12. Hatfield, J.W., Kominers, S.D., Westkamp, A.: Stability, strategy-proofness, and cumulative offer mechanisms. Rev. Econ. Stud. **88**, 1457–1502 (2021)
13. Kelso, A.S., Crawford, V.P.: Job matching, coalition formation, and gross substitutes. Econometrica **50**, 1483–1504 (1982)
14. Klaus, B., Walzl, M.: Stable many-to-many matchings with contracts. J. Math. Econ. **45**, 422–434 (2009)
15. Kominers, S.D.: On the correspondence of contracts to salaries in (many-to-many) matching. Game Econ. Behav. **75**, 984–989 (2012)
16. Konishi, H., Ünver, M.U.: Credible group stability in many-to-many matching problems. J. Econ. Theory **129**, 966–1005 (2006)
17. Roth, A.E.: Stability and polarization of interests in job matching. Econometrica **52**, 47–57 (1984)
18. Roth, A.E., Sotomayor, M.A.O.: Two-sided matching: a study in game-theoretic modelling and analysis. Cambridge University Press, Cambridge (1990)
19. Sotomayor, M.A.O.: Three remarks on the many-to-many stable matching problem. Math. Soc. Sci. **38**, 55–70 (1999)
20. Yenmez, M.B.: A college admissions clearinghouse. J. Econ. Theory **176**, 859–885 (2018)

A Deferred Acceptance Algorithm for Large Marriage Markets

Guilherme Carmona and Krittanai Laohakunakorn

Abstract We consider a version of Gale and Shapley's (College admissions and the stability of marriage 69:9–15) marriage market featuring a continuum of women and men based on Greinecker and Kah (Pairwise stable matching in large economies 89:2929–2974). We define a deferred acceptance algorithm for this setting and show that it terminates after a finite number of iterations to yield a side-optimal stable matching.

Keywords Marriage markets · Stability · Deferred acceptance algorithm

1 Introduction

Matching theory is widely used to address questions such as who gets which jobs, which school places and who marries whom. [14] (GS henceforth) provided the first model of two-sided matching and its key solution concept, stability, which requires that no one is matched with an unacceptable person and that no man and woman who are not matched to each other would both prefer to be.

GS also introduced the so-called deferred acceptance algorithm (DAA henceforth) which converges in finitely many steps to a stable matching. Thus, the DAA establishes the existence of stable matchings and can also be used to compute the side-optimal (i.e., man-optimal and woman-optimal) stable matchings.

There have been many extensions and variations of GS's original work. One such line of research that has been particularly active recently focuses on large matching markets, motivated by the observation that, in reality, many such markets feature a large number of participants.[1] In this paper we extend GS's DAA to the setting of

[1] This literature includes, among others, [1, 2, 5–7, 11–13, 16] and [15].

G. Carmona (✉) · K. Laohakunakorn
School of Economics, University of Surrey, Guildford, UK
e-mail: g.carmona@surrey.ac.uk

K. Laohakunakorn
e-mail: k.laohakunakorn@surrey.ac.uk

M. A. Khan et al. (eds.), *Matching, Dynamics and Games for the Allocation of Resources*, Monographs in Mathematical Economics 7,
https://doi.org/10.1007/978-981-96-6661-4_19

[15] (GK henceforth), which provides a formalization of GS's marriage market for the case where there is a continuum of men and women.

Specifically, we consider a version of GK's setting where there is a finite set of contracts, a finite set of types of men, each having preferences over types of women and contracts, and a finite set of types of women, each having preferences over types of men and contracts; in particular, for each contract, women are indifferent between different men of the same type and vice versa.[2] Instead of having an explicit set of men and women, we only specify the type distribution of each gender. We then define a DAA algorithm for this setting and show that, when preferences are linear orders and each distribution has rational coefficients, it converges in finitely many steps to a stable matching, which is side-optimal.

One interpretation of GK's framework is that there is a continuum of men and women, and the type distributions specify the measure of each type in the population. Another, which is possible when each distribution has rational coefficients, is that each man (resp. woman) type corresponds to a man (resp. woman). Thus, GS's DAA can be applied to the setting we consider, something which is already known since GK, who use [8] version of GS's DAA to show the existence of stable matchings in discrete instances of their framework.

Our version of the DAA has an important advantage over the above as it solves the following dimensionality problem. Suppose that there are 100 women, all of the same type, and 100 men. If all of the men are of the same type, this market can be scaled down to the case where there is just one man and one woman and the stable matching is easy to obtain. If, instead, 99 of the men are of type m_1 and 1 is of type m_2, then we can no longer scale it down. If we represent this marriage market in GS's setting, then, despite there being just one type of woman, each man has to linearly order each of 100 women, each of whom he regards as equivalent. This is in contrast with our algorithm in which we just need the single type of woman to linearly order the two types of men. In short, our DAA does not require that certain ties be broken and, thus, allows for a more parsimonious description of individual preferences. We illustrate this advantage in an example and present a Python code to compute stable allocations using our version of the DAA.

A similar algorithm to ours is the column-greedy algorithm of [3]. The main contribution of our algorithm over theirs is that it allows for contracts.

The paper is organized as follows. The framework we consider is introduced in Sect. 2. Section 3 illustrates our version of the DAA. Its formal definition and our main result showing that it converges in finitely many steps to a side-optimal stable matching is in Sect. 4. The Appendix contains the proof of our main result in Sect. 5.1 and a description of how to use a (rudimentary) Python code to implement our DAA in Sect. 6; the code itself is available online.[3]

---

2 Distributional marriage models with finitely many types have also been considered in [3, 9] and [10].

3 https://drive.google.com/file/d/1LIiaArjO7EPjxkI6vvfXDyXSSJkqd9LF/view.

2 Large Marriage Markets

We consider a simplified version of GK's marriage market framework with finitely many types and contracts. Such a marriage market is defined by the following elements. There are nonempty and finite sets W and M of types describing, respectively, the types of women and men in the marriage market. To represent unmatched individuals, there is a dummy type $\emptyset \notin W \cup M$; let $W_\emptyset = W \cup \{\emptyset\}$ and $M_\emptyset = M \cup \{\emptyset\}$. The distributions of women and men are described by nonzero, finite measures ν_W and ν_M on W and M, respectively.

There is a nonempty and finite set C of contracts. Women have preferences over men (including the "empty man", i.e., being unmatched) and contracts, and analogously for men. Preferences are then described by a linear order $\succ_w$ on $M_\emptyset \times C$ for each $w \in W$ and a linear order $\succ_m$ on $W_\emptyset \times C$ for each $m \in M$.

A *marriage matching* is a Borel measure $\mu \in M(M_\emptyset \times W_\emptyset \times C)$ such that

1. $\mu(M_\emptyset \times \{w\} \times C) = \nu_W(w)$ for each $w \in W$,
2. $\mu(\{m\} \times W_\emptyset \times C) = \nu_M(m)$ for each $m \in M$, and
3. $\mu(\{(m, w, c) : w = m = \emptyset\}) = 0$.[4]

A marriage matching describes, for each $(m, w, c) \in M_\emptyset \times W_\emptyset \times C$, the measure $\mu(m, w, c)$ of people that are matched, whose type profile is (m, w) and whose contract is c; the unmatched are those who are matched with $\emptyset$ so that, in this sense, everyone is matched. Condition 1 then says that the measure of those in a match where the female type is $w \in W$ is, since everyone is matched, exactly the measure of women whose type is w. Condition 2 is the analogous condition for men. Condition 3 says that almost all matches have a woman, a man, or both.

The definition of a stable marriage matching we use is that in [4], which is equivalent to that in GK. For each $m \in M_\emptyset$, define

$$\begin{aligned} T_m(\mu) = \big\{(w, c) \in W \times C :\ & \text{there exists } (m', c') \in M_\emptyset \times C \\ & \text{such that } (m', w, c') \in \operatorname{supp}(\mu) \text{ and } (m, c) \succ_w (m', c')\big\} \\ & \cup (\{\emptyset\} \times C)\,. \end{aligned}$$

The set $T_m(\mu)$ consists of the pairs of women types and contracts $(w, c) \in W_\emptyset \times C$ that a man of type m can target at μ in the sense that either $w = \emptyset$ or there are women of type $w \in W$ that prefers (m, c) to their current match in μ. Analogously, define, for each $w \in W_\emptyset$,

$$\begin{aligned} T_w(\mu) = \big\{(m, c) \in M \times C :\ & \text{there exists } (w', c') \in W_\emptyset \times C \\ & \text{such that } (m, w', c') \in \operatorname{supp}(\mu) \text{ and } (w, c) \succ_m (w', c')\big\} \\ & \cup (\{\emptyset\} \times C)\,. \end{aligned}$$

[4] Throughout the paper, whenever Y is a finite set, $M(Y)$ denotes the set of finite measures on Y and is identified with $\mathbb{R}_+^{|Y|}$. We write $\mu(y)$ for $\mu(\{y\})$ whenever $y \in Y$ and $\mu \in M(Y)$.

Finally, define

$$\begin{aligned}
S_M(\mu) = &\ \{(m, w, c) \in M_\emptyset \times W_\emptyset \times C : \text{ there does not exist } (w', c') \in T_m(\mu) \\
&\ \text{such that } (w', c') \succ_m (w, c)\}, \\
S_W(\mu) = &\ \{(m, w, c) \in M_\emptyset \times W_\emptyset \times C : \text{ there does not exist } (m', c') \in T_w(\mu) \\
&\ \text{such that } (m', c') \succ_w (m, c)\}, \\
S(\mu) = &\ S_M(\mu) \cap S_W(\mu), \text{ and} \\
IR_W(\mu) = &\ \{(m, w, c) \in M_\emptyset \times W_\emptyset \times C : \text{ there does not exist } c' \in C \\
&\ \text{such that } (\emptyset, c') \succ_w (m, c)\}.
\end{aligned}$$

A marriage matching μ is *stable* if $\text{supp}(\mu) \subseteq S(\mu)$. As shown in [4], a matching is stable if and only if $\text{supp}(\mu) \subseteq S_M(\mu) \cap IR_W(\mu)$.

3 An Example

We introduce our version of the DAA using a simple example without contracts. In this example, women prefer men of the same type and men prefer women of a different type. Let $M = \{\alpha, \beta\}$, $W = \{A, B\}$ and preferences be $B \succ_\alpha A \succ_\alpha \emptyset$, $A \succ_\beta B \succ_\beta \emptyset$, $\alpha \succ_A \beta \succ_A \emptyset$ and $\beta \succ_B \alpha \succ_B \emptyset$. Type distributions are $\nu_M(\alpha) = 1$, $\nu_M(\beta) = 2$, $\nu_W(A) = 2$ and $\nu_W(B) = 2$.

The idea of our DAA is analogous to that of GS: Each woman proposes to her favorite man. Each man who receives more than one proposal rejects all but his favorite from among those who have proposed to him and keeps his favorite on a string to allow for the possibility that someone better may come along later.

We start by specifying an initial distribution μ_0 so that every (nonempty) man is matched with the empty woman, $\mu_0(m, \emptyset) = \nu_M(m)$ for each $m \in M$, and no (nonempty) woman is matched, $\mu_0(m, w) = 0$ for each $m \in M_\emptyset$ and $w \in W$. The interpretation is that μ_0 is the distribution of men and women on the string and initially there is no (nonempty) woman on the string. We also specify $\mu_0(\emptyset, \emptyset) = 0$ to satisfy condition 3 in the definition of a matching, and we set $\nu_W(\emptyset) = \nu_M(M) = 3$ which ensures that there is always enough measure of the empty woman to be added to the string with any (nonempty) man.

In the first stage of the DAA, we let the set of rejected women be $R_1 = \{A, B\}$ to reflect that $\sum_{m \in M_\emptyset} \mu_0(m, w) < \nu_W(w)$ for each $w \in W$. Since there have been no proposals yet, we let the set of men who have rejected women of type A be empty, $M_{1,A} = \emptyset$, and likewise for type B, $M_{1,B} = \emptyset$. Women propose to their favorite men: We let $P_w(k)$ be the kth favorite man according to w's preferences, let $r_1(A) = 1$ and $r_1(B) = 1$, and specify that women of type w propose to men of type $P_w(r_1(w))$. Thus, women of type A propose to men of type α and women of type B propose to men of type β. Men can choose between women who have proposed to them, with the convention that the empty woman proposes to each (nonempty) man. Hence, men of type α can choose between women of type $w \in W^*_{1,\alpha} = \{A, \emptyset\}$, and men of type β can choose between women of type $w \in W^*_{1,\beta} = \{B, \emptyset\}$. We also let $W^*_{1,\emptyset} = \emptyset$ to indicate that no woman (nonempty or otherwise) proposed to the empty man. We then allocate all men of type α to women of type A since $A \succ_\alpha \emptyset$, $\mu_1(\alpha, A) = 1$, and all men of type β to women of type B, $\mu_1(\beta, B) = 2$. All the remaining entries of μ_1 are equal to zero.

In stage 2, we obtain that $R_2 = \{A\}$ since $1 = \sum_{m \in M_\emptyset} \mu_0(m, A) < \nu_W(A) = 2$ and $\sum_{m \in M_\emptyset} \mu_0(m, B) = \nu_W(B)$. Some women of type A have been rejected by men of type α, thus $M_{2,A} = \{\alpha\}$; in contrast, no women of type B has been rejected: $M_{2,B} = \emptyset$. We let $r_2(A) = 2 = 1 + |M_{2,A}|$ and $r_2(B) = 1 = 1 + |M_{2,B}|$, hence women of type A now propose to their second favorite type of man, which is β. Proposals are made by the women who were rejected, thus by a measure $\nu^2_W(A) = \nu_W(A) - \sum_{m \in M_\emptyset} \mu_0(m, A) = 1$ of women of type A and a measure $\nu^2_W(B) = 0$ of women of type B. Men of type $m \in M_\emptyset$ can choose from $W^*_{2,\alpha} = \{A, \emptyset\}$, $W^*_{2,\beta} = \{A, B, \emptyset\}$ and $W^*_{2,\emptyset} = \emptyset$; for instance, men of type β can always choose to be unmatched (i.e., choose $\emptyset$), can choose from those women of type B who are on the string with them and also from those women of type A that just proposed to them. We then allocate men of type β as follows: Since A is their favorite type of women,

$$\mu_2(\beta, A) = \min\{\mu_1(\beta, A) + \nu^2_W(A), \nu_M(\beta)\} = 1$$

and

$$\mu_2(\beta, B) = \min\{\mu_1(\beta, B), \nu_M(\beta) - \mu_2(\beta, A)\} = 1;$$

note that $\mu_1(\beta, A) = 0$ is the measure of women of type A who were on the string with men of type β in stage 1 and $\nu^2_W(A) = 1$ is the measure of rejected women of type A who proposed to type β in stage 2; thus men of type β can choose from a measure 1 of women of type A; they can also choose from a measure 1 of women of type B which consist only of those who were on the string with men of type β in stage 1. Nothing changed regarding men of type α, thus $\mu_2(\alpha, A) = 1$.

Continuing to apply the above procedure, we obtain the following.

Stage 3: It follows that $R_3 = \{B\}$, $r_3(B) = 2$, $M_{3,A} = \{\alpha\}$, $M_{3,B} = \{\beta\}$, $r_3(A) = 2$, $r_3(B) = 2$, $\nu^3_W(B) = 1$, $W^*_{3,\alpha} = \{A, B, \emptyset\}$, $W^*_{3,\beta} = \{A, B, \emptyset\}$, $\mu_3(\alpha, B) = 1$, $\mu_3(\beta, A) = 1$ and $\mu_3(\beta, B) = 1$.

Stage 4: We have that $R_4 = \{A\}$, $M_{4,A} = \{\alpha\}$, $M_{4,B} = \{\beta\}$, $r_4(A) = 2$, $r_4(B) = 2$, $\nu^4_W(A) = 1$, $W^*_{4,\alpha} = \{B, \emptyset\}$, $W^*_{4,\beta} = \{A, B, \emptyset\}$, $\mu_4(\alpha, B) = 1$ and $\mu_4(\beta, A) = 2$.

Stage 5: It follows that $R_5 = \{B\}$, $M_{5,A} = \{\alpha\}$, $M_{5,B} = \{\beta\}$, $r_5(A) = 2$, $r_5(B) = 2$, $\nu^5_W(B) = 1$, $W^*_{5,\alpha} = \{B, \emptyset\}$, $W^*_{5,\beta} = \{A, \emptyset\}$, $\mu_5(\alpha, B) = 1$ and $\mu_5(\beta, A) = 2$.

Stage 6: We obtain that $R_6 = \{B\}$, $M_{6,A} = \{\alpha\}$, $M_{6,B} = \{\alpha, \beta\}$, $r_6(A) = 2$, $r_6(B) = 3$, $\nu^6_W(B) = 1$, $W^*_{6,\alpha} = \{B, \emptyset\}$, $W^*_{6,\beta} = \{A, \emptyset\}$, $W^*_{6,\emptyset} = \{B, \emptyset\}$, $\mu_6(\alpha, B) = 1$, $\mu_6(\beta, A) = 2$ and $\mu_6(\emptyset, B) = 1$.

Since $R_7 = \emptyset$, the algorithm stops after 6 iterations and μ_6 is a stable matching.

4 A DAA for Large Marriage Markets

4.1 The Algorithm

The key concept in the construction of the algorithm is what GS call the string, which consists in their setting of the best woman that has proposed to a man. Here we have a distribution of men and women on the string. The initial distribution μ_0 is such that there is no woman on the string, so that $\mu_0(m, \emptyset, c_m) = \nu_M(m)$ for each $m \in M$, where $c_m \in C$ is such that $(\emptyset, c_m) \succ_m (\emptyset, c)$ for each $c \in C \setminus \{c_m\}$, $\mu_0(m, w, c) = 0$ for each $m \in M_\emptyset$, $w \in W$ and $c \in C$ and $\mu_0(\emptyset, \emptyset, c) = 0$ for each $c \in C$. Set $\nu_W(\emptyset) = \nu_M(M)$.

Stage 1: The set of rejected women is

$$R_1 = \{w \in W : \sum_{(m,c) \in M_\emptyset \times C} \mu_0(m, w, c) < \nu_W(w)\} = W$$

to reflect that no woman is on the string and define the new measure of women by setting, for each $w \in W_\emptyset$,

$$\nu_W^1(w) = \nu_W(w) - \sum_{(m,c) \in M_\emptyset \times C} \mu_0(m, w, c).$$

We then have that $\nu_W^1(w) = \nu_W(w)$ for each $w \in W$.

For each $w \in W$, the set of man-contract pairs who have rejected w is

$$M_{1,w} = \emptyset$$

to reflect that no offer and, thus, no rejections have been made yet. Let

$$r_1(w) = |M_{1,w}| + 1 = 1$$

be $1 +$ the number of man-contract pairs that have rejected a woman of type w. The reason why we add 1 to the latter number in the definition of $r_1(w)$ is as follows. For each $w \in W$, order $M_\emptyset \times C$ according to $\succ_w$; thus,

$$M_\emptyset \times C = \{(m_1, c_1), \ldots, (m_{|M_\emptyset \times C|}, c_{|M_\emptyset \times C|})\}$$

with $(m_1, c_1) \succ_w \cdots \succ_w (m_{|M_\emptyset \times C|}, c_{|M_\emptyset \times C|})$ and, for each $1 \leq k \leq |M_\emptyset \times C|$, let $P_w(k) = (m_k, c_k)$ be the kth favorite man-contract pair for any woman of type w. Each $(m, c) \in M_\emptyset \times C$ receives proposals from $w \in R_1$ such that $P_w(1) = P_w(r_1(w)) = (m, c)$.

For each $(m, c) \in M \times C$, let

$$W_{1,m,c} = \{w \in R_1 : P_w(r_1(w)) = (m, c)\} \cup \{\emptyset\}$$

be the set of women types that men of type m have to decide upon (with contract c), and which equals those w that have proposed to match with m under contract c as well as the empty woman type $\emptyset$. In general, men of type m need also consider those women on the string with m under contract c, hence, the set of women types in

$$W^*_{1,m,c} = W_{1,m,c} \cup \text{supp}(\mu_0(m, \cdot, c));$$

since initially there are no women on the string, we have that $\text{supp}(\mu_0(m, \cdot, c))$ equals $\{\emptyset\}$ if $c = c_m$ or $\emptyset$ otherwise; in either case, $W^*_{1,m,c} = W_{1,m,c}$. Let

$$W^*_{1,m} = \cup_{c \in C}(W^*_{1,m,c} \times \{c\})$$

be the set of woman-contract pairs such that the women have proposed to m and order it according to $\succ_m$: $W^*_{1,m} = \{(w_1^{1,m}, c_1^{1,m}), \ldots, (w_{j^{1,m}}^{1,m}, c_{j^{1,m}}^{1,m})\}$ with $j^{1,m} = |W^*_{1,m}|$ and $(w_k^{1,m}, c_k^{1,m}) \succ_m (w_{k+1}^{1,m}, c_{k+1}^{1,m})$ for each $k \in \{1, \ldots, j^{1,m} - 1\}$. The distribution of those on the string is:

$$\mu_1(m, w_1^{1,m}, c_1^{1,m}) = \min\{\mu_0(m, w_1^{1,m}, c_1^{1,m}) + \nu_W^1(w_1^{1,m}) 1_{W_{1,m,c_1^{1,m}}}(w_1^{1,m}), \nu_M(m)\} \text{ and}$$
$$\mu_1(m, w_k^{1,m}, c_k^{1,m}) = \min\{\mu_0(m, w_k^{1,m}, c_k^{1,m}) + \nu_W^1(w_k^{1,m}) 1_{W_{1,m,c_k^{1,m}}}(w_k^{1,m}), \nu_M(m) - \sum_{j<k} \mu_1(m, w_j^{1,m}, c_j^{1,m})\}.$$

In short, men of type m add women outside the string in accordance with their preferences up to capacity.

If $m = \emptyset$, let, for each $c \in C$,

$$W_{1,\emptyset,c} = \{w \in R_1 : P_w(r_1(w)) = (\emptyset, c)\} \text{ and}$$
$$W^*_{1,\emptyset,c} = W_{1,\emptyset,c} \cup \text{supp}(\mu_0(\emptyset, \cdot, c)).$$

For each $w \in W^*_{1,\emptyset,c}$, let

$$\mu_1(\emptyset, w, c) = \mu_0(\emptyset, w, c) + \nu_W^1(w) 1_{W_{1,\emptyset,c}}(w).$$

The case of the empty man type $\emptyset$ differs from the previous case in that there is no limit to the measure of women that can be added to the string with $\emptyset$.

Stage $n > 1$: The set of rejected women is

$$R_n = \{w \in W : \sum_{(m,c) \in M_\emptyset \times C} \mu_{n-1}(m, w, c) < \nu_W(w)\},$$

i.e., the set of women types with some of its members being outside the string. For each $w \in W$, let

$$M_{n,w} = M_{n-1,w} \cup \{(m,c) \in M \times C : \mu_{n-1}(m,w,c) - \mu_{n-2}(m,w,c) < 0\}$$
$$\cup \{(m,c) \in M \times C : w \in W_{n-1,m,c} \text{ and } \mu_{n-1}(m,w,c) - \mu_{n-2}(m,w,c) < \nu_W^{n-1}(w)\}$$

be the set of (m,c) who have rejected w before the start of stage n. This rejection could have happened before stage $n-1$ or it could happen in stage $n-1$ when some women of type w are removed from the string or when some of them outside the string had proposed to some men of type m but were not added to the string with such men. Set

$$r_n(w) = |M_{n,w}| + 1$$

and, for each $w \in W_\emptyset$,

$$\nu_W^n(w) = \nu_W(w) - \sum_{(m,c)\in M_\emptyset \times C} \mu_{n-1}(m,w,c)$$

be the measure of women of type w who are outside the string.

For each $m \in M$ and $c \in C$, let

$$W_{n,m,c} = \{w \in R_n : P_w(r_n(w)) = (m,c)\} \cup \{\emptyset\} \text{ and}$$
$$W^*_{n,m,c} = W_{n,m,c} \cup \text{supp}(\mu_{n-1}(m,\cdot,c)).$$

Order

$$W^*_{n,m} = \cup_{c\in C}(W^*_{n,m,c} \times \{c\})$$

according to $\succ_m$: $W^*_{n,m} = \{(w_1^{n,m}, c_1^{n,m}), \ldots, (w_{j^{n,m}}^{n,m}, c_{j^{n,m}}^{n,m})\}$ with $j^{n,m} = |W^*_{n,m}|$ and $(w_k^{n,m}, c_k^{n,m}) \succ_m (w_{k+1}^{n,m}, c_{k+1}^{n,m})$ for each $k \in \{1,\ldots,j^{n,m}-1\}$. The distribution of those on the string is:

$$\mu_n(m, w_1^{n,m}, c_1^{n,m}) = \min\{\mu_{n-1}(m, w_1^{n,m}, c_1^{n,m}) + \nu_W^n(w_1^{n,m})1_{W_{n,m,c_1^{n,m}}}(w_1^{n,m}),$$
$$\nu_M(m)\} \text{ and}$$
$$\mu_n(m, w_k^{n,m}, c_k^{n,m}) = \min\{\mu_{n-1}(m, w_k^{n,m}, c_k^{n,m}) + \nu_W^n(w_k^{n,m})1_{W_{n,m,c_k^{n,m}}}(w_k^{n,m}),$$
$$\nu_M(m) - \sum_{j<k} \mu_n(m, w_j^{n,m}, c_j^{n,m})\}.$$

If $m = \emptyset$, let, for each $c \in C$,

$$W_{n,\emptyset,c} = \{w \in R_n : P_w(r_n(w)) = (\emptyset, c)\} \text{ and}$$
$$W^*_{n,\emptyset,c} = W_{n,\emptyset,c} \cup \text{supp}(\mu_{n-1}(\emptyset,\cdot,c)).$$

Then, for each $w \in W^*_{n,\emptyset,c}$, let

$$\mu_n(\emptyset, w, c) = \mu_{n-1}(\emptyset, w, c) + \nu^n_W(w) 1_{W_{n,\emptyset,c}}(w).$$

4.2 Convergence of the DAA

Let N be the first n such that $R_n = \emptyset$. Then we say that the algorithm *terminates after $N-1$ iterations*. Our main result shows that when the type distributions have rational coefficients, such an N always exists and μ_{N-1} is a stable matching. In addition, μ_{N-1} is woman-optimal in the sense that, for each $w \in W$ and each stable matching μ',

$$\begin{aligned} &\sum_{(m',c') \in M_\emptyset \times C : (m',c') \succeq_w (m,c)} \mu'(m', w, c') \\ &\leq \sum_{(m',c') \in M_\emptyset \times C : (m',c') \succeq_w (m,c)} \mu_{N-1}(m', w, c') \end{aligned}$$

for each $(m, c) \in M_\emptyset \times C$, where $(m, c) \succeq_w (m', c')$ stands for either $(m, c) = (m', c')$ or $(m, c) \succ_w (m', c')$ (and analogously for $(w, c) \succeq_m (w', c')$).[5]

Theorem 1 *Let E be a marriage market. If preferences are linear orders, $\nu_M(m) \in \mathbb{Q}$ for each $m \in M$ and $\nu_W(w) \in \mathbb{Q}$ for each $w \in W$, then there is $N \in \mathbb{N}$ such that the DAA terminates after $N-1$ iterations and μ_{N-1} is a woman-optimal stable matching.*

Theorem 1 can be used in marriage markets in which preferences are only acyclic. In this case, since M and W are finite, each preference relation can be extended to a linear order to which our DAA applies. Since extending preferences can only make blocking easier, the stable matching for the extended preferences is a stable matching of the original marriage market.

The assumption that the distributions of men and women have rational coefficients is needed to guarantee that the DAA terminates in finitely many steps. Nevertheless, marriage markets whose distributions of men and women do not have rational coefficients can be approximated by a sequence of marriage markets to which Theorem 1 can be applied and, thus, a stable matching for the original marriage market can be obtained as the limit of a sequence of stable matchings for the marriage markets in the sequence.

Our DAA can be easily modified to produce a man-optimal stable matching. All it takes is to reverse the role of men and women in the algorithm, i.e., to make men be the ones who propose.

The proof of Theorem 1 is in Sect. 5.1; here we provide an outline of it.

[5] For a discussion of this notion of woman-optimality, which is defined in analogy to first-order stochastic dominance, see, for example, [9].

Note first that, by construction, $\mu_n(\emptyset, \emptyset, c) = 0$ for each $n \in \mathbb{N}_0$ and $c \in C$ since $\emptyset \notin W^*_{n,\emptyset,c}$; thus, we obtain condition 3 in the definition of a matching.

Condition 2 for a matching is also obtained by construction since the algorithm allocates all men. Indeed, since the measure of the empty woman is effectively unbounded, it can absorb all men of a given type that are not on the string with women that they prefer to the empty woman; thus, we actually have that

$$\sum_{(w,c)\in W_\emptyset\times C:(w,c)\succeq_m(\emptyset,c_m)} \mu_n(m, w, c) = \sum_{(w,c)\in W_\emptyset\times C} \mu_n(m, w, c) = \nu_M(m) \qquad (1)$$

for each $n \in \mathbb{N}_0$ and $m \in M$ (recall that $c_m \in C$ is the best contract for unmatched men of type m).

The marginal condition for women (i.e., condition 1 for a matching) holds when the algorithm stops. Nevertheless, along the sequence, we have that

$$\sum_{(m,c)\in M_\emptyset\times C} \mu_n(m, w, c) \leq \nu_W(w)$$

for each $n \in \mathbb{N}_0$ and $w \in W_\emptyset$. This, in turn, implies that $\nu^n_W(w) \geq 0$ and $\mu_n(m, w, c) \geq 0$ for each $m \in M_\emptyset$, $w \in W_\emptyset$, $c \in C$ and $n \in \mathbb{N}$; hence μ_n is a measure for each $n \in \mathbb{N}$. When the algorithm stops, i.e., when $R_N = \emptyset$, we have from the definition of R_N that $\sum_{(m,c)\in M_\emptyset\times C} \mu_{N-1}(m, w, c) \geq \nu_W(w)$ for all $w \in W$. Combined with the above, this implies that $\sum_{(m,c)\in M_\emptyset\times C} \mu_{N-1}(m, w, c) = \nu_W(w)$ for all $w \in W$.

The algorithm is guaranteed to stop when the type distributions have rational coefficients. The importance of this feature is that there is a common denominator K for $\nu_M(m)$ and $\nu_W(w)$ for each $m \in M$ and $w \in W$; from this, it can be shown that K is in fact a common denominator for $\{\nu^n_W(w)\}_{n=1}^\infty$, and hence $\nu^n_W(w) > 0$ implies $\nu^n_W(w) \geq \frac{1}{K}$.

To see how the above implies that the algorithm stops, note first that $\max_n r_n(w)$ exists since $r_n(w) \in \{1, \ldots, |M_\emptyset \times C|\}$. Thus, $r_n(w)$ is eventually constant, i.e., there is k such that $r_n(w) = r$ for each $n \geq k$. Then for $n \geq k$, we must have $P_w(r) \notin M_{n,w}$. Since $r_{n+1}(w) > r_n(w)$ if $\mu_n(P_w(r), w) < \mu_{n-1}(P_w(r), w) + \nu^n_W(w)$,[6,7] a contradiction, it follows that for $n \geq k$

$$\mu_n(P_w(r), w) \geq \mu_{n-1}(P_w(r), w) + \nu^n_W(w)$$

and, in fact (from the argument in the previous paragraph),

---

[6] The notation $\mu_n(P_w(r), w)$ stands for $\mu_n(m, w, c)$ with $(m, c) = P_w(r)$.

[7] By definition, $r_{n+1}(w) > r_n(w)$ if for some $(m, c) \notin M_{n,w}$, $\mu_n(m, w, c) < \mu_{n-1}(m, w, c)$ or $\mu_n(m, w, c) < \mu_{n-1}(m, w, c) + \nu^n_W(w)$ and $w \in W_{n,m,c}$. But since $P_w(r) \notin M_{n,w}$ and $w \in W_{n,P_w(r)}$ whenever $\nu^n_W(w) > 0$, we have that $r_{n+1}(w) > r_n(w)$ if $\mu_n(P_w(r), w) < \mu_{n-1}(P_w(r), w) + \nu^n_W(w)$.

$$\mu_n(P_w(r), w) \geq \mu_{n-1}(P_w(r), w) + \frac{1}{K}$$

provided that $\nu_W^n(w) > 0$, i.e., $w \in R_n$. Since $\mu_n(P_w(r), w)$ is bounded, it follows that there is N_w such that $w \notin R_n$ for all $n \geq N_w$. Then, $R_n = \emptyset$ for all $n \geq \max_w N_w$. Let N be the first $n \in \mathbb{N}$ such that $R_n = \emptyset$. Then the algorithm stops after $N - 1$ iterations.

We now outline the argument for why $\mu := \mu_{N-1}$ is stable; in Sect. 5.1, we show that in addition μ is woman-optimal. It follows from (1) that μ is individually rational for the men. Individual rationality for the women follows because one can show that $\mu_n(P_w(r), w) = 0$ for each $n \in \mathbb{N}$ and $r > r_w$, where $r_w \in \{1, \ldots, |M_\emptyset \times C|\}$ is such that $P_w(r_w) = (\emptyset, c_w)$, where $c_w \in C$ is the best contract for unmatched women of type w, i.e., $(\emptyset, c_w) \succ_w (\emptyset, c)$ for each $c \in C \setminus \{c_w\}$.

It can be shown that for each $n \in \mathbb{N}$, $m \in M$, $w \in W$ and $c \in C$ such that $(m, w, c) \in \text{supp}(\mu_n)$,

$$\begin{aligned} &\sum_{(w',c'):(w',c') \succeq_m (w,c)} \mu_n(m, w', c') = \nu_M(m) \text{ or} \\ &\sum_{(m',c'):(m',c') \succeq_w (m,c)} \mu_n(m', w, c') = \sum_{(m',c') \in M_\emptyset \times C} \mu_n(m', w, c'). \end{aligned} \tag{2}$$

In other words, men of type m are on the string only with women that are not worse than w or women of type w are on the string only with men that are not worse than m.

This then shows that there are no blocking pairs as follows. Let $m \in M$, $w \in W$ and $c \in C$ be such that $(m, w, c) \in \text{supp}(\mu)$ and there exists $(w', c', \bar{m}, \bar{c})$ such that $\mu(\bar{m}, w', \bar{c}) > 0$, $(w', c') \succ_m (w, c)$ and $(m, c') \succ_{w'} (\bar{m}, \bar{c})$. Since $\mu(m, w, c) > 0$, it follows that $\sum_{(\tilde{w},\tilde{c}):(\tilde{w},\tilde{c}) \succeq_m (w',c')} \mu(m, \tilde{w}, \tilde{c}) < \nu_M(m)$. Using this inequality, it can be shown that $\mu(m, w', c') > 0$. Then (2) and $R_N = \emptyset$ imply that

$$\sum_{(\tilde{m},\tilde{c}):(\tilde{m},\tilde{c}) \succeq_{w'} (m,c')} \mu(\tilde{m}, w', \tilde{c}) = \sum_{(\tilde{m},\tilde{c}) \in M_\emptyset \times C} \mu(\tilde{m}, w', \tilde{c}) = \nu_W(w').$$

Hence, $\mu(\bar{m}, w', \bar{c}) = 0$, a contradiction to $\mu(\bar{m}, w', \bar{c}) > 0$. This then shows that there are no blocking pairs and that μ is stable.

Acknowledgements We wish to thank an anonymous referee for helpful comments. Any remaining errors are, of course, ours.

5 Appendix

5.1 *Proof of Theorem 1*

We establish Theorem 1 in the following lemmas.

Lemma 1 *For each* $n \in \mathbb{N}_0$ *and* $c \in C$, $\mu_n(\emptyset, \emptyset, c) = 0$.

Proof Let $c \in C$. The conclusion is clear when $n = 0$ and, for $n > 0$, it follows because $\emptyset \notin W^*_{n,\emptyset,c}$ for each $n \in \mathbb{N}$. □

Lemma 2 *For each* $n \in \mathbb{N}_0$ *and* $w \in W_\emptyset$, $\sum_{(m,c)\in M_\emptyset \times C} \mu_n(m, w, c) \le \nu_W(w)$.

Proof Let $w \in W_\emptyset$ and note that the conclusion is clear when $n = 0$. Let $n > 0$, $w \in W$ and $(m^*, c^*) = P_w(r_n(w))$; then $1_{W_{n,m,c}}(w) = 0$ for each $(m, c) \neq (m^*, c^*)$. It then follows that

$$\sum_{(m,c)\in M_\emptyset\times C} \mu_n(m, w, c) \le \sum_{(m,c)\in M_\emptyset\times C} \mu_{n-1}(m, w, c) + \nu^n_W(w)$$
$$= \sum_{(m,c)\in M_\emptyset\times C} \mu_{n-1}(m, w, c) + \nu_W(w) - \sum_{(m,c)\in M_\emptyset\times C} \mu_{n-1}(m, w, c) = \nu_W(w).$$

When $w = \emptyset$, using Lemma 1,

$$\sum_{(m,c)\in M_\emptyset\times C} \mu_n(m, \emptyset, c) = \sum_{(m,c)\in M\times C} \mu_n(m, \emptyset, c)$$
$$\le \sum_{m\in M} \nu_M(m) = \nu_M(M) \le \nu_W(\emptyset).$$

□

Lemma 2 implies that $\nu^n_W(w) \ge 0$ for each $w \in W_\emptyset$ and $n \in \mathbb{N}$. This, in turn, implies that $\mu_n(m, w, c) \ge 0$ for each $m \in M_\emptyset$, $w \in W_\emptyset$, $c \in C$ and $n \in \mathbb{N}$.

Write $(m, c) \succeq_w (m', c')$ if either $(m, c) = (m', c')$ or $(m, c) \succ_w (m', c')$ and analogously for $(w, c) \succeq_m (w', c')$.

Lemma 3 *For each* $n \in \mathbb{N}_0$ *and* $m \in M$,

$$\sum_{(w,c)\in W_\emptyset\times C:(w,c)\succeq_m(\emptyset,c_m)} \mu_n(m, w, c) = \sum_{(w,c)\in W_\emptyset\times C} \mu_n(m, w, c) = \nu_M(m).$$

Proof Let $m \in M$ be given and note that the conclusion is clear when $n = 0$. Let $n > 0$ and assume that the conclusion holds for each $j = 0, \ldots, n-1$. We have that $(\emptyset, c_m) \in W^*_{n,m}$ and $\nu^n_W(\emptyset) \ge \nu_M(M) - \sum_{(m',c)\in M_\emptyset\times C} \mu_{n-1}(m', \emptyset, c) = \nu_M(M) - \sum_{(m',c)\in M\times C} \mu_{n-1}(m', \emptyset, c) = \nu_M(M) - \sum_{m'\in M} \mu_{n-1}(m', \emptyset, c_{m'})$ by Lemma 1 and the inductive hypothesis. Thus,

$$\begin{aligned}\mu_{n-1}(m,\emptyset,c_m)+\nu_W^n(\emptyset) &\geq \nu_M(M)-\sum_{m'\in M\setminus\{m\}}\mu_{n-1}(m',\emptyset,c_{m'})\\ &\geq \nu_M(M)-\sum_{m'\in M\setminus\{m\}}\nu_M(m')=\nu_M(m).\end{aligned}$$

Hence, $\mu_n(m,\emptyset,c_m)=\nu_M(m)-\sum_{(w,c)\in W^*_{n,m}:(w,c)\succ_m(\emptyset,c_m)}\mu_n(m,w,c)$ and, therefore, $\sum_{(w,c)\in W_\emptyset\times C:(w,c)\succeq_m(\emptyset,c_m)}\mu_n(m,w)=\nu_M(m)$. □

Lemma 4 *If* $n\in\mathbb{N}$*,* $m\in M_\emptyset$*,* $w\in W$*,* $c\in C$ *and* $(m,w,c)\in\operatorname{supp}(\mu_n)$*, then* $(m,c)\in\{P_w(r):r\in\{1,\ldots,r_n(w)\}\}$*.*

Proof The conclusion clearly holds for μ_1. Suppose that the conclusion holds for $\mu_1,\ldots,\mu_{n-1}$ and that $(m,w,c)\in\operatorname{supp}(\mu_n)$. Then

$$\mu_{n-1}(m,w,c)>0 \text{ or } \nu_W^n(w)1_{W_{n,m,c}}(w)>0.$$

In the former case, $(m,c)=P_w(r)$ for some $r\leq r_{n-1}(w)$ by the inductive hypothesis; hence, $r\leq r_n(w)$ since $M_{n-1,w}\subseteq M_{n,w}$ and, thus, $r_{n-1}(w)\leq r_n(w)$. In the latter case, it follows by $w\in W_{n,m,c}$ that $(m,c)=P_w(r_n(w))$. □

For each $n\in\mathbb{N}$, $w\in W$ and $r\in\{1,\ldots,|M_\emptyset\times C|\}$, let $\mu_n(P_w(r),w)=\mu_n(m,w,c)$ where $(m,c)=P_w(r)$.

Lemma 5 *For each* $n\in\mathbb{N}$ *and* $w\in W$*,* $M_{n,w}=\{P_w(1),\ldots,P_w(r_n(w)-1)\}$ *(with the convention that* $\{P_w(1),P_w(0)\}=\emptyset$*) and* $r_{n+1}(w)-r_n(w)\in\{0,1\}$*.*

Proof The lemma holds for $M_{1,w}$ and $M_{2,w}$. For the latter, note that $M_{2,w}=\{(m,c)\in M\times C:w\in W_{1,m,c}\text{ and }\mu_1(m,w,c)<\nu_W(w)\}$. Since $w\in W_{1,m,c}$ if and only if $P_w(1)=(m,c)$, $M_{2,w}=\{P_w(1)\}$ if $\mu_1(P_w(1),w)<\nu_W(w)$ and $M_{2,w}=\emptyset$ if $\mu_1(P_w(1),w)=\nu_W(w)$. In the former case, $r_2(w)=2$ so $M_{2,w}=\{P_w(1)\}$ and $r_2(w)-r_1(w)=1$. In the latter case, $r_2(w)=1$ so $M_{2,w}=\{P_w(1),P_w(0)\}=\emptyset$ and $r_2(w)-r_1(w)=0$.

Suppose that it holds for $M_{1,w},\ldots,M_{n-1,w}$. Then

$$M_{n-1,w}=\{P_w(1),\ldots,P_w(r_{n-1}(w)-1)\}.$$

If $r>r_{n-1}(w)\geq r_{n-2}(w)$ (since $M_{n-2,w}\subseteq M_{n-1,w}$), then $\mu_{n-2}(P_w(r),w)=0$ by Lemma 4. Hence,

$$\begin{aligned}&\{(m,c)\in M\times C:\mu_{n-1}(m,w,c)<\mu_{n-2}(m,w,c)\}\subseteq\\ &\{P_w(1),\ldots,P_w(r_{n-1}(w)-1),P_w(r_{n-1}(w))\}.\end{aligned}$$

Furthermore, $\{(m,c)\in M\times C:w\in W_{n-1,m,c}\}\subseteq\{P_w(r_{n-1}(w))\}$.

Thus, either $r_n(w)=r_{n-1}(w)$ and

$$M_{n,w}=\{P_w(1),\ldots,P_w(r_{n-1}(w)-1)\}=\{P_w(1),\ldots,P_w(r_n(w)-1)\}$$

or $r_n(w) = r_{n-1}(w) + 1$ and

$$M_{n,w} = \{P_w(1), \ldots, P_w(r_{n-1}(w))\} = \{P_w(1), \ldots, P_w(r_n(w) - 1)\}.$$

Since the conclusion of the lemma holds in both cases, this completes the proof. □

For each $w \in W$, let $c_w \in C$ be such that $(\emptyset, c_w) \succ_w (\emptyset, c)$ for each $c \in C \setminus \{c_w\}$.

Lemma 6 *For each $w \in W$, let $r_w \in \{1, \ldots, |M \times C| + 1\}$ be such that $P_w(r_w) = (\emptyset, c_w)$. Then $\mu_n(P_w(r), w) = 0$ for each $n \in \mathbb{N}$ and $r \in \{r_w + 1, \ldots, |M_\emptyset \times C|\}$.*

Proof Let $w \in W$, $n \in \mathbb{N}$ and $r \in \{r_w + 1, \ldots, |M_\emptyset \times C|\}$. Note first that $r_n(w) \leq r_w$ for each $n \in \mathbb{N}$ since otherwise, by Lemma 5, $(\emptyset, c_w) \in \{P_w(1), \ldots, P_w(r_w)\} \subseteq M_{n,w} \subseteq M \times C$, a contradiction as $\emptyset \notin M$. It then follows that $w \notin W_{n,m,c}$ for each $n \in \mathbb{N}$ and $(m, c) \notin \{P_w(1), \ldots, P_w(r_w)\}$.

It follows from the above that $\mu_n(P_w(r), w) = 0$ for all $n \in \mathbb{N}$. To see this, note that $\mu_0(P_w(r), w) = 0$. Assume that $\mu_{n-1}(P_w(r), w) = 0$. Then $w \notin \operatorname{supp}(\mu_{n-1}(P_w(r), \cdot)) \cup W_{n,P_w(r)} = W^*_{n,P_w(r)}$. Thus, $\mu_n(P_w(r), w) = 0$. □

For each $w \in W$ and $r \in \{1, \ldots, |M_\emptyset \times C|\}$, write $P_w(r) = (m_w(r), c_w(r))$. Also, for each $m \in M$, $w \in W_\emptyset$ and $c \in C$, write $S_m(w, c) = \{(w', c') \in W_\emptyset \times C : (w', c') \succ_m (w, c)\}$ and $S_w(m, c) = \{(m', c') \in M_\emptyset \times C : (m', c') \succ_w (m, c)\}$.

Lemma 7 *For each $n \in \mathbb{N}$ and $w \in W$,*

$$\mu_n(P_w(r), w) = \nu_M(m_w(r)) - \sum_{(w',c') \in S_{m_w(r)}(w, c_w(r))} \mu_n(m_w(r), w', c')$$

if $r < r_n(w)$

$$\begin{aligned}\mu_n(P_w(r_n(w)), w) = \min\{&\mu_{n-1}(P_w(r_n(w)), w) + \nu_W^n(w),\\ &\nu_M(m_w(r_n(w))) - \sum_{(w',c') \in S_{m_w(r_n(w))}(w, c_w(r_n(w)))} \mu_n(m_w(r_n(w)), w', c')\}\end{aligned}$$

and, if $r_n(w) > r_{n-1}(w)$, then

$$\begin{aligned}&\mu_{n-1}(P_w(r_{n-1}(w)), w) =\\ &\nu_M(m_w(r_{n-1}(w))) - \sum_{(w',c') \in S_{m_w(r_{n-1}(w))}(w, c_w(r_{n-1}(w)))} \mu_{n-1}(m_w(r_{n-1}(w)), w', c').\end{aligned}$$

Proof The conclusion holds for $n = 1$ since $r_1(w) = 1$, $w \in W_{1,P_w(1)}$ and

$$\begin{aligned}\mu_1(P_w(1), w) = \min\{&\mu_0(P_w(1), w) + \nu_W^1(w),\\ &\nu_M(m_w(1)) - \sum_{(w',c') \in S_{m_w(1)}(w, c_w(1))} \mu_n(m_w(1), w', c')\}.\end{aligned}$$

Suppose it holds for $\mu_1, \ldots, \mu_{n-1}$ and consider first the case where $r_n(w) = r_{n-1}(w)$. Let $r < r_n(w)$. Then $r < r_{n-1}(w)$ and, by the inductive hypothesis,

$$\mu_{n-1}(P_w(r), w) = \nu_M(m_w(r)) - \sum_{(w',c') \in S_{m_w(r)}(w, c_w(r))} \mu_{n-1}(m_w(r), w', c').$$

Moreover, $P_w(r_n(w)) \neq P_w(r)$, hence $w \notin W_{n, P_w(r)}$. The latter implies that

$$\mu_{n-1}(P_w(r), w) + \nu_W^n(w) 1_{W_{n,P_w(r)}}(w) = \mu_{n-1}(P_w(r), w)$$

and we may assume that $\mu_n(P_w(r), w) = \mu_{n-1}(P_w(r), w)$ since otherwise there is nothing to prove.

If $\mu_{n-1}(P_w(r), w) = 0$, then

$$\sum_{(w',c') \in S_{m_w(r)}(w, c_w(r))} \mu_{n-1}(m_w(r), w', c') = \nu_M(m_w(r)).$$

This then implies that $\text{supp}(\mu_{n-1}(m_w(r), \cdot)) \subseteq S_{m_w(r)}(w, c_w(r))$ and, hence,

$$\text{supp}(\mu_n(m_w(r), \cdot)) \subseteq S_{m_w(r)}(w, c_w(r))$$

as follows: This is clear if there is $(w', c') \in \text{supp}(\mu_{n-1}(m_w(r), \cdot))$ such that

$$\mu_n(m_w(r), w', c') = \nu_M(m_w(r)) - \sum_{(\tilde{w}, \tilde{c}) \in S_{m_w(r)}(w', c')} \mu_n(m_w(r), \tilde{w}, \tilde{c});$$

if this condition does not hold, then

$$\mu_n(m_w(r), w', c') = \mu_{n-1}(m_w(r), w', c') + \nu_W^n(w') 1_{W_{n, m_w(r), c'}}(w')$$

for each $(w', c') \in \text{supp}(\mu_{n-1}(m_w(r), \cdot))$ and it follows by Lemma 3 that

$$\begin{aligned} \nu_M(m_w(r)) &\geq \sum_{(w',c') \in \text{supp}(\mu_{n-1}(m_w(r), \cdot))} \mu_n(m_w(r), w', c') \\ &\geq \sum_{(w',c') \in \text{supp}(\mu_{n-1}(m_w(r), \cdot))} \mu_{n-1}(m_w(r), w', c') = \nu_M(m_w(r)); \end{aligned}$$

thus, indeed, $\text{supp}(\mu_n(m_w(r), \cdot)) \subseteq S_{m_w(r)}(w, c_w(r))$. This, together with Lemma 3, then implies that $\sum_{(w',c') \in S_{m_w(r)}(w, c_w(r))} \mu_n(m_w(r), w', c') = \nu_M(m_w(r))$ and

$$\mu_n(P_w(r), w) = 0 = \nu_M(m_w(r)) - \sum_{(w',c') \in S_{m_w(r)}(w, c_w(r))} \mu_n(m_w(r), w', c').$$

If $\mu_{n-1}(P_w(r), w) > 0$, then

$$\mu_n(P_w(r), w) = \mu_{n-1}(P_w(r), w)$$
$$= \nu_M(m_w(r)) - \sum_{(w',c')\in S_{m_w(r)}(w,c_w(r))} \mu_{n-1}(m_w(r), w', c').$$

Because $\mu_n(P_w(r), w) > 0$, for each $(w', c') \in S_{m_w(r)}(w, c_w(r))$, it cannot be that

$$\mu_n(m_w(r), w', c') = \nu_M(m_w(r)) - \sum_{(\tilde{w},\tilde{c})\in S_{m_w(r)}(w',c')} \mu_n(m_w(r), \tilde{w}, \tilde{c}).$$

Thus, for each $(w', c') \in S_{m_w(r)}(w, c_w(r))$, we have that $\mu_n(m_w(r), w', c') = \mu_{n-1}(m_w(r), w', c') + \nu_W^n(w')1_{W_{n,m_w(r),c'}}(w')$. Hence,

$$\sum_{(w',c')\in S_{m_w(r)}(w,c_w(r))} \mu_n(m_w(r), w', c')$$
$$\geq \sum_{(w',c')\in S_{m_w(r)}(w,c_w(r))} \mu_{n-1}(m_w(r), w', c');$$

if this inequality holds strictly, then

$$\mu_n(P_w(r), w) \leq \nu_M(m_w(r)) - \sum_{(w',c')\in S_{m_w(r)}(w,c_w(r))} \mu_n(m_w(r), w', c')$$
$$< \nu_M(m_w(r)) - \sum_{(w',c')\in S_{m_w(r)}(w,c_w(r))} \mu_{n-1}(m_w(r), w', c') = \mu_{n-1}(P_w(r), w)$$
$$= \mu_n(P_w(r), w),$$

a contradiction. Thus,

$$\sum_{(w',c')\in S_{m_w(r)}(w,c_w(r))} \mu_n(m_w(r), w', c') = \sum_{(w',c')\in S_{m_w(r)}(w,c_w(r))} \mu_{n-1}(m_w(r), w', c')$$

and

$$\mu_n(P_w(r), w) = \nu_M(m_w(r)) - \sum_{(w',c')\in S_{m_w(r)}(w,c_w(r))} \mu_{n-1}(m_w(r), w', c')$$
$$= \nu_M(m_w(r)) - \sum_{(w',c')\in S_{m_w(r)}(w,c_w(r))} \mu_n(m_w(r), w', c')$$

as claimed.

For $r = r_n(w)$, $w \in W_{n,P_w(r_n(w))}$ if and only if $\nu_W^n(w) > 0$, hence

$$\mu_n(P_w(r_n(w)), w) = \min\{\mu_{n-1}(P_w(r_n(w)), w) + \nu_W^n(w),$$

$$\nu_M(m_w(r_n(w))) - \sum_{(w',c')\in S_{m_w(r_n(w))}(w,c_w(r_n(w)))} \mu_n(m_w(r_n(w)), w', c')\}$$

as claimed.

Consider the remaining (by Lemma 5) case where $r_n(w) = r_{n-1}(w) + 1$. When $r < r_{n-1}(w)$ or $r = r_n(w)$, the above argument yields the desired conclusion. Thus, let $r = r_{n-1}(w)$. If

$$\begin{aligned}&\mu_{n-1}(P_w(r_{n-1}(w)), w) = \\ &\nu_M(m_w(r_{n-1}(w))) - \sum_{(w',c')\in S_{m_w(r_{n-1}(w))}(w,c_w(r_{n-1}(w)))} \mu_{n-1}(m_w(r_{n-1}(w)), w', c'),\end{aligned} \quad (3)$$

then the above argument applies as well. Thus, in this case, it suffices to show that (3) holds.

By Lemma 5,

$$M_{n-1,w} = \{P_w(1), \ldots, P_w(r_{n-1}(w) - 1)\} \text{ and } M_{n,w} = \{P_w(1), \ldots, P_w(r_{n-1}(w))\}$$

since $r_n(w) - 1 = r_{n-1}(w)$. For convenience, let $(m, c) = P_w(r_{n-1}(w))$. Since $(m, c) \in M_{n,w} \setminus M_{n-1,w}$, we have that (a) $\mu_{n-1}(m, w, c) < \mu_{n-2}(m, w, c)$ or (b) $w \in W_{n-1,m,c}$ and $\mu_{n-1}(m, w, c) < \mu_{n-2}(m, w, c) + \nu_W^{n-1}(w)$. If (3) does not hold, then

$$\mu_{n-1}(m, w, c) = \mu_{n-2}(m, w, c) + 1_{W_{n-1,m,c}}(w)\nu_W^{n-1}(w)$$

and neither (a) nor (b) can hold. Thus, (3) must hold and the lemma follows. □

Lemma 8 *For each $n \in \mathbb{N}$, $m \in M$, $w \in W$ and $c \in C$, if $(m, w, c) \in \text{supp}(\mu_n)$ then*

$$\sum_{(w',c'):(w',c')\succeq_m(w,c)} \mu_n(m, w', c') = \nu_M(m)$$

or

$$\sum_{(m',c'):(m',c')\succeq_w(m,c)} \mu_n(m', w, c') = \sum_{(m',c')\in M_\emptyset\times C} \mu_n(m', w, c').$$

Proof Let $n \in \mathbb{N}$, $m \in M$, $w \in W$ and $c \in C$ be such that $(m, w, c) \in \text{supp}(\mu_n)$. Then $(m, c) = P_w(r)$ for some $1 \le r \le r_n(w)$ by Lemma 4. The conclusion of the lemma follows immediately from Lemma 7 except when $(m, c) = P_w(r_n(w))$ and $\mu_n(m, w, c) = \mu_{n-1}(m, w, c) + \nu_W^n(w)$.

Hence, suppose that $(m, c) = P_w(r_n(w))$ and $\mu_n(m, w, c) = \mu_{n-1}(m, w, c) + \nu_W^n(w)$. By Lemma 5, there are two possible cases: (a) $r_n(w) = r_{n-1}(w) + 1$ and (b) $r_n(w) = r_{n-1}(w)$.

In case (a), $\mu_{n-1}(P_w(r_n(w)), w) = 0$ since

$$\text{supp}(\mu_{n-1}(\cdot, w)) \subseteq \{P_w(1), \ldots, P_w(r_n(w) - 1)\},$$

thus $\mu_n(P_w(r_n(w)), w) = \nu_W^n(w)$ and

$$\nu_W^n(w) = \nu_W(w) - \sum_{r=1}^{r_{n-1}(w)} \mu_{n-1}(P_w(r), w) = \nu_W(w) - \sum_{r=1}^{r_n(w)-1} \mu_{n-1}(P_w(r), w).$$

Hence,

$$\begin{aligned}\sum_{r=1}^{r_n(w)} \mu_n(P_w(r), w) &= \nu_W(w) - \sum_{r=1}^{r_n(w)-1} \mu_{n-1}(P_w(r), w) \\ &+ \sum_{r=1}^{r_n(w)-1} \left(\nu_M(m_w(r)) - \sum_{(w',c') \in S_{m_w(r)}(w, c_w(r))} \mu_n(m_w(r), w', c') \right).\end{aligned} \tag{4}$$

In case (b),

$$\nu_W^n(w) = \nu_W(w) - \sum_{r=1}^{r_{n-1}(w)} \mu_{n-1}(P_w(r), w) = \nu_W(w) - \sum_{r=1}^{r_n(w)} \mu_{n-1}(P_w(r), w)$$

and, hence,

$$\begin{aligned}&\sum_{r=1}^{r_n(w)} \mu_n(P_w(r), w) = \mu_{n-1}(P_w(r_n(w)), w) + \nu_W(w) - \sum_{r=1}^{r_n(w)} \mu_{n-1}(P_w(r), w) \\ &+ \sum_{r=1}^{r_n(w)-1} \left(\nu_M(m_w(r)) - \sum_{(w',c') \in S_{m_w(r)}(w, c_w(r))} \mu_n(m_w(r), w', c') \right) \\ &= \nu_W(w) - \sum_{r=1}^{r_n(w)-1} \mu_{n-1}(P_w(r), w) \\ &+ \sum_{r=1}^{r_n(w)-1} \left(\nu_M(m_w(r)) - \sum_{(w',c') \in S_{m_w(r)}(w, c_w(r))} \mu_n(m_w(r), w', c') \right).\end{aligned}$$

Thus, (4) holds.

It then follows from $\mu_n(P_w(r_n(w)), w) = \mu_{n-1}(P_w(r_n(w)), w) + \nu_W^n(w)$, (4) and Lemma 7 that

$$\begin{aligned}\sum_{(m',c'):(m',c') \succeq_w (m,c)} \mu_n(m', w, c') &= \sum_{r=1}^{r_n(w)} \mu_n(P_w(r), w) \\ &= \nu_W(w) - \sum_{r=1}^{r_n(w)-1} \mu_{n-1}(P_w(r), w)\end{aligned}$$

$$+\sum_{r=1}^{r_n(w)-1}\left(\nu_M(m_w(r)) - \sum_{(w',c')\in S_{m_w(r)}(w,c_w(r))} \mu_n(m_w(r), w', c')\right)$$

$$= \nu_W(w) - \sum_{r=1}^{r_n(w)-1} \mu_{n-1}(P_w(r), w) + \sum_{r=1}^{r_n(w)-1} \mu_n(P_w(r), w)$$

$$= \nu_W(w) - \sum_{r=1}^{r_n(w)-1} \mu_{n-1}(P_w(r), w) + \sum_{r=1}^{r_n(w)-1} \mu_n(P_w(r), w)$$
$$- \mu_{n-1}(P_w(r_n(w)), w) + \mu_n(P_w(r_n(w)), w) - \nu_W^n(w)$$

$$= \nu_W(w) - \sum_{r=1}^{r_n(w)} \mu_{n-1}(P_w(r), w) + \sum_{r=1}^{r_n(w)} \mu_n(P_w(r), w) - \nu_W^n(w)$$

$$= \sum_{r=1}^{r_n(w)} \mu_n(P_w(r), w) = \sum_{(m',c')\in M_\emptyset\times C} \mu_n(m', w, c').$$

□

Lemma 9 *There exists $\underline{\nu} > 0$ such that if $\nu_W^n(w) > 0$, then $\nu_W^n(w) \geq \underline{\nu}$.*

Proof We have that $\nu_M(m)$ and $\nu_W(w)$ are rational for all $m \in M$ and $w \in W_\emptyset$, and let $K > 0$ be such that $\nu_M(m) = \frac{q_M(m)}{K}$ and $\nu_W(w) = \frac{q_W(w)}{K}$ for each $m \in M$, $w \in W_\emptyset$, and $q_M(w)$, $q_W(w) \in \mathbb{Z}$. We will show that for all $m \in M_\emptyset$, $w \in W_\emptyset$, $c \in C$ and $n \in \mathbb{N}$, $\mu_n(m, w, c) = \frac{q_\mu(n,m,w,c)}{K}$ and $\nu_W^{n+1}(w) = \frac{q_\nu(n+1,w)}{K}$ for some $q_\mu(n, m, w, c)$, $q_\nu(n+1, w) \in \mathbb{Z}$. Thus, $\underline{\nu} = \frac{1}{K}$ proves the lemma.

For any $m \in M_\emptyset$, let $W_{1,m}^* = \{(w_1, c_1), \ldots, (w_{|W_{1,m}^*|}, c_{|W_{1,m}^*|})\}$. Then

$$\mu_1(m, w_1, c) = \min\left\{\frac{q_W(w_1)}{K}, \frac{q_M(m)}{K}\right\} = \frac{q_\mu(1, m, w_1, c_1)}{K},$$

where $q_\mu(1, m, w_1, c_1) \in \mathbb{Z}$. For each $1 < k \leq |W_{1,m}^*|$, assume that for each $j < k$, $\mu_1(m, w_j, c_j) = \frac{q_\mu(1,m,w_j,c_j)}{K}$ for some $q_\mu(1, m, w_j, c_j) \in \mathbb{Z}$. Then

$$\mu_1(m, w_k, c_k) = \min\left\{\frac{q_W(w_k)}{K}, \frac{q_M(m)}{K} - \sum_{j<k} \frac{q_\mu(1, m, w_j, c_j)}{K}\right\}$$
$$= \frac{q_\mu(1, m, w_k, c_k)}{K},$$

where $q_\mu(1, m, w_k, c_k) \in \mathbb{Z}$, since $q_W(w_1) \in \mathbb{Z}$, $q_M(m) \in \mathbb{Z}$, and $q_\mu(1, m, w_j, c_j) \in \mathbb{Z}$ for each $j < k$. Thus, for each $(w, c) \in W_{1,m}^*$, $\mu_1(m, w, c) = \frac{q_\mu(1,m,w,c)}{K}$ for some $q_\mu(1, m, w, c) \in \mathbb{Z}$. For each $(w, c) \notin W_{1,m}^*$, $\mu_1(m, w, c) = \frac{q_\mu(1,m,w,c)}{K}$ for $q_\mu(1, m, w, c) = 0 \in \mathbb{Z}$. Finally, for each $w \in W_\emptyset$, $\nu_W^2(w) = \frac{q_W(w)}{K} - \sum_{(m,c)\in M_\emptyset\times C} \frac{q_\mu(1,m,w,c)}{K} = \frac{q_\nu(2,w)}{K}$ for some $q_\nu(2, w) \in \mathbb{Z}$.

Now suppose that for all $m \in M_\emptyset$, $w \in W_\emptyset$ and $c \in C$, $\mu_{n-1}(m, w, c) = \frac{q_\mu(n-1,m,w,c)}{K}$ and $\nu_W^n(w) = \frac{q_\nu(n,w)}{K}$ for some $q_\mu(n-1, m, w, c), q_\nu(n, w) \in \mathbb{Z}$. For any $m \in M_\emptyset$, let $W_{n,m}^* = \{(w_1, c_1), \ldots, (w_{|W_{n,m}^*|}, c_{|W_{n,m}^*|})\}$. Then

$$\begin{aligned}\mu_n(m, w_1, c_1) &= \min\left\{\frac{q_\mu(n-1, m, w_1, c_1)}{K} + \frac{q_\nu(n, w_1)}{K} 1_{W_{n,m,c_1}}(w_1), \frac{q_M(m)}{K}\right\} \\ &= \frac{q_\mu(n, m, w_1, c_1)}{K},\end{aligned}$$

where $q_\mu(n, m, w_1, c_1) \in \mathbb{Z}$. For each $1 < k \le |W_{n,m}^*|$, assume that for each $j < k$, $\mu_n(m, w_j, c_j) = \frac{q_\mu(n,m,w_j,c_j)}{K}$ for some $q_\mu(n, m, w_j, c_j) \in \mathbb{Z}$. Then

$$\begin{aligned}\mu_n(m, w_k, c_k) &= \min\left\{\frac{q_\mu(n-1, m, w_k, c_k)}{K} + \frac{q_\nu(n, w_k)}{K} 1_{W_{n,m,c_k}}(w_k),\right. \\ &\qquad \left.\frac{q_M(m)}{K} - \sum_{j<k} \frac{q_\mu(n, m, w_j, c_j)}{K}\right\} \\ &= \frac{q_\mu(n, m, w_k, c_k)}{K},\end{aligned}$$

where $q_\mu(n, m, w_k, c_k) \in \mathbb{Z}$, since $q_\mu(n-1, m, w_k, c_k) \in \mathbb{Z}$, $q_\nu(n, w_k) \in \mathbb{Z}$, $q_M(m) \in \mathbb{Z}$, and $q_\mu(n, m, w_j, c_j) \in \mathbb{Z}$ for each $j < k$. Thus, for each $(w, c) \in W_{n,m}^*$, $\mu_n(m, w, c) = \frac{q_\mu(n,m,w,c)}{K}$ for some $q_\mu(n, m, w, c) \in \mathbb{Z}$. For each $(w, c) \notin W_{n,m}^*$, $\mu_n(m, w, c) = \frac{q_\mu(n,m,w,c)}{K}$ for $q_\mu(n, m, w, c) = 0 \in \mathbb{Z}$. Finally, for each $w \in W_\emptyset$, $\nu_W^{n+1}(w) = \frac{q_W(w)}{K} - \sum_{(m,c)\in M_\emptyset \times C} \frac{q_\mu(n,m,w,c)}{K} = \frac{q_\nu(n+1,w)}{K}$ for some $q_\nu(n+1, w) \in \mathbb{Z}$. □

Lemma 10 *For each $w \in W$ and $k \in \mathbb{N}$, if $|\{n > k : w \in R_n\}| > \frac{\nu_W(w)}{\underline{\nu}}$ then $r_{k'}(w) = r_k(w) + 1$ for some $k' > k$.*

Proof Let $w \in W$ and $k \in \mathbb{N}$. Suppose for a contradiction that $r_n(w) = r_k(w)$ for all $n \ge k$ and, for convenience, let $r = r_k(w)$. For each $n > k$, it follows that $P_w(r) \notin M_{n+1,w}$ since $M_{n+1,w} = \{P_w(1), \ldots, P_w(r-1)\}$ by Lemma 5 and, therefore, $\mu_n(P_w(r), w) \ge \mu_{n-1}(P_w(r), w) + \nu_W^n(w)$ since $w \in W_{n,P_w(r)}$ whenever $\nu_W^n(w) > 0$. Hence, $\mu_n(P_w(r), w) \ge \mu_{n-1}(P_w(r), w)$ if $w \notin R_n$ and $\mu_n(P_w(r), w) - \mu_{n-1}(P_w(r), w) \ge \underline{\nu}$ if $w \in R_n$ by Lemma 9 and because $\nu_W^n(w) > 0$ whenever $w \in R_n$.

It then follows by induction that

$$\mu_{k'}(P_w(r), w) \ge \mu_k(P_w(r), w) + |\{k' \ge n > k : w \in R_n\}|\underline{\nu}.$$

Then for k' such that $|\{k' \ge n > k : w \in R_n\}| > \nu_W(w)/\underline{\nu}$, $\mu_{k'}(P_w(r), w) > \mu_k(P_w(r), w) + \nu_W(w)$, a contradiction to Lemma 2. Thus, there exists $k^* > k$ such that $r_{k^*}(w) > r_k(w)$ and it follows by Lemma 5 that $r_{k'}(w) = r_k(w) + 1$ for some $k' > k$. □

Lemma 11 *For each $w \in W$ and $k \in \mathbb{N}$, either there exists $k' > k$ such that $r_{k'}(w) = r_k(w) + 1$, or there exists $N_w \in \mathbb{N}$ such that $w \notin R_n$ for all $n > N_w$.*

Proof By Lemma 10, if there does not exist $k' > k$ such that $r_{k'}(w) = r_k(w) + 1$, then $|\{n > k : w \in R_n\}| \leq \nu_W(w)/\underline{\nu}$. Thus, there must exist some N_w such that $w \notin R_n$ for all $n > N_w$. □

Lemma 12 *For each $w \in W$, there exists $N_w \in \mathbb{N}$ such that $w \notin R_n$ for all $n > N_w$.*

Proof Let $r = \max_{n \in \mathbb{N}} r_n(w)$, which exists since $r_n(w) \in \{1, \ldots, |M_\emptyset \times C|\}$, and $k \in \mathbb{N}$ such that $r_k(w) = r$. It then follows from Lemma 11 that $w \notin R_n$ for all $n > N_w$. □

Lemma 13 *There exists $n \in \mathbb{N}$ such that $R_n = \emptyset$.*

Proof This follows from Lemma 12 (take n to be the max N_w over all w). □

Let $N = \min\{n \in \mathbb{N} : R_n = \emptyset\}$ and define $\mu = \mu_{N-1}$.

Lemma 14 *For each $w \in W$, $\sum_{(m,c) \in M_\emptyset \times C} \mu(m, w, c) = \nu_W(w)$.*

Proof It follows from Lemma 2 and from $R_n = \emptyset$ that

$$\sum_{(m,c) \in M_\emptyset \times C} \mu(m, w, c) = \sum_{(m,c) \in M_\emptyset \times C} \mu_{n-1}(m, w, c) = \nu_W(w).$$

□

Lemma 15 $\text{supp}(\mu) \subseteq S_M(\mu) \cap IR_W(\mu)$.

Proof Let $(m, w, c) \in \text{supp}(\mu)$ and suppose that $(m, w, c) \notin IR_W(\mu)$. Thus, there exists $c' \in C$ such that $(\emptyset, c') \succ_w (m, c)$ and, hence, $w \in W$ and $r > r_w$, where $P_w(r) = (m, c)$. Hence, $\mu(m, w, c) = 0$ by Lemma 6, a contradiction to $\mu(m, w, c) > 0$.

Suppose next that $(m, w, c) \notin S_M(\mu)$. Then $m \in M$. If $(\emptyset, c') \succ_m (w, c)$, then $\mu(m, w, c) = 0$ by Lemma 3, a contradiction to $\mu(m, w, c) > 0$.

Thus, there exists $(w', c', \bar{m}, \bar{c})$ such that $\mu(\bar{m}, w', \bar{c}) > 0$, $(w', c') \succ_m (w, c)$ and $(m, c') \succ_{w'} (\bar{m}, \bar{c})$. Since $\mu(m, w, c) > 0$, it follows that

$$\sum_{(\tilde{w},\tilde{c}):(\tilde{w},\tilde{c}) \succeq_m (w',c')} \mu(m, \tilde{w}, \tilde{c}) < \nu_M(m).$$

Letting $r, \bar{r}$ be such that $P_{w'}(r) = (m, c')$ and $P_{w'}(\bar{r}) = (\bar{m}, \bar{c})$, $(m, c') \succ_{w'} (\bar{m}, \bar{c})$ implies that $r < \bar{r}$ and $\mu(\bar{m}, w', \bar{c}) > 0$ implies that $\bar{r} \leq r_{N-1}(w')$ since $\text{supp}(\mu(w', \cdot)) \subseteq \{P_{w'}(1), \ldots, P_{w'}(r_{N-1}(w'))\}$. Hence, by Lemma 7,

$$\mu(m, w', c') = \nu_M(m) - \sum_{(\tilde{w},\tilde{c}):(\tilde{w},\tilde{c}) \succ_m (w',c')} \mu(m, \tilde{w}, \tilde{c}) > 0.$$

Since $\sum_{(\tilde{w},\tilde{c}):(\tilde{w},\tilde{c})\succeq_m(w',c')}\mu(m,\tilde{w},\tilde{c})<\nu_M(m)$ and $\mu(m,w',c')>0$, Lemmas 8 and 14 then imply that

$$\sum_{(\tilde{m},\tilde{c}):(\tilde{m},\tilde{c})\succeq_{w'}(m,c')}\mu(\tilde{m},w',\tilde{c})=\sum_{(\tilde{m},\tilde{c})\in M_\emptyset\times C}\mu(\tilde{m},w',\tilde{c})=\nu_W(w').$$

Hence, $\mu(\bar{m},w',\bar{c})=0$, a contradiction to $\mu(\bar{m},w',\bar{c})>0$. □

Thus, μ is stable. Lemmas 16 to 18 establish that μ is woman-optimal.

Lemma 16 *For each* $w\in W$ *and* $r'\in\{1,\ldots,|M_\emptyset\times C|\}$, $\mu_n(P_w(r'),w)\leq\nu_W(w)-\sum_{r=1}^{r'-1}\mu_{n-1}(P_w(r),w)$.

Proof If $(w,c_w(r'))\notin W^*_{n,m_w(r')}$, then $\mu_n(P_w(r'),w)=0$. Otherwise, by definition, the distribution must satisfy $\mu_n(P_w(r'),w)\leq\mu_{n-1}(P_w(r'),w)+1_{W_{n,P_w(r')}}\nu_W^n(w)$. By Lemma 2, $\mu_{n-1}(P_w(r'),w)\leq\nu_W(w)-\sum_{r=1}^{r'-1}\mu_{n-1}(P_w(r),w)$. Whenever $w\in W_{n,P_w(r')}$, $r'=r_n(w)$, and by Lemma 4, we have that $\text{supp}(\mu_{n-1}(\cdot,w))\subseteq\{P_w(1),\ldots,P_w(r_{n-1}(w))\}\subseteq\{P_w(1),\ldots,P_w(r_n(w))\}$. Thus, $\nu_W^n(w)=\nu_W(w)-\sum_{r=1}^{r'}\mu_{n-1}(P_w(r),w)$. □

Lemma 17 *For each* $n\in\mathbb{N}$, *for each* $w\in W$, *and for each stable matching* μ':

1. $\sum_{r=1}^{k}\mu_n(P_w(r),w)\geq\sum_{r=1}^{k}\mu'(P_w(r),w)$ *for each* $k\in\{1,\ldots,r_n(w)-1\}$.
2. $\sum_{r=1}^{r_n(w)}\mu_n(P_w(r),w)\geq\sum_{r=1}^{r_n(w)}\mu'(P_w(r),w)$ *if* $r_n(w)<r_{n+1}(w)$.

Proof For $n=1$: It is sufficient to show that $\mu'(P_w(1),w)\leq\mu_1(P_w(1),w)$ for each $w\in W$. Fix $w\in W$, and suppose for a contradiction that $\mu'(P_w(1),w)>\mu_1(P_w(1),w)$. This implies that $\mu_1(P_w(1),w)<\nu_W(w)=\mu_0(P_w(1),w)+\nu_W^1(w)$ and hence by Lemma 7:

$$\mu_1(P_w(1),w)=\nu_M(m_w(1))-\sum_{(\hat{w},\hat{c})\in S_{m_w(1)}(w,c_w(1))}\mu_1(m_w(1),\hat{w},\hat{c}).$$

Suppose that $\mu'(m_w(1),\hat{w},\hat{c})\geq\mu_1(m_w(1),\hat{w},\hat{c})$ for each $(\hat{w},\hat{c})\in S_{m_w(1)}(w,c_w(1))$. Then:

$$\begin{aligned}\nu_M(m_w(1))-\mu'(P_w(1),w)&\geq\sum_{(\hat{w},\hat{c})\in S_{m_w(1)}(w,c_w(1))}\mu'(m_w(1),\hat{w},\hat{c})\\&\geq\sum_{(\hat{w},\hat{c})\in S_{m_w(1)}(w,c_w(1))}\mu_1(m_w(1),\hat{w},\hat{c})\\&=\nu_M(m_w(1))-\mu_1(P_w(1),w),\end{aligned}$$

a contradiction because $\mu_1(P_w(1),w)<\mu'(P_w(1),w)$. Thus, it must be that for some $(\hat{w},\hat{c})\in S_{m_w(1)}(w,c_w(1))$, $0\leq\mu'(m_w(1),\hat{w},\hat{c})<\mu_1(m_w(1),\hat{w},\hat{c})\leq\nu_W(\hat{w})$. Since $0<\mu_1(m_w(1),\hat{w},\hat{c})$, $P_{\hat{w}}(1)=(m_w(1),\hat{c})$ by Lemma 4. Since

$\mu'(m_w(1), \hat{w}, \hat{c}) < \nu_W(\hat{w})$ and $P_{\hat{w}}(1) = (m_w(1), \hat{c})$, $(\hat{w}, \hat{c}) \in T_{m_w(1)}(\mu')$. But $(\hat{w}, \hat{c}) \in S_{m_w(1)}(w, c_w(1))$ and $(P_w(1), w) \in \text{supp}(\mu')$, a contradiction to the stability of μ'.

Now assume that the Lemma holds for $\mu_1, \ldots, \mu_{n-1}$. We need to show that if $k \leq r_n(w) - 1$ or if $k = r_n(w) < r_{n+1}(w)$, then $\sum_{r=1}^{k} \mu'(P_w(r), w) \leq \sum_{r=1}^{k} \mu_n(P_w(r), w)$ for each $w \in W$. First consider $k = 1$. Fix $w \in W$, and suppose for a contradiction that $\mu'(P_w(1), w) > \mu_n(P_w(1), w)$. Since $\mu'(P_w(1), w) \leq \mu_{n-1}(P_w(1), w)$, this implies that $\mu_{n-1}(P_w(1), w) > \mu_n(P_w(1), w)$, and hence by Lemma 7:

$$\mu_n(P_w(1), w) = \nu_M(m_w(1)) - \sum_{(\hat{w},\hat{c}) \in S_{m_w(1)}(w, c_w(1))} \mu_n(m_w(1), \hat{w}, \hat{c}).$$

Suppose that $\mu'(m_w(1), \hat{w}, \hat{c}) \geq \mu_n(m_w(1), \hat{w}, \hat{c})$ for each $(\hat{w}, \hat{c}) \in S_{m_w(1)}(w, c_w(1))$. Then:

$$\begin{aligned} \nu_M(m_w(1)) - \mu'(P_w(1), w) &\geq \sum_{(\hat{w},\hat{c}) \in S_{m_w(1)}(w, c_w(1))} \mu'(m_w(1), \hat{w}, \hat{c}) \\ &\geq \sum_{(\hat{w},\hat{c}) \in S_{m_w(1)}(w, c_w(1))} \mu_n(m_w(1), \hat{w}, \hat{c}) \\ &= \nu_M(m_w(1)) - \mu_n(P_w(1), w), \end{aligned}$$

a contradiction because $\mu_n(P_w(1), w) < \mu'(P_w(1), w)$. Thus, it must be that for some $(\hat{w}, \hat{c}) \in S_{m_w(1)}(w, c_w(1))$, $0 \leq \mu'(m_w(1), \hat{w}, \hat{c}) < \mu_n(m_w(1), \hat{w}, \hat{c}) \leq \nu_W(\hat{w})$. Since $0 < \mu_n(m_w(1), \hat{w}, \hat{c})$, $P_{\hat{w}}(r') = (m_w(1), \hat{c})$ for some $r' \leq r_n(\hat{w})$ by Lemma 4. Note that either $r' \leq r_{n-1}(\hat{w})$ or $r_{n-1}(w) < r_n(w)$. By the inductive hypothesis, $\sum_{r=1}^{r'-1} \mu'(P_{\hat{w}}(r), \hat{w}) \leq \sum_{r=1}^{r'-1} \mu_{n-1}(P_{\hat{w}}(r), \hat{w})$. By Lemma 16, $\mu_n(P_{\hat{w}}(r'), \hat{w}) \leq \nu_W(\hat{w}) - \sum_{r=1}^{r'-1} \mu_{n-1}(P_{\hat{w}}(r), \hat{w})$. Therefore, $\sum_{r=1}^{r'-1} \mu'(P_{\hat{w}}(r), \hat{w}) + \mu'(m_w(1), \hat{w}, \hat{c}) < \sum_{r=1}^{r'-1} \mu_{n-1}(P_{\hat{w}}(r), \hat{w}) + \mu_n(m_w(1), \hat{w}, \hat{c}) \leq \nu_W(\hat{w})$, so $\nu_W(\hat{w}) > \sum_{r=1}^{r'-1} \mu'(P_{\hat{w}}(r), \hat{w}) + \mu'(m_w(1), \hat{w}, \hat{c})$. Thus, $\mu'(P_{\hat{w}}(r), \hat{w}) > 0$ for some $r > r'$, and so $(\hat{w}, \hat{c}) \in T_{m_w(1)}(\mu')$. But $(\hat{w}, \hat{c}) \in S_{m_w(1)}(w, c_w(1))$ and $(P_w(1), w) \in \text{supp}(\mu')$, a contradiction to the stability of μ'.

Now assume that $\sum_{r=1}^{k-1} \mu'(P_w(r), w) \leq \sum_{r=1}^{k-1} \mu_n(P_w(r), w)$ for each $w \in W$. Fix $w \in W$, and suppose for a contradiction that $\sum_{r=1}^{k} \mu'(P_w(r), w) > \sum_{r=1}^{k} \mu_n(P_w(r), w)$. This implies that $\mu_n(P_w(k), w) < \mu'(P_w(k), w)$. If $k < r_n(w)$ or if $k = r_n(w) < r_{n+1}(w)$, then by Lemma 7:

$$\mu_n(P_w(k), w) = \nu_M(m_w(k)) - \sum_{(\hat{w},\hat{c}) \in S_{m_w(k)}(w, c_w(k))} \mu_n(m_w(k), \hat{w}, \hat{c}).$$

If $\mu'(m_w(k), \hat{w}, \hat{c}) \geq \mu_n(m_w(k), \hat{w}, \hat{c})$ for every $(\hat{w}, \hat{c}) \in S_{m_w(k)}(w, c_w(k))$, then:

$$\begin{aligned}
\nu_M(m_w(k)) - \mu'(P_w(k), w) &\geq \sum_{(\hat{w},\hat{c})\in S_{m_w(k)}(w,c_w(k))} \mu'(m_w(k), \hat{w}, \hat{c}) \\
&\geq \sum_{(\hat{w},\hat{c})\in S_{m_w(k)}(w,c_w(k))} \mu_n(m_w(k), \hat{w}, \hat{c}) \\
&= \nu_M(m_w(k)) - \mu_n(P_w(k), w),
\end{aligned}$$

a contradiction because $\mu_n(P_w(k), w) < \mu'(P_w(k), w)$. Thus, it must be that for some $(\hat{w}, \hat{c}) \in S_{m_w(k)}(w, c_w(k))$, $0 \leq \mu'(m_w(k), \hat{w}, \hat{c}) < \mu_n(m_w(k), \hat{w}, \hat{c}) \leq \nu_W(\hat{w})$. Since $0 < \mu_n(m_w(k), \hat{w}, \hat{c})$, $P_{\hat{w}}(r') = (m_w(k), \hat{c})$ for some $r' \leq r_n(\hat{w})$ by Lemma 4. Note that either $r' \leq r_{n-1}(\hat{w})$ or $r_{n-1}(w) < r_n(w)$. By the inductive hypothesis, $\sum_{r=1}^{r'-1} \mu'(P_{\hat{w}}(r), \hat{w}) \leq \sum_{r=1}^{r'-1} \mu_{n-1}(P_{\hat{w}}(r), \hat{w})$. By Lemma 16, $\mu_n(P_{\hat{w}}(r'), \hat{w}) \leq \nu_W(\hat{w}) - \sum_{r=1}^{r'-1} \mu_{n-1}(P_{\hat{w}}(r), \hat{w})$. Then $\sum_{r=1}^{r'-1} \mu'(P_{\hat{w}}(r), \hat{w}) + \mu'(m_w(k), \hat{w}, \hat{c}) < \sum_{r=1}^{r'-1} \mu_{n-1}(P_{\hat{w}}(r), \hat{w}) + \mu_n(m_w(k), \hat{w}, \hat{c}) \leq \nu_W(\hat{w})$, so $\nu_W(\hat{w}) > \sum_{r=1}^{r'-1} \mu'(P_{\hat{w}}(r), \hat{w}) + \mu'(m_w(k), \hat{w}, \hat{c})$. Thus, $\mu'(P_{\hat{w}}(r), \hat{w}) > 0$ for some $r > r'$, and so $(\hat{w}, \hat{c}) \in T_{m_w(k)}(\mu')$. But $(\hat{w}, \hat{c}) \in S_{m_w(k)}(w, c_w(k))$ and $(P_w(k), w) \in \text{supp}(\mu')$, a contradiction to the stability of μ'. □

Lemma 18 *Let N be the first n such that $R_n = \emptyset$. For each $w \in W$ and each stable matching μ',*

$$\begin{aligned}
&\sum_{(m',c')\in M_\emptyset\times C:(m',c')\succeq_w(m,c)} \mu'(m', w, c') \\
&\leq \sum_{(m',c')\in M_\emptyset\times C:(m',c')\succeq_w(m,c)} \mu_{N-1}(m', w, c')
\end{aligned}$$

for each $(m, c) \in M_\emptyset \times C$.

Proof For each $k \in \{1, \ldots, r_{N-1}(w) - 1\}$,

$$\sum_{r=1}^{k} \mu_{N-1}(P_w(r), w) \geq \sum_{r=1}^{k} \mu'(P_w(r), w)$$

for every stable matching μ' by Lemma 17. Also by Lemmas 4 and 14,

$$\sum_{r=1}^{k} \mu_{N-1}(P_w(r), w) = \sum_{r=1}^{r_{N-1}(w)} \mu_{N-1}(P_w(r), w) = \nu_W(w) \geq \sum_{r=1}^{k} \mu'(P_w(r), w)$$

for any $k \in \{r_{N-1}(w), \ldots, |M_\emptyset \times C|\}$. □

6 Python Code

In this section we describe how to use a Python code that implements the DAA defined in Sect. 4; the code itself is available online.[8] It considers the specification of the example in Sect. 3 and here we describe how to modify its preamble, i.e., lines 3–18, to obtain any other example.

We first specify the number of men and women types and the number of contracts:

```
#m is the number of men types
m=2
#w is the number of women types
w=2
#c is the number of contracts
c=1
```

Then we specify preferences in the following form: Using $m = |M|$, $w = |W|$ and $c = |C|$ as above and writing $C = \{c_1, \ldots, c_c\}$, $M_\emptyset = \{m_1, \ldots, m_m, m_{m+1}\}$ and $W_\emptyset = \{w_1, \ldots, w_w, w_{w+1}\}$, where $m_{m+1} = w_{w+1} = \emptyset$, preferences as described by

$$p = [[P_{m_1}(w_1, c_1), \ldots, P_{m_1}(w_1, c_c), \ldots, P_{m_1}(w_{w+1}, c_1), \ldots, P_{m_1}(w_{w+1}, c_c)], \ldots, [P_{m_m}(w_1, c_1), \ldots, P_{m_m}(w_1, c_c), \ldots, P_{m_m}(w_{w+1}, c_1), \ldots, P_{m_m}(w_{w+1}, c_c)]] \text{ and}$$
$$q = [[P_{w_1}(m_1, c_1), \ldots, P_{w_1}(m_1, c_c), \ldots, P_{w_1}(m_{m+1}, c_1), \ldots, P_{w_1}(m_{m+1}, c_c)], \ldots, [P_{w_w}(m_1, c_1), \ldots, P_{w_w}(m_1, c_c), \ldots, P_{w_w}(m_{m+1}, c_1), \ldots, P_{w_w}(m_{m+1}, c_c)]],$$

where $P_w(m, c) = k$ if and only if $P_w(k) = (m, c)$ and analogously for $P_m(w, c)$. In the case of the example:

```
#p describes men's preferences, last entry is the empty woman
p=[[2,1,3],
   [1,2,3]]
#q describes women's preferences
q=[[1,2,3],
   [2,1,3]]
```

The final element in the preamble consists of the type distributions. These are described as

$$dm = [\nu_M(m_1), \ldots, \nu_M(m_m)] \text{ and}$$
$$dw = [\nu_W(w_1), \ldots, \nu_W(w_w)]$$

[8] https://drive.google.com/file/d/1LIiaArjO7EPjxkI6vvfXDyXSSJkqd9LF/view.

in general, and as

```
#distribution of men
dm=[1,2]
#distribution of women
dw=[2,2]
```

in the case of example.

Once the above specifications are made, the code computes a stable matching using our DAA. Its output for the example is:

```
mu is: [[0, 1, 0], [2, 0, 0], [0, 1, 0]]
number of iterations 6
```

In general, the output is a stable matching μ in the form

$$[[\mu(m_1, w_1, c_1), \ldots, \mu(m_1, w_1, c_c), \ldots, \mu(m_1, w_{w+1}, c_1), \ldots, \mu(m_1, w_{w+1}, c_c)], \ldots,$$
$$[\mu(m_{m+1}, w_1, c_1), \ldots, \mu(m_{m+1}, w_1, c_c), \ldots, \mu(m_{m+1}, w_{w+1}, c_1), \ldots,$$
$$\mu(m_{m+1}, w_{w+1}, c_c)]]$$

and $N - 1$, where N is the first $n \in \mathbb{N}$ such that $R_n = \emptyset$.

References

1. Ashlagi, I., Kanoria, Y., Leshno, J.: Unbalanced random matching markets: The stark effect of competition. J. Political Econ. **64**, 69–98 (2017)
2. Azevedo, E., Leshno, J.: A supply and demand framework for two-sided matching markets. J. Political Econ. **124**, 1235–1268 (2016)
3. Baïou, M., Balinski, M.: The stable allocation (or ordinal transportation) problem. Math. Oper. Res. **27**, 662–680 (2002)
4. Carmona, G., Laohakunakorn, K.: Stable matching in large markets with occupational choice. Theor. Econ. **19**, 1261–1304 (2024)
5. Che, Y.K., Tercieux, O.: Efficiency and stability in large matching markets. J. Political Econ. **127**, 2301–2342 (2019)
6. Che, Y.K., Kim, J., Kojima, F.: Stable matching in large economies. Econometrica **87**, 65–110 (2019)
7. Chiappori, P.A., Reny, P.: Matching to share risk. Theoret. Econ. **11**, 227–251 (2016)
8. Crawford, V., Knoer, E.: Job matching with heteregeneous firms and workers. Econometrica **49**, 437–450 (1981)
9. Echenique, F., Lee, S., Yenmez, B.: Existence and testable implications of extreme stable matchings. California Institute of Technology Social Science Working Paper 1337 (2010)
10. Echenique, F., Lee, S., Shum, M., Yenmez, B.: The revealed preference theory of stable and extremal stable matchings. Econometrica **81**, 153–171 (2013)
11. Eeckhout, J., Kircher, P.: Assortative matching with large firms. Econometrica **86**, 85–132 (2018)
12. Fisher, J., Hafalir, I.: Matching with aggregate externalities. Math. Soc. Sci. **81**, 1–7 (2016)

13. Fuentes, M., Tohmé, F.: Stable matching with double infinity of workers and firms. BE J. Theoret. Econ. **19** (2018)
14. Gale, D., Shapley, L.: College admissions and the stability of marriage. Am. Math. Month. **69**, 9–15 (1962)
15. Greinecker, M., Kah, C.: Pairwise stable matching in large economies. Econometrica **89**, 2929–2974 (2021)
16. Nöldeke, G., Samuelson, L.: The implementation duality. Econometrica **86**, 1283–1324 (2018)

[illegible]

Gale-Shapley Algorithm Revisited: Semistability

Vladimir Danilov and Gleb Koshevoy

Abstract We consider the broader problem of the existence of stable systems of contracts (we are not limited to the assumption of finite sets of contracts) between agents of two complementary groups, say workers and firms. To do this, we introduce the concept of semistable pairs. We define a dynamic process (a lá Gale-Shapley algorithm) on the set of semistable pairs and show that for agents's preferences defined by path-independent choice functions, the steady states yield stable sets of contracts. In addition to existence, this process makes it possible to show that the set of stable sets forms a complete lattice. In the Appendix, we discuss Lehmann hyper-orders and establish a bijection between a set of Lehmann hyper-orders and a set of path-independent choice functions.

Keywords Plott choice functions · Stable and semistable pairs · Poset · Lattice · Noetherian order

1 Introduction

Gale and Shapley in their fundamental article [7] introduced three new concepts in matching models. Namely, they introduced the concept of stability, an algorithm that builds a stable matching, and gave a non-trivial example of path-independent preferences that allow an extended algorithm in the one-to-many cases to produce a stable system. The concept of stability (one-to-many, many-to-many, contracts) has been explored in numerous works by David Gale himself and with co-authors [2, 16], as well as in others, among which we highlight Kelso and Crawford [9], Roth [15], Fleiner [6], Hatfield and Milgrom [8]. Gale himself kept coming back to this topic, see Sotomayor's wonderful memoirs [16].

V. Danilov (✉)
Central Institute for Economics and Mathematics RAS, Moscow, Russia
e-mail: vdanilov43@mail.ru

G. Koshevoy
Institute for Information Transmission Problesms RAS, Moscow, Russia

M. A. Khan et al. (eds.), *Matching, Dynamics and Games for the Allocation of Resources*, Monographs in Mathematical Economics 7,
https://doi.org/10.1007/978-981-96-6661-4_20

The main idea of the Gale-Shapley algorithm is that at each step of the procedure, one group of negotiations or transactions makes offers to the other group. For clarity, the former are called *workers* and the latter *firms*. After receiving offers and comparing them with already achieved (intermediate) offers, firms keep the best offers and reject (once and for all) the rest. At each stage, the position of the "receiving" group (firm) improves, so the procedure ends sooner or later. The result is a stable contract system.

We propose a variant of the Gale-Shapley iterative procedure. One of the features is as follows. Usually, in the literature, a variant of the Gale-Shapley algorithm began with some specific initial state, when all "rights" or "opportunities" were given to one group, for example, workers. Gradually, these rights are reduced, passing to firms. We can begin our procedure in any state where workers have more rights than firms. We introduce the notion of a semistable pair for such states and define a dynamic process (a lá Gale-Shapley algorithm) on the set **SSP** of semistable pairs. The fixed points of this process (steady states) exactly correspond to stable systems. The existence of fixed points is based on the monotonicity of the process. This dynamic process provides us not only with the existence of stable states, but also allows us to establish the lattice structure of the set **S** of all stable systems.

How to get semistable pairs different from the usual one constituted from the empty set and the whole set of contracts? We show that it is possible to start with any pair and first apply the inverted above dynamics, gradually strengthening the rights of workers until they exceed the rights of firms. The result is a semistable pair, as close as possible to the original.

It should be noted that the main assumption that ensures the efficiency of the proposed dynamics is the assumption that the preferences of agents are defined (due to the multiplicity of contracts concluded) by path-independent choice functions. Gale and Shapley have already implicitly used this assumption. Such an assumption was formulated more explicitly in [9, 15]. Later it turned out that the desired class of choice functions is exactly the class introduced earlier (and for a different reason) by Plott [14] (which is why we call such choice functions Plott functions). Note that Plott limited himself to a simple example with three alternatives, while Gale and Shapley considered interesting examples of such functions, choices with quotas.

Four papers [2, 3, 6, 12] are related to our work. Fleiner [6] suggested to aggregate the agents of each side, this allows to make a presentation more concise and focus on the essence of the matter, and dealing only with pairs of sets, and his notion of a stable pair of sets influenced our notion of a semistable pair. However, his dynamic process occurs on another set of pairs and the existence of fixed points is based on Tarski's theorem. Alkan and Gale [2] built sequences (of triples, not of pairs) that iteratively were recalculated and in the limit (topological, not transfinite, what we use) give stable outcomes. Their process again starts from a specific initial state. Blair [3] in his Lemma 3.1 gives a hint to the notion of a semistable pair. Finally, Lehmann [12] proves a generalization of 'polarization' of the interests of the groups which we apply in an even more general form (see Proposition 1 and its corollaries) to semistable pairs.

The paper is organized as follows. In Sect. 2, we introduce the general concept of a stable contract system. Although any number of agents is allowed in the model, it can be reduced to the case when there are only two agents. Section 3 is premier on stable and semistable pairs, and a process on semistable pairs is defined. In Sect. 4, we deviate somewhat from the main line and remind the necessary information about Plott functions. In Sect. 5 we study semistable pairs. This is, in a sense, a key step, because the existence of stable contract systems is reduced to the existence of maximum elements in the poset of semistable pairs. In Sect. 6 we study a process of sequential improvement and prove our main result Theorem 1 states that for any semistable pair (Y, Z) there exists a stable set S which is an upper bound to (Y, Z). In Sect. 7 we establish the structural properties of the set of stable sets.

Finally, in the Appendix, we give a bijection between the set of Plott choice functions and the set of Lehmann hyper-orders.

2 Stable Contract Systems

We consider the following general model. There are two complementary sets $\mathcal{F}$ and $\mathcal{W}$ of agents (for example, firms and workers). For each pair of agents of opposite types (f, w), $f \in \mathcal{F}$, $w \in \mathcal{W}$, there is a set of contracts $C(f, w)$ which they can sign. Then, $C(f) := \coprod_{w \in \mathcal{W}} C(f, w)$ is the set of contracts available to sign for the agent f, and $C(w) := \coprod_{f \in \mathcal{F}} C(f, w)$ is the set of contracts available for w.

Which subset of contracts will be signed, depends on the preferences of the agents. Following accepted tradition in the literature, going back to Roth [15], preferences of agents are specified by the corresponding choice functions. This means the following: suppose a set of contracts $X \subset C(f)$ is available to sign for an agent f, then f chooses a subset of $G_f(X) \subseteq X$ to sign. That is the preferences of agent f are given by the choice function $G_f : 2^{C(f)} \to 2^{C(f)}$ that sends X to $G_f(X) \subset X$, $X \subset C(f)$.

Similarly, the preferences of agent w are given by the choice function F_w defined on the set $C(w)$.

Denote by $C = \coprod_f C(f) = \coprod_w C(w)$ the set of all possible contracts. Then, for given collections of preferences of agents $\{G_f, f \in \mathcal{F}\}$ and $\{F_w, w \in \mathcal{W}\}$, which subset $S \subseteq C$ could be signed?

Following the Gale and Shapley stability concept [7], two requirements have to be satisfied for a stable set S of contracts.

Firstly, no one of agents wishes to abandon any of the contracts from the proposed set S. This means that, for any agent f, the equality holds $G_f(S \cap C(f)) = S \cap C(f)$, and, for any agent w, it holds $F_w(S \cap C(w)) = S \cap C(w)$.

Secondly, there is no pair (f, w) which wants to conclude a new contract. That is if c belongs to $C(f, w)$ and does not belong to S, then either c does not belong to $G_f(S \cup c)$, or c does not belongs to $F_w(S \cup c)$.

If both of these conditions are met, the contract system S is called a *stable set*.

It is convenient to rewrite these conditions in a compact form by aggregating all workers into one Worker, and all firms into one Firm. The Worker's preferences are represented by a choice function F defined on the set C being the union of individual ones,

$$F(X) = \coprod_w F_w(X \cap C(w)).$$

Similarly, the preferences of the Firm are given by the choice function G on C,

$$G(X) = \coprod_f G_f(X \cap C(f)).$$

Then, we have in these notations

Definition 1 A subset $S \subseteq C$ is *stable* if

S1. $F(S) = S$, $G(S) = S$;
S2. If $c \notin S$, then either $c \notin F(S \cup c)$, or $c \notin G(S \cup c)$.

Here is a couple of examples.

Example 1 There are two agents. The set C consists of six contracts (depicted below as circles). The choice of the Worker is defined by maximization of u_1 (if the utility is non-negative) and the choice of the Firm is defined by similar maximization of u_2.

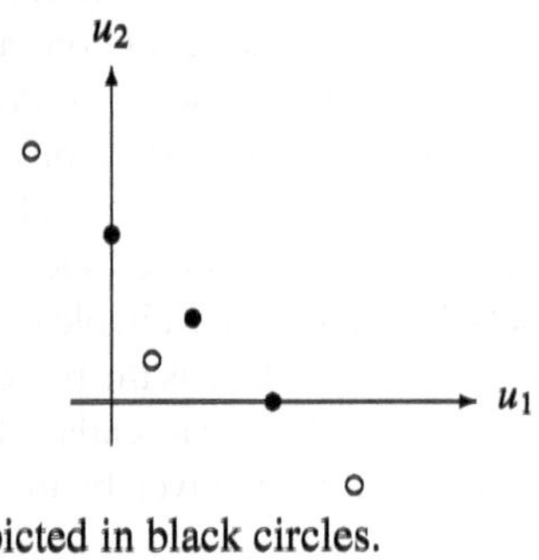

Stable contracts are depicted in black circles.

Example 2 The set C consists of two contracts, a and b. The preferences of the Firm are the choice function G and for the Worker it is F that are specified as follows:

$F(a) = a$, $F(b) = \emptyset$, $F(a, b) = \{a, b\}$;
$G(a) = a$, $G(b) = b$, $G(a, b) = b$.

We claim that there are no stable sets in such a case. In fact, the empty set $\emptyset$ is unstable, since a is better for both. The singleton a is unstable due to the presence of a contract b: $F(a, b)$ contains b, as well as $G(a, b)$. The set $C = \{a, b\}$ is unstable, because the Firm prefers to abandon a. The singleton b is unstable, since the Worker refuses such a contract.

Example 2 shows that for the existence of a stable set, the choice functions F and G have to satisfy certain assumptions.

3 Stable and Semistable Pairs

The central concept of our work is the notion of semistable pairs.

Firstly, following Fleiner [6], we define stable pairs.

Definition 2 A pair (Y, Z) of sets of C is *stable* if

SP1. $Y \cup Z = C$,
SP2. $G(Y) = F(Z)$.

The interpretation of this notion is as follows: we split the set of contracts in two sets Y and Z, and give the set of 'opportunities' Y to the Firm, and give the set Z to the Worker. The Firm and the Worker freely choose contracts from these sets, $G(Y)$ and $F(Z)$, respectively. If their choices coincide, it hints that stability is achieved.

It is not clear how to get stable pairs. We propose an iterative process in the spirit of the Gale-Shapley algorithm. This process will be defined on the set of semistable pairs, we introduce them now.

Definition 3 A pair (Y, Z) of subsets in C is called *semistable* if the following conditions hold

SSP1. $Y \cup Z = C$,
SSP2. $G(Y) \subseteq F(Z)$.

Semistable pairs exist for obvious reasons. For example, the pair $(\emptyset, C)$ and the pair $(F(C), C)$ are semistable. Later we will consider a method for constructing semistable pairs.

Suppose that (Y, Z) is a semistable pair. Let us form a new pair

$$\Phi(Y, Z) = (Y', Z'), \text{ where } Y' = Y \cup F(Z),\ Z' = (Z - F(Z)) \cup G(Y'). \quad (1)$$

That is, for a given pair (Y, Z), we increase Y by adding $F(Z)$, and we throw out from Z those elements of $F(Z)$ that do not belong to $G(Y')$. Transition from a pair of (Y, Z) to a pair of $\Phi(Y, Z)$ implements the basic idea of a general step in the Gale-Shapley algorithm. Firm receives the offer $F(Z)$ from Worker and leaves (for the next step) the subset $G(Y')$, rejecting the set $F(Z) - G(Y')$.

In order to relate stable pairs to stable sets, and to ensure that, for a semistable pair (Y, Z), the pair $\Phi(Y, Z)$ is semistable, and to prove that the steady state of Φ is a stable pair we need to assume that F and G are path-independent choice functions, Plott functions. So we pause to remind a reader of useful facts about Plott functions.

4 Plott Functions

In 1973, Plott [14] invented a notion of path-independence which reflects "rationality" of a choice. Namely, a choice function $F : 2^C \to 2^C$, $F(A) \subseteq A$, is *path-independent* if, for any A and $B \subset C$, it holds

$$F(A \cup B) = F(F(A) \cup B). \tag{2}$$

In [4], we called such choice functions as *Plott functions*.

4.1 Examples and Constructions

Maximization with respect to a Noetherian order $\leq$ on C provides an example of the Plott function. (An order is called *Noetherian* if any increasing chain is stabilized. In other words, any nonempty subset contains a maximum element.) For finite C, a Noetherian order is a usual reflexive, transitive complete binary relation $\prec$ on C. Then the choice of A is the best element of A with respect to $\prec$. Plott by himself showed that such "ordinal" CFs do not exhaust Plott functions.

An interesting example of Plott functions, the quota choice was given by Gale and Shapley [7]. Namely, fix a quota k, and choose from each set at most k best elements wrt $\prec$. Then such a choice function is a Plott function.

4.2 Heredity and Outcast

Path-independence (2) is the equivalent to the union of two conditions, Outcast and Heredity (see for example, [1, 4, 13]).

The *Outcast* condition tells us about the choice intervals $(F(A), A)$.

O Assume that $F(A) \subseteq B \subseteq A$. Then $F(B) = F(A)$.

Aizerman and Malishevskii [1] came up with this name for this property of choice functions, although often it is called "independence of rejected alternatives", since it states that removing elements that are not chosen in A does not change the choice.

The *Heredity* is the monotonicity of complements $A - F(A)$.

H If $A \subseteq B$, then $A \cap F(B) \subseteq F(A)$.

That is, if an element from the smaller set A is chosen from the larger B, then it is also chosen from A.

In economics, Heredity is called as *substitutability* of alternatives. Alkan and Gale in [2] speak about *persistency*.

The Heredity condition is equivalent ([13]) to *subadditivity*.

$$F(A \cup B) \subseteq F(A) \cup F(B). \tag{3}$$

Note that the inclusion (3) holds not only for two terms, but for any: if $(A_i, i \in I)$ is an arbitrary family of subsets in X, then

$$F(\cup_i A_i) \subseteq \cup_i F(A_i).$$

Using **O**, one can obtain a more general form of "path independence": for any families (A_i)

$$F(\cup_i A_i) = F(\cup_i F(A_i)). \tag{4}$$

This also shows that the properties **O** and **H** imply (2).

Note that the union of any family of Plott functions is again a Plott function.

4.3 Conservativity

Note that Plott functions satisfy a stronger condition than Heredity, which we call *conservativity* (Alkan and Gale [2] called it stationarity).

Cons If $G(Y) \subseteq A$, then $G(A) \cap Y \subseteq G(Y)$.

In fact, $G(A \cup Y) = G(A \cup GY) = G(A)$. Applying **H** to the inclusion $A \subseteq A \cup Y$ gives $G(A) \cap Y = G(A \cup Y) \cap Y \subseteq G(Y)$.

The converse is also true: if a choice function G is conservative, it is a Plott function.

The condition **Cons** can be written in the following form, useful for what follows

$$\text{if } G(Y) \subseteq A, \text{ then } G(A) \cap (Y - G(Y)) = \emptyset. \tag{5}$$

In fact, let $a \in G(A) \cap (Y - G(Y))$. Then all the more so $a \in Y$ and by virtue of **Cons** $a \in G(Y)$, which contradicts the fact that $a \in Y - G(Y)$.

4.4 Closure Operator

Using (4), it is easy to understand that among sets with the same choice set $F(A)$, there is the largest one, which we denote as $\sigma_F(A)$. For $x \notin A$

$$x \in \sigma_F(A) \text{ if and only if } x \notin F(A \cup x). \tag{6}$$

Moreover, for any set B between $F(A)$ and $\sigma_F(A)$ we have $F(B) = F(A)$. One can show (see [4, 5, 10]) that σ_F is an anti-exchange closure operator.

The choice function F is determined from its closure σ_F by the formula

$$F(A) = \{a \in A, a \notin \sigma_F(A - a)\}.$$

4.5 The Blair Hyper-Order

Using Plott function F, Blair [3] introduced a remarkable pre-order $\preceq = \preceq_F$ on the set 2^X that allows to compare subsets in X. For subsets A and B

$$A \preceq B \text{ if } F(A \cup B) \subseteq B. \tag{7}$$

For example, $A \preceq B$ if $A \subseteq B$. The (hyper)relation $\preceq$ is reflexive and transitive, although not necessarily antisymmetric. The following statements are equivalent (for $A, B \subseteq X$):

1. $F(A \cup B) = F(B)$,
2. $A \preceq B$,
3. $\sigma_F(A) \subseteq \sigma_F(B)$,
4. $A \subseteq \sigma_F(B)$,
5. $F(A) \subseteq \sigma_F(B)$.

The hyper-order $\preceq$ has another remarkable Union property:

if $(A_i, i \in I)$ is a family of subsets in X, and $A_i \preceq B$ for any i, then $\cup_i A_i \preceq B$. (8)

Indeed, $F((\cup_i A_i) \cup B) = F(\cup_i (A_i \cup B)) = F(\cup_i F(A_i \cup B)) = F(\cup_i F(B)) \subseteq B$.

5 Stable and Semistable Pairs: Continued

In what follows we assume F and G are Plott functions. Firstly we show relation between stable sets and stable pairs. Namely,

Lemma 1 *Let S be a stable set. Then the pair $(\sigma_G(S), \sigma_F(S))$ is a stable pair of sets.*

Proof Since F and G are Plott functions, we have SP2, $S = G(\sigma_G(S)) = F(\sigma_F(S))$. Let us check SP1: suppose c does not belong to $\sigma_G(S)$, that is due to (6) that $c \in G(S \cup c)$. If c does not belong to $\sigma_F(S)$, then $c \in F(S \cup c)$. From the stability of S we get $c \in S$, which contradicts the inclusion $S \subseteq \sigma_G(S)$. □

Vice versa

Lemma 2 *Let (Y, Z) be a stable pair. Then $S = F(Z) = G(Y)$ is a stable set.*

Proof Property S1 follows from the idempotence of F and G: $F(S) = F(F(Z)) = F(Z) = S$ and similarly for G. Let us check S2. $Y \subseteq \sigma_G(S)$ and $Z \subseteq \sigma_F(S)$. So $\sigma_G(S) \cup \sigma_F(S) = C$. Therefore, any blocking contract b ($b \notin S$) lies either in $\sigma_G(S)$ or in $\sigma_F(S)$. In the first case, $b \notin G(S \cup b)$, in the second $b \notin F(S \cup b)$. □

The following proposition is useful for the dynamic iterative process analysis.

Proposition 1 *Let $Y \cup Z = C$, and $G(Y) \subseteq Z$. If $G(Y) \subseteq A$, then $G(A) \subseteq Z$.*

Proof By virtue of (5), $G(A)$ does not intersect $Y - G(Y)$. Suppose that $a \in G(A)$, but does not belong to Z. Since Y and Z cover C, $a \in Y$. But a does not belong to $Y - G(Y)$, so $a \in G(Y) \subseteq Z$. A contradiction. □

Corollary 1 *Let (Y, Z) be a semistable pair, and S be a stable set. If $Y \preceq_G S$, then $S \subseteq Z$.*

Proof Indeed, we have $G(Y) \subseteq F(Z) \subseteq Z$. Set $A = \sigma_G(S)$, then $Y \preceq_G S$ means, that $G(Y) \subseteq A$. By Proposition 1 we get $S = G(A) \subseteq Z$. □

Corollary 2 *([12] Lemma 20) Let S be a stable set and $S \preceq_G A$. Then $G(A) \preceq_F S$. Symmetrically if $S \preceq_F A$, then $F(A) \preceq_G S$.*

Proof Let $Y = \sigma_G(S)$, $Z = \sigma_F(S)$. Then $G(Y) = S$ is contained in Z. $S \preceq_G A$ means that $S = G(Y)$ is a subset of $\sigma_G(A)$. By Proposition 1 we get that $G(\sigma_G(A)) = G(A)$ is a subset of $Z = \sigma_F(S)$. The latter means that $G(A) \preceq_F S$. □

In particular, if A is a stable set, then $S \preceq_G A$ implies $A = G(A) \preceq_F S$ (and vice versa). That is, the hyper-orders $\preceq_G$ and $\preceq_F$ on **S** are opposite to each other.

Proposition 2 *Let $((Y_i, Z_i), i \in I)$ be a family of semistable pairs, $Y = \cup_i Y_i$, $Z = \cap_i Z_i$. Then the pair (Y, Z) is also semistable.*

Proof Obviously, $Y \cup Z = C$, so it remains to check the inclusion $G(Y) \subseteq F(Z)$. Since $G(Y) \subseteq Y$, then due to Proposition 1 $Y \subseteq Z$. Let $y \in G(Y)$. Since $G(Y) \subseteq \cup_i G(Y_i)$, $y \in G(Y_i)$ for some i. Since $G(Y_i) \subseteq F(Z_i)$, we have $y \in F(Z_i)$. Since $y \in Z \subseteq Z_i$, then by virtue of the heredity **H** of F we get $y \in F(Z)$. □

Define the following partial order $\leq$ on the set **SSP** of semistable pairs

$$(Y, Z) \leq (Y', Z'), \text{ if } Y \subseteq Y', Z' \subseteq Z.$$

The minimal element of the poset is the pair $(\emptyset, C)$. Due to Proposition 2 any family of elements of the poset (**SSP**, $\leq$) has the supremum. This implies that the poset (**SSP**. $\leq$) is a complete lattice. In the next section we show that the maximal element of the lattice is a stable pair. This implies the existence of stable sets. Moreover, following Corollary 2 one can build other stable sets. Namely, let (Y, Z) be a semistable pair. If we want to find a stable set S that is not worse than Y ($Y \preceq S$), we must look for it among subsets of Z. Therefore, it is natural to pass from the original problem (with C) to the auxiliary one (with Z contracts). Then the maximal lattice element of **SSP**(Z) will give a stable set S (not worse than Y and contained in Z). This already gives for any family $(S_i, i \in I)$ of stable sets the existence of a stable S such that $S_i \preceq S$ for all i. But it is not yet clear, for example, whether there exists a minimal (in the sense of $\preceq$) stable set. The answer to this question is given in the next section.

6 Process of Sequential Improvement

We defined the mapping Φ by (1). The idea behind this map is that Φ should send **SSP** to itself and has to be monotone with respect to $\leq$.

Firstly, let us check that Φ sends **SSP** to itself.

Lemma 3 *The pair* $\Phi(Y, Z)$ *is semistable.*

Proof Recall that $\Phi(Y, Z) = (Y', Z')$, where $Y' = Y \cup F(Z)$, and $Z' = (Z - F(Z)) \cup G(Y')$. Then obvilously $Y' \cup Z' = C$. So we have to check the inclusion $G(Y') \subseteq F(Z')$. Since $Z' \subseteq Z$, the inclusion $F(Z) \cap Z' \subseteq F(Z')$ holds. From (2) $G(Y') = G(G(Y) \cup F(Z)) = G(F(Z))$, hence $G(Y')$ is a subset of both $F(Z)$ and Z'. □

Transition from a pair (Y, Z) to a pair $\Phi(Y, Z)$ implements the basic idea of a general step in the Gale-Shapley algorithm. Firm receives the offer $F(Z)$ from Worker and leaves (for the next step) the subset $G(Y')$, rejecting the set $F(Z) - G(Y')$. Note that by virtue of Heredity,

$$G(Y') = G(G(Y) \cup F(Z)) = G(F(Z)) \subseteq F(Z).$$

Therefore $Z' \subseteq Z$ and $(Y, Z) \leq \Phi(Y, Z)$.

Thus, the mapping Φ : **SSP** $\to$ **SSP** defines a monotone dynamics on the poset (lattice) (**SSP**, $\leq$). The following statement says that fixed points of the process Φ are exactly stable pairs.

Lemma 4 *The pair* (Y, Z) *is stable if and only if* $(Y, Z) = \Phi(Y, Z)$.

Proof By the definition Φ, a stable pair is stationary. Let us check that if $(Y, Z) = \Phi(Y, Z)$, then the pair (Y, Z) is stable. Since $Y = Y' = Y \cup F(Z)$, we get $F(Z) \subseteq Y$. Since $G(Y) \subseteq F(Z) \subseteq Y$, from the Outcast axiom **O**, we get $G(F(Z)) = G(Y)$. Further, since $Z' = Z$ we have $G(F(Z)) = F(Z)$, so $G(Y) = F(Z)$, that is the pair $(Y.Z)$ is stable. □

In particular, the maximal lattice element **SSP** is stationary with respect to Φ and is (as we stated in the previous section) a stable pair. An advantage of the dynamics Φ is that it allows us to describe all stable pairs. It is natural to construct fixed points of Φ using sequential approximations. Beginning with a semistable pair $P = (Y, Z)$, we define the process $P_1 = \Phi(P)$, $P_2 = \Phi(P_1) = \Phi^2(P)$ and so on. One can hope that 'in the limit' we will get a fixed point of Φ. However, if the set C is infinite, this process may not be completed for a finite number of steps and will have to consider transfinite sequences.

More precisely, with the help of the transfinite induction, for each ordinal number α (that is, a well-ordered set α), we define the pair $\Phi^\alpha(P) = (Y_\alpha, Z_\alpha)$, guided by the following rule. If α is not a limit ordinal number, then there is $\alpha = \beta + 1$, then $\Phi^\alpha(P) = \Phi(\Phi^\beta(P))$. If α is a limit ordinal number, then $\Phi^\alpha(P) = (\cup_{\beta<\alpha} Y_\beta, \cap_{\beta<\alpha} Z_\beta)$. By Proposition 2 it is again a semistable pair.

For a sufficiently large ordinal number α (not exceeding double the cardinality of the set C), this process stabilizes ($\Phi^{\alpha}(P) = \Phi(\Phi^{\alpha}(P))$) and thus give a fixed point Φ. This limiting pair (defined uniquely, as well as the process itself) is denoted as $\Phi^{\infty}(P) = (Y_{\infty}, Z_{\infty})$. Thus we get the mapping

$$\Phi^{\infty} : \mathbf{SSP} \to \mathbf{SP}. \tag{9}$$

For certainty, let $S_{\infty} = G(Y_{\infty}) = F(Z_{\infty})$ be a stable set, corresponding to the pair (Y_{∞}, Z_{∞}). Since $Y \subseteq Y_{\infty}$, we have $Y \preceq S_{\infty}$. Note that S_{∞} is not just better (in the sense of $\preceq$) than Y, it is a minimal stable set, no worse than Y.

Theorem 1 *Let S be a stable set, and $Y \preceq S$. Then $S_{\infty} \preceq S$.*

Proof Consider again the transfinite sequence (Y_{α}). We assert that all $Y_{\alpha} \preceq S$; of course, then S_{∞} contained in Y_{∞}, it will be $\preceq S$. Suppose this is not the case, and α is the first ordinal number for which the statement $Y_{\alpha} \preceq S$ is incorrect. There are two possible cases: α is a limit number or a non-limit one.

1. Let α be limit ordinal; then $Y_{\alpha} = \cup_{\beta<\alpha} Y_{\beta}$. And for any such β there holds $Y_{\beta} \preceq S$. By virtue of (8), Y_{α}, as a union Y_{β}, also $\preceq S$.
2. Let α be non-limit, $\alpha = \beta + 1$. We claim that again $Y_{\alpha} \preceq S$. In the proof, we can assume that $Y_{\beta} = Y$, $Z_{\beta} = Z$, and $Y_{\alpha} = Y' = Y \cup F(Z)$. We claim that $F(Z) \preceq S$; obviously (8) that then also $Y' = Y \cup F(Z) \preceq S$.

From $Y \preceq S$, by Corollary 1, we obtain $S \subseteq Z$. Assume that $F(Z)$ is not less S, i.e., $F(Z)$ is not contained in $\sigma_G(S)$ and z belongs to $F(Z)$ but does not belong to $\sigma_G(S)$ (and, in particular, does not belong to S). This means that $z \in G(S \cup z)$. Due to the stability of S, $z \notin F(S \cup z)$ does not belong to the choice from the larger set $S \cup Z = Z$. That is, $z \notin F(Z)$, contrary to the assumption.

So, we have proved that $Y_{\alpha} \preceq S$ for any α. □

7 Some Consequences

Here are some consequences of the above results.

1. Theorem 1 shows that, for given initial semistable pair (Y, Z), the stable state S_{∞} of the process Φ is not only a stable set, but a minimal stable state among stable sets which dominate Y. In particular, for the minimal semistable pair $(\emptyset, C)$, the corresponding stable set is the worst for Firm and the best for Worker, a fact discovered in [7].

2. *The poset* $(\mathbf{S}, \preceq_G)$ *is a complete lattice.*
   In other words, for any family $(S_i, i \in I)$ of stable sets, there is the least upper bound $S = \vee_i S_i$. Namely, due to Proposition 2 for the family of pairs $(\sigma_G(S_i), \sigma_F(S_i))$, we get the semistability of the pair (Y, Z), where $Y = \cup_i \sigma_G(S_i)$, $Z = \cap_i \sigma_F(S_i)$. It is clear (4) that $S_i \preceq Y_i \preceq Y \preceq Y_{\infty}$. Let now $S = S_{\infty}(Y, Z)$ be

the corresponding limit stable set, then $S_i \preceq_G S$ for any i. On the other hand, let T be a stable set such that $S_i \preceq_G T$ for any i. Then $\sigma_G(S_i) \preceq_G T$, and, due to (4), $Y = \cup_i \sigma_G(S_i) \preceq_G T$. Therefore, due to Theorem 1, $S \preceq_G T$. But this means exactly that S is the least upper bound (the join) for the family (S_i).
Since (due to the item 1 above) there exists the minimal (least) stable set, the poset $(\mathbf{S}, \preceq_G)$ is a complete lattice.

3. Another consequence is related to comparative statics. Suppose that S is a stable set with respect to agents with choice functions F and G. Suppose that the preferences of Worker have changed to a new choice function F', such that $F \leq F'$ (that is, for any $X \subset C$, $F(X) \subset F'(X)$)). In other words, Worker has "weakened" its requirements for the best contracts. Note that if $S \subset C$ is a stable set for agents with choice functions F and G, S remains stable for agents with choice functions F' and G.
For the stable set S, let us consider a pair (Y, Z'), where $Y = \sigma_G(S)$ and $Z' = \sigma'_F(\sigma_F(S))$. We assert that (Y, Z') is a semistable pair with respect to choice functions F' and G. Indeed, since $\sigma_F(S)) \subseteq Z'$ we see that $Y \cup Z' = C$, that is **SSP1** satisfies. **SSP2** satisfies as well,
$$G(Y) = S = F(\sigma_F(S)) \subseteq F'(\sigma_F(S)) = F'(Z').$$
Let S' be the corresponding limit stable for the process Φ' defined with respect to F' and G. Then S' is not worse than the initial $Y = \sigma_G(S)$, $Y \preceq_G (S')$, so that $S \preceq_G S'$. From Corollary 2 we see that $S' = G(S') \preceq_F S$. Thus, changing from F to an upper F', $F \leq F'$, only worsening the outcome for Worker.
Similarly one can can consider switching to preferences (F, G'), where $G' \subseteq G$. One can show that this does not improve the position of Firm.

4. How to build semistable pairs which lead to different stable states? It turns out that the procedure, reminiscent of the Gale-Shapley algorithm, but only "inverted" in a certain sense, allows us to construct a semistable pair close to Z for any set Z (Worker competence). Namely, we take an arbitrary set Z and begin to increase it gradually.
Given the set Z, we first construct the set $Y(Z) = (C - Z) \cup F(Z) = C - (Z - F(Z))$. Note that $Z \cup Y(Z) = C$ and $Z \cap Y(Z) = F(Z)$. Let $\Psi(Z) = Z' \cup G(Y(Z))$. Obviously, $Z \subseteq \Psi(Z)$, so Ψ allows one to define a monotone dynamical process on the set 2^C. A set Z is a fixed point Ψ if $G(Y(Z)) \subseteq Z$, that is, if $G(Y(Z)) \subset F(Z)$. Hence the pair $(Y(Z), Z)$ is semistable.
The Ψ fixed points can be found by using the same methods as in Sect. 6. By iterating over Ψ, we construct an increasing (transfinite) sequence $\Psi^\alpha(Z)$, indexed by ordinal numbers α. Due to monotonicity, it stabilizes on the set $Z^\infty = \Psi^\infty(Z) = \cup_\alpha \Psi^\alpha(Z)$, which is a fixed point of Ψ. Obviously $Z \subseteq \Psi^\infty(Z)$. In fact, the set $\Psi^\infty(Z)$ satisfies not only two conditions: $Z \subseteq \Psi^\infty(Z)$ and the semistability of the pair $(Y(Z^\infty), Z^\infty)$, but minimal under these conditions. Namely, it is true that if Z' is a set such that $Z \subseteq Z'$ and it is such that $G(Y(Z') \subseteq Z')$. Then we claim that in this case $\Psi(Z) \subseteq Z'$.

To do this, it suffices to check that $G(Y) \subseteq Z'$, where Y=Y(Z). Suppose $y \in G(Y)$ but y does not belong to Z'. Then y belongs to $Y' = Y(Z')$. But $Y' \subseteq Y$. Indeed, from the inclusion $Z \subseteq Z'$ we have $Z - F(Z) \subseteq Z' - F(Z')$. Therefore, for their complements we have $Y' \subseteq Y$. From the heredity condition for G we obtain $y \in G(Y') \subseteq F(Z') \subseteq Z'$. A contradiction. Thus, $\Psi(Z) \subseteq Z'$, but then $\Psi(\Psi(Z)) \subseteq Z'$ and so on, for any α we have $\Psi^{\alpha}(Z) \subseteq Z'$, whence $Z^{\infty} \subseteq Z'$. Beginning the process Φ with the semistable pair (Y^{∞}, Z^{∞}), we (temporarily) move away (from the Firm's position) from Y to Y^{∞}, and then we move to the stable pair $\Phi^{\infty}(Y^{\infty}, Z^{\infty})$.

Acknowledgements We thank Arkady Slinko and the referee for useful remarks and suggestions.

8 Appendix. Lehmann Hyper-Orders

For a Plott function G, the Blair hyper-order $\preceq_G$ is defined by

$$A \preceq_G B \text{ if } G(A \cup B) \subseteq B.$$

Lehmann in [11] have defined another (more strong) hyper-order $\prec_G$ by

$$A \prec_G B \text{ if } G(B) \neq \emptyset \text{ and } G(A \cup B) \text{ does not intersect } A.$$

Note that in this case, $G(A \cup B)$ is contained in B and moreover coincides with $G(B)$. In fact, we have $G(A \cup B) = G(G(A \cup B) \cup B) = G(B)$. Thus $A \preceq B$ implies $\preceq B$.

This "strict" hyper-order is interesting because it uniquely defines the initial Plott function G. To do this, we need to restore the set $G(B)$ in terms of $\prec=\prec_G$. It is convenient here to introduce two notions related to the hyper-relation $\prec$. Namely, a set B is called *essential* if $\emptyset \prec B$, and *negligible* otherwise.

If B is negligible, then we have $G(B) = \emptyset$.

If B is essential, then $G(B) = B - A$, where $A = \{a \in B, a \prec_G B\}$. In fact, $\{a\}$ does not intersect with $G(a \cup B) = G(B)$, so a is not contained in $G(B)$. On the other hand, if $b \in G(B)$, then the relation $b \prec_G B$ does not perform.

Below we will get the necessary and sufficient conditions for a hyper-order $\prec$ has the form $\prec_G$ for some Plott function G.

The following are the important properties of the hyper-relation $\prec_G$.

Proposition A1. *The hyper-relation $\prec=\prec_G$ has following properties:*

L0 $\prec$ *is irreflexive.*
L1 Left weakening. *If $A' \subseteq A \prec B$, then $A' \prec B$.*
L2 Union. *Let $(A_i, i \in I)$ be a nonempty family of subsets in C. If $A_i \prec B$ for any $i \in I$, then $\cup_i A_i \prec B$.*

L3 Right strengthening. *If* $A \prec B \subseteq B'$, *then* $A \prec B'$.
L4 Cancelation. *If* $A \prec A \cup B$, *then* $A \prec B$.
L5 Domination. *Any essential set dominates (in the sense of* $\prec$*) any negligible one.*

Proof L0 is obvious.

Let's check L1. We are given that $G(A \cup B) \subseteq B$ and even $G(A \cup B) = G(B)$. Then $G(A' \cup B) = G(A' \cup G(B)) = G(A' \cup G(A \cup B)) = G(A' \cup A \cup B) = G(A \cup B)$ is contained in B.

Check L2. $G((\cup_i A_i) \cup B) = G(\cup_i (A_i \cup B)) = G(\cup_i G(A_i \cup B))$ is contained in $\cup_i G(A_i \cup B)$ and especially in B.

Let's check L3. $G(A \cup B') = G(A \cup B \cup B') = G(G(A \cup B) \cup B') = G(G(B) \cup B')$ is contained in B', since $G(B) \subseteq B \subseteq B'$.

Let's check L4. $G(A \cup B) = G((A \cup B) \cup B) = G(G(A \cup B) \cup B) = G(G(B) \cup B) = G(B)$ and is nonempty.

Let's check L5. Let A be essential, and B be negligible. The essentiality of A means that $G(A) \neq \emptyset$. The neglect of B means that $G(B) = \emptyset$. We have to check that $G(A \cup B) = G(A \cup G(B)) = G(A)$ does not intersect with B. Let's assume that it intersects, and $b \in B$ belongs to $G(A)$. Then b is selected in the larger set $A \cup B$ and belongs to the smaller one B; from the heredity we have $b \in G(B)$. This contradicts to the emptiness of $G(B)$. □

Definition 4 A Lehmann hyper-order is a hyper-relation, which satisfies the properties L0-L5.

Remark. Lehmann [11] called a *qualitative measure* a transitive hyper-relation which satisfies L0-L4 and such that the union of negligible sets is negligible. The next lemma shows that the transitivity is a redundant requirement.

Lemma A2. *A Lehmann hyper-order is transitive.*

Proof Let $A \prec B$ and $B \prec C$. Then $A \prec B \cup C$ (according to L3), as well as $B \prec B \cup C$. Hence by L2, it holds true $A \cup B \prec B \cup C$, and due to L3 we have $A \cup B \prec A \cup B \cup C$. Then from L4 follows $A \cup B \prec C$, and hence from L1 $A \prec C$. □

This lemma justifies the use of the term hyper-order.

Now we construct a mapping from the set of Lehmann hyper-orders to the set of Plott functions.

Let $\prec$ be a Lehmann hyper-order and D be the set of all negligible elements in C. For a set $A \subseteq C$, define

$$L(A) = D \cup \{c \in C, c \prec A\}.$$

Note that, for a negligible set A, we have $L(A) = D$. Because of the union property L2, for an essential set A, it holds that $L(A) \prec A$. Note that $L(A)$ is the largest subset with the property $L(A) \prec A$.

Lemma A3. *If* $A \subseteq L(A) \cup B$, *then* $L(A) \subseteq L(B)$.

Proof For a negligible set A, the statement is obviously true. Let A be essential, then $L(A) \prec A$. Since $A \subseteq L(A) \cup B$, then, from L3, we have $L(A) \prec L(A) \cup B$. From L4, we get $L(A) \prec B$. This implies that B is essential and $L(A) \subseteq L(B)$. □

Now let's us define a choice function $T = T_{\prec}$ by the rule

$$T(A) = A \setminus L(A).$$

Proposition A4. *T is a Plott function.*

Proof Let us verify that T satisfies the heritage and outcast properties.

Heredity: let $A \subseteq B$ and $a \in T(B) \cap A$. Then, suppose on the contrary that $a \notin T(A)$, that is, $a \in L(A)$. If $a \in D$, then $a \in L(B)$, that contradicts to the fact that $a \in T(B)$. If $a \notin D$, then $a \prec A$ and from L3 $a \prec B$, $a \in L(B)$, a contradiction to $a \in T(B)$.

Outcast: let $T(A) \subseteq B \subseteq A$. Then for a negligible set A, we have that B is negligible as well and $T(A) = T(B)$. For an essential set A, from $T(A) \subseteq B$ follows $A \subseteq L(A) \cup A \subseteq L(A) \cup B$. Then, by Lemma A3, $L(A) \subseteq L(B)$. By the same Lemma applied to the inclusion $B \subseteq A$, we have $L(B) \subseteq L(A)$. From that we get $L(A) = L(B)$ and $T(A) = T(B)$. □

Thus, the *operations $G \mapsto \prec_G$ and $\prec \mapsto T$ are mutual inverse and establish a bijection between the set of Plott functions and the set of Lehmann hyper-orders.*

References

1. Aizerman, M.A., Malishevski, A.V.: General theory of best variants choice. IEEE Trans. Autom. Control **AC-26**(5), 1030–1040 (1981)
2. Alkan, A., Gale, D.: Stable schedule matching under revealed preference. J. Econ. Theory **112**(2), 289–306 (2003)
3. Blair, C.: The lattice structure of the set of stable matchings with multiple partners. Math. Oper. Res. **13**, 619–628 (1988)
4. Danilov, V., Koshevoy, G.: Choice functions and extensive operators. Order **26**, 69–94 (2009)
5. Danilov, V., Koshevoy, G., Savaglio, E.: Hyper-relations, choice functions, and orderings of opportunity sets. Soc. Choice Welfare **45**(1), 51–69 (2015)
6. Fleiner, T.: A fixed-point approach to stable mathchings and some applications. Math. Oper. Res. **28**(1), 103–126 (2003)
7. Gale, D., Shapley, L.S.: College admission and stability of marriage. Amer. Math. Montly **69**(1), 9–15 (1962)
8. Hatfield, J.W., Milgrom, P.R.: Matchings with contracts. Am. Econ. Rev. 913–935 (2005)
9. Kelso, A.S., Crawford, V.P.: Job matching, coalition formation and gross substitutes. Econoimetrica **50**, 1483–1593 (1982)
10. Koshevoy, G.A.: Choice functions and abstract convex geometries. Math. Soc. Sci. **38**, 35–44 (1999)
11. Lehmann, D.: Nonmonotonic logics and semantics. J. Logic Comput. **11**(2), 229–256 (2001)
12. Lehmann, D.: Revealed preferences for matching with contracts. arXiv:1908.08823
13. Moulin, H.: Choice functions over a finite set: a summary. Soc. Choice Welf. **2**, 147–160 (1985)

14. Plott, C.R.: Path independence, rationality, and social choice. Econometrica **41**(6), 1075–1091 (1973)
15. Roth, A.E.: Stability and polarization of interests in job mathcing. Econometrica **52**, 47–57 (1984)
16. Sotomayor, M.: My encounters with David Gale. Games Econ. Behav. **66**, 643–646 (2009)

Applying the New Concept of Variances with Uncertainty to Cover the Worst Markets

Shan Li, Xinpeng Li, and George Xianzhi Yuan

Abstract In this paper, we study the general notions of upper and lower variances, which were initially introduced by Peter Walley for the bounded random variables. The properties of upper and lower variances and new formulas for calculation of them are provided. We take US markets for case study, which shows that the upper variance would be a useful tool to provide the risk management in the state or national level for the predication of (financial) systemic risk management by providing worst indicators to prepare the necessary implementation for the worst markets to coming in the practice.

Keywords Market · Random variable · Risk management

1 Introduction

We begin with a simple example to illustrate the idea and purpose of this paper. Let X be a random variable which represents the daily return of one stock in the stock market. We assume that $X \sim N(0.2, 0.4)$ in bull market and $X \sim N(-0.2, 0.4)$ in bear market. In fact, the normal distributions characterize the "risk" in the bull and bear markets, but we don't know the true state of the stock market after one month. This phenomenon is what we called Knightian uncertainty, introduced by Knight [5]: "The practical difference between the two categories, risk and uncertainty, is that in the former the distribution of the outcome in a group of instances is known (either through calculation a priori or from statistics of past experience), while in the

S. Li (✉) · X. Li
Research Center for Mathematics and Interdisciplinary Sciences, Shandong University, Qingdao, China
e-mail: shanlim@foxmail.com

X. Li
e-mail: lixinpeng@sdu.edu.cn

G. X. Yuan
Business School, East China University of Science and Technology, Shanghai, China

M. A. Khan et al. (eds.), *Matching, Dynamics and Games for the Allocation of Resources*, Monographs in Mathematical Economics 7,
https://doi.org/10.1007/978-981-96-6661-4_21

case of uncertainty this is not true." Thus the normal distributions $N(0.2, 0.4)$ and $N(-0.2, 0.4)$ can be estimated from the historical stock data in the bull and bear markets respectively. It is naturally seen that the classical notion of variance is used to measure the "risk" of stock in bull or bear market. The purpose of this paper is to calculate the "variance" when we meet uncertainty, for example, the true state of stock market in the future is unknown. Sublinear expectation theory introduced by Shige Peng (see Peng [8]) is a powerful tool to deal with the problems with uncertainty in finance (see Epstein and Ji [2], Peng et al. [9], etc.), we consider the variances with uncertainty under the framework of sublinear expectation theory.

Peter Walley introduced the notions of upper and lower variances in Appendix G of [11] for the bounded random variables, and pointed out that it is quite easy to calculate the lower variance, but the calculation of upper variance is more difficult. Peng et al. [9] provided a simple method, called max-mean calculation method, to calculate the upper and lower variance based on the historical data with the assumption of mean-certainty. Mean-certainty in our previous example represents that the means of the returns in bull and bear markets should be equal. Li et al. [6] generalize the notions of upper and lower variances and provide the formulas to calculate the upper and lower variances, which allows that the means to be different in bull and bear markets. But the formulas in [6] cannot cover the mean-certainty case in [9]. This paper provides approximate and explicit formulas to calculate upper and lower variances. In particular, we find that the new formula for upper variance covers the corresponding one in [9] and the formula for lower variance in [9] can also be applied to mean-uncertainty case.

One application of upper and lower variance in finance is to use them to calculate the so-called G-VaR. The new Value-at-Risk (VaR) model builds upon the sublinear expectation theory introduced by Shige Peng, which signifies that the volatility of financial returns cannot be characterized by a single distribution but by multiple distributions. Both G-VaR models in mean-certainty case ([9]) and mean-uncertainty case ([6]) perform excellently in the experiments on the NASDAQ Composite Index and S&P 500 Index. This paper provides a new case study of upper and lower variance. By comparing variances of S&P 500 Index with the CBOE Volatility Index (VIX), we show that the upper variance would be a useful tool for risk management by providing worst indicators.

The remainder of this paper is organized as follows. Section 2 provides the definition and properties of upper and lower variances. The estimation of variance with uncertainty is studied in Sect. 3. Section 4 is the case study. All the proofs of properties and theorems are collected in the Appendix.

2 Definition and Properties of Upper and Lower Variances

Let $(\Omega, \mathcal{F})$ be a measurable space, $P_1, \cdots, P_K$ ($K \in \mathbb{N}$) be the probability measures on $(\Omega, \mathcal{F})$. Let $\mathcal{P} = \{P_1, \cdots, P_K\}$ be a finite set, which characterizes Knightian uncertainty or model uncertainty.

The upper expectation of $\mathcal{P}$ is defined as follows: for each random variable X such that $E_P[X]$ exists for each $P \in \mathcal{P}$,

$$\hat{\mathbb{E}}[X] := \max_{P \in \mathcal{P}} E_P[X].$$

It is easy to verify that the upper expectation $\hat{\mathbb{E}}[\cdot]$ corresponding to the family $\mathcal{P}$ is a sublinear expectation, i.e., it satisfies the following properties:

(1) Monotonicity: $\hat{\mathbb{E}}[X] \leq \hat{\mathbb{E}}[Y]$, if $X \leq Y$;
(2) Constant preserving: $\hat{\mathbb{E}}[c] = c$, $\forall c \in \mathbb{R}$;
(3) Sub-additivity: $\hat{\mathbb{E}}[X + Y] \leq \hat{\mathbb{E}}[X] + \hat{\mathbb{E}}[Y]$;
(4) Positive homogeneity: $\hat{\mathbb{E}}[\lambda X] = \lambda \hat{\mathbb{E}}[X]$, $\forall \lambda \geq 0$.

The tripe $(\Omega, \mathcal{F}, \hat{\mathbb{E}})$ is called the sublinear expectation space as in Peng [8]. In fact, the notion of sublinear expectation has closed relations with coherent risk measure in Artzner et al. [1] and coherent prevision in Walley [11].

We also point out that although $\mathcal{P}$ contains finitely many probability measures, but it can cover infinitely many uncertainties. Such assumption of finitely many probability measures can satisfy the requirements in practice. Indeed, let

$$\Lambda = \left\{ (\lambda_1, \cdots, \lambda_K) \in \mathbb{R}^K : \lambda_i \in [0, 1], 1 \leq i \leq K, \sum_{i=1}^{K} \lambda_i = 1 \right\},$$

we consider the convex combination of the probability measures $\{P_i\}_{i=1}^K$

$$P_\lambda = \lambda_1 P_1 + \cdots + \lambda_K P_K,$$

where $\lambda = (\lambda_1, \cdots, \lambda_K) \in \Lambda$. Let

$$\mathcal{P}_\Lambda = \{P_\lambda : \lambda \in \Lambda\}.$$

Clearly, $\mathcal{P}_\Lambda$ is convex. Furthermore, it is also compact with respect to the total variation metric defined by

$$d(P, Q) = \sup_{A \in \mathcal{F}} |P(A) - Q(A)|,$$

since $\mathcal{P}_\Lambda$ is a continuous image of the compact set Λ under the continuous mapping $\varphi : \Lambda \to \mathcal{P}_\Lambda$ defined by

$$\varphi(\lambda_1, \cdots, \lambda_K) = \lambda_1 P_1 + \cdots + \lambda_K P_K.$$

It is easily seen that, if $\{P_i\}_{i=1}^\infty \subset \mathcal{P}_\Lambda$, then there exists a subsequence $\{P_{i_k}\}_{k=1}^\infty$ and $P^* \in \mathcal{P}_\Lambda$, such that, for each random variable X with $\hat{\mathbb{E}}[X] < \infty$,

$$\lim_{k\to\infty} E_{P_{l_k}}[X] = E_{P^*}[X].$$

For each random variable X with $\hat{\mathbb{E}}[X] < \infty$, we have

$$\hat{\mathbb{E}}[X] = \max_{P\in\mathcal{P}_\Lambda} E_P[X] \quad \text{and} \quad E_{P_\lambda}[X] \le \hat{\mathbb{E}}[X], \quad \forall P_\lambda \in \mathcal{P}_\Lambda.$$

This fact illustrates the upper variance of X with uncertainty characterized by $\mathcal{P}$ cannot be defined simply as $\max_{1\le l\le K} V_{P_l}(X)$, where $V_P(X)$ denotes the variance of X under probability measure P.

The following definition in [6] generalizes the original one in [11] from bounded random variables to general random variables.

Definition 1 For each random variable X on sublinear expectation space $(\Omega, \mathcal{F}, \hat{\mathbb{E}})$ with $\hat{\mathbb{E}}[X^2] < \infty$, define the upper variance of X under $\hat{\mathbb{E}}$ to be

$$\overline{V}(X) := \min_{-\hat{\mathbb{E}}[-X]\le\mu\le\hat{\mathbb{E}}[X]} \{\hat{\mathbb{E}}[(X-\mu)^2]\} = \min_{\mu\in\mathbb{R}}\{\hat{\mathbb{E}}[(X-\mu)^2]\},$$

and the lower variance of X to be

$$\underline{V}(X) := \min_{-\hat{\mathbb{E}}[-X]\le\mu\le\hat{\mathbb{E}}[X]} \{-\hat{\mathbb{E}}[-(X-\mu)^2]\} = \min_{\mu\in\mathbb{R}}\{-\hat{\mathbb{E}}[-(X-\mu)^2]\}.$$

The assumption of finitely many probability measures can significantly improve the corresponding results in [6].

Proposition 1 *Let X be a random variable on sublinear expectation space $(\Omega, \mathcal{F}, \hat{\mathbb{E}})$ with $\hat{\mathbb{E}}[X^2] < \infty$. We have*

(i) $\overline{V}(X) = \max_{P\in\mathcal{P}_\Lambda} V_P(X) \ge \max_{P\in\mathcal{P}} V_P(X)$.
Furthermore, there exists $\mu^* \in [-\hat{\mathbb{E}}[-X], \hat{\mathbb{E}}[X]]$ such that $\overline{V}(X) = \hat{\mathbb{E}}[(X-\mu^*)^2]$, and there exists $P_{\lambda^*} \in \mathcal{P}_\Lambda$ such that $E_{P_{\lambda^*}}[X] = \mu^*$ and $\overline{V}(X) = V_{P_{\lambda^*}}(X)$.
(ii) $\underline{V}(X) = \min_{P\in\mathcal{P}_\Lambda} V_P(X) = \min_{P\in\mathcal{P}} V_P(X)$.
Furthermore, there exists $\mu_* \in [-\hat{\mathbb{E}}[-X], \hat{\mathbb{E}}[X]]$ such that $\underline{V}(X) = \hat{\mathbb{E}}[(X-\mu_*)^2]$, and there exists $P_{\lambda_*} \in \mathcal{P}_\Lambda$ such that $E_{P_{\lambda_*}}[X] = \mu_*$ and $\underline{V}(X) = V_{P_{\lambda_*}}(X)$.

The following inequality was given in [11] for the bounded random variables without the proof. It also holds for general random variables and we supplement with the proof in Appendix. This proposition can be used to give the approximate calculation of upper variance.

Proposition 2 *Let X be a random variable on sublinear expectation space $(\Omega, \mathcal{F}, \hat{\mathbb{E}})$ with $\hat{\mathbb{E}}[X^2] < \infty$. We denote $\rho = \frac{1}{2}(\hat{\mathbb{E}}[X] - \hat{\mathbb{E}}[-X])$ and $\Delta = \hat{\mathbb{E}}[X] + \hat{\mathbb{E}}[-X]$. Then*

$$\frac{1}{4}\Delta^2 \le \overline{V}(X) - \underline{V}(X) \le \frac{1}{4}\Delta^2 + \hat{\mathbb{E}}[(X-\rho)^2] + \hat{\mathbb{E}}[-(X-\rho)^2].$$

3 The Estimation of Upper and Lower Variances

To show how upper and lower variances can be calculated, we consider the following example.

Example
Let $\mathcal{P} = \{P_1, P_2\}$ with $P_1 \sim N(0.2, 0.4)$ and $P_2 \sim N(-0.2, 0.4)$. For random variable X under such $\mathcal{P}$, by Proposition 1, we have

$$\underline{V}(X) = \min\{V_{P_1}(X), V_{P_2}(X)\} = 0.4.$$

By Definition 1, we have

$$\begin{aligned}\overline{V}(X) &= \min_{-0.2\leq\mu\leq0.2} \max\{E_{P_1}[(X-\mu)^2], E_{P_2}[(X-\mu)^2]\} \\ &= \min_{-0.2\leq\mu\leq0.2} \max\{\mu^2 - 0.4\mu + 0.44, \mu^2 + 0.4\mu + 0.44\} = 0.44.\end{aligned}$$

By Proposition 1, there exists $P_{\frac{1}{2}} = \frac{1}{2}P_1 + \frac{1}{2}P_2$ such that

$$\overline{V}(X) = V_{P_{\frac{1}{2}}}(X) \text{ and } E_{P_{\frac{1}{2}}}[X] = 0.$$

It is plausible that the stock market will probably be the bull market (or bear market) when the mean of X is close to 0.2 (or -0.2), thus the variance should be close to the corresponding variance of 0.4. But the biggest uncertainty surrounds the zero-mean of X, accompanied by the upper variance of 0.44.

As shown in Walley [11], it is usually quite easy to calculate $\underline{V}(X)$ because P_{λ_*} in Proposition 1 can always be taken to be an extreme point of $\mathcal{P}_\Lambda$. Usually P_{λ^*} is not an extreme point, as shown in the previous example. Thus the calculation of $\overline{V}(X)$ is more difficult. The following theorem provides the explicit formula for the calculation of $\overline{V}(X)$.

Theorem 1 *Let X be a random variable on sublinear expectation space* $(\Omega, \mathcal{F}, \hat{\mathbb{E}})$ *with* $\hat{\mathbb{E}}[X^2] < \infty$ *and* $\mathcal{P} = \{P_1, \cdots, P_K\}$. *Let*

$$\mu_j = E_{P_j}[X] \text{ and } \nu_j = V_{P_j}(X), \quad j = 1, \cdots, K.$$

We have

(i) $\underline{V}(X) = \min_{1\leq j\leq K} \nu_j$.
(ii) $\overline{V}(X) = \min_{\mu\in\Xi} \max_{1\leq j\leq K}\{(\mu-\mu_j)^2 + \nu_j\}$, *where the set* Ξ *is defined by*

$$\Xi = \{\mu_{mn}, 1 \leq m < n \leq K\} \cup \{\mu_i, 1 \leq i \leq K\}.$$

In which,

$$\mu_{mn} = \begin{cases} \frac{1}{2}(\frac{\nu_m - \nu_n}{\mu_m - \mu_n} + \mu_m + \mu_n), & \mu_m \neq \mu_n, \\ \mu_n, & \mu_m = \mu_n. \end{cases}$$

Remark 1 The corresponding results in Li et al. [6] miss the consideration of the set which consists of an axis of symmetry.

Remark 2 The number of elements in the set Ξ is at most $\frac{K(K-1)}{2} + K$. The amounts of calculations will be very large when K is too large. In this case, by Proposition 2, we can approximately calculate $\overline{V}(X)$ by

$$\overline{V}(X) \approx \underline{V}(X) + \frac{1}{4}\Delta^2 + \hat{\mathbb{E}}[(X-\rho)^2] + \hat{\mathbb{E}}[-(X-\rho)^2],$$

since the right hand of the above formula is easy to calculate.

In practice, Peng et al. [9] proposed the max-mean method to calculate the upper and lower variances based on the historical data when the population has mean certainty. The max-mean method is based on the optimal unbiased estimation for maximal distribution (see Jin and Peng [4]) and the law of large numbers (see Peng [8] or Guo and Li [3]). The computation for both upper and lower variances is reasonable to implement as explained by the procedure below.

Actually, let $\{Z_{t-i}\}_{1 \le i \le n}$ be the historical date before time t, which is assumed to be independent and identically distributed (i.i.d.) in the sublinear expectation sense by Peng [8], furthermore, $Z_t \sim N(\mu, [\underline{\sigma}^2, \overline{\sigma}^2])$, here $N(\mu, [\underline{\sigma}^2, \overline{\sigma}^2])$ is G-normally distributed introduced by [8]. In fact, the assumption for G-normal distribution is not necessary as indeed we only require that Z_t are i.i.d. with $\hat{\mathbb{E}}[Z_1] = -\hat{\mathbb{E}}[-Z_1]$. By following Peng et al. [9], the estimation for both upper and lower variances for $\overline{\sigma}^2 = \hat{\mathbb{E}}[(Z_1-\mu)^2]$ and $\underline{\sigma}^2 = -\hat{\mathbb{E}}[-(Z_1-\mu)^2]$ can be given as follows.

First, choose the window of sample L and the numbers of groups K, we firstly calculate the local variance within the same window L,

$$\tilde{\sigma}^2_{t,j} = \frac{\sum_{i=1}^{L}(Z_{t-L+i-j} - \hat{r}_{t,j})^2}{L-1}, \tag{1}$$

where $\hat{r}_{t,j} = \frac{\sum_{i=1}^{L} Z_{t-L+i-j}}{L}$.

Second, the estimators of upper and lower variances are defined by

$$\hat{\underline{\sigma}}^2_t = \min_{0 \le j \le K-1} \tilde{\sigma}^2_{t,j}, \quad \hat{\overline{\sigma}}^2_t = \max_{0 \le j \le K-1} \tilde{\sigma}^2_{t,j}.$$

Indeed, the procedures displayed above can be regarded as that we use date $\{Z_{t-L+i-j}\}_{1 \le i \le L}$ to determine the sample variance $V_{P^{(j)}}(Z_t)$ under the sampling distribution $P^{(j)}$.

Let $\mathcal{P} = \{P^{(1)}, \cdots, P^{(K)}\}$, which characterizes the uncertainty characterized by data. Then we can see that

$$\underline{\hat{\sigma}}_t^2 = \min_{0 \le j \le K-1} V_{P^{(j)}}(Z_t), \quad \hat{\overline{\sigma}}_t^2 = \max_{0 \le j \le K-1} V_{P^{(j)}}(Z_t).$$

It is natural to generalize the above method to the case of mean-uncertainty by Theorem 1.

Thus by assuming that $\{Z_{t-i}\}_{1 \le i \le n}$ are i.i.d. but not necessary to have mean-certainty, with the same notations as above, the estimators of upper and lower variances $\overline{V}(Z_t)$ and $\underline{V}(Z_t)$ are defined by

$$\underline{\hat{V}}_t := \min_{0 \le j \le K-1} \tilde{\sigma}_{t,j}^2, \tag{2}$$

$$\hat{\overline{V}}_t := \min_{\mu \in \Xi} \max_{0 \le j \le K-1} \{(\mu - \hat{r}_{t,j})^2 + \tilde{\sigma}_{t,j}^2\}, \tag{3}$$

where

$$\Xi = \{\mu_{mn}, 1 \le m < n \le K\} \cup \{\hat{r}_{t,j}, 1 \le j \le K\},$$

$$\mu_{mn} = \begin{cases} \frac{1}{2}\left(\frac{\tilde{\sigma}_{t,m}^2 - \tilde{\sigma}_{t,n}^2}{\hat{r}_{t,m} - \hat{r}_{t,n}} + \hat{r}_{t,m} + \hat{r}_{t,n}\right), & \hat{r}_{t,m} \ne \hat{r}_{t,n}, \\ \hat{r}_{t,n}, & \hat{r}_{t,m} = \hat{r}_{t,n}. \end{cases}$$

It is clear that we have

$$\hat{\overline{\sigma}}_t^2 \le \overline{V}_t; \quad \underline{\hat{\sigma}}_t^2 = \underline{V}_t.$$

The calculation formula of lower variance in [9] can be also applied to mean-uncertainty case, but the formula for upper variance in [9] will be underestimated as shown in (i) of Proposition 1.

In practice, we usually firstly choose the window L according to the characteristics of the data. Since upper variance (resp., lower variance) will be increasing (resp. decreasing) when K increases, we then choose the suitable parameter K.

4 Case Study for the Worst Markets by Using Upper Variance

In this section, by applying the above procedure to calculate the upper and lower variances for S&P 500 Index, we show that the upper variances can cover the worst cases indicated by VIX, which is a market index representing the market's expectations for the relative strength of near-term price changes of the S&P 500 Index.

By looking at S&P 500 Index from January 3, 2006 to December 30, 2021 from the US market and using the VIX indicator with equal or bigger than 40% (annualized)

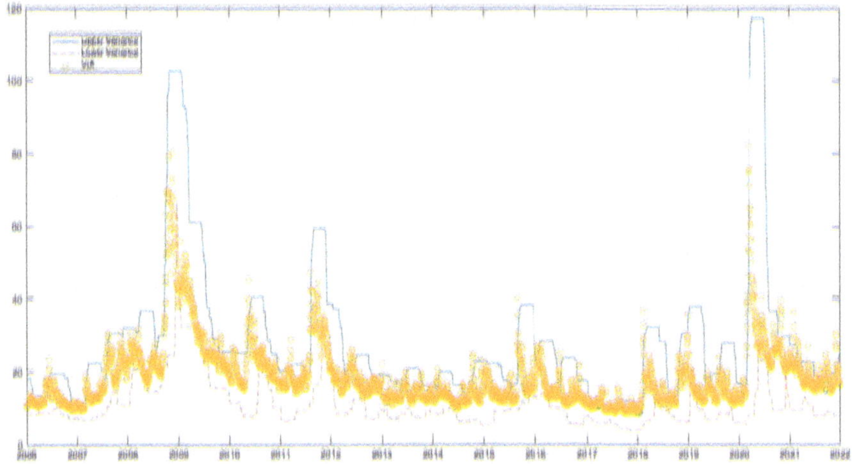

Fig. 1 Upper and lower variance versus VIX

consecutively to classify the "worst market" phase (time period), it is clearly to observe that we have at least two phases being the worst markets, they are:

(i) the time from October 2, 2008 to December 31, 2008 with available VIX data for consecutive 63 days.
(ii) the time from March 6, 2020 to April 13, 2020 with available VIX data for consecutively 26 days.

In our calculation, the daily log-returns of S&P 500 Index is calculated by

$$Z_t = 100(\ln X_t - \ln X_{t-1}),$$

where X_t is the closed price of S&P 500 Index at time t.

We firstly take $L = 22$ (approximately one-month trading data) to calculate the local variance by (1). Then we choose $K = 66$ to reflect 3 months variance uncertainty. Thus we can calculate the upper and lower variance by (3) and (2) respectively. To obtain the annual volatility, we take the square root of the upper and lower variance and then multiply by $\sqrt{365}$.[1]

The results are summarized in Fig. 1. We hope the upper and lower volatilities can cover VIX, i.e., VIX lies between the interval of lower and upper volatilities. In fact, the proposition of such a situation is about 83.32%.

Now back to the two cases for worst markets (i) and (ii) time phases mentioned above, we observed that:

(1) For the worst market phase (i), our upper variance is able to cover almost daily worst VIX from October 2, 2008 to December 31, 2008 with 63 days expect

---

[1] We take 365, rather than 252, due to the Volatility Index Methodology: Cboe Volatility Index, in which the 30-day weighted average of variances is multiplied by $\frac{365}{30}$.

October 10, 2008, which are due to the reason caused during the 2008 global financial crisis by the announcement on September 15, 2008 by Lehman Brothers for its bankruptcy protection.

(2) For the worst market phase (ii), our upper variance is also able to cover almost all daily worst VIX from March 6, 2020 to April 13, 2020 except March 6-16, 2020 (7 trading days) due to the reason of COVID-19's global pandemic starting in early March, 2020.[2]

The case study for worst markets by using upper variance for the US market shows that the way by using upper variance is able to cover the worst case of the markets' unexpected movement in general, which would be used as a useful tool to provide the risk management in the state or national level for the prediction of (financial) systemic risk to prepare necessary implementation for the worst markets to coming in the practice.

Acknowledgements This research is partially supported by National Social Science Foundation of China (No.22AZD044), NSF of Shandong Provence (No.ZR2021MA018), NSF of China (No.11601281, No.U1811462, No.71971031), National Key R&D Program of China (No.2018YFA0703900) and the Young Scholars Program of Shandong University.

5 Appendix

In Appendix, we provide the detailed proof of Proposition 1 and 2 as well as Theorem 1.

Proof of Proposition 1. We firstly prove that there exists $P_{\lambda^*}, P_{\lambda_*} \in \mathcal{P}_\Lambda$ such that

$$\sup_{P\in\mathcal{P}_\Lambda} V_P(X) = V_{P_{\lambda^*}}(X) \text{ and } \inf_{P\in\mathcal{P}} V_P(X) = V_{P_{\lambda_*}}(X). \tag{4}$$

For each $n \in \mathrm{N}$, there exists $P_n \in \mathcal{P}_\Lambda$ such that

$$V_{P_n}(X) \geq \sup_{P\in\mathcal{P}_\Lambda} V_P(X) - \frac{1}{n}.$$

By the compactness of $\mathcal{P}_\Lambda$, there exists subsequence $\{P_{n_k}\}_{k=1}^{\infty}$ and P_{λ^*} such that

$$\lim_{k\to\infty} V_{P_{n_k}}(X) = V_{P_{\lambda^*}}(X),$$

thus

$$V_{P_{\lambda^*}}(X) = \sup_{P\in\mathcal{P}_\Lambda} V_P(X).$$

[2] On March 11, 2020, Dr. Tandesay, director general of the World Health Organization (WHO), said at the 2019 COVID-19 epidemic media briefing that the WHO assessed that COVID-19 could be recognized as a global pandemic.

Similarly, we can prove the existence of $P_{\lambda_*} \in \mathcal{P}_\Lambda$.

For each $P \in \mathcal{P}_\Lambda$, we define the functional $\Pi(\mu, P)$ which is linear on P and convex on μ:

$$\Pi(\mu, P) = E_P[(X-\mu)^2],$$

In order to prove (i), by the minimax theorem (cf. Sion [10], Li and Lin [7] or Yuan [12]), we obtain

$$\begin{aligned}\overline{V}(X) &= \min_{\hat{\mathbb{E}}[-X]\le\mu\le\hat{\mathbb{E}}[X]} \max_{P\in\mathcal{P}_\Lambda} \Pi(\mu, P) \\ &= \max_{P\in\mathcal{P}_\Lambda} \min_{\hat{\mathbb{E}}[-X]\le\mu\le\hat{\mathbb{E}}[X]} \Pi(\mu, P) = \max_{P\in\mathcal{P}_\Lambda} V_P(X).\end{aligned}$$

For (ii), by the definition of $\underline{V}(X)$,

$$\begin{aligned}\underline{V}(X) &= \inf_{\mu\in\mathbb{R}} \inf_{P\in\mathcal{P}_\Lambda} E_P[(X-\mu)^2] = \inf_{P\in\mathcal{P}_\Lambda} \inf_{\mu\in\mathbb{R}} E_P[(X-\mu)^2] \\ &= \inf_{P\in\mathcal{P}_\Lambda} V_P(X) = \min_{P\in\mathcal{P}_\Lambda} V_P(X).\end{aligned}$$

The proof is completed.

Proof of Proposition 2. We firstly prove that

$$\frac{1}{4}\Delta^2 \le \overline{V}(X) - \underline{V}(X) \tag{5}$$

Let $\overline{\mu} = \hat{\mathbb{E}}[X]$ and $\underline{\mu} = -\hat{\mathbb{E}}[-X]$. If $\overline{\mu} = \underline{\mu}$, then $\Delta = 0$. So we only consider the case of $\overline{\mu} > \underline{\mu}$.

There exists $\overline{P}$ and $\underline{P}$ such that

$$E_{\overline{P}}[X] = \overline{\mu}, \quad \text{and} \quad E_{\underline{P}}[X] = \underline{\mu}.$$

It is obvious that

$$\begin{aligned}\overline{V}(X) &\ge \min_{\underline{\mu}\le\mu\le\overline{\mu}} \max\{E_{\overline{P}}[(X-\mu])^2, E_{\underline{P}}[(X-\mu)^2]\} \\ &\ge \min_{\underline{\mu}\le\mu\le\overline{\mu}} \min\{E_{\overline{P}}[(X-\mu])^2, E_{\underline{P}}[(X-\mu)^2]\} \ge \underline{V}(X).\end{aligned}$$

For each probability measure $P \in \mathcal{P}$, we have

$$E_P[(X-\mu)^2] = \mu^2 - 2E_P[X]\mu + E_P[X^2] = (\mu - E_P[X])^2 + V_P(X).$$

Let

$$M(\mu) = \max\{E_{\overline{P}}[(X-\mu])^2, E_{\underline{P}}[(X-\mu)^2]\}.$$

We note that $\overline{\mu} > \underline{\mu}$.

(i) If the minimum of $M(\mu)$ is attained at the intersection of two quadric curves $(\mu - \overline{\mu})^2 + V_{\overline{P}}(X)$ and $(\mu - \underline{\mu})^2 + V_{\underline{P}}(X)$, then we can calculate such point is

$$\mu^* = \frac{1}{2}\left(\frac{V_{\overline{P}}(X) - V_{\underline{P}}(X)}{\overline{\mu} - \underline{\mu}} + \overline{\mu} + \underline{\mu}\right).$$

We have

$$\begin{aligned} M(\mu^*) =& \frac{1}{4}\Delta^2 + \frac{1}{4}\left(\frac{V_{\overline{P}}(X) - V_{\underline{P}}(X)}{\overline{\mu} - \underline{\mu}}\right)^2 + \frac{1}{2}\left(V_{\overline{P}}(X) + V_{\underline{P}}(X)\right) \\ \geq& \frac{1}{4}\Delta^2 + \min\{V_{\overline{P}}(X), V_{\underline{P}}(X)]\} \geq \frac{1}{4}\Delta^2 + \underline{V}(X). \end{aligned}$$

(ii) If the minimum of $M(\mu)$ is attained at $\overline{\mu}$ or $\underline{\mu}$, for example, the corresponding point is denoted by μ^* and $\mu^* = \overline{\mu}$, then

$$\overline{V}(X) \geq E_{\underline{P}}[(X - \overline{\mu}^*)^2] = V_{\underline{P}}(X) + \Delta^2 \geq \underline{V}(X) + \frac{1}{4}\Delta^2.$$

Finally, we can see that (5) holds.

Now we prove that

$$\overline{V}(X) - \underline{V}(X) \leq \frac{1}{4}\Delta^2 + \hat{\mathbb{E}}[(X - \rho)^2] + \hat{\mathbb{E}}[-(X - \rho)^2]. \tag{6}$$

It is clear that

$$\overline{V}(X) \leq \hat{\mathbb{E}}[(X - \rho)^2].$$

We only need to show that

$$-\hat{\mathbb{E}}[-(X - \rho)^2] \leq \underline{V}(X) + \frac{1}{4}\Delta^2. \tag{7}$$

Without loss of generality, we assume that $\underline{\mu} = 0$. In this case, $\Delta = \overline{\mu} = 2\rho$ and there exists $0 \leq \mu_* \leq \overline{\mu}$ such that $\underline{V}(X) = -\hat{\mathbb{E}}[-(X - \mu_*)^2]$. Then we have

$$\begin{aligned} -\hat{\mathbb{E}}[-(X - \rho)^2] &= -\hat{\mathbb{E}}[-(X - \mu_* + \mu_* - \rho)^2] \\ &\leq \underline{V}(X) + (\mu_* - \rho)^2 + 2\hat{\mathbb{E}}[(\mu_* - \rho)(X - \mu_*)]. \end{aligned}$$

If $0 \leq \mu_* \leq \rho$, then we obtain

$$(\mu_* - \rho)^2 + 2\hat{\mathbb{E}}[(\mu_* - \rho)(X - \mu_*)] = (\mu_* - \rho)^2 + 2(\rho - \mu_*)\mu_* = \rho^2 - \mu_*^2 \leq \rho^2 = \frac{1}{4}\Delta^2.$$

If $\rho \leq \mu_* \leq \overline{\mu} = 2\rho$, then we have

$$(\mu_* - \rho)^2 + 2\hat{\mathbb{E}}[(\mu_* - \rho)(X - \mu_*)] = (\mu_* - \rho)(3\rho - \mu_*) \leq \rho^2 = \frac{1}{4}\Delta^2.$$

Thus (6) holds.

Proof of Theorem 1. By Proposition 1, (i) holds.

In order to prove (ii), let

$$\Pi(\mu, P) = E_P[(X - \mu)^2] = \mu^2 - 2E_P[X]\mu + V_P(X) + (E_P[X])^2,$$

which is a linear functional of P and is a quadratic function of μ. Then

$$\begin{aligned}\overline{V}(X) &= \min_{\mu\in\mathbb{R}} \max_{P\in\mathcal{P}} \Pi(\mu, P) = \min_{\mu\in\mathbb{R}} \max_{1\leq j\leq K} \Pi(\mu, P_j) \\ &= \min_{\mu\in\mathbb{R}} \max_{1\leq j\leq K} \{\mu^2 - 2\mu_j\mu + \nu_j + \mu_j^2\} \\ &= \min_{\mu\in\Xi} \max_{1\leq j\leq K} \{(\mu - \mu_j)^2 + \nu_j\}.\end{aligned}$$

The key observation is that the minimum only takes on the intersections of the pair of quadric curves or the axis of symmetry.

References

1. Artzner, P., Delbaen, F., Eber, J., Heath, D.: Coherent measures of risk. Math. Finance **9**, 203–228 (1999)
2. Epstein, L., Ji, S.: Ambiguous volatility and asset pricing in continuous time. Rev. Financ. Stud. **26**(7), 1740–1786 (2013)
3. Guo, X., Li, X.: On the laws of large numbers for pseudo-independent random variables under sublinear expectation. Stat. Probab. Lett. **172**(109042) (2021)
4. Jin, H., Peng, S.: Optimal unbiased estimation for maximal distribution. Probab. Uncertain Quant. Risk **6**(3), 189–198 (2021)
5. Knight, F.: Risk. Cornell University Library, Uncertainty and profit (1921)
6. Li, S., Li, X., Yuan, X.: Upper and lower variances under model uncertainty and their applications in finance. Int. J. Financ. Eng. **9**(1), 2250007 (2022)
7. Li, X., Lin, Y.: Generalized Wasserstein distance and weak convergence of sublinear expectations. J. Theoret. Probab. **30**, 581–593 (2017)
8. Peng, S.: Nonlinear expectations and stochastic calculus under uncertainty. Springer, Berlin (2019)
9. Peng, S., Yang, S., Yao, J.: Improving value-at-risk prediction under model uncertainty. J. Financ. Econ. (2020) https://doi.org/10.1093/jjfinec/nbaa022
10. Sion, M.: On general minimax theorems. Pacific J. Math. **8**, 171–176 (1958)
11. Walley, P.: Statistical reasoning with imprecise probabilities. Chapman & Hall, London (1991)
12. Yuan, X.: KKM theory and applications in nonlinear analysis. Marcel Dekker, Inc (1999)

MIX
Papier aus verantwortungsvollen Quellen
Paper from responsible sources
FSC® C105338

If you have any concerns about our products,
you can contact us on
ProductSafety@springernature.com

In case Publisher is established outside the EU,
the EU authorized representative is:
Springer Nature Customer Service Center GmbH
Europaplatz 3, 69115 Heidelberg, Germany

Printed by Libri Plureos GmbH
in Hamburg, Germany